Student Solutions Guide

to accompany

Chemistry

Seventh Edition

Student Solutions Guide

to accompany

Chemistry

Seventh Edition

Steven S. Zumdahl
Susan Arena Zumdahl

Thomas J. Hummel
Steven S. Zumdahl
Susan Arena Zumdahl

University of Illinois at Urbana-Champaign

HOUGHTON MIFFLIN COMPANY **Boston** **New York**

Publisher and Editor-in-Chief: *Charles Hartford*
Executive Editor: *Richard Stratton*
Development Editor: *Rebecca Berardy Schwartz*
Assistant Editor: *Liz Hogan*
Editorial Assistant: *Brett Pasinella*
Manufacturing Coordinator: *Patricia Cadet*
Senior Marketing Manager: *Katherine Greig*
Market Assistant: *Naveen Hariprasad*

Printed in the U.S.A.

ISBN 13: 978-0-618-52850-9
ISBN 10: 0-618-52850-4

456789-CRS-10 09 08 07

TABLE OF CONTENTS

TO THE STUDENT: HOW TO USE THIS GUIDE

Solutions to odd-numbered chapter questions and exercises are in this manual. This "Solutions Guide" can be very valuable if you use it properly. The way NOT to use it is to look at an exercise in the book and then immediately check the solution, often saying to yourself, "That's easy, I can do it." Developing problem solving skills takes practice. Don't look up a solution to a problem until you have tried to work it on your own. If you are completely stuck, see if you can find a similar problem in the Sample Exercises in the chapter. Only look up the solution as a last resort. If you do this for a problem, look for a similar problem in the end of chapter exercises and try working it. The more problems you do, the easier chemistry becomes. It is also in your self interest to try to work as many problems as possible. Most exams that you will take in chemistry will involve a lot of problem solving. If you have worked several problems similar to the ones on an exam, you will do much better than if the exam is the first time you try to solve a particular type of problem. No matter how much you read and study the text, or how well you think you understand the material, you don't really understand it until you have taken the information in the text and applied the principles to problem solving. You will make mistakes, but the good students learn from their mistakes.

In this manual we have worked problems as in the textbook. We have shown intermediate answers to the correct number of significant figures and used the rounded answer in later calculations. Thus, some of your answers may differ slightly from ours. When we have not followed this convention, we have usually noted this in the solution. The most common exception is when working with the natural logarithm (ln) function, where we usually carried extra significant figures in order to reduce round-off error. In addition, we tried to use constants and conversion factors reported to at least one more significant figure as compared to numbers given in the problem. The practice of carrying one extra significant figure in constants helps minimize round-off error.

We are grateful to Claire O. Szoke for her outstanding effort in preparing the manuscript for this manual. We also thank Linda C. Bush for her careful and thorough accuracy review of the Solutions Guide.

TJH
SSZ
SAZ

Student Solutions Guide

to accompany

Chemistry

Seventh Edition

CHAPTER ONE

CHEMICAL FOUNDATIONS

Questions

17 A law summarizes what happens, e.g., law of conservation of mass in a chemical reaction or the ideal gas law, PV = nRT. A theory (model) is an attempt to explain why something happens. Dalton's atomic theory explains why mass is conserved in a chemical reaction. The kinetic molecular theory explains why pressure and volume are inversely related at constant temperature and moles of gas present as well as explaining the other mathematical relationships summarized in PV = nRT.

19. A qualitative observation expresses what makes something what it is; it does not involve a number, e.g., the air we breathe is a mixture of gases, ice is less dense than water, rotten milk stinks.

The SI units are mass in kilograms, length in meters, and volume in the derived units of m^3. The assumed uncertainty in a number is ±1 in the last significant figure of the number. The precision of an instrument is related to the number of significant figures associated with an experimental reading on that instrument. Different instruments for measuring mass, length, or volume have varying degrees of precision. Some instruments only give a few significant figures for a measurement while others will give more significant figures.

21. Significant figures are the digits we associate with a number. They contain all of the certain digits and the first uncertain digit (the first estimated digit). What follows is one thousand indicated to varying numbers of significant figures: 1000 or 1×10^3 (1 S.F.); 1.0×10^3 (2 S.F.); 1.00×10^3 (3 S.F.); 1000. or 1.000×10^3 (4 S.F.).

To perform the calculation, the addition/subtraction significant figure rule is applied to 1.5 − 1.0. The result of this is the one significant figure answer of 0.5. Next, the multiplication/division rule is applied to 0.5/0.50. A one significant figure number divided by a two significant figure number yields an answer with one significant figure (answer = 1).

23. Straight line equation: y = mx + b where m is the slope of the line and b is the y-intercept. For the T_F vs. T_C plot:

$$T_F = (9/5)T_C + 32$$
$$y = \quad m \quad x + b$$

The slope of the plot is 1.8 (= 9/5) and the y-intercept is 32°F.

For the T_C vs. T_K plot:

$$T_C = T_K - 273$$
$$y = mx + b$$

The slope of the plot is one and the y-intercept is −273°C.

Exercises

Significant Figures and Unit Conversions

25. a. exact b. inexact

c. exact d. inexact (π has an infinite number of decimal places.)

27. a. $\underline{6.07} \times 10^{-15}$; 3 S.F. b. $0.00\underline{3840}$; 4 S.F. c. $\underline{17.00}$; 4 S.F.

d. $\underline{8} \times 10^8$; 1 S.F. e. $\underline{463.8052}$; 7 S.F. f. $\underline{3}00$; 1 S.F.

g. $\underline{301}$; 3 S.F. h. $\underline{300.}$; 3 S.F.

29. When rounding, the last significant figure stays the same if the number after this significant figure is less than 5 and increases by one if the number is greater than or equal to 5.

a. 3.42×10^{-4} b. 1.034×10^4 c. 1.7992×10^1 d. 3.37×10^5

31. For addition and/or subtraction, the result has the same number of decimal places as the number in the calculation with the fewest decimal places. When the result is rounded to the correct number of significant figures, the last significant figure stays the same if the number after this significant figure is less than 5 and increases by one if the number is greater than or equal to 5. The underline shows the last significant figure in the intermediate answers.

a. $212.2 + 26.7 + 402.09 = 640.\underline{9}9 = 641.0$

b. $1.0028 + 0.221 + 0.10337 = 1.32\underline{7}17 = 1.327$

c. $52.331 + 26.01 - 0.9981 = 77.3\underline{4}29 = 77.34$

d. $2.01 \times 10^2 + 3.014 \times 10^3 = 2.01 \times 10^2 + 30.14 \times 10^2 = 32.1\underline{5} \times 10^2 = 3215$

When the exponents are different, it is easiest to apply the addition/subtraction rule when all numbers are based on the same power of 10.

e. $7.255 - 6.8350 = 0.42 = 0.420$ (first uncertain digit is in the third decimal place).

33. a. Here, apply the multiplication/division rule first; then apply the addition/subtraction rule to arrive at the one decimal place answer. We will generally round off at intermediate steps in order to show the correct number of significant figures. However, you should round off at the end of all the mathematical operations in order to avoid round-off error. The best way to do calculations is to keep track of the correct number of significant figures during intermediate steps, but round off at the end. For this problem, we underlined the last significant figure in the intermediate steps.

$$\frac{2.526}{3.1} + \frac{0.470}{0.623} + \frac{80.705}{0.4326} = 0.8\underline{1}48 + 0.75\underline{4}4 + 186.\underline{5}58 = 188.1$$

b. Here, the mathematical operation requires that we apply the addition/subtraction rule first, then apply the multiplication/division rule.

$$\frac{6.404 \times 2.91}{18.7 - 17.1} = \frac{6.404 \times 2.91}{1.\underline{6}} = 12$$

c. $6.071 \times 10^{-5} - 8.2 \times 10^{-6} - 0.521 \times 10^{-4} = 60.71 \times 10^{-6} - 8.2 \times 10^{-6} - 52.1 \times 10^{-6}$
$= 0.\underline{4}1 \times 10^{-6} = 4 \times 10^{-7}$

d. $$\frac{3.8 \times 10^{-12} + 4.0 \times 10^{-13}}{4 \times 10^{12} + 6.3 \times 10^{13}} = \frac{38 \times 10^{-13} + 4.0 \times 10^{-13}}{4 \times 10^{12} + 63 \times 10^{12}} = \frac{4\underline{2} \times 10^{-13}}{6\underline{7} \times 10^{12}} = 6.3 \times 10^{-26}$$

e. $$\frac{9.5 + 4.1 + 2.8 + 3.175}{4} = \frac{19.\underline{5}75}{4} = 4.89 = 4.9$$

Uncertainty appears in the first decimal place. The average of several numbers can only be as precise as the least precise number. Averages can be exceptions to the significant figure rules.

f. $$\frac{8.925 - 8.905}{8.925} \times 100 = \frac{0.02\underline{0}}{8.925} \times 100 = 0.22$$

35. a. $8.43 \text{ cm} \times \frac{1 \text{ m}}{100 \text{ cm}} \times \frac{1000 \text{ mm}}{\text{m}} = 84.3 \text{ mm}$

b. $2.41 \times 10^2 \text{ cm} \times \frac{1 \text{ m}}{100 \text{ cm}} = 2.41 \text{ m}$

c. $294.5 \text{ nm} \times \frac{1 \text{ m}}{1 \times 10^9 \text{ nm}} \times \frac{100 \text{ cm}}{\text{m}} = 2.945 \times 10^{-5} \text{ cm}$

d. $1.445 \times 10^4 \text{ m} \times \frac{1 \text{ km}}{1000 \text{ m}} = 14.45 \text{ km}$

e. $235.3 \text{ m} \times \frac{1000 \text{ mm}}{\text{m}} = 2.353 \times 10^5 \text{ mm}$

f. $903.3 \text{ nm} \times \frac{1 \text{ m}}{1 \times 10^9 \text{ nm}} \times \frac{1 \times 10^6 \text{ μm}}{\text{m}} = 0.9033 \text{ μm}$

37. a. Appropriate conversion factors are found in Appendix 6. In general, the number of significant figures we use in the conversion factors will be one more than the number of significant figures from the numbers given in the problem. This is usually sufficient to avoid round-off error.

$$3.91\text{ kg} \times \frac{1\text{ lb}}{0.4536\text{ kg}} = 8.62\text{ lb};\ 0.62\text{ lb} \times \frac{16\text{ oz}}{\text{lb}} = 9.9\text{ oz}$$

Baby's weight = 8 lb and 9.9 oz or to the nearest ounce, 8 lb and 10. oz.

$$51.4\text{ cm} \times \frac{1\text{ in}}{2.54\text{ cm}} = 20.2\text{ in} \approx 20\ 1/4\text{ in} = \text{baby's height}$$

b. $$25{,}000\text{ mi} \times \frac{1.61\text{ km}}{\text{mi}} = 4.0 \times 10^4\text{ km};\ 4.0 \times 10^4\text{ km} \times \frac{1000\text{ m}}{\text{km}} = 4.0 \times 10^7\text{ m}$$

c. $$V = l \times w \times h = 1.0\text{ m} \times \left(5.6\text{ cm} \times \frac{1\text{ m}}{100\text{ cm}}\right) \times \left(2.1\text{ dm} \times \frac{1\text{ m}}{10\text{ dm}}\right) = 1.2 \times 10^{-2}\text{ m}^3$$

$$1.2 \times 10^{-2}\text{ m}^3 \times \left(\frac{10\text{ dm}}{\text{m}}\right)^3 \times \frac{1\text{ L}}{\text{dm}^3} = 12\text{ L}$$

$$12\text{ L} \times \frac{1000\text{ cm}^3}{\text{L}} \times \left(\frac{1\text{ in}}{2.54\text{ cm}}\right)^3 = 730\text{ in}^3;\ 730\text{ in}^3 \times \left(\frac{1\text{ ft}}{12\text{ in}}\right)^3 = 0.42\text{ ft}^3$$

39. a. $$1.25\text{ mi} \times \frac{8\text{ furlongs}}{\text{mi}} = 10.0\text{ furlongs};\ 10.0\text{ furlongs} \times \frac{40\text{ rods}}{\text{furlong}} = 4.00 \times 10^2\text{ rods}$$

$$4.00 \times 10^2\text{ rods} \times \frac{5.5\text{ yd}}{\text{rod}} \times \frac{36\text{ in}}{\text{yd}} \times \frac{2.54\text{ cm}}{\text{in}} \times \frac{1\text{ m}}{100\text{ cm}} = 2.01 \times 10^3\text{ m}$$

$$2.01 \times 10^3\text{ m} \times \frac{1\text{ km}}{1000\text{ m}} = 2.01\text{ km}$$

b. Let's assume we know this distance to ± 1 yard. First convert 26 miles to yards.

$$26\text{ mi} \times \frac{5280\text{ ft}}{\text{mi}} \times \frac{1\text{ yd}}{3\text{ ft}} = 45{,}760.\text{ yd}$$

$$26\text{ mi} + 385\text{ yd} = 45{,}760.\text{ yd} + 385\text{ yd} = 46{,}145\text{ yards}$$

$$46{,}145\text{ yard} \times \frac{1\text{ rod}}{5.5\text{ yd}} = 8390.0\text{ rods};\ 8390.0\text{ rods} \times \frac{1\text{ furlong}}{40\text{ rods}} = 209.75\text{ furlongs}$$

$$46{,}145\text{ yard} \times \frac{36\text{ in}}{\text{yd}} \times \frac{2.54\text{ cm}}{\text{in}} \times \frac{1\text{ m}}{100\text{ cm}} = 42{,}195\text{ m};\ 42{,}195\text{ m} \times \frac{1\text{ km}}{1000\text{ m}} = 42.195\text{ km}$$

41. a. $1 \text{ troy lb} \times \frac{12 \text{ troy oz}}{\text{troy lb}} \times \frac{20 \text{ pw}}{\text{troy oz}} \times \frac{24 \text{ grains}}{\text{pw}} \times \frac{0.0648 \text{ g}}{\text{grain}} \times \frac{1 \text{ kg}}{1000 \text{ g}} = 0.373 \text{ kg}$

$1 \text{ troy lb} = 0.373 \text{ kg} \times \frac{2.205 \text{ lb}}{\text{kg}} = 0.822 \text{ lb}$

b. $1 \text{ troy oz} \times \frac{20 \text{ pw}}{\text{troy oz}} \times \frac{24 \text{ grains}}{\text{pw}} \times \frac{0.0648 \text{ g}}{\text{grain}} = 31.1 \text{ g}$

$1 \text{ troy oz} = 31.1 \text{ g} \times \frac{1 \text{ carat}}{0.200 \text{ g}} = 156 \text{ carats}$

c. $1 \text{ troy lb} = 0.373 \text{ kg}$; $0.373 \text{ kg} \times \frac{1000 \text{ g}}{\text{kg}} \times \frac{1 \text{ cm}^3}{19.3 \text{ g}} = 19.3 \text{ cm}^3$

43. $\text{warp } 1.71 = \left(5.00 \times \frac{3.00 \times 10^8 \text{ m}}{\text{s}}\right) \times \frac{1.094 \text{ yd}}{\text{m}} \times \frac{60 \text{ s}}{\text{min}} \times \frac{60 \text{ min}}{\text{hr}} \times \frac{1 \text{ knot}}{2000 \text{ yd/hr}}$

$= 2.95 \times 10^9 \text{ knots}$

$\left(5.00 \times \frac{3.00 \times 10^8 \text{ m}}{\text{s}}\right) \times \frac{1 \text{ km}}{1000 \text{ m}} \times \frac{1 \text{ mi}}{1.609 \text{ km}} \times \frac{60 \text{ s}}{\text{min}} \times \frac{60 \text{ min}}{\text{hr}} = 3.36 \times 10^9 \text{ mi/hr}$

45. $\frac{65 \text{ km}}{\text{hr}} \times \frac{0.6214 \text{ mi}}{\text{km}} = 40.4 = 40. \text{ mi/hr}$

To the correct number of significant figures, 65 km/hr does not violate a 40. mi/hr speed limit.

47. $\frac{\$17.25 \text{ U.S.}}{8.21 \text{ gal}} \times \frac{\$1.00 \text{ Canadian}}{\$0.82 \text{ U.S.}} \times \frac{1 \text{ gal}}{3.7854 \text{ L}} = \$0.68 \text{ Canadian/L}$

Temperature

49. a. $T_C = \frac{5}{9}(T_F - 32) = \frac{5}{9}(-459°C - 32) = -273°C$; $T_K = T_C + 273 = -273°C + 273 = 0 \text{ K}$

b. $T_C = \frac{5}{9}(-40.°F - 32) = -40.°C$; $T_K = -40.°C + 273 = 233 \text{ K}$

c. $T_C = \frac{5}{9}(68°F - 32) = 20.°C$; $T_K = 20.°C + 273 = 293 \text{ K}$

d. $T_C = \frac{5}{9}(7 \times 10^7 °F - 32) = 4 \times 10^7 °C$; $T_K = 4 \times 10^7 °C + 273 = 4 \times 10^7 \text{ K}$

51. a. $T_F = \frac{9}{5} \times T_C + 32 = \frac{9}{5} \times 39.2°C + 32 = 102.6°F$ (Note: 32 is exact.)

$T_K = T_C + 273.2 = 39.2 + 273.2 = 312.4$ K

b. $T_F = \frac{9}{5} \times (-25) + 32 = -13°F$; $T_K = -25 + 273 = 248$ K

c. $T_F = \frac{9}{5} \times (-273) + 32 = -459°F$; $T_K = -273 + 273 = 0$ K

d. $T_F = \frac{9}{5} \times 801 + 32 = 1470°F$; $T_K = 801 + 273 = 1074$ K

Density

53. $\text{mass} = 350 \text{ lb} \times \frac{453.6 \text{ g}}{\text{lb}} = 1.6 \times 10^5 \text{ g}$; $V = 1.2 \times 10^4 \text{ in}^3 \times \left(\frac{2.54 \text{ cm}}{\text{in}}\right)^3 = 2.0 \times 10^5 \text{ cm}^3$

$$\text{density} = \frac{\text{mass}}{\text{volume}} = \frac{1 \times 10^5 \text{ g}}{2.0 \times 10^5 \text{ cm}^3} = 0.80 \text{ g/cm}^3$$

Because the material has a density less than water, it will float in water.

55. $V = \frac{4}{3}\pi r^3 = \frac{4}{3} \times 3.14 \times \left(7.0 \times 10^5 \text{ km} \times \frac{1000 \text{ m}}{\text{km}} \times \frac{100 \text{ cm}}{\text{m}}\right)^3 = 1.4 \times 10^{33} \text{ cm}^3$

$$\text{density} = \frac{\text{mass}}{\text{volume}} = \frac{2 \times 10^{36} \text{ kg} \times \frac{1000 \text{ g}}{\text{kg}}}{1.4 \times 10^{33} \text{ cm}^3} = 1.4 \times 10^6 \text{ g/cm}^3 = 1 \times 10^6 \text{ g/cm}^3$$

57. $5.0 \text{ carat} \times \frac{0.200 \text{ g}}{\text{carat}} \times \frac{1 \text{ cm}^3}{3.51 \text{ g}} = 0.28 \text{ cm}^3$

59. $V = 21.6 \text{ mL} - 12.7 \text{ mL} = 8.9 \text{ mL}$; $\text{density} = \frac{33.42 \text{ g}}{8.9 \text{ mL}} = 3.8 \text{ g/mL} = 3.8 \text{ g/cm}^3$

61. a. Both have the same mass of 1.0 kg.

b. 1.0 mL of mercury; Mercury has a greater density than water. Note: $1 \text{ mL} = 1 \text{ cm}^3$

$1.0 \text{ mL} \times \frac{13.6 \text{ g}}{\text{mL}} = 14$ g of mercury; $1.0 \text{ mL} \times \frac{0.998 \text{ g}}{\text{mL}} = 1.0$ g of water

c. Same; Both represent 19.3 g of substance.

$19.3 \text{ mL} \times \frac{0.9982 \text{ g}}{\text{mL}} = 19.3$ g of water; $1.00 \text{ mL} \times \frac{19.32 \text{ g}}{\text{mL}} = 19.3$ g of gold

d. 1.0 L of benzene (880 g vs 670 g)

$$75\text{ mL} \times \frac{8.96\text{ g}}{\text{mL}} = 670\text{ g of copper};\ 1.0\text{ L} \times \frac{1000\text{ mL}}{\text{L}} \times \frac{0.880\text{ g}}{\text{mL}} = 880\text{ g of benzene}$$

63. a. 1.0 kg feather; Feathers are less dense than lead.

b. 100 g water; Water is less dense than gold. c. Same; Both volumes are 1.0 L.

65. $V = 1.00 \times 10^3\text{ g} \times \frac{1\text{ cm}^3}{22.57\text{ g}} = 44.3\text{ cm}^3$

$44.3\text{ cm}^3 = l \times w \times h = 4.00\text{ cm} \times 4.00\text{ cm} \times h,\ h = 2.77\text{ cm}$

Classification and Separation of Matter

67. A gas has molecules that are very far apart from each other while a solid or liquid has molecules that are very close together. An element has the same type of atom, whereas a compound contains two or more different elements. Picture i represents an element that exists as two atoms bonded together (like H_2 or O_2 or N_2). Picture iv represents a compound (like CO, NO, or HF). Pictures iii and iv contain representations of elements that exist as individual atoms (like Ar, Ne, or He).

a. Picture iv represents a gaseous compound. Note that pictures ii and iii also contain a gaseous compound, but they also both have a gaseous element present.

b. Picture vi represents a mixture of two gaseous elements.

c. Picture v represents a solid element.

d. Pictures ii and iii both represent a mixture of a gaseous element and a gaseous compound.

69. Homogeneous: Having visibly indistinguishable parts (the same throughout).
Heterogeneous: Having visibly distinguishable parts (not uniform throughout).

a. heterogeneous (Due to hinges, handles, locks, etc.)

b. homogeneous (hopefully; If you live in a heavily polluted area, air may be heterogeneous.)

c. homogeneous

d. homogeneous (hopefully, if not polluted)

e. heterogeneous

f. heterogeneous

71. A physical change is a change in the state of a substance (solid, liquid and gas are the three states of matter); a physical change does not change the chemical composition of the substance. A chemical change is a change in which a given substance is converted into another substance having a different formula (composition).

a. Vaporization refers to a liquid converting to a gas, so this is a physical change. The formula (composition) of the moth ball does not change.

b. This is a chemical change since hydrofluoric acid (HF) is reacting with glass (SiO_2) to form new compounds which wash away.

c. This is a physical change since all that is happening is the conversion of liquid alcohol to gaseous alcohol. The alcohol formula (C_2H_5OH) does not change.

d. This is a chemical change since the acid is reacting with cotton to form new compounds.

Additional Exercises

73. $15.6\text{ g} \times \frac{1\text{ capsule}}{0.65\text{ g}} = 24\text{ capsules}$

75. $\text{Total volume} = \left(200.\text{ m} \times \frac{100\text{ cm}}{\text{m}}\right) \times \left(300.\text{ m} \times \frac{100\text{ cm}}{\text{m}}\right) \times 4.0\text{ cm} = 2.4 \times 10^9\text{ cm}^3$

$$\text{Vol. of topsoil covered by 1 bag} = \left[10.\text{ ft}^2 \times \left(\frac{12\text{ in}}{\text{ft}}\right)^2 \times \left(\frac{2.54\text{ cm}}{\text{in}}\right)^2\right] \times \left(1.0\text{ in} \times \frac{2.54\text{ cm}}{\text{in}}\right)$$

$$= 2.4 \times 10^4\text{ cm}^3$$

$$2.4 \times 10^9\text{ cm}^3 \times \frac{1\text{ bag}}{2.4 \times 10^4\text{ cm}^3} = 1.0 \times 10^5\text{ bags topsoil}$$

77. $18.5\text{ cm} \times \frac{10.0°\text{F}}{5.25\text{ cm}} = 35.2°\text{F increase}$; $T_{final} = 98.6 + 35.2 = 133.8°\text{F}$

$T_c = 5/9\,(133.8 - 32) = 56.56°\text{C}$

79. a. Volume × density = mass; the orange block is more dense. Because mass (orange) > mass (blue) and because volume (orange) < volume (blue), the density of the orange block must be greater to account for the larger mass of the orange block.

b. Which block is more dense cannot be determined. Because mass (orange) > mass (blue) and because volume (orange) > volume (blue), the density of the orange block may or may not be larger than the blue block. If the blue block is more dense, its density cannot be so large that its mass is larger than the orange block's mass.

c. The blue block is more dense. Because mass (blue) = mass (orange) and because volume (blue) < volume (orange), the density of the blue block must be larger in order to equate the masses.

d. The blue block is more dense. Because mass (blue) > mass (orange) and because the volumes are equal, the density of the blue block must be larger in order to give the blue block the larger mass.

81. $V = V_{final} - V_{initial}$; $d = \dfrac{28.90\text{ g}}{9.8\text{ cm}^3 - 6.4\text{ cm}^3} = \dfrac{28.90\text{ g}}{3.4\text{ cm}^3} = 8.5\text{ g/cm}^3$

$d_{max} = \dfrac{mass_{max}}{V_{min}}$; We get V_{min} from $9.7\text{ cm}^3 - 6.5\text{ cm}^3 = 3.2\text{ cm}^3$.

$$d_{max} = \frac{28.93\text{ g}}{3.2\text{ cm}^3} = \frac{9.0\text{ g}}{\text{cm}^3};\quad d_{min} = \frac{mass_{min}}{V_{max}} = \frac{28.87\text{ g}}{9.9\text{ cm}^3 - 6.3\text{ cm}^3} = \frac{8.0\text{ g}}{\text{cm}^3}$$

The density is: $8.5 \pm 0.5\text{ g/cm}^3$.

Challenge Problems

83. a. $\dfrac{2.70 - 2.64}{2.70} \times 100 = 2\%$ b. $\dfrac{|16.12 - 16.48|}{16.12} \times 100 = 2.2\%$

c. $\dfrac{1.000 - 0.9981}{1.000} \times 100 = \dfrac{0.002}{1.000} \times 100 = 0.2\%$

85. Heavy pennies (old): mean mass = 3.08 ± 0.05 g

Light pennies (new): mean mass = $\dfrac{(2.467 + 2.545 + 2.518)}{3} = 2.51 \pm 0.04$ g

Because we are assuming that volume is additive, let's calculate the volume of 100. g of each type of penny then calculate the density of the alloy. For 100. g of the old pennies, 95 g will be Cu (copper) and 5 g will be Zn (zinc).

$V = 95\text{ g Cu} \times \dfrac{1\text{ cm}^3}{8.96\text{g}} + 5\text{ g Zn} \times \dfrac{1\text{ cm}^3}{7.14\text{g}} = 11.3\text{ cm}^3$ (carrying one extra sig. fig.)

Density of old pennies $= \dfrac{100.\text{ g}}{11.3\text{ cm}^3} = 8.8\text{ g/cm}^3$

For 100. g of new pennies, 97.6 g will be Zn and 2.4 g will be Cu.

$$V = 2.4 \text{ g Cu} \times \frac{1 \text{ cm}^3}{8.96 \text{ g}} + 97.6 \text{ g Zn} \times \frac{1 \text{ cm}^3}{7.14 \text{ g}} = 13.94 \text{ cm}^3 \text{ (carrying one extra sig. fig.)}$$

$$\text{Density of new pennies} = \frac{100. \text{ g}}{13.94 \text{ cm}^3} = 7.17 \text{ g/cm}^3$$

$d = \dfrac{\text{mass}}{\text{volume}}$; Because the volume of both types of pennies are assumed equal, then:

$$\frac{d_{new}}{d_{old}} = \frac{\text{mass}_{new}}{\text{mass}_{old}} = \frac{7.17 \text{ g/cm}^3}{8.8 \text{ g/cm}^3} = 0.81$$

The calculated average mass ratio is: $\dfrac{\text{mass}_{new}}{\text{mass}_{old}} = \dfrac{2.51 \text{ g}}{3.08 \text{ g}} = 0.815$

To the first two decimal places, the ratios are the same. If the assumptions are correct, then we can reasonably conclude that the difference in mass is accounted for by the difference in alloy used.

87. Let x = mass of copper and y = mass of silver.

$105.0 \text{ g} = x + y$ and $10.12 \text{ mL} = \dfrac{x}{8.96} + \dfrac{y}{10.5}$; Solving:

$$\left(10.12 = \frac{x}{8.96} + \frac{105.0 - x}{10.5}\right) \times 8.96 \times 10.5, \quad 952.1 = 10.5\,x + 940.8 - 8.96\,x$$

(carrying 1 extra significant figure)

$$11.3 = 1.54\,x, \quad x = 7.3 \text{ g}; \quad \text{mass \%Cu} = \frac{7.3 \text{ g}}{105.0 \text{ g}} \times 100 = 7.0\% \text{ Cu}$$

89. a. One possibility is that rope B is not attached to anything and rope A and rope C are connected via a pair of pulleys and/or gears.

b. Try to pull rope B out of the box. Measure the distance moved by C for a given movement of A. Hold either A or C firmly while pulling on the other.

Integrative Problems

91. 2.97×10^8 persons $\times$ 0.0100 = 2.97×10^6 persons contributing

$$\frac{\$4.75 \times 10^8}{2.97 \times 10^6 \text{ persons}} = \$160./\text{person}; \quad \frac{\$160.}{\text{person}} \times \frac{20 \text{ nickels}}{\$1} = 3.20 \times 10^3 \text{ nickels/person}$$

$$\frac{\$160.}{\text{person}} \times \frac{1 \text{ pound sterling}}{\$1.869} = 85.6 \text{ pounds sterling/person}$$

93. At 200.0 °F: $T_C = \frac{5}{9}(200.0\ °F - 32\ °F) = 93.33\ °C$; $T_K = 93.33 + 273.15 = 366.48$ K

At −100.0 °F: $T_C = \frac{5}{9}(-100.0\ °F - 32\ °F) = -73.33\ °C$; $T_K = -73.33\ °C + 273.15$

$= 199.82$ K

$\Delta T(°C) = [93.33\ °C - (-73.33\ °C)] = 166.66\ °C$; $\Delta T(K) = [366.48\ K - 199.82\ K]$

$= 166.66$ K

The "300 Club" name only works for the Fahrenheit scale; it does not hold true for the Celsius and Kelvin scales.

CHAPTER TWO

ATOMS, MOLECULES, AND IONS

Questions

15. A compound will always contain the same numbers (and types) of atoms. A given amount of hydrogen will react only with a specific amount of oxygen. Any excess oxygen will remain unreacted.

17. Law of conservation of mass: mass is neither created nor destroyed. The mass before a chemical reaction always equals the mass after a chemical reaction.

    Law of definite proportion: a given compound always contains exactly the same proportion of elements by mass. Water is always 1 g H for every 8 g oxygen.

    Law of multiple proportions: When two elements form a series of compounds, the ratios of the mass of the second element that combine with one gram of the first element can always be reduced to small whole numbers: For CO_2 and CO discussed in section 2.2, the mass ratios of oxygen that react with 1 g of carbon in each compound are in a 2:1 ratio.

19. J. J. Thomson's study of cathode-ray tubes led him to postulate the existence of negatively charged particles which we now call electrons. Ernest Rutherford and his alpha bombardment of metal foil experiments led him to postulate the nuclear atom – an atom with a tiny dense center of positive charge (the nucleus) with electrons moving about the nucleus at relatively large distances away.

21. The number and arrangement of electrons in an atom determines how the atom will react with other atoms. The electrons determine the chemical properties of an atom. The number of neutrons present determines the isotope identity.

23. Statements a and b are true. Counting over in the periodic table, element 118 will be the next noble gas (a nonmetal). For statement c, hydrogen has mostly nonmetallic properties. For statement d, a family of elements is also known as a group of elements. For statement e, two items are incorrect. When a metal reacts with a nonmetal, an ionic compound is produced and the formula of the compound would be AX_2 (alkaline earth metals form +2 ions and halogens form −1 ions in ionic compounds). The correct statement would be: When an alkaline earth metal, A, reacts with a halogen, X, the formula of the ionic compound formed should be AX_2.

Exercises

Development of the Atomic Theory

25. a. The composition of a substance depends on the numbers of atoms of each element making up the compound (on the formula of the compound) and not on the composition of the mixture from which it was formed.

    b. Avogadro's hypothesis implies that volume ratios are equal to molecule ratios at constant temperature and pressure. $H_2(g) + Cl_2(g) \rightarrow 2\ HCl(g)$. From the balanced equation (2 molecules of HCl are produced per molecule of H_2 or Cl_2 reacted), the volume of HCl produced will be twice the volume of H_2 (or Cl_2) reacted.

27. Hydrazine: 1.44×10^{-1} g H/g N; Ammonia: 2.16×10^{-1} g H/g N

    Hydrogen azide: 2.40×10^{-2} g H/g N

    Let's try all of the ratios:

$$\frac{0.144}{0.0240} = 6.00; \quad \frac{0.216}{0.0240} = 9.00; \quad \frac{0.216}{0.144} = 1.50 = \frac{3}{2}$$

    All the masses of hydrogen in these three compounds can be expressed as simple whole number ratios. The g H/g N in hydrazine, ammonia, and hydrogen azide are in the ratios 6:9:1.

29. To get the atomic mass of H to be 1.00, we divide the mass of hydrogen that reacts with 1.00 g of oxygen by 0.126, i.e., $\frac{0.216}{0.216} = 1.00$. To get Na, Mg and O on the same scale, we do the same division.

$$\text{Na: } \frac{2.875}{0.126} = 22.8; \quad \text{Mg: } \frac{1.500}{0.216} = 11.9; \quad \text{O: } \frac{1.00}{0.216} = 7.94$$

	H	O	Na	Mg
Relative Value	1.00	7.94	22.8	11.9
Accepted Value	1.008	16.00	22.99	24.31

    The atomic masses of O and Mg are incorrect; the atomic masses of H and Na are close to the values in the periodic table. Something must be wrong about the assumed formulas of the compounds. It turns out the correct formulas are H_2O, Na_2O, and MgO. The smaller discrepancies result from the error in the atomic mass of H.

The Nature of the Atom

31. Density of hydrogen nucleus (contains one proton only):

$$V_{nucleus} = \frac{4}{3}\pi r^3 = \frac{4}{3}(3.14)(5\times10^{-14}\ cm)^3 = 5 \times 10^{-40}\ cm^3$$

$$d = \frac{1.67\times10^{-24}\ g}{5\times10^{-40}\ cm^3} = 3 \times 10^{15}\ g/cm^3$$

Density of H-atom (contains one proton and one electron):

$$V_{atom} = \frac{4}{3}(3.14)(1\times10^{-8}\ cm)^3 = 4 \times 10^{-24}\ cm^3$$

$$d = \frac{1.67\times10^{-24}\ g + 9\times10^{-28}\ g}{4\times10^{-24}\ cm^3} = 0.4\ g/cm^3$$

33. $5.93\times10^{-18}\ C \times \frac{1\ \text{electron charge}}{1.602\times10^{-19}\ C}$ = 37 negative (electron) charges on the oil drop

35. sodium–Na; radium–Ra; iron–Fe; gold–Au; manganese–Mn; lead–Pb

37. Sn–tin; Pt–platinum; Hg–mercury; Mg–magnesium; K–potassium; Ag–silver

39. a. Metals: Mg, Ti, Au, Bi, Ge, Eu, Am. Nonmetals: Si, B, At, Rn, Br.

b. Si, Ge, B, At. The elements at the boundary between the metals and the nonmetals are: B, Si, Ge, As, Sb, Te, Po, At. Aluminum has mostly properties of metals.

41. a. Six; Be, Mg, Ca, Sr, Ba, Ra

b. Five; O, S, Se, Te, Po

c. Four; Ni, Pd, Pt, Uun

d. Six; He, Ne, Ar, Kr, Xe, Rn

43. a. ${}^{79}_{35}Br$: 35 protons, 79 – 35 = 44 neutrons. Since the charge of the atom is neutral, the number of protons = the number of electrons = 35.

b. ${}^{81}_{35}Br$: 35 protons, 46 neutrons, 35 electrons

c. ${}^{239}_{94}Pu$: 94 protons, 145 neutrons, 94 electrons

d. ${}^{133}_{55}Cs$: 55 protons, 78 neutrons, 55 electrons

e. $^{3}_{1}H$: 1 proton, 2 neutrons, 1 electron

f. $^{56}_{26}Fe$: 26 protons, 30 neutrons, 26 electrons

45. a. Element 8 is oxygen. A = mass number = 9 + 8 = 17; $^{17}_{8}O$

b. Chlorine is element 17. $^{37}_{17}Cl$

c. Cobalt is element 27. $^{60}_{27}Co$

d. Z = 26; A = 26 + 31 = 57; $^{57}_{26}Fe$

e. Iodine is element 53. $^{131}_{53}I$

f. Lithium is element 3. $^{7}_{3}Li$

47. Atomic number = 63 (Eu); Charge = +63 - 60 = +3; Mass number = 63 + 88 = 151;

Symbol: $^{151}_{63}Eu^{3+}$

Atomic number = 50 (Sn); Mass number = 50 + 68 = 118; Net charge = +50 - 48 = +2; The symbol is $^{118}_{50}Sn^{2+}$.

49.

Symbol	Number of protons in nucleus	Number of neutrons in nucleus	Number of electrons	Net charge
$^{238}_{92}U$	92	146	92	0
$^{40}_{20}Ca^{2+}$	20	20	18	2+
$^{51}_{23}V^{3+}$	23	28	20	3+
$^{89}_{39}Y$	39	50	39	0
$^{79}_{35}Br^{-}$	35	44	36	1-
$^{31}_{15}P^{3-}$	15	16	18	3-

51. a. transition metals

b. alkaline earth metals

c. alkali metals

d. noble gases

e. halogens

53. In ionic compounds, metals lose electrons to form cations, and nonmetals gain electrons to form anions. Group 1A, 2A and 3A metals form stable +1, +2 and +3 charged cations, respectively. Group 5A, 6A and 7A nonmetals form −3, −2 and -1 charged anions, respectively.

a. Lose 2 e^- to form Ra^{2+}. b. Lose 3 e^- to form In^{3+}. c. Gain 3 e^- to form P^{3-}.

d. Gain 2 e^- to form Te^{2-}. e. Gain 1 e^- to form Br^-. f. Lose 1 e^- to form Rb^+.

Nomenclature

55. a. sodium bromide b. rubidium oxide

c. calcium sulfide d. aluminum iodide

e. SrF_2 f. Al_2Se_3

g. K_3N h. Mg_3P_2

57. a. cesium fluoride b. lithium nitride

c. silver sulfide (Silver only forms stable +1 ions in compounds so no Roman numerals are needed.)

d. manganese(IV) oxide e. titanium(IV) oxide f. strontium phosphide

59. a. barium sulfite b. sodium nitrite

c. potassium permanganate d. potassium dichromate

61. a. dinitrogen tetroxide b. iodine trichloride

c. sulfur dioxide d. diphosphorus pentasulfide

63. a. copper(I) iodide b. copper(II) iodide c. cobalt(II) iodide

d. sodium carbonate e. sodium hydrogen carbonate or sodium bicarbonate

f. tetrasulfur tetranitride g. sulfur hexafluoride h. sodium hypochlorite

i. barium chromate j. ammonium nitrate

65. a. SF_2 b. SF_6 c. NaH_2PO_4

d. Li_3N e. $Cr_2(CO_3)_3$ f. SnF_2

g. $NH_4C_2H_3O_2$ h. NH_4HSO_4 i. $Co(NO_3)_3$

j. Hg_2Cl_2; Mercury(I) exists as Hg_2^{2+} ions. k. $KClO_3$ l. NaH

67. a. Na_2O b. Na_2O_2 c. KCN

d. $Cu(NO_3)_2$ e. $SeBr_4$ f. HIO_2

g. PbS_2 h. CuCl i. GaAs (Predict Ga^{3+} and As^{3-} ions.)

j. CdSe (Cadmium only forms +2 charged ions in compounds.)

k. ZnS (Zinc only forms +2 charged ions in compounds.)

l. HNO_2 m. P_2O_5

69. a. nitric acid, HNO_3 b. perchloric acid, $HClO_4$ c. acetic acid, $HC_2H_3O_2$

d. sulfuric acid, H_2SO_4 e. phosphoric acid, H_3PO_4

Additional Exercises

71. Yes, 1.0 g H would react with 37.0 g ^{37}Cl and 1.0 g H would react with 35.0 g ^{35}Cl.

No, the mass ratio of H/Cl would always be 1 g H/37 g Cl for ^{37}Cl and 1 g H/35 g Cl for ^{35}Cl. As long as we had pure ^{37}Cl or pure ^{35}Cl, the above ratios will always hold. If we have a mixture (such as the natural abundance of chlorine), the ratio will also be constant as long as the composition of the mixture of the two isotopes does not change.

73. From the Na_2X formula, X has a −2 charge. Since 36 electrons are present, X has 34 p, 79 − 34 = 45 neutrons, and is selenium.

a. True. Nonmetals bond together using covalent bonds and are called covalent compounds.

b. False. The isotope has 34 protons.

c. False. The isotope has 45 neutrons.

d. False. The identity is selenium, Se.

75. a. $Pb(C_2H_3O_2)_2$: lead(II) acetate b. $CuSO_4$: copper(II) sulfate

c. CaO: calcium oxide d. $MgSO_4$: magnesium sulfate

e. $Mg(OH)_2$: magnesium hydroxide f. $CaSO_4$: calcium sulfate

g. N_2O: dinitrogen monoxide or nitrous oxide

77. From the XBr_2 formula, the charge on element X is +2. Therefore, the element has 88 protons, which identifies it as radium, Ra. 230 - 88 = 142 neutrons

79. a. Ca^{2+} and N^{3-}: Ca_3N_2, calcium nitride b. K^+ and O^{2-}: K_2O, potassium oxide

c. Rb^+ and F^-: RbF, rubidium fluoride d. Mg^{2+} and S^{2-}: MgS, magnesium sulfide

e. Ba^{2+} and I^-: BaI_2, barium iodide f. Al^{3+} and Se^{2-}: Al_2Se_3, aluminum selenide

g. Cs^+ and P^{3-}: Cs_3P, cesium phosphide

h. In^{3+} and Br^-: $InBr_3$, indium(III) bromide. In also forms In^+ ions, but one would predict In^{3+} ions from its position in the periodic table.

81. A compound will always have a constant composition by mass. From the initial data given, the mass ratio of H:S:O in sulfuric acid is:

$$\frac{2.02}{2.02} : \frac{32.07}{2.02} : \frac{64.00}{2.02} = 1 : 15.9 : 31.7$$

If we have 7.27 g H, then we will have 7.27 × 15.9 = 116 g S and 7.27 × 31.7 = 230. g O in the second sample of H_2SO_4.

Challenge Problems

83. Copper(Cu), silver (Ag) and gold(Au) make up the coinage metals.

85. Avogadro proposed that equal volumes of gases (at constant temperature and pressure) contain equal numbers of molecules. In terms of balanced equations, Avogadro's hypothesis implies that volume ratios will be identical to molecule ratios. Assuming one molecule of octane reacting, then 1 molecule of C_xH_y produces 8 molecules of CO_2 and 9 molecules of H_2O. $C_xH_y + O_2 \rightarrow 8\ CO_2 + 9\ H_2O$. Because all the carbon in octane ends up as carbon in CO_2, octane contains 8 atoms of C. Similarly, all hydrogen in octane ends up as hydrogen in H_2O, so one molecule of octane contains 9 × 2 = 18 atoms of H. Octane formula = C_8H_{18} and the ratio of C:H = 8:18 or 4:9.

87. Compound I: $\frac{14.0 \text{ g R}}{3.00 \text{ g Q}} = \frac{4.67 \text{ g R}}{1.00 \text{ g Q}}$; Compound II: $\frac{7.00 \text{ g R}}{4.50 \text{ g Q}} = \frac{1.56 \text{ g R}}{1.00 \text{ g Q}}$

The ratio of the masses of R that combine with 1.00 g Q is: $\frac{4.67}{1.56} = 2.99 \approx 3$

As expected from the law of multiple proportions, this ratio is a small whole number.

Because Compound I contains three times the mass of R per gram of Q as compared to Compound II (RQ), the formula of Compound I should be R_3Q.

89. a. Both compounds have C_2H_6O as the formula. Because they have the same formula, their mass percent composition will be identical. However, these are different compounds with different properties since the atoms are bonded together differently. These compounds are called isomers of each other.

b. When wood burns, most of the solid material in wood is converted to gases, which escape. The gases produced are most likely CO_2 and H_2O.

c. The atom is not an indivisible particle, but is instead composed of other smaller particles, e.g., electrons, neutrons, protons.

d. The two hydride samples contain different isotopes of either hydrogen and/or lithium. Although the compounds are composed of different isotopes, their properties are similar because different isotopes of the same element have similar properties (except, of course, their mass).

Integrated Problems

91. The systematic name of Ta_2O_5 is tantalum(V) oxide. Tantalum is a transition metal and requires a Roman numeral. Sulfur is in the same group as oxygen and its most common ion is S^{2-}. Therefore, the formula of the sulfur analogue would be Ta_2S_5.

Total number of protons in Ta_2O_5:

Ta, Z = 73, so 73 protons × 2 = 146 protons; O, Z = 8, so 8 protons × 5 = 40 protons

Total protons = 186 protons

Total number of protons in Ta_2S_5:

Ta, Z = 73, so 73 protons × 2 = 146 protons; S, Z = 16, so 16 protons × 5 = 80 protons

Total protons = 226 protons

Proton difference between Ta_2S_5 and Ta_2O_5: 226 protons – 186 protons = 40 protons

93. Number of electrons in the unknown ion:

$$2.55 \times 10^{-26}\ \text{g} \times \frac{1\ \text{kg}}{1000\ \text{g}} \times \frac{1\ \text{electron}}{9.11 \times 10^{-31}\ \text{kg}} = 28\ \text{electrons}$$

Number of protons in the unknown ion:

$$5.34 \times 10^{-23}\ \text{g} \times \frac{1\ \text{kg}}{1000\ \text{g}} \times \frac{1\ \text{proton}}{1.67 \times 10^{-27}\ \text{kg}} = 32\ \text{protons}$$

Therefore this ion has 32 protons and 28 electrons. This is element number 32, germanium (Ge). The charge is +4 because four electrons have been lost from a neutral germanium atom.

The number of electrons in the unknown atom:

$$3.92 \times 10^{-26}\ \text{g} \times \frac{1\ \text{kg}}{1000\ \text{g}} \times \frac{1\ \text{electron}}{9.11 \times 0^{-31}\ \text{kg}} = 43\ \text{electrons}$$

In a neutral atom, the number of protons and electrons is the same. Therefore, this is element 43, technetium (Tc).

The number of neutrons in the technetium atom:

$$9.35 \times 10^{-23}\ \text{g} \times \frac{1\ \text{kg}}{1000\ \text{g}} \times \frac{1\ \text{proton}}{1.67 \times 10^{-27}\ \text{kg}} = 56\ \text{neutrons}$$

The mass number is the sum of the protons and neutrons. In this atom, the mass number is 43 protons + 56 neutrons = 99. Thus, this atom and its mass number is ^{99}Tc.

CHAPTER THREE

STOICHIOMETRY

Questions

19.

isotope	mass	abundance
^{12}C	$12.000\overline{0}$ amu	98.89%
^{13}C	13.034 amu	1.11%

$$\text{average mass} = 0.9889\,(12.000\overline{0}) + 0.0111(13.034) = 12.01 \text{ amu}$$

From the relative abundances, there would be 9889 atoms of ^{12}C and 111 atoms of ^{13}C in the 10,000 atom sample. The average mass of carbon is independent of the sample size; it will always be 12.01 amu.

$$\text{total mass} = 10{,}000 \text{ atoms} \times \frac{12.01 \text{ amu}}{\text{atom}} = 1.201 \times 10^5 \text{ amu}$$

For one mol of carbon (6.0221×10^{23} atoms C), the average mass would still be 12.01 amu.

The number of ^{12}C atoms would be $0.9889\,(6.0221 \times 10^{23}) = 5.955 \times 10^{23}$ atoms ^{12}C and the number of ^{13}C atoms would be $0.0111\,(6.0221 \times 10^{23}) = 6.68 \times 10^{21}$ atoms ^{13}C.

$$\text{total mass} = 6.0221 \times 10^{23} \text{ atoms} \times \frac{12.01 \text{ amu}}{\text{atom}} = 7.233 \times 10^{24} \text{ amu}$$

$$\text{total mass in g} = 6.0221 \times 10^{23} \text{ atoms} \times \frac{12.01 \text{ amu}}{\text{atom}} \times \frac{1 \text{ g}}{6.0221 \times 10^{23} \text{ amu}} = 12.01 \text{ g/mol}$$

By using the carbon-12 standard to define the relative masses of all of the isotopes as well as to define the number of things in a mole, then each element's average atomic mass in units of grams is the mass of a mole of that element as it is found in nature.

21.

$$\text{Avogadro's number of dollars} = \frac{6.022 \times 10^{23} \text{ dollars}}{\text{mol dollars}}$$

$$\frac{1 \text{ mol dollars} \times \dfrac{6.022 \times 10^{23} \text{ dollars}}{\text{mol dollars}}}{6 \times 10^9 \text{ people}} = 1 \times 10^{14} \text{ dollars/person}$$

1 trillion = 1,000,000,000,000 = 1×10^{12}; Each person would have 100 trillion dollars.

23. The mass percent of a compound is a constant no matter what amount of substance is present. Compounds always have constant composition.

25. The specific information needed is mostly the coefficients in the balanced equation and the molar masses of the reactants and products. For percent yield, we would need the actual yield of the reaction and the amounts of reactants used.

a. mass of CB produced = 1.00×10^4 molecules A_2B_2 ×

$$\frac{1 \text{ mol } A_2B_2}{6.022 \times 10^{23} \text{ molecules } A_2B_2} \times \frac{2 \text{ mol CB}}{1 \text{ mol } A_2B_2} \times \frac{\text{molar mass of CB}}{\text{mol CB}}$$

b. atoms of A produced = $1.00 \times 10^4 \text{ molecules } A_2B_2 \times \frac{2 \text{ atoms A}}{1 \text{ molecule } A_2B_2}$

c. mol of C reacted = $1.00 \times 10^4 \text{ molecules } A_2B_2 \times \frac{1 \text{ mol } A_2B_2}{6.022 \times 10^{23} \text{ molecules } A_2B_2} \times \frac{2 \text{ mol C}}{1 \text{ mol } A_2B_2}$

d. $\% \text{ yield} = \frac{\text{actual mass}}{\text{theoretical mass}} \times 100$; The theoretical mass of CB produced was calculated in part a. If the actual mass of CB produced is given, then the percent yield can be determined for the reaction using the % yield equation.

Exercises

Atomic Masses and the Mass Spectrometer

27. A = 0.0140(203.973) + 0.2410(205.9745) + 0.2210(206.9759) + 0.5240(207.9766)

A = 2.86 + 49.64 + 45.74 + 109.0 = 207.2 amu; From the periodic table, the element is Pb.

29. Let A = mass of ^{185}Re:

186.207 = 0.6260(186.956) + 0.3740(A), 186.207 − 117.0 = 0.3740(A)

$A = \frac{69.2}{0.3740} = 185$ amu (A = 184.95 amu without rounding to proper significant figures.)

31. There are three peaks in the mass spectrum, each 2 mass units apart. This is consistent with two isotopes, differing in mass by two mass units. The peak at 157.84 corresponds to a Br_2 molecule composed of two atoms of the lighter isotope. This isotope has mass equal to 157.84/2 or 78.92. This corresponds to ^{79}Br. The second isotope is ^{81}Br with mass equal to 161.84/2 = 80.92. The peaks in the mass spectrum correspond to $^{79}Br_2$, $^{79}Br^{81}Br$, and $^{81}Br_2$ in order of increasing mass. The intensities of the highest and lowest mass tell us the two

isotopes are present in about equal abundance. The actual abundance is 50.69% ^{79}Br and 49.31% ^{81}Br. The calculation of the abundance from the mass spectrum is beyond the scope of this text.

Moles and Molar Masses

33. When more than one conversion factor is necessary to determine the answer, we will usually put all the conversion factors into one calculation instead of determining intermediate answers. This method reduces round-off error and is a time saver.

$$500.\text{ atoms Fe} \times \frac{1\text{ mol Fe}}{6.022\times10^{23}\text{ atoms Fe}} \times \frac{55.85\text{ g Fe}}{\text{mol Fe}} = 4.64\times10^{-20}\text{ g Fe}$$

35. $$1.00\text{ carat} \times \frac{0.200\text{ g C}}{\text{carat}} \times \frac{1\text{ mol C}}{12.01\text{ g C}} \times \frac{6.022\times10^{23}\text{ atoms C}}{\text{mol C}} = 1.00\times10^{22}\text{ atoms C}$$

37. Al_2O_3: 2(26.98) + 3(16.00) = 101.96 g/mol

Na_3AlF_6: 3(22.99) + 1(26.98) + 6(19.00) = 209.95 g/mol

39. a. The formula is NH_3. 14.01 g/mol + 3(1.008 g/mol) = 17.03 g/mol

b. The formula is N_2H_4. 2(14.01) + 4(1.008) = 32.05 g/mol

c. $(NH_4)_2Cr_2O_7$: 2(14.01) + 8(1.008) + 2(52.00) + 7(16.00) = 252.08 g/mol

41. a. $$1.00\text{ g }NH_3 \times \frac{1\text{ mol }NH_3}{17.03\text{ g }NH_3} = 0.0587\text{ mol }NH_3$$

b. $$1.00\text{ g }N_2H_4 \times \frac{1\text{ mol }N_2H_4}{32.05\text{ g }N_2H_4} = 0.0312\text{ mol }N_2H_4$$

c. $$1.00\text{ g }(NH_4)_2Cr_2O_7 \times \frac{1\text{ mol }(NH_4)_2Cr_2O_7}{252.08\text{ g }(NH_4)_2Cr_2O_7} = 3.97\times10^{-3}\text{ mol }(NH_4)_2Cr_2O_7$$

43. a. $$5.00\text{ mol }NH_3 \times \frac{17.03\text{ g }NH_4}{\text{mol }NH_3} = 85.2\text{ g }NH_3$$

b. $$5.00\text{ mol }N_2H_4 \times \frac{32.05\text{ g }N_2H_4}{\text{mol }N_2H_4} = 160.\text{ g }N_2H_4$$

c. $$5.00\text{ mol }(NH_4)_2Cr_2O_7 \times \frac{252.08\text{ g }(NH_4)_2Cr_2O_7}{1\text{ mol }(NH_4)_2Cr_2O_7} = 1260\text{ g }(NH_4)_2Cr_2O_7$$

45. Chemical formulas give atom ratios as well as mol ratios.

a. $5.00 \text{ mol } NH_3 \times \frac{1 \text{ mol N}}{\text{mol } NH_3} \times \frac{14.01 \text{ g N}}{\text{mol N}} = 70.1 \text{ g N}$

b. $5.00 \text{ mol } N_2H_4 \times \frac{2 \text{ mol N}}{\text{mol } N_2H_4} \times \frac{14.01 \text{ g N}}{\text{mol N}} = 140. \text{ g N}$

c. $5.00 \text{ mol } (NH_4)_2Cr_2O_7 \times \frac{2 \text{ mol N}}{\text{mol } (NH_4)_2Cr_2O_7} \times \frac{14.01 \text{ g N}}{\text{mol N}} = 140. \text{ g N}$

47. a. $1.00 \text{ g } NH_3 \times \frac{1 \text{ mol } NH_3}{17.03 \text{ g } NH_3} \times \frac{6.022 \times 10^{23} \text{ molecules } NH_3}{\text{mol } NH_3} = 3.54 \times 10^{22} \text{ molecules } NH_3$

b. $1.00 \text{ g } N_2H_4 \times \frac{1 \text{ mol } N_2H_4}{32.05 \text{ g } N_2H_4} \times \frac{6.022 \times 10^{23} \text{ molecules } N_2H_4}{\text{mol } N_2H_4}$

$= 1.88 \times 10^{22} \text{ molecules } N_2H_4$

c. $1.00 \text{ g } (NH_4)_2Cr_2O_7 \times \frac{1 \text{ mol } (NH_4)_2Cr_2O_7}{252.08 \text{ g } (NH_4)_2Cr_2O_7}$

$\times \frac{6.022 \times 10^{23} \text{ formula units } (NH_4)_2Cr_2O_7}{\text{mol } (NH_4)_2Cr_2O_7} = 2.39 \times 10^{21} \text{ formula units } (NH_4)_2Cr_2O_7$

49. Using answers from Exercise 47:

a. $3.54 \times 10^{22} \text{ molecules } NH_3 \times \frac{1 \text{ atom N}}{\text{molecule } NH_3} = 3.54 \times 10^{22} \text{ atoms N}$

b. $1.88 \times 10^{22} \text{ molecules } N_2H_4 \times \frac{2 \text{ atoms N}}{\text{molecule } N_2H_4} = 3.76 \times 10^{22} \text{ atoms N}$

c. $2.39 \times 10^{21} \text{ formula units } (NH_4)_2Cr_2O_7 \times \frac{2 \text{ atoms N}}{\text{formula unit } (NH_4)_2Cr_2O_7}$

$= 4.78 \times 10^{21} \text{ atoms N}$

51. Molar mass of $C_6H_8O_6 = 6(12.01) + 8(1.008) + 6(16.00) = 176.12$ g/mol

$500.0 \text{ mg} \times \frac{1 \text{ g}}{1000 \text{ mg}} \times \frac{1 \text{ mol}}{176.12 \text{ g}} = 2.839 \times 10^{-3} \text{ mol}$

$2.839 \times 10^{-3} \text{ mol} \times \frac{6.022 \times 10^{23} \text{ molecules}}{\text{mol}} = 1.710 \times 10^{21} \text{ molecules}$

53. a. $150.0 \text{ g } Fe_2O_3 \times \dfrac{1 \text{ mol}}{159.70 \text{ g}} = 0.9393 \text{ mol } Fe_2O_3$

b. $10.0 \text{ mg } NO_2 \times \dfrac{1 \text{ g}}{1000 \text{ mg}} \times \dfrac{1 \text{ mol}}{46.01 \text{ g}} = 2.17 \times 10^{-4} \text{ mol } NO_2$

c. $1.5 \times 10^{16} \text{ molecules } BF_3 \times \dfrac{1 \text{ mol}}{6.02 \times 10^{23} \text{ molecules}} = 2.5 \times 10^{-8} \text{ mol } BF_3$

55. a. A chemical formula gives atom ratios as well as mole ratios. We will use both ideas to show how these conversion factors can be used.

Molar mass of $C_2H_5O_2N = 2(12.01) + 5(1.008) + 2(16.00) + 14.01 = 75.07$ g/mol

$$5.00 \text{ g } C_2H_5O_2N \times \frac{1 \text{ mol } C_2H_5O_2N}{75.07 \text{ g } C_2H_5O_2N} \times \frac{6.022 \times 10^{23} \text{ molecules } C_2H_5O_2N}{\text{mol } C_2H_5O_2N} \times$$

$$\frac{1 \text{ atom N}}{\text{molecule } C_2H_5O_2N} = 4.01 \times 10^{22} \text{ atoms N}$$

b. Molar mass of $Mg_3N_2 = 3(24.31) + 2(14.01) = 100.95$ g/mol

$$5.00 \text{ g } Mg_3N_2 \times \frac{1 \text{ mol } Mg_3N_2}{100.95 \text{ g } Mg_3N_2} \times \frac{6.022 \times 10^{23} \text{ formula units } Mg_3N_2}{\text{mol } Mg_3N_2} \times \frac{2 \text{ atoms N}}{\text{mol } Mg_3N_2}$$

$$= 5.97 \times 10^{22} \text{ atoms N}$$

c. Molar mass of $Ca(NO_3)_2 = 40.08 + 2(14.01) + 6(16.00) = 164.10$ g/mol

$$5.00 \text{ g } Ca(NO_3)_2 \times \frac{1 \text{ mol } Ca(NO_3)_2}{164.10 \text{ g } Ca(NO_3)_2} \times \frac{2 \text{ mols N}}{\text{mol } Ca(NO_3)_2} \times \frac{6.022 \times 10^{23} \text{ atoms N}}{\text{mol N}}$$

$$= 3.67 \times 10^{22} \text{ atoms N}$$

d. Molar mass of $N_2O_4 = 2(14.01) + 4(16.00) = 92.02$ g/mol

$$5.00 \text{ g } N_2O_4 \times \frac{1 \text{ mol } N_2O_4}{92.02 \text{ g } N_2O_4} \times \frac{2 \text{ mol N}}{\text{mol } N_2O_4} \times \frac{6.022 \times 10^{23} \text{ atoms N}}{\text{mol N}}$$

$$= 6.54 \times 10^{22} \text{ atoms N}$$

57. a. $$14 \text{ mol C}\left(\frac{12.01 \text{ g}}{\text{mol C}}\right) + 18 \text{ mol H}\left(\frac{1.008 \text{ g}}{\text{mol H}}\right) + 2 \text{ mol N}\left(\frac{14.01 \text{ g}}{\text{mol N}}\right) + 5 \text{ mol O}\left(\frac{16.00 \text{ g}}{\text{mol O}}\right)$$

$$= 294.30 \text{ g/mol}$$

b. $10.0 \text{ g aspartame} \times \dfrac{1 \text{ mol}}{294.30 \text{ g}} = 3.40 \times 10^{-2} \text{ mol}$

c. $1.56 \text{ mol} \times \frac{294.30 \text{ g}}{\text{mol}} = 459 \text{ g}$

d. $5.0 \text{ mg} \times \frac{1 \text{ g}}{1000 \text{ mg}} \times \frac{1 \text{ mol}}{294.30 \text{ g}} \times \frac{6.02 \times 10^{23} \text{ molecules}}{\text{mol}} = 1.0 \times 10^{19} \text{ molecules}$

e. The chemical formula tells us that 1 molecule of aspartame contains two atoms of N. The chemical formula also says that 1 mol of aspartame contains two mol of N.

$$1.2 \text{ g aspartame} \times \frac{1 \text{ mol aspartame}}{294.30 \text{ g aspartame}} \times \frac{2 \text{ mol N}}{\text{mol aspartame}} \times \frac{6.02 \times 10^{23} \text{ atoms N}}{\text{mol N}}$$

$$= 4.9 \times 10^{21} \text{ atoms of nitrogen}$$

f. $1.0 \times 10^9 \text{ molecules} \times \frac{1 \text{ mol}}{6.02 \times 10^{23} \text{ molecules}} \times \frac{294.30 \text{ g}}{\text{mol}} = 4.9 \times 10^{-13} \text{ g or } 490 \text{ fg}$

g. $1 \text{ molecule aspartame} \times \frac{1 \text{ mol}}{6.022 \times 10^{23} \text{ molecules}} \times \frac{294.30 \text{ g}}{\text{mol}} = 4.887 \times 10^{-22} \text{ g}$

Percent Composition

59. a. $C_3H_4O_2$: Molar mass = 3(12.01) + 4(1.008) + 2(16.00) = 36.03 + 4.032 + 32.00 = 72.06 g/mol

$$\%C = \frac{36.03 \text{ g C}}{72.06 \text{ g compound}} \times 100 = 50.00\% \text{ C}; \quad \%H = \frac{4.032 \text{ g H}}{72.06 \text{ g compound}} \times 100 = 5.595\% \text{ H}$$

$$\%O = 100.00 - (50.00 + 5.595) = 44.41\% \text{ O or } \%O = \frac{32.00 \text{ g}}{72.06 \text{ g}} \times 100 = 44.41\% \text{ O}$$

b. $C_4H_6O_2$: Molar mass = 4(12.01) + 6(1.008) + 2(16.00) = 48.04 + 6.048 + 32.00 = 86.09 g/mol

$$\%C = \frac{48.04 \text{ g}}{86.09 \text{ g}} \times 100 = 55.80\% \text{ C}; \quad \%H = \frac{6.048 \text{ g}}{86.09 \text{ g}} \times 100 = 7.025\% \text{ H}$$

$$\%O = 100.00 - (55.80 + 7.025) = 37.18\% \text{ O}$$

c. C_3H_3N: Molar mass = 3(12.01) + 3(1.008) + 1(14.01) = 36.03 + 3.024 + 14.01 = 53.06 g/mol

$$\%C = \frac{36.03 \text{ g}}{53.06 \text{ g}} \times 100 = 67.90\% \text{ C}; \quad \%H = \frac{3.024 \text{ g}}{53.06 \text{ g}} \times 100 = 5.699\% \text{ H}$$

$$\%N = \frac{14.01 \text{ g}}{53.06 \text{ g}} \times 100 = 26.40\% \text{ N or } \%N = 100.00 - (67.90 + 5.699) = 26.40\% \text{ N}$$

61. a. NO: $\%N = \frac{14.01 \text{ g N}}{30.01 \text{ g NO}} \times 100 = 46.68\% \text{ N}$

b. NO_2: $\%N = \frac{14.01 \text{ g N}}{46.01 \text{ g NO}_2} \times 100 = 30.45\% \text{ N}$

c. N_2O_4: $\%N = \frac{28.02 \text{ g N}}{92.02 \text{ g N}_2\text{O}_4} \times 100 = 30.45\% \text{ N}$

d. N_2O: $\%N = \frac{28.02 \text{ g N}}{44.02 \text{ g N}_2\text{O}} \times 100 = 63.65\% \text{ N}$

The order from lowest to highest mass percentage of nitrogen is: $NO_2 = N_2O_4 < NO < N_2O$.

63. There are many valid methods to solve this problem. We will assume 100.00 g of compound, then determine from the information in the problem how many mol of compound equals 100.00 g of compound. From this information, we can determine the mass of one mol of compound (the molar mass) by setting up a ratio. Assuming 100.00 g cyanocobalamin:

$$\text{mol cyanocobalamin} = 4.34 \text{ g Co} \times \frac{1 \text{ mol Co}}{58.93 \text{ g Co}} \times \frac{1 \text{ mol cyanocobalamin}}{\text{mol Co}}$$

$$= 7.36 \times 10^{-2} \text{ mol cyanocobalamin}$$

$$\frac{x \text{ g cyanocobalamin}}{1 \text{ mol cyanocobalamin}} = \frac{100.00 \text{ g}}{7.36 \times 10^{-2} \text{ mol}}, \quad x = \text{molar mass} = 1360 \text{ g/mol}$$

Empirical and Molecular Formulas

65. a. Molar mass of CH_2O = $1 \text{ mol C} \left(\frac{12.01 \text{ g C}}{\text{mol C}}\right) + 2 \text{ mol H} \left(\frac{1.008 \text{ g H}}{\text{mol H}}\right)$

$$+ 1 \text{ mol O} \left(\frac{16.00 \text{ g O}}{\text{mol O}}\right) = 30.03 \text{ g/mol}$$

$\%C = \frac{12.01 \text{ g C}}{30.03 \text{ g CH}_2\text{O}} \times 100 = 39.99\% \text{ C}$; $\%H = \frac{2.016 \text{ g H}}{30.03 \text{ g CH}_2\text{O}} \times 100 = 6.713\% \text{ H}$

$\%O = \frac{16.00 \text{ g O}}{30.03 \text{ g CH}_2\text{O}} \times 100 = 53.28\% \text{ O}$ or $\%O = 100.00 - (39.99 + 6.713) = 53.30\%$

b. Molar Mass of $C_6H_{12}O_6$ = 6(12.01) + 12(1.008) + 6(16.00) = 180.16 g/mol

$\%C = \frac{76.06 \text{ g C}}{180.16 \text{ g C}_6\text{H}_{12}\text{O}_6} \times 100 = 40.00\%$; $\%H = \frac{12.(1.008) \text{ g}}{180.16 \text{ g}} \times 100 = 6.714\%$

$\%O = 100.00 - (40.00 + 6.714) = 53.29\%$

c. Molar mass of $HC_2H_3O_2$ = 2(12.01) + 4(1.008) + 2(16.00) = 60.05 g/mol

$$\%C = \frac{24.02 \text{ g}}{60.05 \text{ g}} \times 100 = 40.00\%; \quad \%H = \frac{4.032 \text{ g}}{60.05 \text{ g}} \times 100 = 6.714\%$$

$$\%O = 100.00 - (40.00 + 6.714) = 53.29\%$$

67. a. The molecular formula is N_2O_4. The smallest whole number ratio of the atoms (the empirical formula) is NO_2.

b. Molecular formula: C_3H_6; empirical formula = CH_2

c. Molecular formula: P_4O_{10}; empirical formula = P_2O_5

d. Molecular formula: $C_6H_{12}O_6$; empirical formula = CH_2O

69. Out of 100.00 g of adrenaline, there are:

$$56.79 \text{ g C} \times \frac{1 \text{ mol C}}{12.01 \text{ g C}} = 4.729 \text{ mol C}; \quad 6.56 \text{ g H} \times \frac{1 \text{ mol H}}{1.008 \text{ g H}} = 6.51 \text{ mol H}$$

$$28.37 \text{ g O} \times \frac{1 \text{ mol O}}{16.00 \text{ g O}} = 1.773 \text{ mol O}; \quad 8.28 \text{ g N} \times \frac{1 \text{ mol N}}{14.01 \text{ g N}} = 0.591 \text{ mol N}$$

Dividing each mol value by the smallest number:

$$\frac{4.729}{0.591} = 8.00; \quad \frac{6.51}{0.591} = 11.0; \quad \frac{1.773}{0.591} = 3.00; \quad \frac{0.591}{0.591} = 1.00$$

This gives adrenaline an empirical formula of $C_8H_{11}O_3N$.

71. Compound I: mass O = 0.6498 g Hg_xO_y − 0.6018 g Hg = 0.0480 g O

$$0.6018 \text{ g Hg} \times \frac{1 \text{ mol Hg}}{200.6 \text{ g Hg}} = 3.000 \times 10^{-3} \text{ mol Hg}$$

$$0.0480 \text{ g O} \times \frac{1 \text{ mol O}}{16.00 \text{ g O}} = 3.00 \times 10^{-3} \text{ mol O}$$

The mol ratio between Hg and O is 1:1, so the empirical formula of compound I is HgO.

Compound II: mass Hg = 0.4172 g Hg_xO_y − 0.016 g O = 0.401 g Hg

$$0.401 \text{ g Hg} \times \frac{1 \text{ mol Hg}}{200.6 \text{ g Hg}} = 2.00 \times 10^{-3} \text{ mol Hg}; \quad 0.016 \text{ g O} \times \frac{1 \text{ mol O}}{16.00 \text{ g O}}$$

$$= 1.0 \times 10^{-3} \text{ mol O}$$

The mol ratio between Hg and O is 2:1, so the empirical formula is Hg_2O.

73. Out of 100.0 g, there are:

$$69.6\text{ g S} \times \frac{1\text{ mol S}}{32.07\text{ g S}} = 2.17\text{ mol S}; \; 30.4\text{ g N} \times \frac{1\text{ mol N}}{14.01\text{ g N}} = 2.17\text{ mol N}$$

Empirical formula is SN since mol values are in a 1:1 mol ratio.

The empirical formula mass of SN is ~ 46 g. Because 184/46 = 4.0, the molecular formula is S_4N_4.

75. Assuming 100.00 g of compound (mass hydrogen = 100.00 g - 49.31 g C - 43.79 g O = 6.90 g H):

$$49.31\text{ g C} \times \frac{1\text{ mol C}}{12.01\text{ g C}} = 4.106\text{ mol C}; \; 6.90\text{ g H} \times \times \frac{1\text{ mol H}}{1.008\text{ g H}} = 6.85\text{ mol H}$$

$$43.79\text{ g O} \times \frac{1\text{ mol O}}{16.00\text{ g O}} = 2.737\text{ mol O}$$

Dividing all mole values by 2.737 gives:

$$\frac{4.106}{2.737} = 1.500; \; \frac{6.85}{2.737} = 2.50; \; \frac{2.737}{2.737} = 1.000$$

Since a whole number ratio is required, the empirical formula is $C_3H_5O_2$.

The empirical formula mass is: 3(12.01) + 5(1.008) +2(16.00) = 73.07 g/mol

$$\frac{\text{molar mass}}{\text{empirical formula mass}} = \frac{146.1}{73.07} = 1.999; \; \text{molecular formula} = (C_3H_5O_2)_2 = C_6H_{10}O_4$$

77. When combustion data are given, it is assumed that all the carbon in the compound ends up as carbon in CO_2 and all the hydrogen in the compound ends up as hydrogen in H_2O. In the sample of propane combusted, the moles of C and H are:

$$\text{mol C} = 2.641\text{ g } CO_2 \times \frac{1\text{ mol } CO_2}{44.01\text{ g } CO_2} \times \frac{1\text{ mol C}}{\text{mol } CO_2} = 0.06001\text{ mol C}$$

$$\text{mol H} = 1.442\text{ g } H_2O \times \frac{1\text{ mol } H_2O}{18.02\text{ g } H_2O} \times \frac{2\text{ mol H}}{\text{mol } H_2O} = 0.1600\text{ mol H}$$

$$\frac{\text{mol H}}{\text{mol C}} = \frac{0.1600}{0.06001} = 2.666$$

Multiplying this ratio by three gives the empirical formula of C_3H_8.

79. The combustion data allow determination of the amount of hydrogen in cumene. One way to determine the amount of carbon in cumene is to determine the mass percent of hydrogen in the compound from the data in the problem; then determine the mass percent of carbon by difference (100.0 - mass %H = mass %C).

$$42.8 \text{ mg } H_2O \times \frac{1 \text{ g}}{1000 \text{ mg}} \times \frac{2.016 \text{ g H}}{18.02 \text{ g } H_2O} \times \frac{1000 \text{ mg}}{\text{g}} = 4.79 \text{ mg H}$$

$$\%H = \frac{4.79 \text{ mg H}}{47.6 \text{ mg cumene}} \times 100 = 10.1\% \text{ H}; \ \%C = 100.0 - 10.1 = 89.9\% \text{ C}$$

Now solve this empirical formula problem. Out of 100.0 g cumene, we have:

$$89.9 \text{ g C} \times \frac{1 \text{ mol C}}{12.01 \text{ g C}} = 7.49 \text{ mol C}; \ 10.1 \text{ g H} \times \frac{1 \text{ mol H}}{1.008 \text{ g H}} = 10.0 \text{ mol H}$$

$\frac{10.0}{7.49} = 1.34 \approx \frac{4}{3}$, i.e., mol H to mol C are in a 4:3 ratio. Empirical formula = C_3H_4

Empirical formula mass $\approx 3(12) + 4(1) = 40$ g/mol

The molecular formula is $(C_3H_4)_3$ or C_9H_{12} since the molar mass will be between 115 and 125 g/mol (molar mass $\approx 3 \times 40$ g/mol = 120 g/mol).

Balancing Chemical Equations

81. When balancing reactions, start with elements that appear in only one of the reactants and one of the products, then go on to balance the remaining elements.

a. $C_6H_{12}O_6(s) + O_2(g) \rightarrow CO_2(g) + H_2O(g)$

Balance C atoms: $C_6H_{12}O_6 + O_2 \rightarrow 6\ CO_2 + H_2O$

Balance H atoms: $C_6H_{12}O_6 + O_2 \rightarrow 6\ CO_2 + 6\ H_2O$

Lastly, balance O atoms: $C_6H_{12}O_6(s) + 6\ O_2(g) \rightarrow 6\ CO_2(g) + 6\ H_2O(g)$

b. $Fe_2S_3(s) + HCl(g) \rightarrow FeCl_3(s) + H_2S(g)$

Balance Fe atoms: $Fe_2S_3 + HCl \rightarrow 2\ FeCl_3 + H_2S$

Balance S atoms: $Fe_2S_3 + HCl \rightarrow 2\ FeCl_3 + 3\ H_2S$

There are 6 H and 6 Cl on right, so balance with 6 HCl on left:

$Fe_2S_3(s) + 6\ HCl(g) \rightarrow 2\ FeCl_3(s) + 3\ H_2S(g)$.

c. $CS_2(l) + NH_3(g) \rightarrow H_2S(g) + NH_4SCN(s)$

C and S balanced; balance N:

$$CS_2 + 2\ NH_3 \rightarrow H_2S + NH_4SCN$$

H is also balanced. So: $CS_2(l) + 2\ NH_3(g) \rightarrow H_2S(g) + NH_4SCN(s)$

83. a. $3\ Ca(OH)_2(aq) + 2\ H_3PO_4(aq) \rightarrow 6\ H_2O(l) + Ca_3(PO_4)_2(s)$

b. $Al(OH)_3(s) + 3\ HCl(aq) \rightarrow AlCl_3(aq) + 3\ H_2O(l)$

c. $2\ AgNO_3(aq) + H_2SO_4(aq) \rightarrow Ag_2SO_4(s) + 2\ HNO_3(aq)$

85. a. The formulas of the reactants and products are $C_6H_6(l) + O_2(g) \rightarrow CO_2(g) + H_2O(g)$. To balance this combustion reaction, notice that all of the carbon in C_6H_6 has to end up as carbon in CO_2 and all of the hydrogen in C_6H_6 has to end up as hydrogen in H_2O. To balance C and H, we need 6 CO_2 molecules and 3 H_2O molecules for every 1 molecule of C_6H_6. We do oxygen last. Because we have 15 oxygen atoms in 6 CO_2 molecules and 3 H_2O molecules, we need 15/2 O_2 molecules in order to have 15 oxygen atoms on the reactant side.

$C_6H_6(l) + \frac{15}{2}\ O_2(g) \rightarrow 6\ CO_2(g) + 3\ H_2O(g)$; Multiply by two to give whole numbers.

$$2\ C_6H_6(l) + 15\ O_2(g) \rightarrow 12\ CO_2(g) + 6\ H_2O(g)$$

b. The formulas of the reactants and products are $C_4H_{10}(g) + O_2(g) \rightarrow CO_2(g) + H_2O(g)$.

$C_4H_{10}(g) + \frac{13}{2}\ O_2(g) \rightarrow 4\ CO_2(g) + 5\ H_2O(g)$; Multiply by two to give whole numbers.

$$2\ C_4H_{10}(g) + 13\ O_2(g) \rightarrow 8\ CO_2(g) + 10\ H_2O(g)$$

c. $C_{12}H_{22}O_{11}(s) + 12\ O_2(g) \rightarrow 12\ CO_2(g) + 11\ H_2O(g)$

d. $2\ Fe(s) + \frac{3}{2}\ O_2(g) \rightarrow Fe_2O_3(s)$; For whole numbers: $4\ Fe(s) + 3\ O_2(g) \rightarrow 2\ Fe_2O_3(s)$

e. $2\ FeO(s) + \frac{1}{2}\ O_2(g) \rightarrow Fe_2O_3(s)$; For whole numbers, multiply by two.

$$4\ FeO(s) + O_2(g) \rightarrow 2\ Fe_2O_3(s)$$

87. a. $SiO_2(s) + C(s) \rightarrow Si(s) + CO(g)$

Balance oxygen atoms: $SiO_2 + C \rightarrow Si + 2\ CO$

Balance carbon atoms: $SiO_2(s) + 2\ C(s) \rightarrow Si(s) + 2\ CO(g)$

b. $SiCl_4(l) + Mg(s) \rightarrow Si(s) + MgCl_2(s)$

Balance Cl atoms: $SiCl_4 + Mg \rightarrow Si + 2\ MgCl_2$

Balance Mg atoms: $SiCl_4(l) + 2\ Mg(s) \rightarrow Si(s) + 2\ MgCl_2(s)$

c. $Na_2SiF_6(s) + Na(s) \rightarrow Si(s) + NaF(s)$

Balance F atoms: $Na_2SiF_6 + Na \rightarrow Si + 6\ NaF$

Balance Na atoms: $Na_2SiF_6(s) + 4\ Na(s) \rightarrow Si(s) + 6\ NaF(s)$

Reaction Stoichiometry

89. The stepwise method to solve stoichiometry problems is outlined in the text. Instead of calculating intermediate answers for each step, we will combine conversion factors into one calculation. This practice reduces round-off error and saves time.

$Fe_2O_3(s) + 2\ Al(s) \rightarrow 2\ Fe(l) + Al_2O_3(s)$

$$15.0\text{ g Fe} \times \frac{1\text{ mol Fe}}{55.85\text{ g Fe}} = 0.269\text{ mol Fe};\ 0.269\text{ mol Fe} \times \frac{2\text{ mol Al}}{2\text{ mol Fe}} \times \frac{26.98\text{ g Al}}{\text{mol Al}} = 7.26\text{ g Al}$$

$$0.269\text{ mol Fe} \times \frac{1\text{ mol Fe}_2\text{O}_3}{2\text{ mol Fe}} \times \frac{159.70\text{ g Fe}_2\text{O}_3}{\text{mol Fe}_2\text{O}_3} = 21.5\text{ g Fe}_2\text{O}_3$$

$$0.269\text{ mol Fe} \times \frac{1\text{ mol Al}_2\text{O}_3}{2\text{ mol Fe}} \times \frac{101.96\text{ g Al}_2\text{O}_3}{\text{mol Al}_2\text{O}_3} = 13.7\text{ g Al}_2\text{O}_3$$

91. $$1.000\text{ kg Al} \times \frac{1000\text{ g Al}}{\text{kg Al}} \times \frac{1\text{ mol Al}}{26.98\text{ g Al}} \times \frac{3\text{ mol NH}_4\text{ClO}_4}{3\text{ mol Al}} \times \frac{117.49\text{ g NH}_4\text{ClO}_4}{\text{mol NH}_4\text{ClO}_4} = 4355\text{ g}$$

93. $$1.0 \times 10^4\text{ kg waste} \times \frac{3.0\text{ g NH}_4^+}{100\text{ kg waste}} \times \frac{1000\text{ g}}{\text{kg}} \times \frac{1\text{ mol NH}_4^+}{18.04\text{ g NH}_4^+} \times \frac{1\text{ mol C}_5\text{H}_7\text{O}_2\text{N}}{55\text{ mol NH}_4^+} \times$$

$$\frac{113.12\text{ g C}_5\text{H}_7\text{O}_2\text{N}}{\text{mol C}_5\text{H}_7\text{O}_2\text{N}} = 3.4 \times 10^4\text{ g tissue if all NH}_4^+\text{ converted}$$

Since only 95% of the NH_4^+ ions react:

mass of tissue = $(0.95)(3.4 \times 10^4\text{ g}) = 3.2 \times 10^4$ g or 32 kg bacterial tissue

95. a. $$1.00 \times 10^2\text{ g C}_7\text{H}_6\text{O}_3 \times \frac{1\text{ mol C}_7\text{H}_6\text{O}_3}{138.12\text{ g C}_7\text{H}_6\text{O}_3} \times \frac{1\text{ mol C}_4\text{H}_6\text{O}_3}{1\text{ mol C}_7\text{H}_6\text{O}_3} \times \frac{102.09\text{ g C}_4\text{H}_6\text{O}_3}{1\text{ mol C}_4\text{H}_6\text{O}_3}$$

$$= 73.9\text{ g C}_4\text{H}_6\text{O}_3$$

b. $1.00 \times 10^2 \text{ g } C_7H_6O_3 \times \frac{1 \text{ mol } C_7H_6O_3}{138.12 \text{ g } C_7H_6O_3} \times \frac{1 \text{ mol } C_9H_8O_4}{1 \text{ mol } C_7H_6O_3} \times \frac{180.15 \text{ g } C_9H_8O_4}{\text{mol } C_9H_8O_4}$

$= 1.30 \times 10^2$ g aspirin

Limiting Reactants and Percent Yield

97. The product formed in the reaction is NO_2; the other species present in the product representation is excess O_2. Therefore, NO is the limiting reactant. In the pictures, 6 NO molecules react with 3 O_2 molecules to form 6 NO_2 molecules.

$$6\ NO(g) + 3\ O_2(g) \rightarrow 6\ NO_2(g)$$

For smallest whole numbers, the balanced reaction is:

$$2\ NO(g) + O_2(g) \rightarrow 2\ NO_2(g)$$

99. $1.50 \text{ g } BaO_2 \times \frac{1 \text{ mol } BaO_2}{169.3 \text{ g } BaO_2} = 8.86 \times 10^{-3} \text{ mol } BaO_2$

$25.0 \text{ mL} \times \frac{0.0272 \text{ g HCl}}{\text{mL}} \times \frac{1 \text{ mol HCl}}{36.46 \text{ g HCl}} = 1.87 \times 10^{-2} \text{ mol HCl}$

The required mole ratio from the balanced reaction is 2 mol HCl to 1 mol BaO_2. The actual ratio is:

$$\frac{1.87 \times 10^{-2} \text{ mol HCl}}{8.86 \times 10^{-3} \text{ mol } BaO_2} = 2.11$$

Because the actual mole ratio is larger than the required mole ratio, the denominator (BaO_2) is the limiting reagent.

$8.86 \times 10^{-3} \text{ mol } BaO_2 \times \frac{1 \text{ mol } H_2O_2}{\text{mol } BaO_2} \times \frac{34.02 \text{ g } H_2O_2}{\text{mol } H_2O_2} = 0.301 \text{ g } H_2O_2$

The amount of HCl reacted is:

$$8.86 \times 10^{-3} \text{ mol } BaO_2 \times \frac{2 \text{ mol HCl}}{\text{mol } BaO_2} = 1.77 \times 10^{-2} \text{ mol HCl}$$

excess mol HCl = 1.87×10^{-2} mol − 1.77×10^{-2} mol = 1.0×10^{-3} mol HCl

mass of excess HCl = $1.0 \times 10^{-3} \text{ mol HCl} \times \frac{36.46 \text{ g HCl}}{\text{mol HCl}} = 3.6 \times 10^{-2} \text{ g HCl}$

101. An alternative method to solve limiting reagent problems is to assume each reactant is limiting and calculate how much product could be produced from each reactant. The reactant that produces the smallest amount of product will run out first and is the limiting reagent.

$$5.00 \times 10^6 \text{ g } NH_3 \times \frac{1 \text{ mol } NH_3}{17.03 \text{ g } NH_3} \times \frac{2 \text{ mol HCN}}{2 \text{ mol } NH_3} = 2.94 \times 10^5 \text{ mol HCN}$$

$$5.00 \times 10^6 \text{ g } O_2 \times \frac{1 \text{ mol } O_2}{32.00 \text{ g } O_2} \times \frac{2 \text{ mol HCN}}{3 \text{ mol } O_2} = 1.04 \times 10^5 \text{ mol HCN}$$

$$5.00 \times 10^6 \text{ g } CH_4 \times \frac{1 \text{ mol } CH_4}{16.04 \text{ g } CH_4} \times \frac{2 \text{ mol HCN}}{2 \text{ mol } CH_4} = 3.12 \times 10^5 \text{ mol HCN}$$

O_2 is limiting because it produces the smallest amount of HCN. Although more product could be produced from NH_3 and CH_4, only enough O_2 is present to produce 1.04×10^5 mol HCN.

The mass of HCN produced is:

$$1.04 \times 10^5 \text{ mol HCN} \times \frac{27.03 \text{ g HCN}}{\text{mol HCN}} = 2.81 \times 10^6 \text{ g HCN}$$

$$5.00 \times 10^6 \text{ g } O_2 \times \frac{1 \text{ mol } O_2}{32.00 \text{ g } O_2} \times \frac{6 \text{ mol } H_2O}{3 \text{ mol } O_2} \times \frac{18.02 \text{ g } H_2O}{1 \text{ mol } H_2O} = 5.63 \times 10^6 \text{ g } H_2O$$

103. $C_7H_6O_3 + C_4H_6O_3 \rightarrow C_9H_8O_4 + HC_2H_3O_2$

$$1.50 \text{ g } C_7H_6O_3 \times \frac{1 \text{ mol } C_7H_6O_3}{138.12 \text{ g } C_7H_6O_3} = 1.09 \times 10^{-2} \text{ mol } C_7H_6O_3$$

$$2.00 \text{ g } C_4H_6O_3 \times \frac{1 \text{ mol } C_4H_6O_3}{102.09 \text{ g } C_4H_6O_3} = 1.96 \times 10^{-2} \text{ mol } C_4H_6O_3$$

$C_7H_6O_3$ is the limiting reagent because the actual moles of $C_7H_6O_3$ present are below the required 1:1 mol ratio. The theoretical yield of aspirin is:

$$1.09 \times 10^{-2} \text{ mol } C_7H_6O_3 \times \frac{1 \text{ mol } C_9H_8O_4}{\text{mol } C_7H_6O_3} \times \frac{180.15 \text{ g } C_9H_8O_4}{\text{mol } C_9H_8O_4} = 1.96 \text{ g } C_9H_8O_4$$

$$\% \text{ yield} = \frac{1.50 \text{ g}}{1.96 \text{ g}} \times 100 = 76.5\%$$

105. $$2.50 \text{ metric tons } Cu_3FeS_3 \times \frac{1000 \text{ kg}}{\text{metric ton}} \times \frac{1000 \text{ g}}{\text{kg}} \times \frac{1 \text{ mol } Cu_3FeS_3}{342.71 \text{ g}} \times \frac{3 \text{ mol Cu}}{1 \text{ mol } Cu_3FeS_3} \times \frac{63.55 \text{ g}}{\text{mol Cu}}$$

$$= 1.39 \times 10^6 \text{ g Cu (theoretical)}$$

$$1.39 \times 10^6 \text{ g Cu (theoretical)} \times \frac{86.3 \text{ g Cu (actual)}}{100. \text{ g Cu (theoretical)}} = 1.20 \times 10^6 \text{ g Cu} = 1.20 \times 10^3 \text{ kg Cu}$$

$$= 1.20 \text{ metric tons Cu (actual)}$$

Additional Exercises

107. molar mass XeF_n = $\dfrac{0.368 \text{ g } XeF_n}{9.03\times10^{20} \text{ molecules } XeF_n \times \dfrac{1 \text{ mol } XeF_n}{6.022\times10^{23} \text{ molecules}}}$ = 245 g/mol

245 g = 131.3 g + n(19.00 g), n = 5.98; formula = XeF_6

109. $2\,H_2(g) + O_2(g) \rightarrow 2\,H_2O(g)$

a. 50 molecules $H_2 \times \dfrac{1 \text{ molecule } O_2}{2 \text{ molecules } H_2}$ = 25 molecules O_2

Stoichiometric mixture. Neither is limiting.

b. 100 molecules $H_2 \times \dfrac{1 \text{ molecule } O_2}{2 \text{ molecules } H_2}$ = 50 molecules O_2;

O_2 is limiting since only 40 molecules O_2 are present.

c. From b, 50 molecules of O_2 will react completely with 100 molecules of H_2. We have 100 molecules (an excess) of O_2. So, H_2 is limiting.

d. 0.50 mol $H_2 \times \dfrac{1 \text{ mol } O_2}{2 \text{ mol } H_2}$ = 0.25 mol O_2; H_2 is limiting because 0.75 mol O_2 are present.

e. 0.80 mol $H_2 \times \dfrac{1 \text{ mol } O_2}{2 \text{ mol } H_2}$ = 0.40 mol O_2; H_2 is limiting because 0.75 mol O_2 are present.

f. 1.0 g $H_2 \times \dfrac{1 \text{ mol } H_2}{2.016 \text{ g } H_2} \times \dfrac{1 \text{ mol } O_2}{2 \text{ mol } H_2}$ = 0.25 mol O_2

Stoichiometric mixture, neither is limiting.

g. 5.00 g $H_2 \times \dfrac{1 \text{ mol } H_2}{2.016 \text{ g } H_2} \times \dfrac{1 \text{ mol } O_2}{2 \text{ mol } H_2} \times \dfrac{32.00 \text{ g } O_2}{\text{mol } O_2}$ = 39.7 g O_2

H_2 is limiting because 56.00 g O_2 are present.

111. Empirical formula mass = 12.01 + 1.008 = 13.02 g/mol; Because 104.14/13.02 = 7.998 ≈ 8, the molecular formula for styrene is $(CH)_8 = C_8H_8$.

2.00 g $C_8H_8 \times \dfrac{1 \text{ mol } C_8H_8}{104.14 \text{ g } C_8H_8} \times \dfrac{8 \text{ mol H}}{\text{mol } C_8H_8} \times \dfrac{6.002\times10^{23} \text{ atoms H}}{\text{mol H}}$ = 9.25×10^{22} atoms H

113. $17.3\text{ g H} \times \dfrac{1\text{ mol H}}{1.008\text{ g H}} = 17.2\text{ mol H}$; $82.7\text{ g C} \times \dfrac{1\text{ mol C}}{12.01\text{ g C}} = 6.89\text{ mol C}$

$\dfrac{17.2}{6.89} = 2.50$; The empirical formula is C_2H_5.

The empirical formula mass is ~29 g, so two times the empirical formula would put the compound in the correct range of the molar mass. Molecular formula = $(C_2H_5)_2 = C_4H_{10}$

$$2.59 \times 10^{23}\text{ atoms H} \times \frac{1\text{ molecule } C_4H_{10}}{10\text{ atoms H}} \times \frac{1\text{ mol } C_4H_{10}}{6.022\times10^{23}\text{ molecules}} = 4.30 \times 10^{-2}\text{ mol } C_4H_{10}$$

$$4.30 \times 10^{-2}\text{ mol } C_4H_{10} \times \frac{58.12\text{ g}}{\text{mol } C_4H_{10}} = 2.50\text{ g } C_4H_{10}$$

115. Mass of H_2O = 0.755 g $CuSO_4 \bullet xH_2O$ - 0.483 g $CuSO_4$ = 0.272 g H_2O

$$0.483\text{ g } CuSO_4 \times \frac{1\text{ mol } CuSO_4}{159.62\text{ g } CuSO_4} = 0.00303\text{ mol } CuSO_4$$

$$0.272\text{ g } H_2O \times \frac{1\text{ mol } H_2O}{18.02\text{ g } H_2O} = 0.0151\text{ mol } H_2O$$

$$\frac{0.0151\text{ mol } H_2O}{0.00303\text{ g } CuSO_4} = \frac{4.98\text{ mol } H_2O}{1\text{ mol } CuSO_4}$$; Compound formula = $CuSO_4 \bullet 5\,H_2O$, $x = 5$

117. $$1.20\text{ g } CO_2 \times \frac{1\text{ mol } CO_2}{44.01\text{ g}} \times \frac{1\text{ mol C}}{\text{mol } CO_2} \times \frac{1\text{ mol } C_{24}H_{30}N_3O}{24\text{ mol C}} \times \frac{376.51\text{ g}}{\text{mol } C_{24}H_{30}N_3O}$$

$$= 0.428\text{ g } C_{24}H_{30}N_3O$$

$$\frac{0.428\text{ g } C_{24}H_{30}N_3O}{1.00\text{ g sample}} \times 100 = 42.8\%\ C_{24}H_{30}N_3O$$

119. $126\text{ g } B_5H_9 \times \dfrac{1\text{ mol}}{63.12\text{ g}} = 2.00\text{ mol } B_5H_9$; $192\text{ g } O_2 \times \dfrac{1\text{ mol}}{32.00\text{ g}} = 6.00\text{ mol } O_2$

$$\frac{\text{mol } O_2}{\text{mol } B_5H_9}\text{ (actual)} = \frac{6.00}{2.00} = 3.00$$

The required mol O_2 to mol B_5H_9 ratio is 12/2 = 6. The actual mole ratio is less than the required mole ratio, thus the numerator (O_2) is limiting.

$$6.00\text{ mol } O_2 \times \frac{9\text{ mol } H_2O}{12\text{ mol } O_2} \times \frac{18.02\text{ g } H_2O}{\text{mol } H_2O} = 81.1\text{ g } H_2O$$

121. $$453\text{ g Fe} \times \frac{1\text{ mol Fe}}{55.85\text{ g Fe}} \times \frac{1\text{ mol Fe}_2\text{O}_3}{2\text{ mol Fe}} \times \frac{159.70\text{ g Fe}_2\text{O}_3}{\text{mol Fe}_2\text{O}_3} = 648\text{ g Fe}_2\text{O}_3$$

$$\text{mass \%Fe}_2\text{O}_3 = \frac{648\text{ g Fe}_2\text{O}_3}{752\text{ g ore}} \times 100 = 86.2\%$$

123. Assuming one mol of vitamin A (286.4 g vitamin A):

$$\text{mol C} = 286.4\text{ g vitamin A} \times \frac{0.8386\text{ g C}}{\text{g vitamin A}} \times \frac{1\text{ mol C}}{12.01\text{ g C}} = 20.00\text{ mol C}$$

$$\text{mol H} = 286.4\text{ g vitamin A} \times \frac{0.1056\text{ g H}}{\text{g vitamin A}} \times \frac{1\text{ mol H}}{1.008\text{ g H}} = 30.00\text{ mol H}$$

Since one mol of vitamin A contains 20 mol C and 30 mol H, the molecular formula of vitamin A is $C_{20}H_{30}E$. To determine E, let's calculate the molar mass of E.

$$286.4\text{ g} = 20(12.01) + 30(1.008) + \text{molar mass E},\ \ \text{molar mass E} = 16.0\text{ g/mol}$$

From the periodic table, E = oxygen and the molecular formula of vitamin A is $C_{20}H_{30}O$.

Challenge Problems

125. First, we will determine composition in mass percent. We assume all the carbon in the 0.213 g CO_2 came from 0.157 g of the compound and that all the hydrogen in the 0.0310 g H_2O came from the 0.157 g of the compound.

$$0.213\text{ g CO}_2 \times \frac{12.01\text{ g C}}{44.01\text{ g CO}_2} = 0.0581\text{ g C};\ \ \%\text{C} = \frac{0.0581\text{ g C}}{0.1571\text{ g compound}} \times 100 = 37.0\%\text{ C}$$

$$0.0310\text{ g H}_2\text{O} \times \frac{2.016\text{ g H}}{18.02\text{ g H}_2\text{O}} = 3.47 \times 10^{-3}\text{ g H};\ \ \%\text{H} = \frac{3.47\times10^{-3}\text{ g}}{0.157\text{ g}} = 2.21\%\text{ H}$$

We get %N from the second experiment:

$$0.0230\text{ g NH}_3 \times \frac{14.01\text{ g N}}{17.03\text{ g NH}_3} = 1.89 \times 10^{-2}\text{ g N}$$

$$\%\text{N} = \frac{1.89\times10^{-2}\text{ g}}{0.103\text{ g}} \times 100 = 18.3\%\text{ N}$$

The mass percent of oxygen is obtained by difference:

$$\%\text{O} = 100.00 - (37.0 + 2.21 + 18.3) = 42.5\%$$

So out of 100.00 g of compound, there are:

$$37.0 \text{ g C} \times \frac{1 \text{ mol C}}{12.01 \text{ g C}} = 3.08 \text{ mol C}; \quad 2.21 \text{ g H} \times \frac{1 \text{ mol H}}{1.008 \text{ g H}} = 2.19 \text{ mol H}$$

$$18.3 \text{ g N} \times \frac{1 \text{ mol N}}{14.01 \text{ g N}} = 1.31 \text{ mol N}; \quad 42.5 \text{ g O} \times \frac{1 \text{ mol O}}{16.00 \text{ g O}} = 2.66 \text{ mol O}$$

The last, and often the hardest part, is to find simple whole number ratios. Divide all mole values by the smallest number:

$$\frac{3.08}{1.31} = 2.35; \quad \frac{2.19}{1.31} = 1.67; \quad \frac{1.31}{1.31} = 1.00; \quad \frac{2.66}{1.31} = 2.03$$

Multiplying all these ratios by 3 gives an empirical formula of $C_7H_5N_3O_6$.

127. The two relevant equations are:

$$4 \text{ FeO(s)} + O_2\text{(g)} \longrightarrow 2 \text{ Fe}_2O_3\text{(s) and } 4 \text{ Fe}_3O_4\text{(s)} + O_2\text{(g)} \longrightarrow 6 \text{ Fe}_2O_3\text{(s)}$$

Let x = mass FeO, so 5.430 - x = mass Fe_3O_4. The mol of each are:

$$\text{moles FeO} = \frac{x}{71.85} \text{ and moles Fe}_3O_4 = \frac{5.430 - x}{231.55}$$

Thus, moles Fe_2O_3 is:

$$\left(\frac{x}{71.85} \times \frac{2 \text{ moles Fe}_2O_3}{4 \text{ moles FeO}}\right) + \left(\frac{5.430 - x}{231.35} \times \frac{6 \text{ moles Fe}_2O_3}{4 \text{ moles Fe}_3O_4}\right)$$

and mass Fe_2O_3 is:

$$159.70 \text{ g/mol} \left[\left(\frac{x}{71.85} \times \frac{2}{4}\right) + \left(\frac{5.430 - x}{231.35} \times \frac{6}{4}\right)\right] = 5.779 \text{ g}$$

Solving: $x = 2.10$ g; Thus, the mixture is $\frac{2.10 \text{ g}}{5.430 \text{ g}} \times 100 = 38.7\%$ FeO by mass.

129. The two relevant equations are:

$$\text{Zn(s)} + 2 \text{ HCl(aq)} \rightarrow \text{ZnCl}_2\text{(aq)} + H_2\text{(g) and Mg(s)} + 2 \text{ HCl(aq)} \rightarrow \text{MgCl}_2\text{(aq)} + H_2\text{(g)}$$

Let x = mass Mg, so 10.00 − x = mass Zn.

From the balanced equations, moles H_2 = moles Zn + moles Mg.

$$\text{mol } H_2 = 0.5171 \text{ g } H_2 \times \frac{1 \text{ mol } H_2}{2.016 \text{ g } H_2} = 0.2565 \text{ mol } H_2$$

Thus, $0.2565 = \frac{x}{24.31} + \frac{10.00 - x}{65.38}$; Solving, $x = 4.008$ g Mg.

$$\frac{4.008 \text{ g}}{10.00 \text{ g}} \times 100 = 40.08\% \text{ Mg}$$

131. We know water is a product, so one of the elements in the compound is hydrogen.

$X_aH_b + O_2 \rightarrow H_2O + ?$

To balance the H atoms, the mole ratio between $X_aH_b : H_2O = \frac{2}{b}$.

$$\text{mol compound} = \frac{1.39 \text{ g}}{62.09 \text{ g/mol}} = 0.0224 \text{ mol}; \ \text{mol } H_2O = \frac{1.21 \text{ g}}{18.02 \text{ g/mol}} = 0.0671 \text{ mol}$$

$\frac{2}{b} = \frac{0.0224}{0.0671}$ and b = 6; X_aH_6 has a molar mass of 62.09 g/mol.

62.09 = a × molar mass of X + 6 × 1.008, a × molar mass of X = 56.04

Some possible identities for X could be Fe (a = 1), Si (a = 2), N (a = 4), Li (a = 8). N fits the data best so N_4H_6 is the formula.

133. Total mass of copper used:

$$10{,}000 \text{ boards} \times \frac{(8.0 \text{ cm} \times 16.0 \text{ cm} \times 0.060 \text{ cm})}{\text{board}} \times \frac{8.96 \text{ g}}{\text{cm}^3} = 6.9 \times 10^5 \text{ g Cu}$$

Amount of Cu removed = 0.80 × 6.9 × 10^5 g = 5.5 × 10^5 g Cu

$$5.5 \times 10^5 \text{ g Cu} \times \frac{1 \text{ mol Cu}}{63.55 \text{ g Cu}} \times \frac{1 \text{ mol Cu(NH}_3)_4\text{Cl}_2}{\text{mol Cu}} \times \frac{202.59 \text{ g Cu(NH}_3)_4\text{Cl}_2}{\text{mol Cu(NH}_3)_4\text{Cl}_2}$$
$$= 1.8 \times 10^6 \text{ g Cu(NH}_3)_4\text{Cl}_2$$

$$5.5 \times 10^5 \text{ g Cu} \times \frac{1 \text{ mol Cu}}{63.55 \text{ g Cu}} \times \frac{4 \text{ mol NH}_3}{\text{mol Cu}} \times \frac{17.03 \text{ g NH}_3}{\text{mol NH}_3} = 5.9 \times 10^5 \text{ g NH}_3$$

135. 10.00 g XCl_2 + excess $Cl_2 \rightarrow$ 12.55 g XCl_4; 2.55 g Cl reacted with XCl_2 to form XCl_4. XCl_4 contains 2.55 g Cl and 10.00 g XCl_2. From mol ratios, 10.00 g XCl_2 must also contain 2.55 g Cl; mass X in XCl_2 = 10.00 - 2.55 = 7.45 g X.

$$2.55 \text{ g Cl} \times \frac{1 \text{ mol Cl}}{35.45 \text{ g Cl}} \times \frac{1 \text{ mol XCl}_2}{2 \text{ mol Cl}} \times \frac{1 \text{ mol X}}{\text{mol XCl}_2} = 3.60 \times 10^{-2} \text{ mol X}$$

So, 3.60×10^{-2} mol X has a mass equal to 7.45 g X. The molar mass of X is:

$$\frac{7.45 \text{ g X}}{3.60 \times 10^{-2} \text{ mol X}} = 207 \text{ g/mol X; Atomic mass} = 207 \text{ amu so X is Pb.}$$

137. Consider the case of aluminum plus oxygen. Aluminum forms Al^{3+} ions; oxygen forms O^{2-} anions. The simplest compound of the two elements is Al_2O_3. Similarly, we would expect the formula of any group 6A element with Al to be Al_2X_3. Assuming this, out of 100.00 g of compound there are 18.56 g Al and 81.44 g of the unknown element, X. Let's use this information to determine the molar mass of X which will allow us to identify X from the periodic table.

$$18.56 \text{ g Al} \times \frac{1 \text{ mol Al}}{26.98 \text{ g Al}} \times \frac{3 \text{ mol X}}{2 \text{ mol Al}} = 1.032 \text{ mol X}$$

81.44 g of X must contain 1.032 mol of X.

$$\text{The molar mass of X} = \frac{81.44 \text{ g X}}{1.032 \text{ mol X}} = 78.91 \text{ g/mol X.}$$

From the periodic table, the unknown element is selenium and the formula is Al_2Se_3.

139. The balanced equations are:

$$4\ NH_3(g) + 5\ O_2(g) \rightarrow 4\ NO(g) + 6\ H_2O(g) \text{ and } 4\ NH_3(g) + 7\ O_2(g) \rightarrow 4\ NO_2(g) + 6\ H_2O(g)$$

Let 4x = number of mol of NO formed, and let 4y = number of mol of NO_2 formed. Then:

$$4x\ NH_3 + 5x\ O_2 \rightarrow 4x\ NO + 6x\ H_2O \text{ and } 4y\ NH_3 + 7y\ O_2 \rightarrow 4y\ NO_2 + 6y\ H_2O$$

All the NH_3 reacted, so $4x + 4y = 2.00$. $10.00 - 6.75 = 3.25$ mol O_2 reacted, so $5x + 7y = 3.25$.

Solving by the method of simultaneous equations:

$$\begin{array}{r} 20x + 28y = 13.0 \\ -20x - 20y = -10.0 \\ \hline 8y = 3.0, \end{array} \quad y = 0.38;\ 4x + 4 \times 0.38 = 2.00,\ x = 0.12$$

mol NO = $4x = 4 \times 0.12 = 0.48$ mol NO formed

Integrative Problems

141. a. 1.05×10^{-20} g Fe $\times \frac{1 \text{ mol Fe}}{55.85 \text{ g Fe}} \times \frac{6.022 \times 10^{23} \text{ atoms Fe}}{\text{mol Fe}} = 113$ atoms Fe

b. The total number of platinum atoms is $14 \times 20 = 280$ atoms (exact number). The mass of these atoms is:

$$280 \text{ atoms Pt} \times \frac{1 \text{ mol Pt}}{6.022 \times 10^{23} \text{ atoms Pt}} \times \frac{195.1 \text{ g Pt}}{\text{mol Pt}} = 9.071 \times 10^{-20} \text{ g Pt}$$

c. 9.071×10^{-20} g Ru $\times \frac{1 \text{ mol Ru}}{101.1 \text{ g Ru}} \times \frac{6.022 \times 10^{23} \text{ atoms Ru}}{\text{mol Ru}} = 540.3 = 540$ atoms Ru

143. $$\text{molar mass } X_2 = \frac{0.105 \text{ g}}{8.92 \times 10^{20} \text{ molecules} \times \frac{1 \text{ mol}}{6.022 \times 10^{23} \text{ molecules}}} = 70.9 \text{ g/mol}$$

The mass of X = 1/2(70.9 g/mol) = 35.5 g/mol. This is the element chlorine.

Assuming 100.00 g of MX_3 compound:

$$54.47 \text{ g Cl} \times \frac{1 \text{ mol}}{35.45 \text{ g}} = 1.537 \text{ mol Cl}$$

$$1.537 \text{ mol Cl} \times \frac{1 \text{ mol M}}{3 \text{ mol Cl}} = 0.5123 \text{ mol M}$$

$$\text{molar mass M} = \frac{45.53 \text{ g M}}{0.5123 \text{ mol M}} = 88.87 \text{ g/mol M}$$

M is the element yttrium (Y) and the name of YCl_3 is yttrium(III) chloride.

The balanced equation is: $2\text{ Y} + 3\text{ Cl}_2 \rightarrow 2\text{ YCl}_3$

Assuming Cl_2 is limiting:

$$1.00 \text{ g Cl}_2 \times \frac{1 \text{ mol Cl}_2}{70.90 \text{ g Cl}_2} \times \frac{2 \text{ mol YCl}_3}{3 \text{ mol Cl}_2} \times \frac{195.26 \text{ g YCl}_3}{1 \text{ mol YCl}_3} = 1.84 \text{ g YCl}_3$$

Assuming Y is limiting:

$$1.00 \text{ g Y} \times \frac{1 \text{ mol Y}}{88.91 \text{ g Y}} \times \frac{2 \text{ mol YCl}_3}{2 \text{ mol Y}} \times \frac{195.26 \text{ g YCl}_3}{1 \text{ mol YCl}_3} = 2.20 \text{ g YCl}_3$$

Cl_2 is the limiting reagent and the theoretical yield is 1.84 g YCl_3.

CHAPTER FOUR

TYPES OF CHEMICAL REACTIONS AND SOLUTION STOICHIOMETRY

Questions

9. a. Polarity is a term applied to covalent compounds. Polar covalent compounds have an unequal sharing of electrons in bonds that results in unequal charge distribution in the overall molecule. Polar molecules have a partial negative end and a partial positive end. These are not full charges like in ionic compounds, but are charges much smaller in magnitude. Water is a polar molecule and dissolves other polar solutes readily. The oxygen end of water (the partial negative end of the polar water molecule) aligns with the partial positive end of the polar solute while the hydrogens of water (the partial positive end of the polar water molecule) align with the partial negative end of the solute. These opposite charge attractions stabilize polar solutes in water. This process is called hydration. Nonpolar solutes do not have permanent partial negative and partial positive ends; nonpolar solutes are not stabilized in water and do not dissolve.

   b. KF is a soluble ionic compound so it is a strong electrolyte. KF(aq) actually exists as separate hydrated K^+ ions and hydrated F^- ions in solution: $C_6H_{12}O_6$ is a polar covalent molecule that is a nonelectrolyte. $C_6H_{12}O_6$ is hydrated as described in part a.

   c. RbCl is a soluble ionic compound so it exists as separate hydrated Rb^+ ions and hydrated Cl^- ions in solution. AgCl is an insoluble ionic compound so the ions stay together in solution and fall to the bottom of the container as a precipitate.

   d. HNO_3 is a strong acid and exists as separate hydrated H^+ ions and hydrated NO_3^- ions in solution. CO is a polar covalent molecule and is hydrated as explained in part a.

11. Use the solubility rules in Table 4.1. Some soluble bromides by rule 2 would be NaBr, KBr, and NH_4Br (there are others). The insoluble bromides by rule 3 would be AgBr, $PbBr_2$, and Hg_2Br_2. Similar reasoning is used for the other parts to this problem.

    Sulfates: Na_2SO_4, K_2SO_4, and $(NH_4)_2SO_4$ (and others) would be soluble and $BaSO_4$, $CaSO_4$, and $PbSO_4$ (or Hg_2SO_4) would be insoluble.

    Hydroxides: NaOH, KOH, $Ca(OH)_2$ (and others) would be soluble and $Al(OH)_3$, $Fe(OH)_3$, and $Cu(OH)_2$ (and others) would be insoluble.

    Phosphates: Na_3PO_4, K_3PO_4, $(NH_4)_3PO_4$ (and others) would be soluble and Ag_3PO_4, $Ca_3(PO_4)_2$, and $FePO_4$ (and others) would be insoluble.

    Lead: $PbCl_2$, $PbBr_2$, PbI_2, $Pb(OH)_2$, $PbSO_4$, and PbS (and others) would be insoluble. $Pb(NO_3)_2$ would be a soluble Pb^{2+} salt.

13. The Brønsted-Lowry definitions are best for our purposes. An acid is a proton donor and a base is a proton acceptor. A proton is an H^+ ion. Neutral hydrogen has 1 electron and 1 proton, so an H^+ ion is just a proton. An acid-base reaction is the transfer of an H^+ ion (a proton) from an acid to a base.

15. a. The species reduced is the element that gains electrons. The reducing agent causes reduction to occur by itself being oxidized. The reducing agent generally refers to the entire formula of the compound/ion that contains the element oxidized.

b. The species oxidized is the element that loses electrons. The oxidizing agent causes oxidation to occur by itself being reduced. The oxidizing agent generally refers to the entire formula of the compound/ion that contains the element reduced.

c. For simple binary ionic compounds, the actual charge on the ions are the oxidation states. For covalent compounds, nonzero oxidation states are imaginary charges the elements would have if they were held together by ionic bonds (assuming the bond is between two different nonmetals). Nonzero oxidation states for elements in covalent compounds are not actual charges. Oxidation states for covalent compounds are a bookkeeping method to keep track of electrons in a reaction.

Exercises

Aqueous Solutions: Strong and Weak Electrolytes

17. a. $NaBr(s) \rightarrow Na^+(aq) + Br^-(aq)$

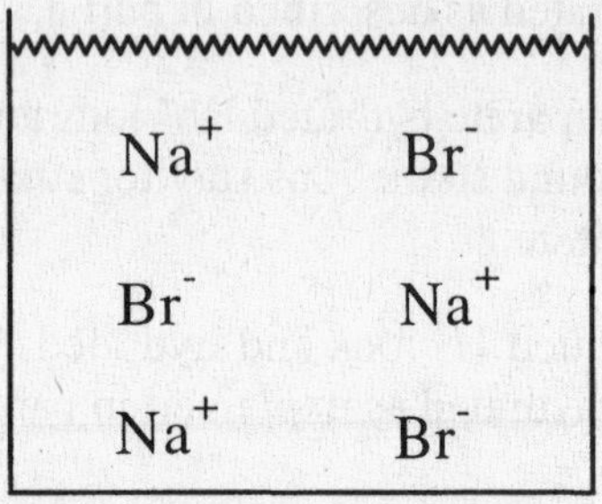

Na^+ Br^-
Br^- Na^+
Na^+ Br^-

Your drawing should show equal number of Na^+ and Br^- ions.

b. $MgCl_2(s) \rightarrow Mg^{2+}(aq) + 2\ Cl^-(aq)$

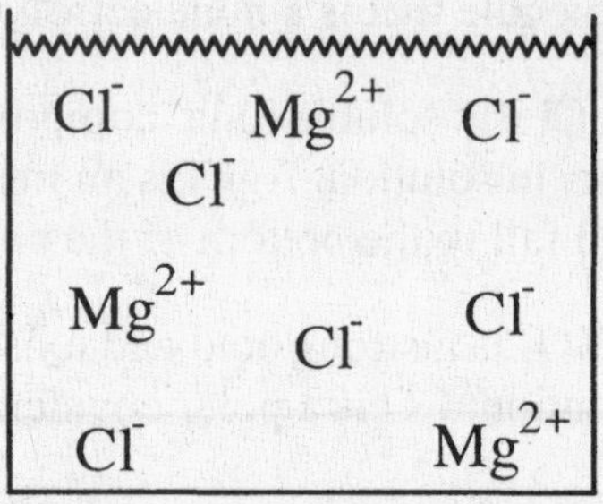

Cl^- Mg^{2+} Cl^-
Cl^-
Mg^{2+} Cl^- Cl^-
Cl^- Mg^{2+}

Your drawing should show twice the number of Cl^- ions as Mg^{2+} ions.

c. $Al(NO_3)_3(s) \rightarrow Al^{3+}(aq) + 3\ NO_3^-(aq)$

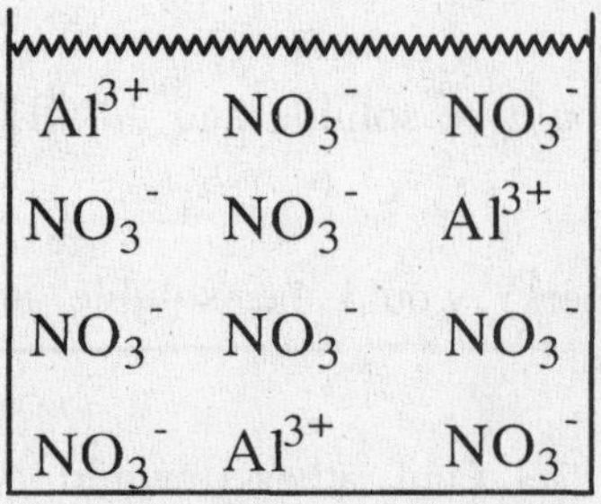

Al^{3+} NO_3^- NO_3^-
NO_3^- NO_3^- Al^{3+}
NO_3^- NO_3^- NO_3^-
NO_3^- Al^{3+} NO_3^-

d. $(NH_4)_2SO_4(s) \rightarrow 2\ NH_4^+(aq) + SO_4^{2-}(aq)$

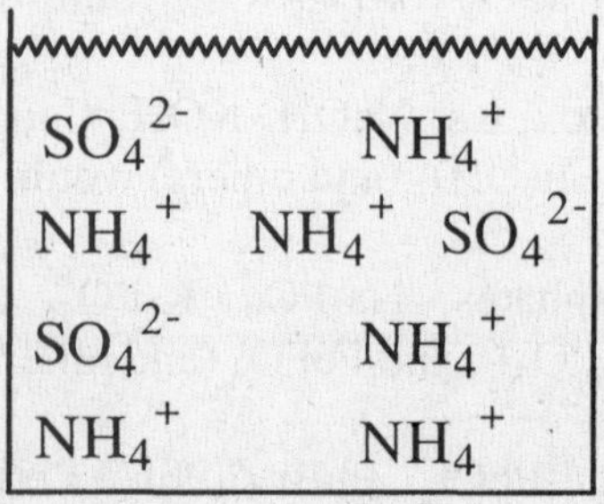

SO_4^{2-} NH_4^+
NH_4^+ NH_4^+ SO_4^{2-}
SO_4^{2-} NH_4^+
NH_4^+ NH_4^+

For e-i, your drawings should show equal numbers of the cations and anions present as each salt is a 1:1 salt. The ions present are listed in the following dissolution reactions.

e. $NaOH(s) \rightarrow Na^+(aq) + OH^-(aq)$

f. $FeSO_4(s) \rightarrow Fe^{2+}(aq) + SO_4^{2-}(aq)$

g. $KMnO_4(s) \rightarrow K^+(aq) + MnO_4^-(aq)$

h. $HClO_4(aq) \rightarrow H^+(aq) + ClO_4^-(aq)$

i. $NH_4C_2H_3O_2(s) \rightarrow NH_4^+(aq) + C_2H_3O_2^-(aq)$

19. $CaCl_2(s) \rightarrow Ca^{2+}(aq) + 2\ Cl^-(aq)$

Solution Concentration: Molarity

21. a. $5.623\text{ g }NaHCO_3 \times \dfrac{1\text{ mol }NaHCO_3}{84.01\text{ g }NaHCO_3} = 6.693 \times 10^{-2}\text{ mol }NaHCO_3$

$$M = \frac{6.693\times10^{-2}\text{ mol}}{250.0\text{ mL}} \times \frac{1000\text{ mL}}{\text{L}} = 0.2677\ M\ NaHCO_3$$

b. $0.1846\text{ g }K_2Cr_2O_7 \times \dfrac{1\text{ mol }K_2Cr_2O_7}{294.20\text{ g }K_2Cr_2O_7} = 6.275 \times 10^{-4}\text{ mol }K_2Cr_2O_7$

$$M = \frac{6.275\times10^{-4}\text{ mol}}{500.0\times10^{-3}\text{ L}} = 1.255 \times 10^{-3}\ M\ K_2Cr_2O_7$$

c. $0.1025\text{ g Cu} \times \dfrac{1\text{ mol Cu}}{63.55\text{ g Cu}} = 1.613 \times 10^{-3}\text{ mol Cu} = 1.613 \times 10^{-3}\text{ mol }Cu^{2+}$

$$M = \frac{1.613\times10^{-2}\text{ mol }Cu^{2+}}{200.0\text{ mL}} \times \frac{1000\text{ mL}}{\text{L}} = 8.065 \times 10^{-3}\ M\ Cu^{2+}$$

23. a. $M_{Ca(NO_3)_2} = \dfrac{0.100\text{ mol }Ca(NO_3)_2}{0.100\text{ L}} = 1.00\ M$

$Ca(NO_3)_2(s) \rightarrow Ca^{2+}(aq) + 2\ NO_3^-(aq)$; $M_{ca^{2+}} = 1.00\ M$; $M_{NO_3^-} = 2(1.00) = 2.00\ M$

b. $M_{Na_2SO_4} = \dfrac{2.5\text{ mol }Na_2SO_4}{1.25\text{ L}} = 2.0\ M$

$Na_2SO_4(s) \rightarrow 2\ Na^+(aq) + SO_4^{2-}(aq)$; $M_{Na^+} = 2(2.0) = 4.0\ M$; $M_{SO_4^{2-}} = 2.0\ M$

c. $5.00\text{ g } NH_4Cl \times \dfrac{1\text{ mol } NH_4Cl}{53.49\text{ g } NH_4Cl} = 0.0935\text{ mol } NH_4Cl$

$$M_{NH_4Cl} = \frac{0.0935\text{ mol } NH_4Cl}{0.5000\text{ L}} = 0.187\ M$$

$NH_4Cl(s) \rightarrow NH_4^+(aq) + Cl^-(aq);\ M_{NH_4^+} = M_{Cl^-} = 0.187\ M$

d. $1.00\text{ g } K_3PO_4 \times \dfrac{1\text{ mol } K_3PO_4}{212.27\text{ g}} = 4.71 \times 10^{-3}\text{ mol } K_3PO_4$

$$M_{K_3PO_4} = \frac{4.71\times10^{-3}\text{ mol}}{0.2500\text{ L}} = 0.0188\ M$$

$K_3PO_4(s) \rightarrow 3\ K^+(aq) + PO_4^{3-}(aq);\ M_{K^+} = 3(0.0188) = 0.0564\ M;\ M_{PO_4^{3-}} = 0.0188\ M$

25. mol solute = volume (L) $\times \left(\dfrac{\text{mol}}{\text{L}}\right)$; $AlCl_3(s) \rightarrow Al^{3+}(aq) + 3\ Cl^-(aq)$

$$\text{mol } Cl^- = 0.1000\text{ L} \times \frac{0.30\text{ mol } AlCl_3}{\text{L}} \times \frac{3\text{ mol } Cl^-}{\text{mol } AlCl_3} = 9.0 \times 10^{-2}\text{ mol } Cl^-$$

$MgCl_2(s) \rightarrow Mg^{2+}(aq) + 2\ Cl^-(aq)$

$$\text{mol } Cl^- = 0.0500\text{ L} \times \frac{0.60\text{ mol } MgCl_2}{\text{L}} \times \frac{2\text{ mol } Cl^-}{\text{mol } MgCl_2} = 6.0 \times 10^{-2}\text{ mol } Cl^-$$

$NaCl(s) \rightarrow Na^+(aq) + Cl^-(aq)$

$$\text{mol } Cl^- = 0.2000\text{ L} \times \frac{0.40\text{ mol NaCl}}{\text{L}} \times \frac{1\text{ mol } Cl^-}{\text{mol NaCl}} = 8.0 \times 10^{-2}\text{ mol } Cl^-$$

100.0 mL of 0.30 *M* $AlCl_3$ contains the most moles of Cl^- ions.

27. Molar mass of NaOH = 22.99 + 16.00 + 1.008 = 40.00 g/mol

$$\text{Mass NaOH} = 0.2500\text{ L} \times \frac{0.400\text{ mol NaOH}}{\text{L}} \times \frac{40.00\text{ g NaOH}}{\text{mol NaOH}} = 4.00\text{ g NaOH}$$

29. a. $2.00\text{ L} \times \dfrac{0.250\text{ mol NaOH}}{\text{L}} \times \dfrac{40.00\text{ g NaOH}}{\text{mol NaOH}} = 20.0\text{ g NaOH}$

Place 20.0 g NaOH in a 2 L volumetric flask; add water to dissolve the NaOH, and fill to the mark with water, mixing several times along the way.

b. $2.00\text{ L} \times \frac{0.250\text{ mol NaOH}}{\text{L}} \times \frac{1\text{ L stock}}{1.00\text{ mol NaOH}} = 0.500\text{ L}$

Add 500. mL of 1.00 M NaOH stock solution to a 2 L volumetric flask; fill to the mark with water, mixing several times along the way.

c. $2.00\text{ L} \times \frac{0.100\text{ mol } K_2CrO_4}{\text{L}} \times \frac{194.20\text{ g } K_2CrO_4}{\text{mol } K_2CrO_4} = 38.8\text{ g } K_2CrO_4$

Similar to the solution made in part a, instead using 38.8 g K_2CrO_4.

d. $2.00\text{ L} \times \frac{0.100\text{ mol } K_2CrO_4}{\text{L}} \times \frac{1\text{ L stock}}{1.75\text{ mol } K_2CrO_4} = 0.114\text{ L}$

Similar to the solution made in part b, instead using 114 mL of the 1.75 M K_2CrO_4 stock solution.

31. $10.8\text{ g } (NH_4)_2SO_4 \times \frac{1\text{ mol}}{132.15\text{ g}} = 8.17 \times 10^{-2}\text{ mol } (NH_4)_2SO_4$

$$\text{Molarity} = \frac{8.17 \times 10^{-2}\text{ mol}}{100.0\text{ mL}} \times \frac{1000\text{ mL}}{\text{L}} = 0.817\ M\ (NH_4)_2SO_4$$

Moles of $(NH_4)_2SO_4$ in final solution:

$$10.00 \times 10^{-3}\text{ L} \times \frac{0.817\text{ mol}}{\text{L}} = 8.17 \times 10^{-3}\text{ mol}$$

$$\text{Molarity of final solution} = \frac{8.17 \times 10^{-3}\text{ mol}}{(10.00 + 50.00)\text{ mL}} \times \frac{1000\text{ mL}}{\text{L}} = 0.136\ M\ (NH_4)_2SO_4$$

$(NH_4)_2SO_4(s) \rightarrow 2\ NH_4^+(aq) + SO_4^{2-}(aq)$; $M_{NH_4^+} = 2(0.136) = 0.272\ M$; $M_{SO_4^{2-}} = 0.136\ M$

33. $\text{Stock solution} = \frac{10.0\text{ mg}}{500.0\text{ mL}} = \frac{10.0 \times 10^{-3}\text{ g}}{500.0\text{ mL}} = \frac{2.00 \times 10^{-5}\text{ g steroid}}{\text{mL}}$

$$100.0 \times 10^{-6}\text{ L stock} \times \frac{1000\text{ mL}}{\text{L}} \times \frac{2.00 \times 10^{-5}\text{ g steroid}}{\text{mL}} = 2.00 \times 10^{-6}\text{ g steroid}$$

This is diluted to a final volume of 100.0 mL.

$$\frac{2.00 \times 10^{-6}\text{ g steroid}}{100.0\text{ mL}} \times \frac{1000\text{ mL}}{\text{L}} \times \frac{1\text{ mol steroid}}{336.43\text{ g steroid}} = 5.94 \times 10^{-8}\ M\text{ steroid}$$

Precipitation Reactions

35. The solubility rules referenced in the following answers are outlined in Table 4.1 of the text.

a. Soluble: most nitrate salts are soluble (Rule 1).

b. Soluble: most chloride salts are soluble except for Ag^+, Pb^{2+}, and Hg_2^{2+} (Rule 3).

c. Soluble: most sulfate salts are soluble except for $BaSO_4$, $PbSO_4$, Hg_2SO_4, and $CaSO_4$ (Rule 4.)

d. Insoluble: most hydroxide salts are only slightly soluble (Rule 5).
Note: we will interpret the phrase "slightly soluble" as meaning insoluble and the phrase "marginally soluble" as meaning soluble. So the marginally soluble hydroxides $Ba(OH)_2$, $Sr(OH)_2$, and $Ca(OH)_2$ will be assumed soluble unless noted otherwise.

e. Insoluble: most sulfide salts are only slightly soluble (Rule 6). Again, "slightly soluble" is interpreted as "insoluble" in problems like these.

f. Insoluble: Rule 5 (see answer d).

g. Insoluble: most phosphate salts are only slightly soluble (Rule 6).

37. In these reactions, soluble ionic compounds are mixed together. To predict the precipitate, switch the anions and cations in the two reactant compounds to predict possible products; then use the solubility rules in Table 4.1 to predict if any of these possible products are insoluble (are the precipitate). Note that the phrase "slightly soluble" in Table 4.1 is interpreted to mean insoluble and the phrase "marginally soluble" is interpreted to mean soluble.

a. Possible products = $FeCl_2$ and K_2SO_4; Both salts are soluble so no precipitate forms.

b. Possible products = $Al(OH)_3$ and $Ba(NO_3)_2$; precipitate = $Al(OH)_3(s)$

c. Possible products = $CaSO_4$ and NaCl; precipitate = $CaSO_4(s)$

d. Possible products = KNO_3 and NiS; precipitate = NiS(s)

39. For the following answers, the balanced formula equation is first, followed by the complete ionic equation, then the net ionic equation.

a. No reaction occurs since all possible products are soluble salts.

b. $2\ Al(NO_3)_3(aq) + 3\ Ba(OH)_2(aq) \rightarrow 2\ Al(OH)_3(s) + 3\ Ba(NO_3)_2(aq)$

$$2\ Al^{3+}(aq) + 6\ NO_3^-(aq) + 3\ Ba^{2+}(aq) + 6\ OH^-(aq) \rightarrow 2\ Al(OH)_3(s) + 3\ Ba^{2+}(aq) + 6\ NO_3^-(aq)$$

$$Al^{3+}(aq) + 3\ OH^-(aq) \rightarrow Al(OH)_3(s)$$

c. $CaCl_2(aq) + Na_2SO_4(aq) \rightarrow CaSO_4(s) + 2\ NaCl(aq)$

$$Ca^{2+}(aq) + 2\ Cl^-(aq) + 2\ Na^+(aq) + SO_4^{2-}(aq) \rightarrow CaSO_4(s) + 2\ Na^+(aq) + 2\ Cl^-(aq)$$

$$Ca^{2+}(aq) + SO_4^{2-}(aq) \rightarrow CaSO_4(s)$$

d. $K_2S(aq) + Ni(NO_3)_2(aq) \rightarrow 2\ KNO_3(aq) + NiS(s)$

$2\ K^+(aq) + S^{2-}(aq) + Ni^{2+}(aq) + 2\ NO_3^-(aq) \rightarrow 2\ K^+(aq) + 2\ NO_3^-(aq) + NiS(s)$

$Ni^{2+}(aq) + S^{2-}(aq) \rightarrow NiS(s)$

41. a. When $CuSO_4(aq)$ is added to $Na_2S(aq)$, the precipitate that forms is CuS(s). Therefore, Na^+ (the grey spheres) and SO_4^{2-} (the blue-green spheres) are the spectator ions.

$CuSO_4(aq) + Na_2S(aq) \rightarrow CuS(s) + Na_2SO_4(aq)$; $Cu^{2+}(aq) + S^{2-}(aq) \rightarrow CuS(s)$

b. When $CoCl_2(aq)$ is added to NaOH(aq), the precipitate that forms is $Co(OH)_2(s)$. Therefore, Na^+ (the grey spheres) and Cl^- (the green spheres) are the spectator ions.

$CoCl_2(aq) + 2\ NaOH(aq) \rightarrow Co(OH)_2(s) + 2\ NaCl(aq)$

$Co^{2+}(aq) + 2\ OH^-(aq) \rightarrow Co(OH)_2(s)$

c. When $AgNO_3(aq)$ is added to KI(aq), the precipitate that forms is AgI(s). Therefore, K^+ (the red spheres) and NO_3^- ((the blue spheres) are the spectator ions.

$AgNO_3(aq) + KI(aq) \rightarrow AgI(s) + KNO_3(aq)$; $Ag^+(aq) + I^-$ ((aq) $\rightarrow AgI(s)$

43. a. $(NH_4)_2SO_4(aq) + Ba(NO_3)_2(aq) \rightarrow 2\ NH_4NO_3(aq) + BaSO_4(s)$

$Ba^{2+}(aq) + SO_4^{2-}(aq) \rightarrow BaSO_4(s)$

b. $Pb(NO_3)_2(aq) + 2\ NaCl(aq) \rightarrow PbCl_2(s) + 2\ NaNO_3(aq)$

$Pb^{2+}(aq) + 2\ Cl^-(aq) \rightarrow PbCl_2(s)$

c. Potassium phosphate and sodium nitrate are both soluble in water. No reaction occurs.

d. No reaction occurs because all possible products are soluble.

e. $CuCl_2(aq) + 2\ NaOH(aq) \rightarrow Cu(OH)_2(s) + 2\ NaCl(aq)$

$Cu^{2+}(aq) + 2\ OH^-(aq) \rightarrow Cu(OH)_2(s)$

45. Because a precipitate formed with Na_2SO_4, the possible cations are Ba^{2+}, Pb^{2+}, Hg_2^{2+}, and Ca^{2+} (from the solubility rules). Because no precipitate formed with KCl, Pb^{2+} and Hg_2^{2+} cannot be present. Because both Ba^{2+} and Ca^{2+} form soluble chlorides and soluble hydroxides, both of these cations could be present. Therefore, the cations could be Ba^{2+} and Ca^{2+} (by the solubility rules in Table 4.1). For students who do a more rigorous study of solubility, Sr^{2+} could also be a possible cation (it forms an insoluble sulfate salt while the chloride and hydroxide salts of strontium are soluble).

47. $2\ AgNO_3(aq) + Na_2CrO_4(aq) \rightarrow Ag_2CrO_4(s) + 2\ NaNO_3(aq)$

$$0.0750\ L \times \frac{0.100\ mol\ AgNO_3}{L} \times \frac{1\ mol\ Na_2CrO_4}{2\ mol\ AgNO_3} \times \frac{161.98\ g\ Na_2CrO_4}{mol\ Na_2CrO_4} = 0.607\ g\ Na_2CrO_4$$

49. $Al(NO_3)_3(aq) + 3\ KOH(aq) \rightarrow Al(OH)_3(s) + 3\ KNO_3(aq)$

$$0.0500\ L \times \frac{0.200\ mol\ Al(NO_3)_3}{L} = 0.0100\ mol\ Al(NO_3)_3$$

$$0.2000\ L \times \frac{0.100\ mol\ KOH}{L} = 0.0200\ mol\ KOH$$

From the balanced equation, 3 mol of KOH are required to react with 1 mol of $Al(NO_3)_3$ (3:1 mole ratio). The actual KOH to $Al(NO_3)_3$ mole ratio present is 0.0200/0.0100 = 2 (2:1). Since the actual mole ratio present is less than the required mole ratio, KOH (the numerator) is the limiting reagent.

$$0.0200\ mol\ KOH \times \frac{1\ mol\ Al(OH)_3}{3\ mol\ KOH} \times \frac{78.00\ g\ Al(OH)_3}{mol\ Al(OH)_3} = 0.520\ g\ Al(OH)_3$$

51. $2\ AgNO_3(aq) + CaCl_2(aq) \rightarrow 2\ AgCl(s) + Ca(NO_3)_2(aq)$

$$mol\ AgNO_3 = 0.1000\ L \times \frac{0.20\ mol\ AgNO_3}{L} = 0.020\ mol\ AgNO_3$$

$$mol\ CaCl_2 = 0.1000\ L \times \frac{0.15\ mol\ CaCl_2}{L} = 0.015\ mol\ CaCl_2$$

The required mol $AgNO_3$ to mol $CaCl_2$ ratio is 2:1 (from the balanced equation). The actual mole ratio present is 0.020/0.015 = 1.3 (1.3:1). Therefore, $AgNO_3$ is the limiting reagent.

$$mass\ AgCl = 0.020\ mol\ AgNO_3 \times \frac{1\ mol\ AgCl}{1\ mol\ AgNO_3} \times \frac{143.4\ g\ AgCl}{mol\ AgCl} = 2.9\ g\ AgCl$$

The net ionic equation is: $Ag^+(aq) + Cl^-(aq) \rightarrow AgCl(s)$. The ions remaining in solution are the unreacted Cl^- ions and the spectator ions, NO_3^- and Ca^{2+} (all of the Ag^+ is used up in forming AgCl). The moles of each ion present initially (before reaction) can be easily determined from the moles of each reactant. 0.020 mol $AgNO_3$ dissolves to form 0.020 mol Ag^+ and 0.020 mol NO_3^-. 0.015 mol $CaCl_2$ dissolves to form 0.015 mol Ca^{2+} and 2(0.015) = 0.030 mol Cl^-.

mol unreacted Cl^- = 0.030 mol Cl^- initially − 0.020 mol Cl^- reacted = 0.010 mol Cl^- unreacted

$$M_{Cl^-} = \frac{0.010\ mol\ Cl^-}{total\ volume} = \frac{0.010\ mol\ Cl^-}{0.1000\ L + 0.1000\ L} = 0.050\ M\ Cl^-$$

The molarity of the spectator ions are:

$$M_{NO_3^-} = \frac{0.20 \text{ mol } NO_3^-}{0.2000 \text{ L}} = 0.10\ M\ NO_3^-;\ M_{Ca^{2+}} = \frac{0.015 \text{ mol } Ca^{2+}}{0.2000 \text{ L}} = 0.075\ M\ Ca^{2+}$$

53. $M_2SO_4(aq) + CaCl_2(aq) \rightarrow CaSO_4(s) + 2\ MCl(aq)$

$$1.36 \text{ g } CaSO_4 \times \frac{1 \text{ mol } CaSO_4}{136.15 \text{ g } CaSO_4} \times \frac{1 \text{ mol } M_2SO_4}{\text{mol } CaSO_4} = 9.99 \times 10^{-3} \text{ mol } M_2SO_4$$

From the problem, 1.42 g M_2SO_4 was reacted so:

$$\text{molar mass} = \frac{1.42 \text{ g } M_2SO_4}{9.99 \times 10^{-3} \text{ mol } M_2SO_4} = 142 \text{ g/mol}$$

142 amu = 2(atomic mass M) + 32.07 + 4(16.00), atomic mass M = 23 amu

From periodic table, M = Na (sodium).

Acid-Base Reactions

55. All the bases in this problem are ionic compounds containing OH^-. The acids are either strong or weak electrolytes. The best way to determine if an acid is a strong or weak electrolyte is to memorize all the strong electrolytes (strong acids). Any other acid you encounter that is not a strong acid will be a weak electrolyte (a weak acid) and the formula should be left unaltered in the complete ionic and net ionic equations. The strong acids to recognize are HCl, HBr, HI, HNO_3, $HClO_4$ and H_2SO_4. For the following answers, the order of the equations are formula, complete ionic, and net ionic.

a. $2\ HClO_4(aq) + Mg(OH)_2(s) \rightarrow 2\ H_2O(l) + Mg(ClO_4)_2(aq)$

$2\ H^+(aq) + 2\ ClO_4^-(aq) + Mg(OH)_2(s) \rightarrow 2\ H_2O(l) + Mg^{2+}(aq) + 2\ ClO_4^-(aq)$

$2\ H^+(aq) + Mg(OH)_2(s) \rightarrow 2\ H_2O(l) + Mg^{2+}(aq)$

b. $HCN(aq) + NaOH(aq) \rightarrow H_2O(l) + NaCN(aq)$

$HCN(aq) + Na^+(aq) + OH^-(aq) \rightarrow H_2O(l) + Na^+(aq) + CN^-(aq)$

$HCN(aq) + OH^-(aq) \rightarrow H_2O(l) + CN^-(aq)$

c. $HCl(aq) + NaOH(aq) \rightarrow H_2O(l) + NaCl(aq)$

$H^+(aq) + Cl^-(aq) + Na^+(aq) + OH^-(aq) \rightarrow H_2O(l) + Na^+(aq) + Cl^-(aq)$

$H^+(aq) + OH^-(aq) \rightarrow H_2O(l)$

57. All the acids in this problem are strong electrolytes (strong acids). The acids to recognize as strong electrolytes are HCl, HBr, HI, HNO_3, $HClO_4$ and H_2SO_4.

a. $KOH(aq) + HNO_3(aq) \rightarrow H_2O(l) + KNO_3(aq)$

$K^+(aq) + OH^-(aq) + H^+(aq) + NO_3^-(aq) \rightarrow H_2O(l) + K^+(aq) + NO_3^-(aq)$

$OH^-(aq) + H^+(aq) \rightarrow H_2O(l)$

b. $Ba(OH)_2(aq) + 2\ HCl(aq) \rightarrow 2\ H_2O(l) + BaCl_2(aq)$

$Ba^{2+}(aq) + 2\ OH^-(aq) + 2\ H^+(aq) + 2\ Cl^-(aq) \rightarrow 2\ H_2O(l) + Ba^{2+}(aq) + 2\ Cl^-(aq)$

$2\ OH^-(aq) + 2\ H^+(aq) \rightarrow 2\ H_2O(l)$ or $OH^-(aq) + H^+(aq) \rightarrow H_2O(l)$

c. $3\ HClO_4(aq) + Fe(OH)_3(s) \rightarrow 3\ H_2O(l) + Fe(ClO_4)_3(aq)$

$3\ H^+(aq) + 3\ ClO_4^-(aq) + Fe(OH)_3(s) \rightarrow 3\ H_2O(l) + Fe^{3+}(aq) + 3\ ClO_4^-(aq)$

$3\ H^+(aq) + Fe(OH)_3(s) \rightarrow 3\ H_2O(l) + Fe^{3+}(aq)$

59. If we begin with 50.00 mL of 0.200 *M* NaOH, then:

$$50.00 \times 10^{-3}\ L \times \frac{0.200\ mol}{L} = 1.00 \times 10^{-2}\ \text{mol NaOH is to be neutralized.}$$

a. $NaOH(aq) + HCl(aq) \rightarrow NaCl(aq) + H_2O(l)$

$$1.00 \times 10^{-2}\ \text{mol NaOH} \times \frac{1\ \text{mol HCl}}{\text{mol NaOH}} \times \frac{1\ L}{0.100\ mol} = 0.100\ L \text{ or } 100.\ mL$$

b. $HNO_3(aq) + NaOH(aq) \rightarrow H_2O(l) + NaNO_3(aq)$

$$1.00 \times 10^{-2}\ \text{mol NaOH} \times \frac{1\ \text{mol } HNO_3}{\text{mol NaOH}} \times \frac{1\ L}{0.150\ \text{mol } HNO_3} = 6.67 \times 10^{-2}\ L \text{ or } 66.7\ mL$$

c. $HC_2H_3O_2(aq) + NaOH(aq) \rightarrow H_2O(l) + NaC_2H_3O_2(aq)$

$$1.00 \times 10^{-2}\ \text{mol NaOH} \times \frac{1\ \text{mol } HC_2H_3O_2}{\text{mol NaOH}} \times \frac{1\ L}{0.200\ \text{mol } HC_2H_3O_2} = 5.00 \times 10^{-2}\ L$$

$$= 50.0\ mL$$

61. $Ba(OH)_2(aq) + 2\ HCl(aq) \rightarrow BaCl_2(aq) + 2\ H_2O(l)$; $H^+(aq) + OH^-(aq) \rightarrow H_2O(l)$

$$75.0 \times 10^{-3}\ L \times \frac{0.250\ \text{mol HCl}}{L} = 1.88 \times 10^{-2}\ \text{mol HCl} = 1.88 \times 10^{-2}\ \text{mol } H^+ +$$

$$1.88 \times 10^{-2}\ \text{mol } Cl^-$$

$$225.0 \times 10^{-3}\,\text{L} \times \frac{0.0550\text{ mol Ba(OH)}_2}{\text{L}} = 1.24 \times 10^{-2}\text{ mol Ba(OH)}_2$$

$$= 1.24 \times 10^{-2}\text{ mol Ba}^{2+} + 2.48 \times 10^{-2}\text{ mol OH}^-$$

The net ionic equation requires a 1:1 mole ratio between OH^- and H^+. The actual mol OH^- to mol H^+ ratio is greater than 1:1 so OH^- is in excess.

Since 1.88×10^{-2} mol OH^- will be neutralized by the H^+, we have $(2.48 - 1.88) \times 10^{-2} = 0.60 \times 10^{-2}$ mol OH^- remaining in excess.

$$M_{OH^-} = \frac{\text{mol OH}^-\text{ excess}}{\text{total volume}} = \frac{6.0 \times 10^{-3}\text{ mol OH}^-}{0.750\text{ L} + 0.2250\text{ L}} = 2.0 \times 10^{-2}\,M\text{ OH}^-$$

63. $HCl(aq) + NaOH(aq) \rightarrow H_2O(l) + NaCl(aq)$

$$24.16 \times 10^{-3}\text{ L NaOH} \times \frac{0.106\text{ mol NaOH}}{\text{L NaOH}} \times \frac{1\text{ mol HCl}}{\text{mol NaOH}} = 2.56 \times 10^{-3}\text{ mol HCl}$$

$$\text{Molarity of HCl} = \frac{2.56 \times 10^{-3}\text{ mol}}{25.00 \times 10^{-3}\text{ L}} = 0.102\,M\text{ HCl}$$

65. KHP is a monoprotic acid: $NaOH(aq) + KHP(aq) \rightarrow H_2O(l) + NaKP(aq)$

$$\text{Mass KHP} = 0.02046\text{ L NaOH} \times \frac{0.1000\text{ mol NaOH}}{\text{L NaOH}} \times \frac{1\text{ mol KHP}}{\text{mol NaOH}} \times \frac{204.22\text{ g KHP}}{\text{mol KHP}}$$

$$= 0.4178\text{ g KHP}$$

Oxidation-Reduction Reactions

67. Apply the rules in Table 4.2.

a. $KMnO_4$ is composed of K^+ and MnO_4^- ions. Assign oxygen a value of −2, which gives manganese a +7 oxidation state since the sum of oxidation states for all atoms in MnO_4^- must equal the −1 charge on MnO_4^-. K, +1; O, −2; Mn, +7.

b. Assign O a −2 oxidation state, which gives nickel a +4 oxidation state. Ni, +4; O, −2.

c. $Na_4Fe(OH)_6$ is composed of Na^+ cations and $Fe(OH)_6^{4-}$ anions. $Fe(OH)_6^{4-}$ is composed of an iron cation and 6 OH^- anions. For an overall anion charge of −4, iron must have a +2 oxidation state. As is usually the case in compounds, assign O a −2 oxidation state and H a +1 oxidation state. Na, +1; Fe, +2; O, −2; H, +1.

d. $(NH_4)_2HPO_4$ is made of NH_4^+ cations and HPO_4^{2-} anions. Assign +1 as the oxidation state of H and −2 as the oxidation state of O. In NH_4^+, $x + 4(+1) = +1$, $x = -3$ = oxidation state of N. In HPO_4^{2-}, $+1 + y + 4(-2) = -2$, $y = +5$ = oxidation state of P.

e. O, −2; P, +3

f. O, −2; Fe, + 8/3

g. O, −2; F, −1; Xe, +6

h. F, −1; S, +4

i. O, −2; C, +2

j. H, +1; O, −2; C, 0

69. a. −3 b. −3 c. $2(x) + 4(+1) = 0,\ x = -2$
d. +2 e. +1 f. +4
g. +3 h. +5 i. 0

71. To determine if the reaction is an oxidation-reduction reaction, assign oxidation numbers. If the oxidation numbers change for some elements, then the reaction is a redox reaction. If the oxidation numbers do not change, then the reaction is not a redox reaction. In redox reactions, the species oxidized (called the reducing agent) shows an increase in oxidation numbers and the species reduced (called the oxidizing agent) shows a decrease in oxidation numbers.

	Redox?	Oxidizing Agent	Reducing Agent	Substance Oxidized	Substance Reduced
a.	Yes	Ag^+	Cu	Cu	Ag^+
b.	No	–	–	–	–
c.	No	–	–	–	–
d.	Yes	$SiCl_4$	Mg	Mg	$SiCl_4$ (Si)
e.	No	–	–	–	–

In b, c, and e, no oxidation numbers change.

73. Use the method of half-reactions described in Section 4.10 of the text to balance these redox reactions. The first step always is to separate the reaction into the two half-reactions, then balance each half-reaction separately.

a. $Zn \rightarrow Zn^{2+} + 2\,e^-$ $2e^- + 2\,HCl \rightarrow H_2 + 2\,Cl^-$

Adding the two balanced half-reactions:

$$Zn(s) + 2\,HCl(aq) \rightarrow H_2(g) + Zn^{2+}(aq) + 2\,Cl^-(aq)$$

$2\,Cl^-(aq)$

b. $3\,I^- \rightarrow I_3^- + 2e^-$

$ClO^- \rightarrow Cl^-$

$2e^- + 2H^+ + ClO^- \rightarrow Cl^- + H_2O$

Adding the two balanced half-reactions so electrons cancel:

$$3\,I^-(aq) + 2\,H^+(aq) + ClO^-(aq) \rightarrow I_3^-(aq) + Cl^-(aq) + H_2O(l)$$

c. $As_2O_3 \rightarrow H_3AsO_4$

$As_2O_3 \rightarrow 2\,H_3AsO_4$

Left 3 − O; Right 8 − O

Right hand side has 5 extra O.

$NO_3^- \rightarrow NO + 2\,H_2O$

$4\,H^+ + NO_3^- \rightarrow NO + 2\,H_2O$

$(3\,e^- + 4\,H^+ + NO_3^- \rightarrow NO + 2\,H_2O) \times 4$

Balance the oxygen atoms first using H_2O, then balance H using H^+, and finally balance charge using electrons. This gives:

$$(5\ H_2O + As_2O_3 \rightarrow 2\ H_3AsO_4 + 4\ H^+ + 4\ e^-) \times 3$$

Common factor is a transfer of 12 e^-. Add half-reactions so electrons cancel.

$$12\ e^- + 16\ H^+ + 4\ NO_3^- \rightarrow 4\ NO + 8\ H_2O$$
$$15\ H_2O + 3\ As_2O_3 \rightarrow 6\ H_3AsO_4 + 12\ H^+ + 12\ e^-$$

$$7\ H_2O(l) + 4\ H^+(aq) + 3\ As_2O_3(s) + 4\ NO_3^-(aq) \rightarrow 4\ NO(g) + 6\ H_3AsO_4(aq)$$

d. $(2\ Br^- \rightarrow Br_2 + 2\ e^-) \times 5$

$$MnO_4^- \rightarrow Mn^{2+} + 4\ H_2O$$
$$(5\ e^- + 8\ H^+ + MnO_4^- \rightarrow Mn^{2+} + 4\ H_2O) \times 2$$

Common factor is a transfer of 10 e^-.

$$10\ Br^- \rightarrow 5\ Br_2 + 10\ e^-$$
$$10\ e^- + 16\ H^+ + 2\ MnO_4^- \rightarrow 2\ Mn^{2+} + 8\ H_2O$$

$$16\ H^+(aq) + 2\ MnO_4^-(aq) + 10\ Br^-(aq) \rightarrow 5\ Br_2(l) + 2\ Mn^{2+}(aq) + 8\ H_2O(l)$$

e. $CH_3OH \rightarrow CH_2O$

$(CH_3OH \rightarrow CH_2O + 2\ H^+ + 2\ e^-) \times 3$

$$Cr_2O_7^{2-} \rightarrow 2\ Cr^{3+}$$
$$14\ H^+ + Cr_2O_7^{2-} \rightarrow 2\ Cr^{3+} + 7\ H_2O$$
$$6\ e^- + 14\ H^+ + Cr_2O_7^{2-} \rightarrow 2\ Cr^{3+} + 7\ H_2O$$

Common factor is a transfer of 6 e^-.

$$3\ CH_3OH \rightarrow 3\ CH_2O + 6\ H^+ + 6\ e^-$$
$$6\ e^- + 14\ H^+ + Cr_2O_7^{2-} \rightarrow 2\ Cr^{3+} + 7\ H_2O$$

$$8\ H^+(aq) + 3\ CH_3OH(aq) + Cr_2O_7^{2-}(aq) \rightarrow 2\ Cr^{3+}(aq) + 3\ CH_2O(aq) + 7\ H_2O(l)$$

75. Use the same method as with acidic solutions. After the final balanced equation, convert H^+ to OH^- as described in section 4.10 of the text. The extra step involves converting H^+ into H_2O by adding equal moles of OH^- to each side of the reaction. This converts the reaction to a basic solution while still keeping it balanced.

a.

$$Al \rightarrow Al(OH)_4^-$$
$$4\ H_2O + Al \rightarrow Al(OH)_4^- + 4\ H^+$$
$$4\ H_2O + Al \rightarrow Al(OH)_4^- + 4\ H^+ + 3\ e^-$$

$$MnO_4^- \rightarrow MnO_2$$
$$3\ e^- + 4\ H^+ + MnO_4^- \rightarrow MnO_2 + 2\ H_2O$$

$$4\ H_2O + Al \rightarrow Al(OH)_4^- + 4\ H^+ + 3\ e^-$$
$$3\ e^- + 4\ H^+ + MnO_4^- \rightarrow MnO_2 + 2\ H_2O$$

$$2\ H_2O(l) + Al(s) + MnO_4^-(aq) \rightarrow Al(OH)_4^-(aq) + MnO_2(s)$$

H^+ doesn't appear in the final balanced reaction, so we are done.

b. $Cl_2 \rightarrow Cl^-$ $Cl_2 \rightarrow OCl^-$

$2\ e^- + Cl_2 \rightarrow 2\ Cl^-$ $2\ H_2O + Cl_2 \rightarrow 2\ OCl^- + 4\ H^+ + 2\ e^-$

$$2\ e^- + Cl_2 \rightarrow 2\ Cl^-$$
$$2\ H_2O + Cl_2 \rightarrow 2\ OCl^- + 4\ H^+ + 2\ e^-$$
$$\overline{2\ H_2O + 2\ Cl_2 \rightarrow 2\ Cl^- + 2\ OCl^- + 4\ H^+}$$

Now convert to a basic solution. Add 4 OH^- to both sides of the equation. The 4 OH^- will react with the 4 H^+ on the product side to give 4 H_2O. After this step, cancel identical species on both sides (2 H_2O). Applying these steps gives: $4\ OH^- + 2\ Cl_2 \rightarrow 2\ Cl^- + 2\ OCl^- + 2\ H_2O$, which can be further simplified to:

$$2\ OH^-(aq) + Cl_2(g) \rightarrow Cl^-(aq) + OCl^-(aq) + H_2O(l)$$

c. $NO_2^- \rightarrow NH_3$ $Al \rightarrow AlO_2^-$

$6\ e^- + 7\ H^+ + NO_2^- \rightarrow NH_3 + 2\ H_2O$ $(2\ H_2O + Al \rightarrow AlO_2^- + 4\ H^+ + 3\ e^-) \times 2$

Common factor is a transfer of 6 e^-.

$$6e^- + 7\ H^+ + NO_2^- \rightarrow NH_3 + 2\ H_2O$$
$$4\ H_2O + 2\ Al \rightarrow 2\ AlO_2^- + 8\ H^+ + 6\ e^-$$
$$\overline{OH^- + 2\ H_2O + NO_2^- + 2\ Al \rightarrow NH_3 + 2\ AlO_2^- + H^+ + OH^-}$$

Reducing gives: $OH^-(aq) + H_2O(l) + NO_2^-(aq) + 2\ Al(s) \rightarrow NH_3(g) + 2\ AlO_2^-(aq)$

77. $NaCl + H_2SO_4 + MnO_2 \rightarrow Na_2SO_4 + MnCl_2 + Cl_2 + H_2O$

We could balance this reaction by the half-reaction method or by inspection. Let's try inspection. To balance Cl^-, we need 4 NaCl:

$$4\ NaCl + H_2SO_4 + MnO_2 \rightarrow Na_2SO_4 + MnCl_2 + Cl_2 + H_2O$$

Balance the Na^+ and SO_4^{2-} ions next:

$$4\ NaCl + 2\ H_2SO_4 + MnO_2 \rightarrow 2\ Na_2SO_4 + MnCl_2 + Cl_2 + H_2O$$

On the left side: 4−H and 10−O; On the right side: 8−O not counting H_2O

We need 2 H_2O on the right side to balance H and O:

$$4\ NaCl(aq) + 2\ H_2SO_4(aq) + MnO_2(s) \rightarrow 2\ Na_2SO_4(aq) + MnCl_2(aq) + Cl_2(g) + 2\ H_2O(l)$$

Additional Exercises

79. Only statement b is true. A concentrated solution can also contain a nonelectrolyte dissolved in water, e.g., concentrated sugar water. Acids are either strong or weak electrolytes. Some ionic compounds are not soluble in water so they are not labeled as a specific type of electrolyte.

81. There are other possible correct choices for most of the following answers. We have listed only three possible reactants in each case.

a. $AgNO_3$, $Pb(NO_3)_2$, and $Hg_2(NO_3)_2$ would form precipitates with the Cl^- ion.
$Ag^+(aq) + Cl^-(aq) \rightarrow AgCl(s)$; $Pb^{2+}(aq) + 2\ Cl^-(aq) \rightarrow PbCl_2(s)$;
$Hg_2^{2+}(aq) + 2\ Cl^-(aq) \rightarrow Hg_2Cl_2(s)$

b. Na_2SO_4, Na_2CO_3, and Na_3PO_4 would form precipitates with the Ca^{2+} ion.
$Ca^{2+}(aq) + SO_4^{2-}(aq) \rightarrow CaSO_4(s)$; $Ca^{2+}(aq) + CO_3^{2-}(aq) \rightarrow CaCO_3(s)$
$3\ Ca^{2+}(aq) + 2\ PO_4^{3-}(aq) \rightarrow Ca_3(PO_4)_2(s)$

c. NaOH, Na_2S, and Na_2CO_3 would form precipitates with the Fe^{3+} ion.
$Fe^{3+}(aq) + 3\ OH^-(aq) \rightarrow Fe(OH)_3(s)$; $2\ Fe^{3+}(aq) + 3\ S^{2-}(aq) \rightarrow Fe_2S_3(s)$;
$2\ Fe^{3+}(aq) + 3\ CO_3^{2-}(aq) \rightarrow Fe_2(CO_3)_3(s)$

d. $BaCl_2$, $Pb(NO_3)_2$, and $Ca(NO_3)_2$ would form precipitates with the SO_4^{2-} ion.
$Ba^{2+}(aq) + SO_4^{2-}(aq) \rightarrow BaSO_4(s)$; $Pb^{2+}(aq) + SO_4^{2-}(aq) \rightarrow PbSO_4(s)$;
$Ca^{2+}(aq) + SO_4^{2-}(aq) \rightarrow CaSO_4(s)$

e. Na_2SO_4, NaCl, and NaI would form precipitates with the Hg_2^{2+} ion.
$Hg_2^{2+}(aq) + SO_4^{2-}(aq) \rightarrow Hg_2SO_4(s)$; $Hg_2^{2+}(aq) + 2\ Cl^-(aq) \rightarrow Hg_2Cl_2(s)$;
$Hg_2^{2+}(aq) + 2\ I^-(aq) \rightarrow Hg_2I_2(s)$

f. NaBr, Na_2CrO_4, and Na_3PO_4 would form precipitates with the Ag^+ ion.
$Ag^+(aq) + Br^-(aq) \rightarrow AgBr(s)$; $2\ Ag^+(aq) + CrO_4^{2-}(aq) \rightarrow Ag_2CrO_4(s)$;
$3\ Ag^+(aq) + PO_4^{3-}(aq) \rightarrow Ag_3PO_4(s)$

83. $XCl_2(aq) + 2\ AgNO_3(aq) \rightarrow 2\ AgCl(s) + X(NO_3)_2(aq)$

$$1.38\ \text{g AgCl} \times \frac{1\ \text{mol}}{143.4\ \text{g}} \times \frac{1\ \text{mol XCl}_2}{2\ \text{mol AgCl}} = 4.81 \times 10^{-3}\ \text{mol XCl}_2$$

$$\frac{1.00\ \text{g XCl}_2}{4.91 \times 10^{-3}\ \text{mol XCl}_2} = 208\ \text{g/mol}; \qquad x + 2(35.45) = 208,\ \ x = 137\ \text{g/mol}$$

The metal X is barium (Ba).

85. All the sulfur in $BaSO_4$ came from the saccharin. The conversion from $BaSO_4$ to saccharin utilizes the molar masses of each compound.

$$0.5032 \text{ g BaSO}_4 \times \frac{32.07 \text{ g S}}{233.4 \text{ g BaSO}_4} \times \frac{183.19 \text{ g saccharin}}{32.07 \text{ g S}} = 0.3949 \text{ g saccharin}$$

$$\frac{\text{Avg. mass}}{\text{Tablet}} = \frac{0.3949 \text{ g}}{10 \text{ tablets}} = \frac{3.949 \times 10^{-2} \text{ g}}{\text{tablet}} = \frac{39.49 \text{ mg}}{\text{tablet}}$$

$$\text{Avg. mass \%} = \frac{0.3949 \text{ g saccharin}}{0.5894 \text{ g}} \times 100 = 67.00\% \text{ saccharin by mass}$$

87. $Cr(NO_3)_3(aq) + 3\ NaOH(aq) \rightarrow Cr(OH)_3(s) + 3\ NaNO_3(aq)$

$$2.06 \text{ g Cr(OH)}_3 \times \frac{1 \text{ mol Cr(OH)}_3}{103.02 \text{ g}} \times \frac{3 \text{ mol NaOH}}{1 \text{ mol Cr(OH)}_3} = 6.00 \times 10^{-2} \text{ mol NaOH to form ppt}$$

$NaOH(aq) + HCl(aq) \rightarrow NaCl(aq) + H_2O(l)$

$$0.1000 \text{ L} \times \frac{0.400 \text{ mol HCl}}{\text{L}} \times \frac{1 \text{ mol NaOH}}{\text{mol HCl}} = 4.00 \times 10^{-2} \text{ mol NaOH to react with HCl}$$

$$M_{\text{NaOH}} = \frac{6.00 \times 10^{-2} \text{ mol} + 4.00 \times 10^{-2} \text{ mol}}{0.0500 \text{ L}} = 2.00\ M \text{ NaOH}$$

89. $HC_2H_3O_2(aq) + NaOH(aq) \rightarrow H_2O(l) + NaC_2H_3O_2(aq)$

a. $$16.58 \times 10^{-3} \text{ L soln} \times \frac{0.5062 \text{ mol NaOH}}{\text{L soln}} \times \frac{1 \text{ mol acetic acid}}{\text{mol NaOH}}$$

$$= 8.393 \times 10^{-3} \text{ mol acetic acid}$$

$$\text{Concentration of acetic acid} = \frac{8.393 \times 10^{-3} \text{ mol}}{0.01000 \text{ L}} = 0.8393\ M$$

b. If we have 1.000 L of solution: total mass = 1000. mL × $\frac{1.006 \text{ g}}{\text{mL}}$ = 1006 g

$$\text{Mass of HC}_2\text{H}_3\text{O}_2 = 0.8393 \text{ mol} \times \frac{60.05 \text{ g}}{\text{mol}} = 50.40 \text{ g}$$

$$\text{Mass \% acetic acid} = \frac{50.40 \text{ g}}{1006 \text{ g}} \times 100 = 5.010\%$$

91. Let HA = unknown acid; $HA(aq) + NaOH(aq) \rightarrow NaA(aq) + H_2O(l)$

$$\text{mol HA present} = 0.0250 \text{ L} \times \frac{0.500 \text{ mol NaOH}}{\text{L}} \times \frac{1 \text{ mol HA}}{1 \text{ mol NaOH}} = 0.0125 \text{ mol HA}$$

$$\frac{x \text{ g HA}}{\text{mol HA}} = \frac{2.20 \text{ g HA}}{0.0125 \text{ mol HA}} \quad x = \text{molar mass of HA} = 176 \text{ g/mol}$$

Empirical formula weight ≈ 3(12) + 4(1) + 3(16) = 88 g/mol

176/88 = 2.0, so the molecular formula is $(C_3H_4O_3)_2 = C_6H_8O_6$.

93. Strong bases contain the hydroxide ion, OH^-. The reaction that occurs is $H^+ + OH^- \rightarrow H_2O$.

$$0.0120\ \text{L} \times \frac{0.150\ \text{mol H}^+}{\text{L}} \times \frac{1\ \text{mol OH}^-}{\text{mol H}^+} = 1.80 \times 10^{-3}\ \text{mol OH}^-$$

The 30.0 mL of the unknown strong base contains 1.80×10^{-3} mol OH^-, so:

$$\frac{1.8 \times 10^{-3}\ \text{mol OH}^-}{0.0300\ \text{L}} = 0.0600\ \text{M OH}^-$$

The unknown base concentration is one-half the concentration of OH^- ions produced from the base, so the base must contain 2 OH^- in each formula unit. The three soluble strong bases that have 2 OH^- ions in the formula are $Ca(OH)_2$, $Sr(OH)_2$, and $Ba(OH)_2$. These are all possible identities for the strong base.

95. $Mn + HNO_3 \rightarrow Mn^{2+} + NO_2$

$Mn \rightarrow Mn^{2+} + 2\ e^-$

$HNO_3 \rightarrow NO_2$

$HNO_3 \rightarrow NO_2 + H_2O$

$(e^- + H^+ + HNO_3 \rightarrow NO_2 + H_2O) \times 2$

$$Mn \rightarrow Mn^{2+} + 2\ e^-$$
$$2\ e^- + 2\ H^+ + 2\ HNO_3 \rightarrow 2\ NO_2 + 2\ H_2O$$

$$2\ H^+(aq) + Mn(s) + 2\ HNO_3(aq) \rightarrow Mn^{2+}(aq) + 2\ NO_2(g) + 2\ H_2O(l)$$

$Mn^{2+} + IO_4^- \rightarrow MnO_4^- + IO_3^-$

$(4\ H_2O + Mn^{2+} \rightarrow MnO_4^- + 8\ H^+ + 5\ e^-) \times 2$ $(2\ e^- + 2\ H^+ + IO_4^- \rightarrow IO_3^- + H_2O) \times 5$

$$8\ H_2O + 2\ Mn^{2+} \rightarrow 2\ MnO_4^- + 16\ H^+ + 10\ e^-$$
$$10\ e^- + 10\ H^+ + 5\ IO_4^- \rightarrow 5\ IO_3^- + 5\ H_2O$$

$$3\ H_2O(l) + 2\ Mn^{2+}(aq) + 5\ IO_4^-(aq) \rightarrow 2\ MnO_4^-(aq) + 5\ IO_3^-(aq) + 6\ H^+(aq)$$

Challenge Problems

97. a. $0.308\ \text{g AgCl} \times \frac{35.45\ \text{g Cl}}{143.4\ \text{g AgCl}} = 0.0761\ \text{g Cl}$; $\%Cl = \frac{0.761\ \text{g}}{0.256\ \text{g}} \times 100 = 29.7\%\ \text{Cl}$

Cobalt(III) oxide, Co_2O_3: 2(58.93) + 3(16.00) = 165.86 g/mol

$$0.145\ \text{g Co}_2\text{O}_3 \times \frac{117.86\ \text{g Co}}{165.86\ \text{g Co}_2\text{O}_3} = 0.103\ \text{g Co};\ \%Co = \frac{0.103\ \text{g}}{0.416\ \text{g}} \times 100 = 24.8\%\ \text{Co}$$

The remainder, 100.0 - (29.7 + 24.8) = 45.5%, is water.

Assuming 100.0 g of compound:

$$45.5 \text{ g } H_2O \times \frac{2.016 \text{ g H}}{18.02 \text{ g } H_2O} = 5.09 \text{ g H; \%H} = \frac{5.09 \text{ g H}}{100.0 \text{ g compound}} \times 100 = 5.09\% \text{ H}$$

$$45.5 \text{ g } H_2O \times \frac{16.00 \text{ g O}}{18.02 \text{ g } H_2O} = 40.4 \text{ g O; \%O} = \frac{40.4 \text{ g O}}{100.0 \text{ g compound}} \times 100 = 40.4\% \text{ O}$$

The mass percent composition is 24.8% Co, 29.7% Cl, 5.09% H and 40.4% O.

b. Out of 100.0 g of compound, there are:

$$24.8 \text{ g Co} \times \frac{1 \text{ mol Co}}{58.93 \text{ g Co}} = 0.421 \text{ mol Co; } 29.7 \text{ g Cl} \times \frac{1 \text{ mol Cl}}{35.45 \text{ g Cl}} = 0.838 \text{ mol Cl}$$

$$5.09 \text{ g H} \times \frac{1 \text{ mol H}}{1.008 \text{ g H}} = 5.05 \text{ mol H; } 40.4 \text{ g O} \times \frac{1 \text{ mol O}}{16.00 \text{ g O}} = 2.53 \text{ mol O}$$

Dividing all results by 0.421, we get $CoCl_2 \bullet 6H_2O$ for the formula.

c. $CoCl_2 \bullet 6H_2O(aq) + 2\ AgNO_3(aq) \rightarrow 2\ AgCl(s) + Co(NO_3)_2(aq) + 6\ H_2O(l)$

$CoCl_2 \bullet 6H_2O(aq) + 2\ NaOH(aq) \rightarrow Co(OH)_2(s) + 2\ NaCl(aq) + 6\ H_2O(l)$

$Co(OH)_2 \rightarrow Co_2O_3$ This is an oxidation-reduction reaction. Thus, we also need to include an oxidizing agent. The obvious choice is O_2.

$4\ Co(OH)_2(s) + O_2(g) \rightarrow 2\ Co_2O_3(s) + 4\ H_2O(l)$

99. a. $2\ AgNO_3(aq) + K_2CrO_4(aq) \rightarrow Ag_2CrO_4(s) + 2\ KNO_3(aq)$

Molar mass: 169.9 g/mol 194.20 g/mol 331.8 g/mol

The molar mass of Ag_2CrO_4 is 331.8 g/mol, so one mol of precipitate was formed.

We have equal masses of $AgNO_3$ and K_2CrO_4. Since the molar mass of $AgNO_3$ is less than that of K_2CrO_4, then we have more mol of $AgNO_3$ present. However, we will not have twice the mol of $AgNO_3$ present as compared to K_2CrO_4 as required by the balanced reaction; this is because the molar mass of $AgNO_3$ is nowhere near one-half the molar mass of K_2CrO_4. Therefore, $AgNO_3$ is limiting.

$$\text{mass } AgNO_3 = 1.000 \text{ mol } Ag_2CrO_4 \times \frac{2 \text{ mol } AgNO_3}{\text{mol } Ag_2CrO_4} \times \frac{169.9 \text{ g}}{\text{mol } AgNO_3} = 339.8 \text{ g } AgNO_3$$

Because equal masses of reactants are present, 339.8 g K_2CrO_4 were present initially.

$$M_{K^+} = \frac{\text{mol } K^+}{\text{total volume}} = \frac{339.8 \text{ g } K_2CrO_4 \times \frac{1 \text{ mol } K_2CrO_4}{194.20 \text{ g}} \times \frac{2 \text{ mol } K^+}{\text{mol } K_2CrO_4}}{0.5000 \text{ L}} = 7.000\ M\ K^+$$

b. $\text{mol } CrO_4^{2-} = 339.8 \text{ g } K_2CrO_4 \times \frac{1 \text{ mol } K_2CrO_4}{194.20 \text{ g}} \times \frac{1 \text{ mol } CrO_4^{2-}}{\text{mol } K_2CrO_4} = 1.750 \text{ mol } CrO_4^{2-}$
present initially

$$\text{mol } CrO_4^{2-} \text{ in } = 1.000 \text{ mol } Ag_2CrO_4 \times \frac{1 \text{ mol } CrO_4^{2-}}{\text{mol } Ag_2CrO_4} = 1.000 \text{ mol } CrO_4^{2-} \text{ precipitate}$$

$$M_{CrO_4^{2-}} = \frac{\text{excess mol } CrO_4^{2-}}{\text{total volume}} = \frac{1.750 \text{ mol} - 1.000 \text{ mol}}{0.5000 \text{ L} + 0.5000 \text{ L}} = \frac{0.750 \text{ mol}}{1.0000 \text{ L}} = 0.750\ M$$

101. $0.298 \text{ g } BaSO_4 \times \frac{96.07 \text{ g } SO_4^{2-}}{233.4 \text{ g } BaSO_4} = 0.123 \text{ g } SO_4^{2-}$; $\% \text{ sulfate} = \frac{0.123 \text{ g } SO_4^{2-}}{0.205 \text{ g}} = 60.0\%$

Assume we have 100.0 g of the mixture of Na_2SO_4 and K_2SO_4. There are:

$$60.0 \text{ g } SO_4^{2-} \times \frac{1 \text{ mol}}{96.07 \text{ g}} = 0.625 \text{ mol } SO_4^{2-}$$

There must be $2 \times 0.625 = 1.25$ mol of +1 cations to balance the −2 charge of SO_4^{2-}.

Let x = number of moles of K^+ and y = number of moles of Na^+; then $x + y = 1.25$.

The total mass of Na^+ and K^+ must be 40.0 g in the assumed 100.0 g of mixture. Setting up an equation:

$$x \text{ mol } K^+ \times \frac{39.10 \text{ g}}{\text{mol}} + y \text{ mol } Na^+ \times \frac{22.99 \text{ g}}{\text{mol}} = 40.0 \text{ g}$$

So, we have two equations with two unknowns: $x + y = 1.25$ and $39.10\,x + 22.99\,y = 40.0$

$x = 1.25 - y$, so $39.10(1.25 - y) + 22.99\,y = 40.0$

$48.9 - 39.10\,y + 22.99\,y = 40.0$, $-16.11\,y = -8.9$

$y = 0.55$ mol Na^+ and $x = 1.25 - 0.55 = 0.70$ mol K^+

Therefore:

$$0.70 \text{ mol } K^+ \times \frac{1 \text{ mol } K_2SO_4}{2 \text{ mol } K^+} = 0.35 \text{ mol } K_2SO_4;\ 0.35 \text{ mol } K_2SO_4 \times \frac{174.27 \text{ g}}{\text{mol}} = 61 \text{ g } K_2SO_4$$

We assumed 100.0 g, therefore the mixture is 61% K_2SO_4 and 39% Na_2SO_4.

103. $Pb^{2+}(aq) + 2\,Cl^-(aq) \rightarrow PbCl_2(s)$
3.407 g

$$3.407 \text{ g } PbCl_2 \times \frac{1 \text{ mol } PbCl_2}{278.1 \text{ g } PbCl_2} \times \frac{1 \text{ mol } Pb^{2+}}{1 \text{ mol } PbCl_2} = 0.01225 \text{ mol } Pb^{2+}$$

$$\frac{0.01225 \text{ mol}}{2.00 \times 10^{-3} \text{ L}} = 6.13\ M\ Pb^{2+} \text{ (evaporated concentration)}$$

$$\text{original concentration} = \frac{0.0800 \text{ L} \times 6.13 \text{ mol/L}}{0.100 \text{ L}} = 4.90\ M$$

105. 0.2750 L × 0.300 mol/L = 0.0825 mol H^+; Let y = volume (L) delivered by Y and z = volume (L) delivered by Z.

$$H^+(aq) + OH^-(aq) \rightarrow H_2O(l); \quad \underbrace{y(0.150 \text{ mol/L}) + z(0.250 \text{ mol/L})}_{\text{mol } OH^-} = 0{,}9725 \text{ mol } H^+$$

0.2750 L + y + z = 0.655 L, y + z = 0.380, z = 0.380 − y

y(0.150) + (0.380 − y)(0.250) = 0.0825, Solving: y = 0.125 L, z = 0.255 L

$$\text{flow rates: Y} \rightarrow \frac{125 \text{ mL}}{60.65 \text{ min}} = 2.06 \text{ mL/min and Z} \rightarrow \frac{255 \text{ mL}}{60.65 \text{ min}} = 4.20 \text{ mL/min}$$

107. $2\ H_3PO_4(aq) + 3\ Ba(OH)_2(aq) \rightarrow 6\ H_2O(l) + Ba_3(PO_4)_2(s)$

$$0.01420 \text{ L} \times \frac{0.141 \text{ mol } H_3PO_4}{\text{L}} \times \frac{3 \text{ mol } Ba(OH)_2}{2 \text{ mol } H_3PO_4} \times \frac{1 \text{ L } Ba(OH)_2}{0.0521 \text{ mol } Ba(OH)_2} = 0.0576 \text{ L}$$

$$= 57.6 \text{ mL } Ba(OH)_2$$

109. a. $MgO(s) + 2\ HCl(aq) \rightarrow MgCl_2(aq) + H_2O(l)$

$Mg(OH)_2(s) + 2\ HCl(aq) \rightarrow MgCl_2(aq) + 2\ H_2O(l)$

$Al(OH)_3(s) + 3\ HCl(aq) \rightarrow AlCl_3(aq) + 3\ H_2O(l)$

b. Let's calculate the number of moles of HCl neutralized per gram of substance. We can get these directly from the balanced equations and the molar masses of the substances.

$$\frac{2 \text{ mol HCl}}{\text{mol MgO}} \times \frac{1 \text{ mol MgO}}{40.31 \text{ g MgO}} = \frac{4.962 \times 10^{-2} \text{ mol HCl}}{\text{g MgO}}$$

$$\frac{2 \text{ mol HCl}}{\text{mol } Mg(OH)_2} \times \frac{1 \text{ mol } Mg(OH)_2}{58.33 \text{ g } Mg(OH)_2} = \frac{3.429 \times 10^{-2} \text{ mol HCl}}{\text{g } Mg(OH)_2}$$

$$\frac{3 \text{ mol HCl}}{\text{mol } Al(OH)_3} \times \frac{1 \text{ mol } Al(OH)_3}{78.00 \text{ g } Al(OH)_3} = \frac{3.846 \times 10^{-2} \text{ mol HCl}}{\text{g } Al(OH)_3}$$

Therefore, one gram of magnesium oxide would neutralize the most 0.10 *M* HCl.

111. $\text{mol } C_6H_8O_7 = 0.250 \text{ g } C_6H_8O_7 \times \frac{1 \text{ mol } C_6H_8O_7}{192.12 \text{ g } C_6H_8O_7} = 1.30 \times 10^{-3} \text{ mol } C_6H_8O_7$

Let H_xA represent citric acid where x is the number of acidic hydrogens. The balanced neutralization reaction is:

$$H_xA(aq) + x\, OH^-(aq) \rightarrow x\, H_2O(l) + A^{x-}(aq)$$

$$\text{mol } OH^- \text{ reacted} = 0.0372 \text{ L} \times \frac{0.105 \text{ mol } OH^-}{\text{L}} = 3.91 \times 10^{-3} \text{ mol } OH^-$$

$$x = \frac{\text{mol } OH^-}{\text{mol citric acid}} = \frac{3.91 \times 10^{-3} \text{ mol}}{1.30 \times 10^{-3} \text{ mol}} = 3.01$$

Therefore, the general acid formula for citric acid is H_3A, meaning that citric acid has three acidic hydrogens per citric acid molecule (citric acid is a triprotic acid).

113. $\text{mol KHP used} = 0.4016 \text{ g} \times \frac{1 \text{ mol}}{204.22 \text{ g}} = 1.967 \times 10^{-3} \text{ mol KHP}$

One mole of NaOH reacts completely with one mole of KHP, therefore the NaOH solution contains 1.967×10^{-3} mol NaOH.

$$\text{Molarity of NaOH} = \frac{1.967 \times 10^{-3} \text{ mol}}{25.06 \times 10^{-3} \text{ L}} = \frac{7.849 \times 10^{-2} \text{ mol}}{\text{L}}$$

$$\text{Maximum molarity} = \frac{1.967 \times 10^{-3} \text{ mol}}{25.01 \times 10^{-3} \text{ L}} = \frac{7.865 \times 10^{-2} \text{ mol}}{\text{L}}$$

$$\text{Minimum molarity} = \frac{1.967 \times 10^{-3} \text{ mol}}{25.11 \times 10^{-3} \text{ L}} = \frac{7.834 \times 10^{-2} \text{ mol}}{\text{L}}$$

We can express this as 0.07849 ± 0.00016 *M*. An alternative is to express the molarity as 0.0785 ± 0.0002 *M*. The second way shows the actual number of significant figures in the molarity. The advantage of the first method is that it shows that we made all of our individual measurements to four significant figures.

Integrative Problems

115. $3\, (NH_4)_2CrO_4(aq) + 2\, Cr(NO_2)_3(aq) \rightarrow 6\, NH_4NO_2(aq) + Cr_2(CrO_4)_3(s)$

$$0.203 \text{ L} \times \frac{0.307 \text{ mol}}{\text{L}} = 6.23 \times 10^{-2} \text{ mol } (NH_4)_2CrO_4$$

$$0.137 \text{ L} \times \frac{0.269 \text{ mol}}{\text{L}} = 3.69 \times 10^{-2} \text{ mol } Cr(NO_2)_3$$

$\frac{0.0623 \text{ mol}}{0.0369 \text{ mol}}$ = 1.69 (actual); The balanced reaction requires a 3/2 = 1.5 mol ratio between

$(NH_4)_2CrO_4$ and $Cr(NO_2)_3$. Actual > required, so $Cr(NO_2)_3$ (the denominator) is limiting.

$$3.69 \times 10^{-2} \text{ mol } Cr(NO_2)_3 \times \frac{1 \text{ mol } Cr_2(CrO_4)_3}{2 \text{ mol } Cr(NO_2)_3} \times \frac{452.00 \text{ g } Cr_2(CrO_4)_3}{1 \text{ mol } Cr_2(CrO_4)_3} = 8.34 \text{ g } Cr_2(CrO_4)_3$$

$$0.880 = \frac{\text{actual yield}}{8.34 \text{ g}}, \text{ actual yield} = (8.34 \text{ g})(0.880) = 7.34 \text{ g } Cr_2(CrO_4)_3 \text{ isolated}$$

117. X^{2-} contains 36 electrons, so X^{2-} has 34 protons which identifies X as selenium (Se). The name of H_2Se would be hydroselenic acid following the conventions described in Chapter 2.

$$H_2Se(aq) + 2\ OH^-(aq) \rightarrow Se^{2-}(aq) + 2\ H_2O(l)$$

$$0.0356 \text{ L} \times \frac{0.175 \text{ mol } OH^-}{\text{L}} \times \frac{1 \text{ mol } H_2Se}{2 \text{ mol } OH^-} \times \frac{80.98 \text{ g } H_2Se}{\text{mol } H_2Se} = 0.252 \text{ g } H_2Se$$

CHAPTER FIVE

GASES

Questions

17. The column of water would have to be 13.6 times taller than a column of mercury. When the pressure of the column of liquid standing on the surface of the liquid is equal to the pressure of air on the rest of the surface of the liquid, then the height of the column of liquid is a measure of atmospheric pressure. Because water is 13.6 times less dense than mercury, the column of water must be 13.6 times longer than that of mercury to match the force exerted by the columns of liquid standing on the surface.

19. The P versus 1/V plot is incorrect. The plot should be linear with positive slope and a y-intercept of zero. PV = k so P = k(1/V). This is in the form of the straight-line equation y = mx + b. The y-axis is pressure and the x-axis is 1/V.

21. d = (molar mass) P/RT; Density is directly proportional to the molar mass of a gas. Helium, with the smallest molar mass of all the noble gases, will have the smallest density.

23. No; At any nonzero Kelvin temperature, there is a distribution of kinetic energies. Similarly, there is a distribution of velocities at any nonzero Kelvin temperature. The reason there is a distribution of energies at any specific temperature is because there is a distribution of gas particle velocities at any T.

25. $2\ NH_3(g) \rightarrow N_2(g) + 3\ H_2(g)$: As reactants are converted into products, we go from 2 moles of gaseous reactants to 4 moles of gaseous products (1 mol N_2 + 3 mol H_2). Because the mol of gas doubles as reactants are converted into products, the volume of the gases will double (at constant P and T).

$PV = nRT$, $P = \left(\frac{RT}{V}\right) n$ = (constant) n; Pressure is directly related to n at constant T and V.

As the reaction occurs, the moles of gas will double so the pressure will double. Because 1 mol of N_2 is produced for every 2 mol of NH_3 reacted, $P_{N_2} = 1/2\ P^{o}_{NH_3}$. Due to the 3 to 2 mol ratio in the balanced equation, $P_{H_2} = 3/2\ P^{o}_{NH_3}$.

Note: $P_{tot} = P_{H_2} + P_{N_2} = 3/2\ P^{o}_{NH_3} + 1/2\ P^{o}_{NH_3} = 2\ P^{o}_{NH_3}$. As said earlier, the total pressure will double from the initial pressure of NH_3 as the reactants are completely converted into products.

Exercises

Pressure

27. a. $4.8\text{ atm} \times \dfrac{760\text{ mm Hg}}{\text{atm}} = 3.6 \times 10^3\text{ mm Hg}$ b. $3.6 \times 10^3\text{ mm Hg} \times \dfrac{1\text{ torr}}{\text{mm Hg}}$

$= 3.6 \times 10^3\text{ torr}$

c. $4.8\text{ atm} \times \dfrac{1.013 \times 10^5\text{ Pa}}{\text{atm}} = 4.9 \times 10^5\text{ Pa}$ d. $4.8\text{ atm} \times \dfrac{14.7\text{ psi}}{\text{atm}} = 71\text{ psi}$

29. $6.5\text{ cm} \times \dfrac{10\text{ mm}}{\text{cm}} = 65\text{ mm Hg} = 65\text{ torr}$; $65\text{ torr} \times \dfrac{1\text{ atm}}{760\text{ torr}} = 8.6 \times 10^{-2}\text{ atm}$

$8.6 \times 10^{-2}\text{ atm} \times \dfrac{1.013 \times 10^5\text{ Pa}}{\text{atm}} = 8.7 \times 10^3\text{ Pa}$

31. If the levels of Hg in each arm of the manometer are equal, the pressure in the flask is equal to atmospheric pressure. When they are unequal, the difference in height in mm will be equal to the difference in pressure in mm Hg between the flask and the atmosphere. Which level is higher will tell us whether the pressure in the flask is less than or greater than atmospheric.

a. $P_{flask} < P_{atm}$; $P_{flask} = 760. - 118 = 642\text{ mm Hg} = 642\text{ torr}$; $642\text{ torr} \times \dfrac{1\text{ atm}}{760\text{ torr}}$

$= 0.845\text{ atm}$

$0.845\text{ atm} \times \dfrac{1.013 \times 10^5\text{ Pa}}{\text{atm}} = 8.56 \times 10^4\text{ Pa}$

b. $P_{flask} > P_{atm}$; $P_{flask} = 760.\text{ torr} + 215\text{ torr} = 975\text{ torr}$; $975\text{ torr} \times \dfrac{1\text{ atm}}{760\text{ torr}} = 1.28\text{ atm}$

$1.28\text{ atm} \times \dfrac{1.013 \times 10^5\text{ Pa}}{\text{atm}} = 1.30 \times 10^5\text{ Pa}$

c. $P_{flask} = 635 - 118 = 517\text{ torr}$; $P_{flask} = 635 + 215 = 850.\text{ torr}$

Gas Laws

33. At constant n and T, PV = nRT = constant, $P_1V_1 = P_2V_2$; At sea level, P = 1.00 atm = 760. mm Hg.

$$V_2 = \frac{P_1V_1}{P_2} = \frac{760.\text{ mm} \times 2.0\text{ L}}{500.\text{ mm Hg}} = 3.0\text{ L}$$

The balloon will burst at this pressure since the volume must expand beyond the 2.5 L limit of the balloon.

Note: To solve this problem, we did not have to convert the pressure units into atm; the units of mm Hg canceled each other. In general, only convert units if you have to. Whenever the gas constant R is not used to solve a problem, pressure and volume units must only be consistent, and not necessarily in units of atm and L. The exception is temperature which must always be converted to the Kelvin scale.

35. At constant T and P, Avogadro's law holds.

$$\frac{V_1}{n_1} = \frac{V_2}{n_2},\ n_2 = \frac{V_2 n_1}{V_1} = \frac{20.\ \text{L} \times 0.50\ \text{mol}}{11.2\ \text{L}} = 0.89\ \text{mol}$$

As expected, as V increases, n increases.

37. a. $$PV = nRT,\ V = \frac{nRT}{P} = \frac{2.00\ \text{mol} \times \frac{0.08206\ \text{L atm}}{\text{mol K}} \times (155 + 273)\ \text{K}}{5.00\ \text{atm}} = 14.0\ \text{L}$$

b. $$PV = nRT,\ n = \frac{PV}{RT} = \frac{0.300\ \text{atm} \times 2.00\ \text{L}}{\frac{0.08206\ \text{L atm}}{\text{mol K}} \times 155\ \text{K}} = 4.72\ 10^{-2}\ \text{mol}$$

c. $$PV = nRT,\ T = \frac{PV}{nR} = \frac{4.47\ \text{atm} \times 25.0\ \text{L}}{2.01\ \text{mol} \times \frac{0.08206\ \text{L atm}}{\text{mol K}}} = 678\ \text{K} = 405°\text{C}$$

d. $$PV = nRT,\ P = \frac{nRT}{V} = \frac{10.5\ \text{mol} \times \frac{0.08206\ \text{L atm}}{\text{mol K}} \times (273 + 75)\ \text{K}}{2.25\ \text{L}} = 133\ \text{atm}$$

39. $$n = \frac{PV}{RT} = \frac{135\ \text{atm} \times 200.0\ \text{L}}{\frac{0.08206\ \text{L atm}}{\text{mol K}} \times (273 + 24)\ \text{K}} = 1.11 \times 10^3\ \text{mol}$$

For He: $$1.11 \times 10^3\ \text{mol} \times \frac{4.003\ \text{g He}}{\text{mol}} = 4.44 \times 10^3\ \text{g He}$$

For H_2: $$1.11 \times 10^3\ \text{mol} \times \frac{2.016\ \text{g He}}{\text{mol}} = 2.24 \times 10^3\ \text{g}\ H_2$$

41. a. $$PV = nRT;\ 175\ \text{g Ar} \times \frac{1\ \text{mol Ar}}{39.95\ \text{g Ar}} = 4.38\ \text{mol Ar}$$

$$T = \frac{PV}{nR} = \frac{10.0\ \text{atm} \times 2.50\ \text{L}}{4.38\ \text{mol} \times \frac{0.08206\ \text{L atm}}{\text{mol K}}} = 69.6\ \text{K}$$

b. $$PV = nRT,\ P = \frac{nRT}{V} = \frac{4.38\ \text{mol} \times \frac{0.08206\ \text{L atm}}{\text{mol K}} \times 255\ \text{K}}{2.50\ \text{L}} = 32.3\ \text{atm}$$

43. For a gas at two conditions: $\frac{P_1V_1}{n_1T_1} = \frac{P_2V_2}{n_2T_2}$

Because V is constant: $\frac{P_1}{n_1T_1} = \frac{P_2}{n_2T_2}$, $n_2 = \frac{n_1P_2T_1}{P_1T_2}$

$$n_2 = \frac{1.50\ \text{mol} \times 800.\ \text{torr} \times 298\ \text{K}}{400.\ \text{torr} \times 323\ \text{K}} = 2.77\ \text{mol}$$

mol of gas added = $n_2 - n_1 = 2.77 - 1.50 = 1.27$ mol

For two condition problems, units for P and V just need to be the same units for both conditions, not necessarily atm and L. The unit conversions from other P or V units would cancel when applied to both conditions. However, temperature always must be converted to the Kelvin scale. The temperature conversions between other units and Kelvin will not cancel each other.

45. At two conditions: $\frac{P_1V_1}{n_1T_1} = \frac{P_2V_2}{n_2T_2}$; All gases follow the ideal gas law. The identity of the gas in container B is unimportant as long as we know the mol of gas.

$$\frac{P_B}{P_A} = \frac{V_A\, n_b T_b}{V_B\, n_a\, T_a} = \frac{1.0\ \text{L} \times 2.0\ \text{mol} \times 560.\ \text{K}}{2.0\ \text{L} \times 1.0\ \text{mol} \times 280.\ \text{K}} = 2.0$$

The pressure of the gas in container B is twice the pressure of the gas in container A.

47. $\frac{PV}{T} = nR = \text{constant},\ \frac{P_1V_1}{T_1} = \frac{P_2V_2}{T_2}$

$$P_2 = \frac{P_1V_1T_2}{V_2T_1} = 7.10.\ \text{torr} \times \frac{5.0 \times 10^2\ \text{mL}}{25\ \text{mL}} \times \frac{(273 + 820.)\ \text{K}}{(273 + 30.)\ \text{K}} = 5.1 \times 10^4\ \text{torr}$$

49. $PV = nRT$, n is constant. $\frac{PV}{T} = nR = \text{constant}$, $\frac{P_1V_1}{T_1} = \frac{P_2V_2}{T_2}$, $V_2 = \frac{V_1P_1T_2}{V_2T_1}$

$$V_2 = 1.00\text{ L} \times \frac{760.\text{torr}}{220.\text{ torr}} \times \frac{(273-31)\text{ K}}{(273+23)\text{ K}} = 2.82\text{ L};\ \Delta V = 2.82 - 1.00 = 1.82\text{ L}$$

Gas Density, Molar Mass, and Reaction Stoichiometry

51. STP: T = 273 K and P = 1.00 atm; At STP, the molar volume of a gas is 22.42 L.

$$2.00\text{ L } O_2 \times \frac{1\text{ mol } O_2}{22.42\text{ L}} \times \frac{4\text{ mol Al}}{3\text{ mol } O_2} \times \frac{26.98\text{ g Al}}{\text{mol Al}} = 3.21\text{ g Al}$$

53. $2\ NaN_3(s) \rightarrow 2\ Na(s) + 3\ N_2(g)$

$$n_{N_2} = \frac{PV}{RT} = \frac{1.00\text{ atm} \times 70.0\text{ L}}{\frac{0.08206\text{ L atm}}{\text{mol K}} \times 273\text{ K}} = 3.12\text{ mol } N_2 \text{ needed to fill air bag.}$$

$$\text{mass } NaN_3 \text{ reacted} = 3.12\text{ mol } N_2 \times \frac{2\text{ mol } NaN_3}{3\text{ mol } N_2} \times \frac{65.02\text{ g } NaN_3}{\text{mol } NaN_3} = 135\text{ g } NaN_3$$

55. $$n_{H_2} = \frac{PV}{RT} = \frac{1.0\text{ atm} \times \left[4800\text{ m}^3 \times \left(\frac{100\text{ cm}}{\text{m}}\right)^3 \times \frac{1\text{ L}}{1000\text{ cm}^3}\right]}{\frac{0.08206\text{ L atm}}{\text{mol K}} \times 273\text{ K}} = 2.1 \times 10^5\text{ mol}$$

2.1×10^5 mol H_2 are in the balloon. This is 80.% of the total amount of H_2 that had to be generated:

$$0.80(\text{total mol } H_2) = 2.1 \times 10^5,\ \text{total mol } H_2 = 2.6 \times 10^5\text{ mol } H_2$$

$$2.6 \times 10^5\text{ mol } H_2 \times \frac{1\text{ mol Fe}}{\text{mol } H_2} \times \frac{55.85\text{ g Fe}}{\text{mol Fe}} = 1.5 \times 10^7\text{ g Fe}$$

$$2.6 \times 10^5\text{ mol } H_2 \times \frac{1\text{ mol } H_2SO_4}{\text{mol } H_2} \times \frac{98.09\text{ g } H_2SO_4}{\text{mol } H_2SO_4} \times \frac{100\text{ g reagent}}{98\text{ g } H_2SO_4}$$

$$= 2.6 \times 10^7\text{ g of 98\% sulfuric acid}$$

57. $CH_3OH + 3/2\ O_2 \rightarrow CO_2 + 2\ H_2O$ or $2\ CH_3OH(l) + 3\ O_2(g) \rightarrow 2\ CO_2(g) + 4\ H_2O(g)$

$$50.0\text{ mL} \times \frac{0.850\text{ g}}{\text{mL}} \times \frac{1\text{ mol}}{32.04\text{ g}} = 1.33\text{ mol } CH_3OH(l) \text{ available}$$

$$n_{O_2} = \frac{PV}{RT} = \frac{2.00 \text{ atm} \times 22.8 \text{ L}}{\frac{0.08206 \text{ L atm}}{\text{mol K}} \times 300. \text{ K}} = 1.85 \text{ mol } O_2 \text{ available}$$

$$1.33 \text{ mol } CH_3OH \times \frac{3 \text{ mol } O_2}{2 \text{ mol } CH_3OH} = 2.00 \text{ mol } O_2$$

2.00 mol O_2 are required to react completely with all of the CH_3OH available. We only have 1.85 mol O_2, so O_2 is limiting.

$$1.85 \text{ mol } O_2 \times \frac{4 \text{ mol } H_2O}{3 \text{ mol } O_2} = 2.47 \text{ mol } H_2O$$

59. a. $CH_4(g) + NH_3(g) + O_2(g) \rightarrow HCN(g) + H_2O(g)$; Balancing H first, then O, gives:

$$CH_4 + NH_3 + \frac{3}{2}O_2 \rightarrow HCN + 3\ H_2O \text{ or } 2\ CH_4(g) + 2\ NH_3(g) + 3\ O_2(g) \rightarrow 2\ HCN(g) + 6\ H_2O(g)$$

b. $PV = nRT$, T and P constant; $\frac{V_1}{n_1} = \frac{V_2}{n_2}$, $\frac{V_1}{V_2} = \frac{n_1}{n_2}$

The volumes are all measured at constant T and P, so the volumes of gas present are directly proportional to the moles of gas present (Avogadro's law). Because Avogadro's law applies, the balanced reaction gives mole relationships as well as volume relationships. Therefore, 2 L of CH_4, 2 L of NH_3 and 3 L of O_2 are required by the balanced equation for the production of 2 L of HCN. The actual volume ratio is 20.0 L CH_4:20.0 L NH_3:20.0 L O_2 (or 1:1:1). The volume of O_2 required to react with all of the CH_4 and NH_3 present is 20.0 L ×(3/2) = 30.0 L. Since only 20.0 L of O_2 are present, O_2 is the limiting reagent. The volume of HCN produced is:

$$20.0 \text{ L } O_2 \times \frac{2 \text{ L HCN}}{3 \text{ L } O_2} = 13.3 \text{ L HCN}$$

61. molar mass = $\frac{dRT}{P}$ where d = density of gas in units of g/L

$$\text{molar mass} = \frac{3.164 \text{ g/L} \times \frac{0.08206 \text{ L atm}}{\text{mol K}} \times 273.2 \text{ K}}{1.000 \text{ atm}} = 70.98 \text{ g/mol}$$

The gas is diatomic, so the atomic mass = 70.93/2 = 35.47. This is chlorine and the identity of the gas is Cl_2.

63. $$d_{UF_6} = \frac{P \times (\text{molar mass})}{RT} = \frac{\left(745 \text{ torr} \times \frac{1 \text{ atm}}{760 \text{ torr}}\right) \times 352.0 \text{ g/mol}}{\frac{0.08206 \text{ L atm}}{\text{mol K}} \times 333 \text{ K}} = 12.6 \text{ g/L}$$

Partial Pressure

65. $$P_{CO_2} = \frac{nRT}{V} = \frac{\left(7.8\text{ torr} \times \frac{1\text{ mol}}{44.01\text{ g}}\right) \times \frac{0.08206\text{ L atm}}{\text{mol K}} \times 300.\text{ K}}{4.0\text{ L}} = 1.1\text{ atm}$$

With air present, the partial pressure of CO_2 will still be 1.1 atm. The total pressure will be the sum of the partial pressures, $P_{total} = P_{CO_2} + P_{air}$.

$$P_{total} = 1.1\text{ atm} + \left(740\text{ torr} \times \frac{1\text{ atm}}{760\text{ torr}}\right) = 1.1 + 0.97 = 2.1\text{ atm}$$

67. Use the relationship $P_1V_1 = P_2V_2$ for each gas, since T and n for each gas are constant.

For H_2: $P_2 = \frac{P_1V_1}{V_2} = 475\text{ torr} \times \frac{2.00\text{ L}}{3.00\text{ L}} = 317\text{ torr}$

For N_2: $P_2 = 0.200\text{ atm} \times \frac{1.00\text{ L}}{3.00\text{ L}} = 0.0667\text{ atm}$; $0.0667\text{ atm} \times \frac{760\text{ torr}}{\text{atm}} = 50.7\text{ torr}$

$P_{total} = P_{H_2} + P_{N_2} = 317 + 50.7 = 368\text{ torr}$

69. a. mol fraction $CH_4 = \chi_{CH_4} = \frac{P_{CH_4}}{P_{total}} = \frac{0.175\text{ atm}}{0.175\text{ atm} + 0.250\text{ atm}} = 0.412$

$\chi_{O_2} = 1.000 - 0.412 = 0.588$

b. $PV = nRT$, $n_{total} = \frac{P_{total} \times V}{RT} = \frac{0.425\text{ atm} \times 10.5\text{ L}}{\frac{0.08206\text{ L atm}}{\text{mol K}} \times 338\text{ K}} = 0.161\text{ mol}$

c. $\chi_{CH_4} = \frac{n_{CH_4}}{n_{total}}$, $n_{CH_4} = \chi_{CH_4} \times n_{total} = 0.412 \times 0.161\text{ mol} = 6.63 \times 10^{-2}\text{ mol } CH_4$

$6.63 \times 10^{-2}\text{ mol } CH_4 \times \frac{16.04\text{ g } CH_4}{\text{mol } CH_4} = 1.06\text{ g } CH_4$

$n_{O_2} = 0.588 \times 0.161\text{ mol} = 9.47 \times 10^{-2}\text{ mol } O_2$; $9.47 \times 10^{-2}\text{ mol } O_2 \times \frac{32.00\text{ g } O_2}{\text{mol } O_2}$

$= 3.03\text{ g } O_2$

71. $P_{TOT} = P_{H_2} + P_{H_2O}$, $1.032 \text{ atm} = P_{H_2} + 32 \text{ torr} \times \frac{1 \text{ atm}}{760 \text{ torr}}$, $P_{H_2} = 1.032 - 0.042 = 0.990$ atm

$$n_{H_2} = \frac{P_{H_2}V}{RT} = \frac{0.990 \text{ atm} \times 0.240 \text{ L}}{\frac{0.08206 \text{ L atm}}{\text{mol K}} \times 303 \text{ K}} = 9.56 \times 10^{-3} \text{ mol } H_2$$

$$9.56 \times 10^{-3} \text{ mol } H_2 \times \frac{1 \text{ mol Zn}}{\text{mol } H_2} \times \frac{65.38 \text{ g Zn}}{\text{mol Zn}} = 0.625 \text{ g Zn}$$

73. $2 NaClO_3(s) \rightarrow 2 NaCl(s) + 3 O_2(g)$

$P_{total} = P_{O_2} + P_{H_2O}$, $P_{O_2} = P_{total} - P_{H_2O} = 734 \text{ torr} - 19.8 \text{ torr} = 714$ torr

$$n_{O_2} = \frac{P_{O_2} \times V}{RT} = \frac{\left(714 \text{ torr} \times \frac{1 \text{ atm}}{760 \text{ torr}}\right) \times 0.0572 \text{ L}}{\frac{0.08206 \text{ L atm}}{\text{mol K}} \times (273 + 22) \text{ K}} = 2.22 \times 10^{-3} \text{ mol } O_2$$

$$\text{Mass } NaClO_3 \text{ decomposed} = 2.22 \times 10^{-3} \text{ mol } O_2 \times \frac{2 \text{ mol } NaClO_3}{3 \text{ mol } O_2} \times \frac{106.44 \text{ g } NaClO_3}{\text{mol } NaClO_3}$$

$$= 0.158 \text{ g } NaClO_3$$

$$\text{Mass \% } NaClO_3 = \frac{0.158 \text{ g}}{0.8765 \text{ g}} \times 100 = 18.0\%$$

75. $2 HN_3(g) \rightarrow 3 N_2(g) + H_2(g)$; At constant V and T, P is directly proportional to n. In the reaction, we go from 2 moles of gaseous reactants to 4 moles of gaseous products. Since moles doubled, the final pressure will double ($P_{tot} = 6.0$ atm). Similarly, from the 2:1 mole ratio between HN_3 and H_2, the partial pressure of H_2 will be 3.0/2 = 1.5 atm. The partial pressure of N_2 will be 3/2 (3.0 atm) = 4.5 atm. This is from the 2:3 mole ratio between HN_3 and N_2.

Kinetic Molecular Theory and Real Gases

77. $KE_{avg} = 3/2\ RT$; The average kinetic energy depends only on temperature. At each temperature, CH_4 and N_2 will have the same average KE. For energy units of joules (J), use R = 8.3145 J/mol•K. To determine average KE per molecule, divide by Avogadro's number, 6.022×10^{23} molecules/mol.

$$\text{at 273 K: } KE_{avg} = \frac{3}{2} \times \frac{8.3145 \text{ J}}{\text{mol K}} \times 273 \text{ K} = 3.40 \times 10^3 \text{ J/mol} = 5.65 \times 10^{-21} \text{ J/molecule}$$

$$\text{at 546 K: } KE_{avg} = \frac{3}{2} \times \frac{8.3145 \text{ J}}{\text{mol K}} \times 546 \text{ K} = 6.81 \times 10^3 \text{ J/mol} = 1.13 \times 10^{-20} \text{ J/molecule}$$

79. $u_{rms} = \left(\frac{3RT}{M}\right)^{1/2}$, $R = \frac{8.3145 \text{ J}}{\text{mol K}}$ and M = molar mass in kg = 1.604×10^{-2} kg/mol for CH_4

For CH_4 at 273 K: $u_{rms} = \left(\frac{\frac{3 \times 8.3145 \text{ J}}{\text{mol K}} \times 273 \text{ K}}{1.604 \times 10^{-2} \text{ kg/mol}}\right)^{1/2} = 652$ m/s

Similarly, u_{rms} for CH_4 at 546 K is 921 m/s.

For N_2 at 273 K: $u_{rms} = \left(\frac{\frac{3 \times 8.3145 \text{ J}}{\text{mol K}} \times 273 \text{ K}}{2.802 \times 10^{-2} \text{ kg/mol}}\right)^{1/2} = 493$ m/s

Similarly, for N_2 at 546 K, u_{rms} = 697 m/s.

81.

	a	b	c	d
avg. KE	inc	dec	same (KE $\propto$ T)	same
avg. velocity	inc	dec	same ($\frac{1}{2}$ mv^2 = KE $\propto$ T)	same
coll. freq wall	inc	dec	inc	inc

Average kinetic energy and average velocity depend on T. As T increases, both average kinetic energy and average velocity increase. At constant T, both average kinetic energy and average velocity are constant. The collision frequency is proportional to the average velocity (as velocity increases it takes less time to move to the next collision) and to the quantity n/V (as molecules per volume increase, collision frequency increases).

83. a. They will all have the same average kinetic energy since they are all at the same temperature.

b. Flask C; H_2 has the smallest molar mass. At constant T, the lightest molecules are the fastest (on the average). This must be true in order for the average kinetic energies to be constant.

85. Graham's law of effusion: $\frac{\text{Rate}_1}{\text{Rate}_2} = \left(\frac{M_2}{M_1}\right)^{1/2}$

Let Freon-12 = gas 1 and Freon-11 = gas 2:

$$\frac{1.07}{1.00} = \left(\frac{137.4}{M_1}\right)^{1/2}, \quad 1.14 = \frac{137.4}{M_1}, \quad M_1 = 121 \text{ g/mol}$$

The molar mass of CF_2Cl_2 is equal to 121 g/mol, so Freon-12 is CF_2Cl_2.

87. $\dfrac{\text{Rate}_1}{\text{Rate}_2} = \left(\dfrac{M_2}{M_1}\right)^{1/2}$, $\dfrac{\text{Rate}(^{12}C^{17}O)}{\text{Rate}(^{12}C^{18}O)} = \left(\dfrac{30.0}{29.0}\right)^{1/2} = 1.02$

$$\frac{\text{Rate}(^{12}C^{16}O)}{\text{Rate}(^{12}C^{18}O)} = \left(\frac{30.0}{28.0}\right)^{1/2} = 1.04$$

The relative rates of effusion of $^{12}C^{16}O$: $^{12}C^{17}O$: $^{12}C^{18}O$ are 1.04: 1.02: 1.00.

Advantage: CO_2 isn't as toxic as CO.

Major disadvantages of using CO_2 instead of CO:

1. Can get a mixture of oxygen isotopes in CO_2.

2. Some species, e.g., $^{12}C^{16}O^{18}O$ and $^{12}C^{17}O_2$, would effuse (gaseously diffuse) at about the same rate since the masses are about equal. Thus, some species cannot be separated from each other.

89. a. $P = \dfrac{nRT}{V} = \dfrac{0.5000 \text{ mol} \times \dfrac{0.08206 \text{ L atm}}{\text{mol K}} \times (25.0 + 273.2) \text{ K}}{1.0000 \text{ L}} = 12.24 \text{ atm}$

b. $\left[P + a\left(\dfrac{n}{V}\right)^2\right] \times (V - nb) = nRT$; For N_2: $a = 1.39 \text{ atm L}^2/\text{mol}^2$ and $b = 0.0391$ L/mol

$$\left[P + 1.39\left(\frac{0.5000}{1.0000}\right)^2 \text{atm}\right] \times (1.0000 \text{ L} - 0.5000 \times 0.0391 \text{ L}) = 12.24 \text{ L atm}$$

$$(P + 0.348 \text{ atm}) \times (0.9805 \text{ L}) = 12.24 \text{ L atm}$$

$$P = \frac{12.24 \text{ L atm}}{0.9805 \text{ L}} - 0.348 \text{ atm} = 12.48 - 0.348 = 12.13 \text{ atm}$$

c. The ideal gas law is high by 0.11 atm or $\dfrac{0.11}{12.13} \times 100 = 0.91\%$.

Atmospheric Chemistry

91. $\chi_{He} = 5.24 \times 10^{-6}$ from Table 5.4. $P_{He} = \chi_{He} \times P_{total} = 5.24 \times 10^{-6} \times 1.0 \text{ atm} = 5.2 \times 10^{-6}$ atm

$$\frac{n}{V} = \frac{P}{RT} = \frac{5.2 \times 10^{-6} \text{ atm}}{\dfrac{0.08206 \text{ L atm}}{\text{mol K}} \times 298 \text{ K}} = 2.1 \times 10^{-7} \text{ mol He/L}$$

$$\frac{2.1\times 10^{-7}\ \text{atm}}{\text{L}} \times \frac{1\ \text{L}}{1000\ \text{cm}^3} \times \frac{6.022\times 10^{23}\ \text{atoms}}{\text{mol}} = 1.3 \times 10^{14}\ \text{atoms He/cm}^3$$

93. $N_2(g) + O_2(g) \rightarrow 2\ NO(g)$, automobile combustion or formed by lightning

$2\ NO(g) + O_2(g) \rightarrow 2\ NO_2(g)$, reaction with atmospheric O_2

$2\ NO_2(g) + H_2O(l) \rightarrow HNO_3(aq) + HNO_2(aq)$, reaction with atmospheric H_2O

$S(s) + O_2(g) \rightarrow SO_2(g)$, combustion of coal

$2\ SO_2(g) + O_2(g) \rightarrow 2SO_3(g)$, reaction with atmospheric O_2

$H_2O(l) + SO_3(g) \rightarrow H_2SO_4(aq)$, reaction with atmospheric H_2O

Additional Exercises

95. a. $PV = nRT$

$PV = \text{constant}$

b. $P = \left(\frac{nR}{V}\right) \times T$

$P = \text{constant} \times T$

c. $T = \left(\frac{P}{nR}\right) \times V$

$T = \text{constant} \times V$

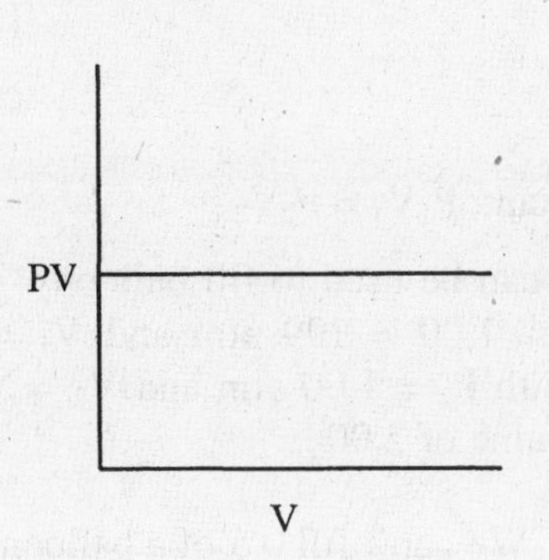

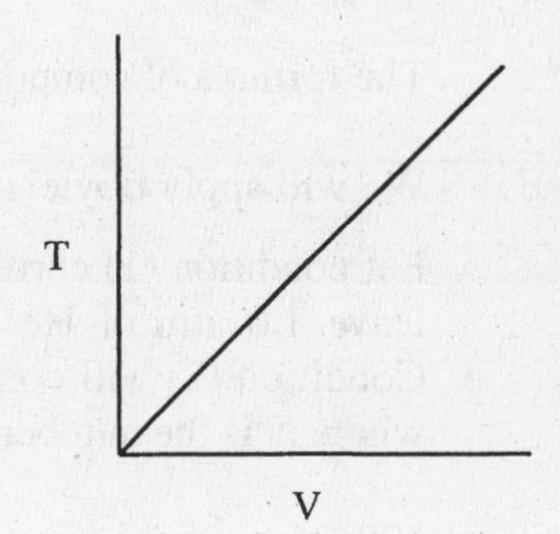

d. $PV = nRT$

$PV = \text{constant}$

e. $P = \frac{nRT}{V}$

$P = \text{constant} \times \frac{1}{V}$

f. $\frac{PV}{T} = nR$

$\frac{PV}{T} = \text{constant}$

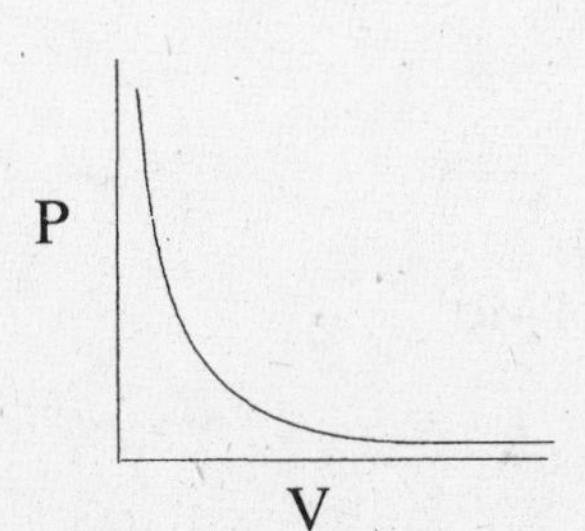

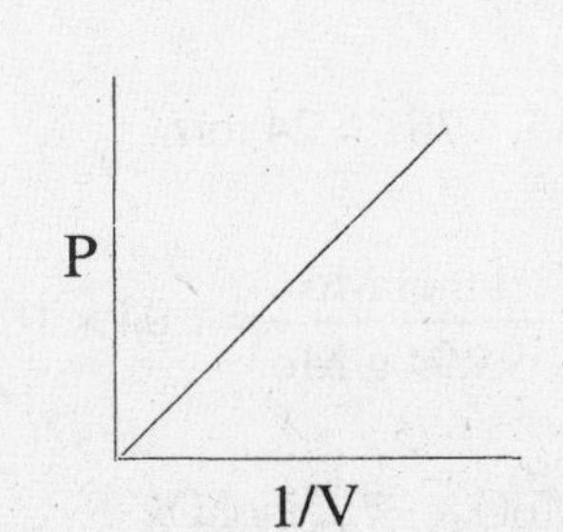

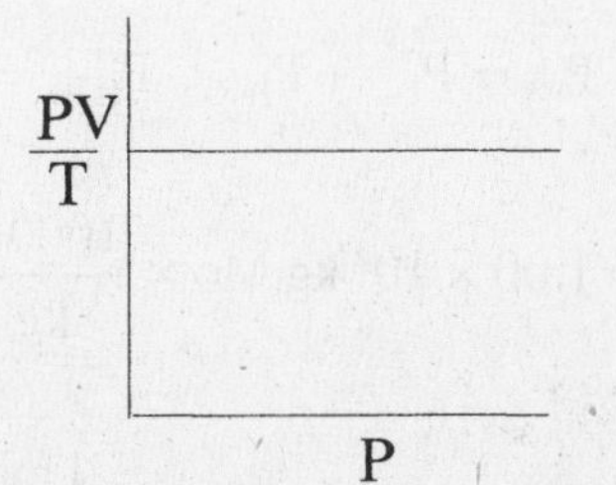

97. $14.1 \times 10^2 \text{ in Hg} \bullet \text{in}^3 \times \frac{2.54\text{ cm}}{\text{in}} \times \frac{10\text{ mm}}{1\text{ cm}} \times \frac{1\text{ atm}}{760\text{ mm}} \times \left(\frac{2.54\text{ cm}}{\text{in}}\right)^3 \times \frac{1\text{ L}}{1000\text{ cm}^3}$

$= 0.772$ atm L

Boyle's law: PV = k where k = nRT; From Sample Exercise 5.3, the k values are around 22 atm L. Because k = nRT, we can assume that Boyle's data and sample Exercise 5.3 data were taken at different temperatures and/or had different sample sizes (different mol).

99. $Mn(s) + x\,HCl(g) \rightarrow MnCl_x(s) + \frac{x}{2}\,H_2(g)$

$$n_{H_2} = \frac{PV}{RT} = \frac{0.951\text{ atm} \times 3.22\text{ L}}{\frac{0.08206\text{ L atm}}{\text{mol K}} \times 373\text{ K}} = 0.100\text{ mol } H_2$$

$$\text{mol Cl in compound} = \text{mol HCl} = 0.100\text{ mol } H_2 \times \frac{x\text{ mol Cl}}{\frac{x}{2}\text{ mol } H_2} = 0.200\text{ mol Cl}$$

$$\frac{\text{mol Cl}}{\text{mol Mn}} = \frac{0.200\text{ mol Cl}}{2.747\text{ g Mn} \times \frac{1\text{ mol Mn}}{54.94\text{ g Mn}}} = \frac{0.200\text{ mol Cl}}{0.05000\text{ mol Mn}} = 4.00$$

The formula of compound is $MnCl_4$.

101. We will apply Boyle's law to solve. PV = nRT = contstant, $P_1V_1 = P_2V_2$

Let condition (1) correspond to He from the tank that can be used to fill balloons. We must leave 1.0 atm of He in the tank, so P_1 = 200. atm - 1.00 = 199 atm and V_1 = 15.0 L. Condition (2) will correspond to the filled balloons with P_2 = 1.00 atm and V_2 = N(2.00 L) where N is the number of filled balloons, each at a volume of 2.00 L.

199 atm × 15.0 L = 1.00 atm × N(2.00 L), N = 1492.5; We can't fill 0.5 of a balloon, so N = 1492 balloons or to 3 significant figures, 1490 balloons.

103. For O_2, n and T are constant, so $P_1V_1 = P_2V_2$.

$$P_1 = \frac{P_2V_2}{V_1} = 785\text{ torr} \times \frac{1.94\text{ L}}{2.00\text{ L}} = 761\text{ torr} = P_{O_2}$$

$P_{tot} = P_{O_2} + P_{H_2O}$, $P_{H_2O} = 785 - 761 = 24$ torr

105. $1.00 \times 10^3\text{ kg Mo} \times \frac{1000\text{ g}}{\text{kg}} \times \frac{1\text{ mol Mo}}{95.94\text{ g Mo}} = 1.04 \times 10^4\text{ mol Mo}$

$$1.04 \times 10^4\text{ mol Mo} \times \frac{1\text{ mol } MoO_3}{\text{mol Mo}} \times \frac{7/2\text{ mol } O_2}{\text{mol } MoO_3} = 3.64 \times 10^4\text{ mol } O_2$$

$$V_{O_2} = \frac{n_{O_2}RT}{P} = \frac{3.64 \times 10^4 \text{ mol} \times \frac{0.08206 \text{ L atm}}{\text{mol K}} \times 290.\text{ K}}{1.00 \text{ atm}} = 8.66 \times 10^5 \text{ L of } O_2$$

$$8.66 \times 10^5 \text{ L } O_2 \times \frac{100 \text{ L air}}{21 \text{ L } O_2} = 4.1 \times 10^6 \text{ L air}$$

$$1.04 \times 10^4 \text{ mol Mo} \times \frac{3 \text{ mol } H_2}{\text{mol Mo}} = 3.12 \times 10^4 \text{ mol } H_2$$

$$V_{H_2} = \frac{3.12 \times 10^4 \text{ mol} \times \frac{0.08206 \text{ L atm}}{\text{mol K}} \times 290.\text{ K}}{1.00 \text{ atm}} = 7.42 \times 10^5 \text{ L of } H_2$$

107. $750.\text{ mL juice} \times \frac{12 \text{ mL } C_2H_5OH}{100 \text{ mL juice}} = 90.\text{ mL } C_2H_5OH$ present

$$90.\text{ mL } C_2H_5OH \times \frac{0.79 \text{ g } C_2H_5OH}{\text{mL } C_2H_5OH} \times \frac{1 \text{ mol } C_2H_5OH}{46.07 \text{ g } C_2H_5OH} \times \frac{2 \text{ mol } CO_2}{2 \text{ mol } C_2H_5OH} = 1.5 \text{ mol } CO_2$$

The CO_2 will occupy (825 − 750. =) 75 mL not occupied by the liquid (headspace).

$$P_{CO_2} = \frac{n_{CO_2} \times RT}{V} = \frac{1.5 \text{ mol} \times \frac{0.08206 \text{ L atm}}{\text{mol K}} \times 298 \text{ K}}{75 \times 10^{-3} \text{ L}} = 490 \text{ atm}$$

Actually, enough CO_2 will dissolve in the wine to lower the pressure of CO_2 to a much more reasonable value.

109. $P_{total} = P_{N_2} + P_{H_2O}$, $P_{N_2} = 726 \text{ torr} - 23.8 \text{ torr} = 702 \text{ torr} \times \frac{1 \text{ atm}}{760 \text{ torr}} = 0.924 \text{ atm}$

$$PV = nRT,\ n_{N_2} = \frac{P_{N_2} \times V}{RT} = \frac{0.924 \text{ atm} \times 31.9 \times 10^{-3} \text{ L}}{\frac{0.08206 \text{ L atm}}{\text{mol K}} \times 298 \text{ K}} = 1.20 \times 10^{-3} \text{ mol } N_2$$

$$\text{Mass of N in compound} = 1.20 \times 10^{-3} \text{ mol} \times \frac{28.02 \text{ g } N_2}{\text{mol}} = 3.36 \times 10^{-2} \text{ g}$$

$$\% \text{ N} = \frac{3.36 \times 10^{-2} \text{ g}}{0.253 \text{ g}} \times 100 = 13.3\% \text{ N}$$

111. $0.2766 \text{ g } CO_2 \times \frac{12.01 \text{ g C}}{44.01 \text{ g } CO_2} = 7.548 \times 10^{-2} \text{ g C}; \ \% \text{ C} = \frac{7.548 \times 10^{-2} \text{ g}}{0.1023 \text{ g}} \times 100 = 73.78\% \text{ C}$

$$0.0991 \text{ g } H_2O \times \frac{2.016 \text{ g H}}{18.02 \text{ g } H_2O} = 1.11 \times 10^{-2} \text{ g H}; \ \% \text{ H} = \frac{1.11 \times 10^{-2} \text{ g}}{0.1023 \text{ g}} \times 100 = 10.9\% \text{ H}$$

$$PV = nRT, \ n_{N_2} = \frac{PV}{RT} = \frac{1.00 \text{ atm} \times 27.6 \times 10^{-3} \text{ L}}{\frac{0.08206 \text{ L atm}}{\text{mol K}} \times 273 \text{ K}} = 1.23 \times 10^{-3} \text{ mol } N_2$$

$$1.23 \times 10^{-3} \text{ mol } N_2 \times \frac{28.02 \text{ g } N_2}{\text{mol } N_2} = 3.45 \times 10^{-2} \text{ g nitrogen}$$

$$\% \text{ N} = \frac{3.45 \times 10^{-2} \text{ g}}{0.4831 \text{ g}} \times 100 = 7.14\% \text{ N}$$

$$\% \text{ O} = 100.00 - (73.78 + 10.9 + 7.14) = 8.2\% \text{ O}$$

Out of 100.00 g of compound, there are:

$$73.78 \text{ g C} \times \frac{1 \text{ mol}}{12.01 \text{ g}} = 6.143 \text{ mol C}; \ 7.14 \text{ g N} \times \frac{1 \text{ mol}}{14.01 \text{ g}} = 0.510 \text{ mol N}$$

$$10.9 \text{ g H} \times \frac{1 \text{ mol}}{1.008 \text{ g}} = 10.8 \text{ mol H}; \ 8.2 \text{ g O} \times \frac{1 \text{ mol}}{16.00 \text{ g}} = 0.51 \text{ mol O}$$

Dividing all values by 0.51 gives an empirical formula of $C_{12}H_{21}NO$.

$$\text{Molar mass} = \frac{dRT}{P} = \frac{\frac{4.02 \text{ g}}{\text{L}} \times \frac{0.08206 \text{ L atm}}{\text{mol K}} \times 400. \text{ K}}{256 \text{ torr} \times \frac{1 \text{ atm}}{760 \text{ torr}}} = 392 \text{ g/mol}$$

Empirical formula mass of $C_{12}H_{21}NO \approx 195$ g/mol; $\frac{392}{195} \approx 2$

Thus, the molecular formula is $C_{24}H_{42}N_2O_2$.

113. The van der Waals constant b is a measure of the size of the molecule. Thus, C_3H_8 should have the largest value of b since it has the largest molar mass (size). The values of a are: H_2, 0.244 L^2 atm/mol^2; CO_2, 3.59; N_2, 1.39; CH_4, 2.25. Because a is a measure of interparticle attractions, the attractions are greatest for CO_2.

Challenge Problems

115. $BaO(s) + CO_2(g) \rightarrow BaCO_3(s)$; $CaO(s) + CO_2(g) \rightarrow CaCO_3(s)$

$$n_i = \frac{P_i V}{RT} = \text{initial moles of } CO_2 = \frac{\frac{750.}{760}\text{ atm} \times 1.50\text{ L}}{\frac{0.08206\text{ L atm}}{\text{mol K}} \times 303.2\text{ K}} = 0.0595\text{ mol } CO_2$$

$$n_f = \frac{P_f V}{RT} = \text{final moles of } CO_2 = \frac{\frac{230.}{760}\text{ atm} \times 1.50\text{ L}}{\frac{0.08206\text{ L atm}}{\text{mol K}} \times 303.2\text{ K}} = 0.0182\text{ mol } CO_2$$

$0.0595 - 0.0182 = 0.0413$ mol CO_2 reacted

Since each metal reacts 1:1 with CO_2, the mixture contains 0.0413 mol of BaO and CaO. The molar masses of BaO and CaO are 153.3 g/mol and 56.08 g/mol, respectively.

Let x = mass of BaO and y = mass of CaO, so:

$$x + y = 5.14\text{ g and } \frac{x}{153.3} + \frac{y}{56.08} = 0.0413\text{ mol}$$

Solving by simultaneous equations:

$$\begin{aligned} x + 2.734\,y &= 6.33 \\ -x \quad -y &= -5.14 \\ \hline 1.734\,y &= 1.19 \end{aligned}$$

$y = 0.686$ g CaO and $5.14 - y = x = 4.45$ g BaO

$$\%BaO = \frac{4.45\text{ g BaO}}{5.14\text{ g}} \times 100 = 86.6\%\text{ BaO}; \quad \%CaO = 100.0 - 86.6 = 13.4\%\text{ CaO}$$

117. Assuming 1.000 L of the hydrocarbon (C_xH_y), then the volume of products will be 4.000 L and the mass of products ($H_2O + CO_2$) will be:

$$1.391\text{ g/L} \times 4.000\text{ L} = 5.564\text{ g products}$$

$$\text{moles } C_xH_y = n_{C_xH_y} = \frac{PV}{RT} = \frac{0.959\text{ atm} \times 1.000\text{ L}}{\frac{0.08206\text{ L atm}}{\text{mol K}} \times 298\text{ K}} = 0.0392\text{ mol}$$

$$\text{moles products} = n_p = \frac{PV}{RT} = \frac{1.51\text{ atm} \times 4.000\text{ L}}{\frac{0.08206\text{ L atm}}{\text{mol K}} \times 375\text{ K}} = 0.196\text{ mol}$$

C_xH_y + oxygen $\rightarrow$ x CO_2 + y/2 H_2O

Setting up two equations:

$$0.0392x + 0.0392(y/2) = 0.196 \text{ (mol of products)}$$

$$0.0392x(44.01 \text{ g/mol}) + 0.0392(y/2)(18.02 \text{ g/mol}) = 5.564 \text{ g (mass of products)}$$

Solving: x = 2 and y = 6, so the formula of the hydrocarbon is C_2H_6.

119. a. The reaction is: $CH_4(g) + 2\ O_2(g) \rightarrow CO_2(g) + 2\ H_2O(g)$

$$PV = nRT, \quad \frac{PV}{n} = RT = \text{constant}, \quad \frac{P_{CH_4}V_{CH_4}}{n_{CH_4}} = \frac{P_{air}V_{air}}{n_{air}}$$

The balanced equation requires 2 mol O_2 for every mol of CH_4 that reacts. For three times as much oxygen, we would need 6 mol O_2 per mol of CH_4 reacted ($n_{O_2} = 6\, n_{CH_4}$). Air is 21% mol percent O_2, so $n_{O_2} = 0.21\ n_{air}$. Therefore, the moles of air we would need to delivery the excess O_2 are:

$$n_{O_2} = 0.21\ n_{air} = 6\, n_{CH_4}, \quad n_{air} = 29\, n_{CH_4}, \quad \frac{n_{air}}{n_{CH_4}} = 29$$

In one minute:

$$V_{air} = V_{CH_4} \times \frac{n_{air}}{n_{CH_4}} \times \frac{P_{CH_4}}{P_{air}} = 200.\ L \times 29 \times \frac{1.50 \text{ atm}}{1.00 \text{ atm}} = 8.7 \times 10^3 \text{ L air/min}$$

b. If x moles of CH_4 were reacted, then 6 x mol O_2 were added, producing 0.950 x mol CO_2 and 0.050 x mol of CO. In addition, 2 x mol H_2O must be produced to balance the hydrogens.

$$CH_4(g) + 2\ O_2(g) \rightarrow CO_2(g) + 2\ H_2O(g); \quad CH_4(g) + 3/2\ O_2(g) \rightarrow CO(g) + 2\ H_2O(g)$$

Amount O_2 reacted:

$$0.950\ x \text{ mol } CO_2 \times \frac{2 \text{ mol } O_2}{\text{mol } CO_2} = 1.90\ x \text{ mol } O_2$$

$$0.050\ x \text{ mol CO} \times \frac{1.5 \text{ mol } O_2}{\text{mol CO}} = 0.075\ x \text{ mol } O_2$$

Amount of O_2 left in reaction mixture = 6.00 x - 1.90 x - 0.075 x = 4.03 x mol O_2

$$\text{Amount of } N_2 = 6.00\ x \text{ mol } O_2 \times \frac{79 \text{ mol } N_2}{21 \text{ mol } O_2} = 22.6\ x \approx 23\ x \text{ mol } N_2$$

The reaction mixture contains:

0.950 x mol CO_2 + 0.050 x mol CO + 4.03 x mol O_2 + 2.00 x mol H_2O +

23 x mol N_2 = 30. x total mol of gas

$$\chi_{CO} = \frac{0.050\,x}{30.\,x} = 0.017; \quad \chi_{CO_2} = \frac{0.950\,x}{30.\,x} = 0.032; \quad \chi_{O_2} = \frac{4.03\,x}{30.\,x} = 0.13$$

$$\chi_{H_2O} = \frac{2.00\,x}{30.\,x} = 0.067; \quad \chi_{N_2} = \frac{23\,x}{30.\,x} = 0.77$$

121. a. Volume of hot air: $V = \frac{4}{3}\pi r^3 = \frac{4}{3}\pi(2.50\text{ m})^3 = 65.4\text{ m}^3$

(Note: radius = diameter/2 = 5.00/2 = 2.50 m)

$$65.4\text{ m}^3\left(\frac{10\text{ dm}}{\text{m}}\right)^3 \times \frac{1\text{ L}}{\text{dm}^3} = 6.54 \times 10^4\text{ L}$$

$$n = \frac{PV}{RT} = \frac{\left(754\text{ torr} \times \frac{1\text{ atm}}{760\text{ torr}}\right) \times 6.54 \times 10^4\text{ L}}{\frac{0.08206\text{ L atm}}{\text{mol K}} \times (273 + 65)\text{ K}} = 2.31 \times 10^3\text{ mol air}$$

$$\text{Mass of hot air} = 2.31 \times 10^3\text{ mol} \times \frac{29.0\text{ g}}{\text{mol}} = 6.70 \times 10^4\text{ g}$$

Mass of air displaced:

$$n = \frac{PV}{RT} = \frac{\frac{745}{760}\text{ atm} \times 6.54 \times 10^4\text{ L}}{\frac{0.08206\text{ L atm}}{\text{mol K}} \times (273 + 21)\text{ K}} = 2.66 \times 10^3\text{ mol air}$$

$$\text{Mass} = 2.66 \times 10^3\text{ mol} \times \frac{29.0\text{ g}}{\text{mol}} = 7.71 \times 10^4\text{ g of air displaced}$$

Lift = 7.71×10^4 g − 6.70×10^4 g = 1.01×10^4 g

b. Mass of air displaced is the same, 7.71×10^4 g. Moles of He in balloon will be the same as moles of air displaced, 2.66×10^3 mol, since P, V and T are the same.

$$\text{Mass of He} = 2.66 \times 10^3\text{ mol} \times \frac{4.003\text{ g}}{\text{mol}} = 1.06 \times 10^4\text{ g}$$

Lift = 7.71×10^4g − 1.06×10^4 g = 6.65×10^4 g

c. Mass of hot air:

$$n = \frac{PV}{RT} = \frac{\frac{630.}{760}\text{atm} \times 6.54 \times 10^4\ \text{L}}{\frac{0.08206\ \text{L atm}}{\text{mol K}} \times 338\ \text{K}} = 1.95 \times 10^3\ \text{mol air}$$

$$1.95 \times 10^3\ \text{mol} \times \frac{29.0\ \text{g}}{\text{mol}} = 5.66 \times 10^4\ \text{g of hot air}$$

Mass of air displaced:

$$n = \frac{PV}{RT} = \frac{\frac{630.}{760}\text{atm} \times 6.54 \times 10^4\ \text{L}}{\frac{0.08206\ \text{L atm}}{\text{mol K}} \times 294\ \text{K}} = 2.25 \times 10^3\ \text{mol air}$$

$$2.25 \times 10^3\ \text{mol} \times \frac{29.0\ \text{g}}{\text{mol}} = 6.53 \times 10^4\ \text{g of air displaced}$$

Lift = 6.53×10^4 g − 5.66×10^4 g = 8.7×10^3 g

123. a. Average molar mass of air = 0.790 × 28.02 g/mol + 0.210 × 32.00 g/mol = 28.9 g/mol; molar mass of helium = 4.003 g/mol

A given volume of air at a given set of conditions has a larger density than helium at those conditions. We need to heat the air to greater than 25°C to lower the air density (by driving air out of the hot air balloon) until the density is the same as that for helium (at 25°C and 1.00 atm).

b. To provide the same lift as the helium balloon (assume V = 1.00 L), the mass of air in the hot air balloon (V = 1.00 L) must be the same as that in the helium balloon. Let MM = molar mass:

$$P \bullet MM = dRT,\ \ \text{mass} = \frac{MM \bullet PV}{RT},\ \ \text{Solving: mass He} = 0.164\ \text{g}$$

$$\text{mass air} = 0.164\ \text{g} = \frac{28.9\ \text{g/mol} \times 1.00\ \text{atm} \times 1.00\ \text{L}}{\frac{0.08206\ \text{L atm}}{\text{mol K}} \times T},\ \ T = 2150\ \text{K (a very high temp)}$$

125. a. If we have 1.0×10^6 L of air, then there are $3.0 \times 10^{-}$ L of CO.

$$P_{CO} = \chi_{CO} \times P_{total};\ \ \chi_{CO} = \frac{V_{CO}}{V_{total}}\ \text{since}\ V \propto n;\ \ P_{CO} = \frac{3.0 \times 10^2\ \text{L}}{1.0 \times 10^6\ \text{L}} \times 628\ \text{torr} = 0.19\ \text{torr}$$

b. $n_{CO} = \frac{P_{CO} \times V}{RT}$; Assuming 1.0 cm^3 of air = 1.0 mL = 1.0×10^{-3} L:

$$n_{CO} = \frac{\frac{0.19}{760}\text{ atm} \times 1.0 \times 10^{-3}\text{ L}}{\frac{0.08206\text{ L atm}}{\text{mol K}} \times 273\text{ K}} = 1.1 \times 10^{-8}\text{ mol CO}$$

$$1.1 \times 10^{-8}\text{ mol} \times \frac{6.022 \times 10^{23}\text{ molecules}}{\text{mol}} = 6.6 \times 10^{15}\text{ molecules CO in the 1.0 cm}^3\text{ of air}$$

Integrative Problems

127. The redox equation must be balanced first using the half-reaction method. The two half-reactions to balance are: $NO_3^- \rightarrow NO$ and $UO^{2+} \rightarrow UO_2^{2+}$; The balanced equation is:

$$2\ H^+(aq) + 2\ NO_3^-(aq) + 3\ UO^{2+}(aq) \rightarrow 3\ UO_2^{2+}(aq) + 2\ NO(g) + H_2O(l)$$

$$n_{NO} = \frac{PV}{RT} = \frac{1.5\text{ atm} \times 0.255\text{ L}}{\frac{0.08206\text{ L atm}}{\text{mol K}} \times 302\text{ K}} = 0.015\text{ mol NO}$$

$$0.015\text{ mol NO} \times \frac{3\text{ mol }UO^{2+}}{2\text{ mol NO}} = 0.023\text{ mol }UO^{2+}$$

129. ThF_4, 232.0 + 4(19.00) = 308.0 g/mL

$$d = \frac{\text{molar mass} \times P}{RT} = \frac{308.0\text{ g/mol} \times 2.5\text{ atm}}{\frac{0.08206\text{ L atm}}{\text{mol K}} \times (1680 + 273)\text{ K}} = 4.8\text{ g/L}$$

The gas with the lower mass will effuse faster. Molar mass of ThF_4 = 308.0 g/mol; molar mass of UF_3 = 238.0 + 3(19.00) = 295.0. Therefore, UF_3 will effuse faster.

$$\frac{\text{rate of effusion of }UF_3}{\text{rate of effusion of }ThF_4} = \sqrt{\frac{\text{molar mass of }ThF_4}{\text{molar mass of }UF_3}} = \sqrt{\frac{308.0\text{ g/mol}}{295.0\text{ g/mol}}} = 1.02$$

UF_3 effuses 1.02 times faster than ThF_4.

CHAPTER SIX

THERMOCHEMISTRY

Questions

9. Path-dependent functions for a trip from Chicago to Denver are those quantities that depend on the route taken. One can fly directly from Chicago to Denver or one could fly from Chicago to Atlanta to Los Angeles and then to Denver. Some path-dependent quantities are miles traveled, fuel consumption of the airplane, time traveling, airplane snacks eaten, etc. State functions are path independent; they only depend on the initial and final states. Some state functions for an airplane trip from Chicago to Denver would be longitude change, latitude change, elevation change, and overall time zone change.

11. $2\ C_8H_{18}(l) + 25\ O_2(g) \rightarrow 16\ CO_2(g) + 18\ H_2O(g)$; All combustion reactions are exothermic; they all release heat to the surroundings so q is negative. To determine the sign of w, concentrate on the moles of gaseous reactants versus the moles of gaseous products. In this combustion reaction, we go from 25 moles of reactant gas molecules to 16 + 18 = 34 moles of product gas molecules. As reactants are converted to products, an expansion will occur. When a gas expands, the system does work on the surroundings and w is negative.

13. $CH_4(g) + 2\ O_2(g) \rightarrow CO_2(g) + 2\ H_2O(l)$ $\qquad \Delta H = -891\ kJ$

$CH_4(g) + 2\ O_2(g) \rightarrow CO_2(g) + 2\ H_2O(g)$ $\qquad \Delta H = -803\ kJ$

$H_2O(l) + 1/2\ CO_2(g) \rightarrow 1/2\ CH_4(g) + O_2(g)$ $\qquad \Delta H_1 = -1/2(-891\ kJ)$

$1/2\ CH_4(g) + 2\ O_2(g) \rightarrow 1/2\ CO_2(g) + H_2O(g)$ $\qquad \Delta H_2 = 1/2(-803\ kJ)$

$H_2O(l) \rightarrow H_2O(g)$ $\qquad \Delta H = \Delta H_1 + \Delta H_2 = 44\ kJ$

The enthalpy of vaporization of water is 44 kJ/mol.

15. Fossil fuels contain carbon; the incomplete combustion of fossil fuels produces CO(g) instead of $CO_2(g)$. This occurs when the amount of oxygen reacting is not sufficient to convert all the carbon to CO_2. Carbon monoxide is a poisonous gas to humans.

Exercises

Potential and Kinctic Energy

17. $KE = \frac{1}{2}mv^2$; Convert mass and velocity to SI units. $1\ J = \frac{1\ kg\ m^2}{s^2}$

$$\text{Mass} = 5.25\ oz \times \frac{1\ lb}{16\ oz} \times \frac{1\ kg}{2.205\ lb} = 0.149\ kg$$

$$\text{Velocity} = \frac{1.0\times10^2 \text{ mi}}{\text{hr}} \times \frac{1 \text{ hr}}{60 \text{ min}} \times \frac{1 \text{ min}}{60 \text{ s}} \times \frac{1760 \text{ yd}}{\text{mi}} \times \frac{1 \text{ m}}{1.094 \text{ yd}} = \frac{45 \text{ m}}{\text{s}}$$

$$\text{KE} = \frac{1}{2}\text{mv}^2 = \frac{1}{2}\times 0.149 \text{ kg} \times \left(\frac{45 \text{ m}}{\text{s}}\right)^2 = 150 \text{ J}$$

19. $$\text{KE} = \frac{1}{2}\text{mv}^2 = \frac{1}{2}\times 2.0 \text{ kg} \times \left(\frac{1.0 \text{ m}}{\text{s}}\right)^2 = 1.0 \text{ J};\ \text{KE} = \frac{1}{2}\text{mv}^2 = \frac{1}{2}\times 1.0 \text{ kg} \times \left(\frac{2.0 \text{ m}}{\text{s}}\right)^2 = 2.0 \text{ J}$$

The 1.0 kg object with a velocity of 2.0 m/s has the greater kinetic energy.

Heat and Work

21 a. $\Delta E = q + w = -47 \text{ kJ} + 88 \text{ kJ} = 41 \text{ kJ}$

b. $\Delta E = 82 - 47 = 35 \text{ kJ}$ c. $\Delta E = 47 + 0 = 47 \text{ kJ}$

d. When the surroundings deliver work to the system, $w > 0$. This is the case for a.

23. $\Delta E = q + w$; Work is done by the system on the surroundings in a gas expansion; w is negative.

300. J = q − 75 J, q = 375 J of heat transferred to the system

25. $w = -P\Delta V$; We need the final volume of the gas. Since T and n are constant, $P_1V_1 = P_2V_2$.

$$V_2 = \frac{V_1 P_1}{P_2} = \frac{10.0 \text{ L } (15.0 \text{ atm})}{2.00 \text{ atm}} = 75.0 \text{ L}$$

$$w = -P\Delta V = -2.00 \text{ atm } (75.0 \text{ L} - 10.0 \text{ L}) = -130. \text{ L atm} \times \frac{101.3 \text{ J}}{\text{L atm}} \times \frac{1 \text{ kJ}}{1000 \text{ J}}$$

$$= -13.2 \text{ kJ} = \text{work}$$

27. In this problem q = w = −950. J

$$-950. \text{ J} \times \frac{1 \text{ L atm}}{101.3 \text{ J}} = -9.38 \text{ L atm of work done by the gases.}$$

$$w = -P\Delta V,\ -9.38 \text{ L atm} = \frac{-650.}{760} \text{ atm} \times (V_f - 0.040 \text{ L}),\ V_f - 0.040 = 11.0 \text{ L},\ V_f = 11.0 \text{ L}$$

29. $$q = \text{molar heat capacity} \times \text{mol} \times \Delta T = \frac{20.8 \text{ J}}{^\circ\text{C mol}} \times 39.1 \text{ mol} \times (38.0 - 0.0)\ ^\circ\text{C} = 30{,}900 \text{ J}$$

$$= 30.9 \text{ kJ}$$

$$w = -P\Delta V = -1.00 \text{ atm} \times (998 \text{ L} - 876 \text{ L}) = -122 \text{ L atm} \times \frac{101.3 \text{ J}}{\text{L atm}} = -12{,}400 \text{ J} = -12.4 \text{ kJ}$$

$$\Delta E = q + w = 30.9 \text{ kJ} + (-12.4 \text{ kJ}) = 18.5 \text{ kJ}$$

Properties of Enthalpy

31. This is an endothermic reaction so heat must be absorbed in order to convert reactants into products. The high temperature environment of internal combustion engines provides the heat.

33. a. Heat is absorbed from the water (it gets colder) as KBr dissolves, so this is an endothermic process.

    b. Heat is released as CH_4 is burned, so this is an exothermic process.

    c. Heat is released to the water (it gets hot) as H_2SO_4 is added, so this is an exothermic process.

    d. Heat must be added (absorbed) to boil water, so this is an endothermic process.

35. $4\ Fe(s) + 3\ O_2(g) \rightarrow 2\ Fe_2O_3(s)$ $\Delta H = -1652$ kJ; Note that 1652 kJ of heat are released when 4 mol Fe react with 3 mol O_2 to produce 2 mol Fe_2O_3.

    a. $4.00 \text{ mol Fe} \times \frac{-1652 \text{ kJ}}{4 \text{ mol Fe}} = -1650 \text{ kJ}$; 1650 kJ of heat released

    b. $1.00 \text{ ml } Fe_2O_3 \times \frac{-1652 \text{ kJ}}{2 \text{ mol } Fe_2O_3} = -826 \text{ kJ}$; 826 kJ of heat released

    c. $1.00 \text{ g Fe} \times \frac{1 \text{ mol Fe}}{55.85 \text{ g}} \times \frac{-1652 \text{ kJ}}{4 \text{ mol Fe}} = -7.39 \text{ kJ}$; 7.39 kJ of heat released

    d. $10.0 \text{ g Fe} \times \frac{1 \text{ mol Fe}}{55.85 \text{ g}} = 0.179 \text{ mol Fe}$; $2.00 \text{ g } O_2 \times \frac{1 \text{ mol } O_2}{32.00 \text{ g}} = 0.0625 \text{ mol } O_2$

    0.179 mol Fe/0.0625 mol O_2 = 2.86; The balanced equation requires a 4 mol Fe/3 mol O_2 = 1.33 mol ratio. O_2 is limiting since the actual mol Fe/mol O_2 ratio is greater than the required mol ratio.

    $0.0625 \text{ mol } O_2 \times \frac{-1652 \text{ kJ}}{3 \text{ mol } O_2} = -34.4 \text{ kJ}$; 34.4 kJ of heat released

37. From Sample Exercise 6.3, $q = 1.3 \times 10^8$ J. Since the heat transfer process is only 60.% efficient, the total energy required is: $1.3 \times 10^8 \text{ J} \times \frac{100. \text{ J}}{60. \text{ J}} = 2.2 \times 10^8 \text{ J}$

    $$\text{mass } C_3H_8 = 2.2 \times 10^8 \text{ J} \times \frac{1 \text{ mol } C_3H_8}{2221 \times 10^3 \text{ J}} \times \frac{44.09 \text{ g } C_3H_8}{\text{mol } C_3H_8} = 4.4 \times 10^3 \text{ g } C_3H_8$$

39. When a liquid is converted into gas, there is an increase in volume. The 2.5 kJ/mol quantity is the work done by the vaporization process in pushing back the atmosphere.

Calorimetry and Heat Capacity

41. Specific heat capacity is defined as the amount of heat necessary to raise the temperature of one gram of substance by one degree Celsius. Therefore, $H_2O(l)$ with the largest heat capacity value requires the largest amount of heat for this process. The amount of heat for $H_2O(l)$ is:

$$\text{energy} = s \times m \times \Delta T = \frac{4.18\ \text{J}}{\text{g °C}} \times 25.0\ \text{g} \times (37.0°\text{C} - 15.0°\text{C}) = 2.30 \times 10^3\ \text{J}$$

The largest temperature change when a certain amount of energy is added to a certain mass of substance will occur for the substance with the smallest specific heat capacity. This is Hg(l), and the temperature change for this process is:

$$\Delta T = \frac{\text{energy}}{s \times m} = \frac{10.7\ \text{kJ} \times \frac{1000\ \text{J}}{\text{kJ}}}{\frac{0.14\ \text{J}}{\text{g °C}} \times 550.\ \text{g}} = 140°\text{C}$$

43. $s = \text{specific heat capacity} = \dfrac{q}{m \times \Delta T} = \dfrac{133\ \text{J}}{5.00\ \text{g} \times (55.1 - 25.2)°\text{C}} = 0.890\ \text{J/°C•g}$

From Table 6.1, the substance is aluminum.

45. | Heat loss by hot water | = | Heat gain by cooler water |

The magnitude of heat loss and heat gain are equal in calorimetry problems. The only difference is the sign (positive or negative). To avoid sign errors, keep all quantities positive and, if necessary, deduce the correct signs at the end of the problem. Water has a specific heat capacity = s = 4.18 J/°C•g = 4.18 J/K•g (ΔT in °C = ΔT in K).

$$\text{Heat loss by hot water} = s \times m \times \Delta T = \frac{4.18\ \text{J}}{\text{g K}} \times 50.0\ \text{g} \times (330.\ \text{K} - T_f)$$

$$\text{Heat gain by cooler water} = \frac{4.18\ \text{J}}{\text{g K}} \times 30.0\ \text{g} \times (T_f - 280.\ \text{K});$$ Heat loss = Heat gain, so:

$$\frac{209\ \text{J}}{\text{K}} \times (330.\ \text{K} - T_f) = \frac{125\ \text{J}}{\text{K}} \times (T_f - 280.\ \text{K}),\quad 6.90 \times 10^4 - 209\ T_f = 125\ T_f - 3.50 \times 10^4$$

$$334\ T_f = 1.040 \times 10^5,\quad T_f = 311\ \text{K}$$

Note that the final temperature is closer to the temperature of the more massive hot water, which is as it should be.

47. Heat loss by Al + heat loss by Fe = heat gain by water; Keeping all quantities positive to avoid sign error:

$$\frac{0.89\text{ J}}{\text{g °C}} \times 5.00\text{ g Al} \times (100.0°\text{C} - T_f) + \frac{0.45\text{ J}}{\text{g °C}} \times 10.00\text{ g Fe} \times (100.0 - T_f)$$

$$= \frac{4.18\text{ J}}{\text{g °C}} \times 97.3\text{ g H}_2\text{O} \times (T_f - 22.0°\text{C})$$

$4.5(100.0 - T_f) + 4.5(100.0 - T_f) = 407(T_f - 22.0)$, $450 - 4.5\,T_f + 450 - 4.5\,T_f = 407\,T_f - 8950$

$416\,T_f = 9850$, $T_f = 23.7°\text{C}$

49. Heat gain by water = heat loss by metal = $s \times m \times \Delta T$ where s = specific heat capacity.

$$\text{Heat gain} = \frac{4.18\text{ J}}{\text{g °C}} \times 150.0\text{ g} \times (18.3°\text{C} - 15.0°\text{C}) = 2100\text{ J}$$

A common error in calorimetry problems is sign errors. Keeping all quantities positive helps eliminate sign errors.

$$\text{heat loss} = 2100\text{ J} = s \times 150.0\text{ g} \times (75.0°\text{C} - 18.3°\text{C}),\ s = \frac{2100\text{ J}}{150.0\text{ g} \times 56.7\text{ °C}} = 0.25\text{ J/g•°C}$$

51. 50.0×10^{-3} L × 0.100 mol/L = 5.00×10^{-3} mol of both $AgNO_3$ and HCl are reacted. Thus, 5.00×10^{-3} mol of AgCl will be produced since there is a 1:1 mole ratio between reactants.

Heat lost by chemicals = Heat gained by solution

$$\text{Heat gain} = \frac{4.18\text{ J}}{\text{g °C}} \times 100.0\text{ g} \times (23.40 - 22.60)°\text{C} = 330\text{ J}$$

Heat loss = 330 J; This is the heat evolved (exothermic reaction) when 5.00×10^{-3} mol of AgCl is produced. So q = −330 J and ΔH (heat per mol AgCl formed) is negative with a value of:

$$\Delta H = \frac{-330\text{ J}}{5.00 \times 10^{-3}\text{ mol}} \times \frac{1\text{ kJ}}{1000\text{ J}} = -66\text{ kJ/mol}$$

Note: Sign errors are common with calorimetry problems. However, the correct sign for ΔH can easily be determined from the ΔT data, i.e., if ΔT of the solution increases, then the reaction is exothermic since heat was released, and if ΔT of the solution decreases, then the reaction is endothermic since the reaction absorbed heat from the water. For calorimetry problems, keep all quantities positive until the end of the calculation, then decide the sign for ΔH. This will help eliminate sign errors.

53. Since ΔH is exothermic, the temperature of the solution will increase as $CaCl_2(s)$ dissolves. Keeping all quantities positive:

$$\text{Heat loss as } CaCl_2 \text{ dissolves} = 11.0 \text{ g } CaCl_2 \times \frac{1 \text{ mol } CaCl_2}{110.98 \text{ g } CaCl_2} \times \frac{81.5 \text{ kJ}}{\text{mol } CaCl_2} = 8.08 \text{ kJ}$$

$$\text{Heat gain by solution} = 8.08 \times 10^3 \text{ J} = \frac{4.18 \text{ J}}{\text{g °C}} \times (125 + 11.0) \text{ g} \times (T_f - 25.0\text{°C})$$

$$T_f - 25.0\text{°C} = \frac{8.08 \times 10^3}{4.18 \times 136} = 14.2\text{°C}, \quad T_f = 14.2\text{°C} + 25.0\text{°C} = 39.2\text{°C}$$

55. a. $\text{heat gain by calorimeter} = \text{heat loss by } CH_4 = 6.79 \text{ g } CH_4 \times \frac{1 \text{ mol } CH_4}{16.04 \text{ g}} \times \frac{802 \text{ kJ}}{\text{mol}} = 340. \text{ kJ}$

$$\text{heat capacity of calorimeter} = \frac{340. \text{ kJ}}{10.8\text{°C}} = 31.5 \text{ kJ/°C}$$

b. $\text{heat loss by } C_2H_2 = \text{heat gain by calorimeter} = 16.9\text{°C} \times \frac{31.5 \text{ kJ}}{\text{°C}} = 532 \text{ kJ}$

$$\Delta E_{comb} = \frac{-532 \text{ kJ}}{12.6 \text{ g } C_2H_2} \times \frac{26.04 \text{ g}}{\text{mol } C_2H_2} = -1.10 \times 10^3 \text{ kJ/mol}$$

Hess's Law

57. Information given:

$$C(s) + O_2(g) \rightarrow CO_2(g) \qquad \Delta H = -393.7 \text{ kJ}$$
$$CO(g) + 1/2\, O_2(g) \rightarrow CO_2(g) \qquad \Delta H = -283.3 \text{ kJ}$$

Using Hess's Law:

$$2\, C(s) + 2\, O_2(g) \rightarrow 2\, CO_2(g) \qquad \Delta H_1 = 2(-393.7 \text{ kJ})$$
$$2\, CO_2(g) \rightarrow 2\, CO(g) + O_2(g) \qquad \Delta H_2 = -2(-283.3 \text{ kJ})$$

$$2\, C(s) + O_2(g) \rightarrow 2\, CO(g) \qquad \Delta H = \Delta H_1 + \Delta H_2 = -220.8 \text{ kJ}$$

Note: The enthalpy change for a reaction that is reversed is the negative quantity of the enthalpy change for the original reaction. If the coefficients in a balanced reaction are multiplied by an integer, the value of ΔH is multiplied by the same integer while the sign stays the same.

59.

$$2\, N_2(g) + 6\, H_2(g) \rightarrow 4\, NH_3(g) \qquad \Delta H = -4(46 \text{ kJ})$$
$$6\, H_2O(g) \rightarrow 6\, H_2(g) + 3\, O_2(g) \qquad \Delta H = -3(-484 \text{ kJ})$$

$$2\, N_2(g) + 6\, H_2O(g) \rightarrow 3\, O_2(g) + 4\, NH_3(g) \qquad \Delta H = 1268 \text{ kJ}$$

No, since the reaction is very endothermic (requires a lot of heat), it would not be a practical way of making ammonia due to the high energy costs.

61.

$NO + O_3 \rightarrow NO_2 + O_2$	$\Delta H = -199$ kJ
$3/2\ O_2 \rightarrow O_3$	$\Delta H = -1/2(-427$ kJ)
$O \rightarrow 1/2\ O_2$	$\Delta H = -1/2(495$ kJ)
$NO(g) + O(g) \rightarrow NO_2(g)$	$\Delta H = -233$ kJ

63.

$CaC_2 \rightarrow Ca + 2\ C$	$\Delta H = -(-62.8$ kJ)
$CaO + H_2O \rightarrow Ca(OH)_2$	$\Delta H = -653.1$ kJ
$2\ CO_2 + H_2O \rightarrow C_2H_2 + 5/2\ O_2$	$\Delta H = -(-1300.$ kJ)
$Ca + 1/2\ O_2 \rightarrow CaO$	$\Delta H = -635.5$ kJ
$2\ C + 2\ O_2 \rightarrow 2\ CO_2$	$\Delta H = 2(-393.5$ kJ)
$CaC_2(s) + 2\ H_2O(l) \rightarrow Ca(OH)_2(aq) + C_2H_2(g)$	$\Delta H = -713$ kJ

Standard Enthalpies of Formation

65. The change in enthalpy that accompanies the formation of one mole of a compound from its elements, with all substances in their standard states, is the standard enthalpy of formation for a compound. The reactions that refer to ΔH_f° are:

$Na(s) + 1/2\ Cl_2(g) \rightarrow NaCl(s)$; $H_2(g) + 1/2\ O_2(g) \rightarrow H_2O(l)$

$6\ C(\text{graphite, s}) + 6\ H_2(g) + 3\ O_2(g) \rightarrow C_6H_{12}O_6(s)$

$Pb(s) + S(\text{rhombic, s}) + 2\ O_2(g) \rightarrow PbSO_4(s)$

67. In general: $\Delta H^\circ = \sum n_p \Delta H^\circ_{f,\ products} - \sum n_r \Delta H^\circ_{f,\ reactants}$ and all elements in their standard state have $\Delta H_f^\circ = 0$ by definition.

a. The balanced equation is: $2\ NH_3(g) + 3\ O_2(g) + 2\ CH_4(g) \rightarrow 2\ HCN(g) + 6\ H_2O(g)$

$$\Delta H^\circ = [\ 2 \text{ mol HCN} \times \Delta H^\circ_{f,\ HCN} + 6 \text{ mol } H_2O(g) \times \Delta H^\circ_{f,\ H_2O}\ ] - [2 \text{ mol } NH_3 \times \Delta H^\circ_{f,\ NH_3} + 2 \text{ mol } CH_4 \times \Delta H^\circ_{f,\ CH_4}\ ]$$

$$\Delta H^\circ = [2(135.1) + 6(-242)] - [2(-46) + 2(-75)] = -940. \text{ kJ}$$

b. $Ca_3(PO_4)_2(s) + 3\ H_2SO_4(l) \rightarrow 3\ CaSO_4(s) + 2\ H_3PO_4(l)$

$$\Delta H^\circ = \left[3 \text{ mol } CaSO_4\left(\frac{-1433 \text{ kJ}}{\text{mol}}\right) + 2 \text{ mol } H_3PO_4(l)\left(\frac{-1267 \text{ kJ}}{\text{mol}}\right)\right]$$

$$- \left[1 \text{ mol } Ca_3(PO_4)_2\left(\frac{-4126 \text{ kJ}}{\text{mol}}\right) + 3 \text{ mol } H_2SO_4(l)\left(\frac{-814 \text{ kJ}}{\text{mol}}\right)\right]$$

$$\Delta H^\circ = -6833 \text{ kJ} - (-6568 \text{ kJ}) = -265 \text{ kJ}$$

c. $NH_3(g) + HCl(g) \rightarrow NH_4Cl(s)$

$$\Delta H° = [1 \text{ mol } NH_4Cl \times \Delta H°_{f,\,NH_4Cl}] - [1 \text{ mol } NH_3 \times \Delta H°_{f,\,NH_3} + 1 \text{ mol } HCl \times \Delta H°_{f,\,HCl}]$$

$$\Delta H° = \left[1 \text{ mol}\left(\frac{-314 \text{ kJ}}{\text{mol}}\right)\right] - \left[1 \text{ mol}\left(\frac{-46 \text{ kJ}}{\text{mol}}\right) + 1 \text{ mol}\left(\frac{-92 \text{ kJ}}{\text{mol}}\right)\right]$$

$$\Delta H° = -314 \text{ kJ} + 138 \text{ kJ} = -176 \text{ kJ}$$

69. a. $4\ NH_3(g) + 5\ O_2(g) \rightarrow 4\ NO(g) + 6\ H_2O(g);\quad \Delta H° = \sum n_p \Delta H°_{f,\text{ products}} - \sum n_r \Delta H°_{f,\text{ reactants}}$

$$\Delta H° = \left[4 \text{ mol}\left(\frac{90. \text{ kJ}}{\text{mol}}\right) + 6 \text{ mol}\left(\frac{-242 \text{ kJ}}{\text{mol}}\right)\right] - \left[4 \text{ mol}\left(\frac{-46 \text{ kJ}}{\text{mol}}\right)\right] = -908 \text{ kJ}$$

$2\ NO(g) + O_2(g) \rightarrow 2\ NO_2(g)$

$$\Delta H° = \left[2 \text{ mol}\left(\frac{34 \text{ kJ}}{\text{mol}}\right)\right] - \left[2 \text{ mol}\left(\frac{90. \text{ kJ}}{\text{mol}}\right)\right] = -112 \text{ kJ}$$

$3\ NO_2(g) + H_2O(l) \rightarrow 2\ HNO_3(aq) + NO(g)$

$$\Delta H° = \left[2 \text{ mol}\left(\frac{-207 \text{ kJ}}{\text{mol}}\right) + 1 \text{ mol}\left(\frac{90. \text{ kJ}}{\text{mol}}\right)\right] - \left[3 \text{ mol}\left(\frac{34 \text{ kJ}}{\text{mol}}\right) + 1 \text{ mol}\left(\frac{-286 \text{ kJ}}{\text{mol}}\right)\right]$$

$$= -140. \text{ kJ}$$

Note: All $\Delta H°_f$ values are assumed ± 1 kJ.

b. $12\ NH_3(g) + 15\ O_2(g) \rightarrow 12\ NO(g) + 18\ H_2O(g)$
$12\ NO(g) + 6\ O_2(g) \rightarrow 12\ NO_2(g)$
$12\ NO_2(g) + 4\ H_2O(l) \rightarrow 8\ HNO_3(aq) + 4\ NO(g)$
$4\ H_2O(g) \rightarrow 4\ H_2O(l)$

$12\ NH_3(g) + 21\ O_2(g) \rightarrow 8\ HNO_3(aq) + 4\ NO(g) + 14\ H_2O(g)$

The overall reaction is exothermic since each step is exothermic.

71. $3\ Al(s) + 3\ NH_4ClO_4(s) \rightarrow Al_2O_3(s) + AlCl_3(s) + 3\ NO(g) + 6\ H_2O(g)$

$$\Delta H° = \left[6 \text{ mol}\left(\frac{-242 \text{ kJ}}{\text{mol}}\right) + 3 \text{ mol}\left(\frac{90. \text{ kJ}}{\text{mol}}\right) + 1 \text{ mol}\left(\frac{-704 \text{ kJ}}{\text{mol}}\right) + 1 \text{ mol}\left(\frac{-1676 \text{ kJ}}{\text{mol}}\right)\right]$$

$$- \left[3 \text{ mol}\left(\frac{-295 \text{ kJ}}{\text{mol}}\right)\right] = -2677 \text{ kJ}$$

73. $2\ ClF_3(g) + 2\ NH_3(g) \rightarrow N_2(g) + 6\ HF(g) + Cl_2(g) \quad \Delta H° = -1196\ kJ$

$$\Delta H° = [6\ \Delta H°_{f,\,HF}] - [2\ \Delta H°_{f,\,ClF_3} + 2\ \Delta H°_{f,\,NH_3}]$$

$$-1196\ kJ = 6\ mol\left(\frac{-271\ kJ}{mol}\right) - 2\ \Delta H°_{f,\,ClF_3} - 2\ mol\left(\frac{-46\ kJ}{mol}\right)$$

$$-1196\ kJ = -1626\ kJ - 2\ \Delta H°_{f,\,ClF_3} + 92\ kJ,\ \Delta H°_{f,\,ClF_3} = \frac{(-1626 + 92 + 1196)\ kJ}{2\ mol} = \frac{-169\ kJ}{mol}$$

Energy Consumption and Sources

75. $C_2H_5OH(l) + 3\ O_2(g) \rightarrow 2\ CO_2(g) + 3\ H_2O(l)$

$$\Delta H° = [2\ (-393.5\ kJ) + 3(-286\ kJ)] - (-278\ kJ) = -1367\ kJ/mol\ ethanol$$

$$\frac{-1367\ kJ}{mol} \times \frac{1\ mol}{46.07\ g} = -29.67\ kJ/g$$

77. $C_3H_8(g) + 5\ O_2(g) \rightarrow 3\ CO_2(g) + 4\ H_2O(l)$

$$\Delta H° = [3(-393.5\ kJ) + 4(-286\ kJ)] - [-104\ kJ] = -2221\ kJ/mol\ C_3H_8$$

$$\frac{-2221\ kJ}{mol} \times \frac{1\ mol}{44.09\ g} = \frac{-50.37\ kJ}{mol}$$ vs. −47.7 kJ/g for octane (Sample Exercise 6.11)

The fuel values are close. An advantage of propane is that it burns more cleanly. The boiling point of propane is −42°C. Thus, it is more difficult to store propane and there are extra safety hazards associated with using high pressure compressed gas tanks.

79. The molar volume of a gas at STP is 22.42 L (from Chapter 5).

$$4.19 \times 10^6\ kJ \times \frac{1\ mol\ CH_4}{891\ kJ} \times \frac{22.42\ L\ CH_4}{mol\ CH_4} = 1.05 \times 10^5\ L\ CH_4$$

Additional Exercises

81. a. $2\ SO_2(g) + O_2(g) \rightarrow 2\ SO_3(g)$ ($w = -P\Delta V$); Because the volume of the piston apparatus decreased as reactants were converted to products, w is positive ($w > 0$).

b. $COCl_2(g) \rightarrow CO(g) + Cl_2(g)$; Because the volume increased, w is negative ($w < 0$).

c. $N_2(g) + O_2(g) \rightarrow 2\ NO(g)$; Because the volume did not change, no PV work is done ($w = 0$).

In order to predict the sign of w for a reaction, compare the coefficients of all the product gases in the balanced equation to the coefficients of all the reactant gases. When a balanced reaction has more mol of product gases than mol of reactant gases (as in b), the reaction will expand in volume (ΔV positive), and the system does work on the surroundings. When a balanced reaction has a decrease in the mol of gas from reactants to products (as in a), the reaction will contract in volume (ΔV negative), and the surroundings will do compression work on the system. When there is no change in the mol of gas from reactants to products (as in c), $\Delta V = 0$ and $w = 0$.

83. $\Delta E_{overall} = \Delta E_{step\ 1} + \Delta E_{step\ 2}$; This is a cyclic process which means that the overall initial state and final state are the same. Since ΔE is a state function, $\Delta E_{overall} = 0$ and $\Delta E_{step\ 1} = -\Delta E_{step\ 2}$.

$$\Delta E_{step\ 1} = q + w = 45\ J + (-10.\ J) = 35\ J$$

$$\Delta E_{step\ 2} = -\Delta E_{step\ 1} = -35\ J = q + w,\ \ -35\ J = -60\ J + w,\ \ w = 25\ J$$

85. $HCl(aq) + NaOH(aq) \rightarrow H_2O(l) + NaCl(aq)\ \ \Delta H = -56\ kJ$

$$0.2000\ L \times \frac{0.400\ mol\ HCl}{L} = 8.00 \times 10^{-2}\ mol\ HCl$$

$$0.1500\ L \times \frac{0.500\ mol\ NaOH}{L} = 7.50 \times 10^{-2}\ mol\ NaOH$$

Because the balanced reaction requires a 1:1 mole ratio between HCl and NaOH, and because fewer moles of NaOH are actually present as compared to HCl, NaOH is the limiting reagent.

$$7.50 \times 10^{-2}\ mol\ NaOH \times \frac{-56\ kJ}{mol\ NaOH} = -4.2\ kJ;$$ 4.2 kJ of heat is released.

87. $q_{surr} = q_{solution} + q_{cal}$; We normally assume q_{cal} is zero (no heat gain/loss by the calorimeter). However, if the calorimeter has a nonzero heat capacity, then some of the heat absorbed by the endothermic reaction came from the calorimeter. If we ignore q_{cal}, then q_{surr} is too small giving a calculated ΔH value which is less positive (smaller) than it should be.

89. Heat released = 1.056 g × 26.42 kJ/g = 27.90 kJ = Heat gain by water and calorimeter

$$\text{Heat gain} = 27.90\ kJ = \frac{4.18\ J}{kg\ ^\circ C} \times 0.987\ kg \times \Delta T + \frac{6.66\ kJ}{^\circ C} \times \Delta T$$

$$27.90 = (4.13 + 6.66)\ \Delta T = 10.79\ \Delta T,\ \ \Delta T = 2.586^\circ C$$

$$2.586^\circ C = T_f - 23.32^\circ C,\ \ T_f = 25.91^\circ C$$

91. a. $\Delta H^\circ = 3\ mol\ (227\ kJ/mol) - 1\ mol\ (49\ kJ/mol) = 632\ kJ$

b. Since 3 $C_2H_2(g)$ is higher in energy than $C_6H_6(l)$, acetylene will release more energy per gram when burned in air.

93. a. $C_2H_4(g) + O_3(g) \rightarrow CH_3CHO(g) + O_2(g)$, $\Delta H° = -166$ kJ − [143 kJ + 52 kJ] = −361 kJ

b. $O_3(g) + NO(g) \rightarrow NO_2(g) + O_2(g)$, $\Delta H° = 34$ kJ − [90. kJ + 143 kJ] = −199 kJ

c. $SO_3(g) + H_2O(l) \rightarrow H_2SO_4(aq)$, $\Delta H° = -909$ kJ − [−396 kJ + (−286 kJ)] = −227 kJ

d. $2\,NO(g) + O_2(g) \rightarrow 2\,NO_2(g)$, $\Delta H° = 2(34)$ kJ − 2(90.) kJ = −112 kJ

Challenge Problems

95. a. $C_{12}H_{22}O_{11}(s) + 12\,O_2(g) \rightarrow 12\,CO_2(g) + 11\,H_2O(l)$

b. A bomb calorimeter is at constant volume, so heat released = $q_v = \Delta E$:

$$\Delta E = \frac{-24.00\text{ kJ}}{1.46\text{ g}} \times \frac{342.30\text{ g}}{\text{mol}} = -5630\text{ kJ/mol } C_{12}H_{22}O_{11}$$

c. PV = nRT; At constant P and T, $P\Delta V = RT\Delta n$ where Δn = mol gaseous products − mol gaseous reactants.

$\Delta H = \Delta E + P\Delta V = \Delta E + RT\Delta n$

For this reaction, $\Delta n = 12 - 12 = 0$, so $\Delta H = \Delta E = -5630$ kJ/mol.

97. Energy used in 8.0 hours = 40. kWh = $\frac{40.0\text{ kJ hr}}{\text{s}} \times \frac{3600\text{ s}}{\text{hr}} = 1.4 \times 10^5$ kJ

Energy from the sun in 8.0 hours = $\frac{1.0\text{ kJ}}{\text{s m}^2} \times \frac{60\text{ s}}{\text{min}} \times \frac{60\text{ min}}{\text{hr}} \times 8.0\text{ hr} = 2.9 \times 10^4\text{ kJ/m}^2$

Only 13% of the sunlight is converted into electricity:

$0.13 \times (2.9 \times 10^4\text{ kJ/m}^2) \times \text{Area} = 1.4 \times 10^5$ kJ, Area = 37 m^2

99. $400\text{ kcal} \times \frac{4.18\text{ kJ}}{\text{kcal}} = 1.7 \times 10^3\text{ kJ} \approx 2 \times 10^3\text{ kJ}$

$$PE = mgz = \left(180\text{ lb} \times \frac{1\text{ kg}}{2.205\text{ lb}}\right) \times \frac{9.81\text{ m}}{\text{s}^2} \times \left(8\text{ in} \times \frac{2.54\text{ cm}}{\text{in}} \times \frac{1\text{ m}}{100\text{ cm}}\right) = 160\text{ J} \approx 200\text{ J}$$

200 J of energy are needed to climb one step. The total number of steps to climb are:

$$2 \times 10^6\text{ J} \times \frac{1\text{ step}}{200\text{ J}} = 1 \times 10^4\text{ steps}$$

101. There are five parts to this problem. We need to calculate:

1. q required to heat $H_2O(s)$ from −30.°C to 0°C; use the specific heat capacity of $H_2O(s)$

2. q required to convert 1 mol $H_2O(s)$ at 0°C into 1 mol $H_2O(l)$ at 0°C; use ΔH_{fusion}

3. q required to heat $H_2O(l)$ from 0°C to 100.°C; use the specific heat capacity of $H_2O(l)$

4. q required to convert 1 mol $H_2O(l)$ at 100.°C into 1 mol $H_2O(g)$ at 100.°C; use $\Delta H_{vaporization}$

5. q required to heat $H_2O(g)$ from 100.°C to 140.°C; use the specific heat capacity of $H_2O(g)$

We will sum up the heat required for all five parts and this will be the total amount of heat required to convert 1.00 mol of $H_2O(s)$ at −30.°C to $H_2O(g)$ at 140.°C. ($q_{total} = q_1 + q_2 + q_3 + q_4 + q_5$). The molar mass of H_2O is 18.02 g/mol.

$q_1 = 2.03\ J/°C{\bullet}g \times 18.02\ g \times [0 - (-30.)]°C = 1.1 \times 10^3\ J$

$q_2 = 1.00\ mol \times 6.02 \times 10^3\ J/mol = 6.02 \times 10^3\ J$

$q_3 = 4.18\ J/°C{\bullet}g \times 18.02\ g \times (100. - 0)°C = 7.53 \times 10^3\ J$

$q_4 = 1.00\ mol \times 40.7 \times 10^4\ J/mol = 4.07 \times 10^4\ J$

$q_5 = 2.02\ J/°C{\bullet}g \times 18.02\ g \times (140. - 100.) = 1.5 \times 10^3\ J$

$q_{total} = q_1 + q_2 + q_3 + q_4 + q_5 = 5.68 \times 10^4\ J = 56.9\ kJ$

Integrative Problems

103. $N_2(g) + 2\ O_2(g) \rightarrow 2\ NO_2(g) \qquad \Delta H = 67.7\ kJ$

$$n_{N_2} = \frac{PV}{RT} = \frac{3.50\ atm \times 0.250\ L}{\frac{0.08206 \times L\ atm}{mol\ K} \times 373\ K} = 2.86 \times 10^{-2}\ mol\ N_2$$

$$n_{O_2} = \frac{PV}{RT} = \frac{3.50\ atm \times 0.450\ L}{\frac{0.08206 \times L\ atm}{mol\ K} \times 373\ K} = 5.15 \times 10^{-2}\ mol\ O_2$$

The balanced equation requires a 2:1 O_2 to N_2 mole ratio. The actual mole ratio is $5.15 \times 10^{-2} / 2.86 \times 10^{-2} = 1.80$; Because the actual mole ratio < required mole ratio, O_2 in the numerator is limiting.

$$5.15 \times 10^{-2}\ \text{mol}\ O_2 \times \frac{2\ \text{mol}\ NO_2}{2\ \text{mol}\ O_2} = 5.15 \times 10^{-2}\ \text{mol}\ NO_2$$

$$5.15 \times 10^{-2}\ \text{mol}\ NO_2 \times \frac{67.7\ \text{kJ}}{2\ \text{mol}\ NO_2} = 1.74\ \text{kJ}$$

105. heat loss by U = heat gain by heavy water; vol of cube = (cube edge)^3

$$\text{mass of heavy water} = 1.00 \times 10^3\ \text{mL} \times \frac{1.11\ \text{g}}{\text{mL}} = 1110\ \text{g}$$

$$\text{heat gain by heavy water} = \frac{4.211\ \text{J}}{\text{g}\ ^\circ\text{C}} \times 1110\ \text{g} \times (28.5 - 25.5)^\circ\text{C} = 1.4 \times 10^4\ \text{J}$$

$$\text{heat loss by U} = 1.4 \times 10^4\ \text{J} = \frac{0.117\ \text{J}}{\text{g}\ ^\circ\text{C}} \times \text{mass} \times (200.0 - 28.5)^\circ\text{C},\quad \text{mass} = 7.0 \times 10^2\ \text{g U}$$

$$7.0 \times 10^2\ \text{g U} \times \frac{1\ \text{cm}^3}{19.05\ \text{g}} = 37\ \text{cm}^3;\ \text{cube edge} = (37\ \text{cm}^3)^{1/3} = 3.3\ \text{cm}$$

CHAPTER SEVEN

ATOMIC STRUCTURE AND PERIODICITY

Questions

15. The equations relating the terms are $\nu\lambda = c$, $E = h\nu$, and $E = hc/\lambda$. From the equations, wavelength and frequency are inversely related, photon energy and frequency are directly related, and photon energy and wavelength are inversely related. The unit of 1 Joule (J) = 1 $kg\ m^2/s^2$. This is why you must change mass units to kg when using the deBroglie equation.

17. Sample Exercise 7.3 calculates the deBroglie wavelength of a ball and of an electron. The ball has a wavelength on the order of 10^{-34} m. This is incredibly short and, as far as the wave-particle duality is concerned, the wave properties of large objects are insignificant. The electron, with its tiny mass, also has a short wavelength; on the order of 10^{-10} m. However, this wavelength is significant as it is on the same order as the spacing between atoms in a typical crystal. For very tiny objects like electrons, the wave properties are important. The wave properties must be considered, along with the particle properties, when hypothesizing about the electron motion in an atom.

19. For the radial probability distribution, the space around the hydrogen nucleus is cut-up into a series of thin spherical shells. When the total probability of finding the electron in each spherical shell is plotted versus the distance from the nucleus, we get the radial probability distribution graph. The plot initially shows a steady increase with distance from the nucleus, reaches a maximum, then shows a steady decrease. Even though it is likely to find an electron near the nucleus, the volume of the spherical shell close to the nucleus is tiny, resulting in a low radial probability. The maximum radial probability distribution occurs at a distance of 5.29×10^{-2} nm from the nucleus; the electron is most likely to be found in the volume of the shell centered at this distance from the nucleus. The 5.29×10^{-2} nm distance is the exact radius of innermost ($n = 1$) orbit in the Bohr model.

21. If one more electron is added to a half-filled subshell, electron-electron repulsions will increase since two electrons must now occupy the same atomic orbital. This may slightly decrease the stability of the atom. Hence, half-filled subshells minimize electron-electron repulsions.

23. The valence electrons are strongly attracted to the nucleus for elements with large ionization energies. One would expect these species to readily accept another electron and have very exothermic electron affinities. The noble gases are an exception; they have a large IE but have an endothermic EA. Noble gases have a stable arrangement of electrons. Adding an electron disrupts this stable arrangement, resulting in unfavorable electron affinities.

25. For hydrogen and one-electron ions (hydrogenlike ions), all atomic orbitals with the same n value have the same energy. For polyatomic atoms/ions, the energy of the atomic orbitals also depends on ℓ. Because there are more nondegenerate energy levels for polyatomic atoms/ions as compared to hydrogen, there are many more possible electronic transitions resulting in more complicated line spectra.

27. Yes, the maximum number of unpaired electrons in any configuration corresponds to a minimum in electron-electron repulsions.

29. Ionization energy is for removal of the electron from the atom in the gas phase. The work function is for the removal of an electron from the solid.

$M(g) \rightarrow M^+(g) + e^-$ ionization energy; $M(s) \rightarrow M^+(s) + e^-$ work function

Exercises

Light and Matter

31. $\nu\lambda = c,\ \nu = \dfrac{c}{\lambda} = \dfrac{2.998\times10^8\ m/s}{660\ nm \times \dfrac{1\ m}{1\times10^9\ nm}} = 4.5 \times 10^{14}\ s^{-1}$

33. $\nu = \dfrac{c}{\lambda} = \dfrac{3.00\times10^8\ m/s}{1.0\times10^{-2}\ m} = 3.0 \times 10^{10}\ s^{-1}$

$E = h\nu = 6.63 \times 10^{-34}\ J\ s \times 3.0 \times 10^{10}\ s^{-1} = 2.0 \times 10^{-23}$ J/photon

$$\frac{2.0\times10^{-23}\ J}{photon} \times \frac{6.01\times10^{23}\ photons}{mol} = 12\ J/mol$$

35. The wavelength is the distance between consecutive wave peaks. Wave a shows 4 wavelengths and wave b shows 8 wavelengths.

Wave a: $\lambda = \dfrac{1.6\times10^{-3}\ m}{4} = 4.0 \times 10^{-4}\ m$

Wave b: $\lambda = \dfrac{1.6\times10^{-3}\ m}{8} = 2.0 \times 10^{-4}\ m$

Wave a has the longer wavelength. Frequency and photon energy are both inversely proportional to wavelength, thus wave b will have the higher frequency and larger photon energy because it has the shorter wavelength.

$$\nu = \frac{c}{\lambda} = \frac{3.00\times10^8\ m/s}{2.0\times10^{-4}\ m} = 1.5 \times 10^{12}\ s^{-1}$$

$$E = \frac{hc}{\lambda} = \frac{6.63\times10^{-34}\ \text{J s}\times3.00\times10^{8}\ \text{m/s}}{2.0\times10^{-4}\ \text{m}} = 9.9\times10^{-22}\ \text{J}$$

Both waves are examples of electromagnetic radiation, so both waves travel at the same speed, c, the speed of light. From Figure 7.2 of the text, both of these waves represent infrared electromagnetic radiation.

37. $$E_{photon} = \frac{hc}{\lambda} = \frac{6.626\times10^{-34}\ \text{J s}\times2.998\times10^{8}\ \text{m/s}}{150.\ \text{nm}\times\frac{1\ \text{m}}{1\times10^{9}\ \text{nm}}} = 1.32\times10^{-18}\ \text{J}$$

$$1.98\times10^{5}\ \text{J}\times\frac{1\ \text{photon}}{1.32\times10^{-18}\ \text{J}}\times\frac{1\ \text{atom C}}{\text{photon}} = 1.50\times10^{23}\ \text{atoms C}$$

39. The energy needed to remove a single electron is:

$$\frac{279.7\ \text{kJ}}{\text{mol}}\times\frac{1\ \text{mol}}{6.0221\times10^{23}} = 4.645\times10^{-22}\ \text{kJ} = 4.645\times10^{-19}\ \text{J}$$

$$E = \frac{hc}{\lambda},\ \lambda = \frac{hc}{E} = \frac{6.6261\times10^{-34}\ \text{J s}\times2.9979\times10^{8}\ \text{m/s}}{4.645\times10^{-19}\ \text{J}} = 4.277\times10^{-7}\ \text{m} = 427.7\ \text{nm}$$

41. a. 10.% of speed of light = $0.10\times3.00\times10^{8}$ m/s = 3.0×10^{7} m/s

$$\lambda = \frac{h}{mv},\ \lambda = \frac{6.63\times10^{-34}\ \text{J s}}{9.11\times10^{-31}\ \text{kg}\times3.0\times10^{7}\ \text{m/s}} = 2.4\times10^{-11}\ \text{m} = 2.4\times10^{-2}\ \text{nm}$$

Note: For units to come out, the mass must be in kg since $1\ \text{J} = \frac{1\ \text{kg m}^2}{\text{s}^2}$

b. $$\lambda = \frac{h}{mv} = \frac{6.63\times10^{-34}\ \text{J s}}{0.055\ \text{kg}\times35\ \text{m/s}} = 3.4\times10^{-34}\ \text{m} = 3.4\times10^{-25}\ \text{nm}$$

This number is so small that it is insignificant. We cannot detect a wavelength this small. The meaning of this number is that we do not have to worry about the wave properties of large objects.

43. $$\lambda = \frac{h}{mv},\ m = \frac{h}{\lambda v} = \frac{6.63\times10^{-34}\ \text{J s}}{1.5\times10^{-15}\ \text{m}\times(0.90\times3.00\times10^{8}\ \text{m/s})} = 1.6\times10^{-27}\ \text{kg}$$

This particle is probably a proton or a neutron.

Hydrogen Atom: The Bohr Model

45. For the H atom (Z = 1): $E_n = -2.178 \times 10^{-18}\ J/n^2$; For a spectral transition, $\Delta E = E_f - E_i$:

$$\Delta E = -2.178 \times 10^{-18}\ J\left(\frac{1}{n_f^2} - \frac{1}{n_i^2}\right)$$

where n_i and n_f are the levels of the initial and final states, respectively. A positive value of ΔE always corresponds to an absorption of light, and a negative value of ΔE always corresponds to an emission of light.

a. $$\Delta E = -2.178 \times 10^{-18}\ J\left(\frac{1}{2^2} - \frac{1}{3^2}\right) = -2.178 \times 10^{-18}\ J\left(\frac{1}{4} - \frac{1}{9}\right)$$

$$\Delta E = -2.178 \times 10^{-18}\ J \times (0.2500 - 0.1111) = -3.025 \times 10^{-19}\ J$$

The photon of light must have precisely this energy (3.025×10^{-19} J).

$$|\Delta E| = E_{photon} = h\nu = \frac{hc}{\lambda} \text{ or } \lambda = \frac{hc}{|\Delta E|} = \frac{6.6261 \times 10^{-34}\ J\,s \times 2.9979 \times 10^8\ m/s}{3.025 \times 10^{-19}\ J}$$

$$= 6.567 \times 10^{-7}\ m = 656.7\ nm$$

From Figure 7.2, this is visible electromagnetic radiation (red light).

b. $$\Delta E = -2.178 \times 10^{-18}\ J\left(\frac{1}{2^2} - \frac{1}{4^2}\right) = -4.084 \times 10^{-19}\ J$$

$$\lambda = \frac{hc}{|\Delta E|} = \frac{6.6261 \times 10^{-34}\ J\,s \times 2.9979 \times 10^8\ m/s}{4.084 \times 10^{-19}\ J} = 4.864 \times 10^{-7}\ m = 486.4\ nm$$

This is visible electromagnetic radiation (green-blue light).

c. $$\Delta E = -2.178 \times 10^{-18}\ J\left(\frac{1}{1^2} - \frac{1}{2^2}\right) = -1.634 \times 10^{-18}\ J$$

$$\lambda = \frac{6.6261 \times 10^{-34}\ J\,s \times 2.9979 \times 10^8\ m/s}{1.634 \times 10^{-18}\ J} = 1.216 \times 10^{-7}\ m = 121.6\ nm$$

This is ultraviolet electromagnetic radiation.

47.

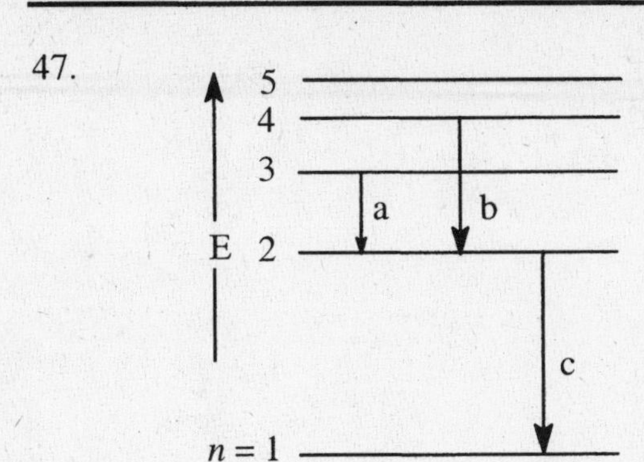

a. 3 → 2

b. 4 → 2

c. 2 → 1

Energy levels are not to scale.

49. $\Delta E = -2.178 \times 10^{-18}\ J\left(\frac{1}{n_f^2} - \frac{1}{n_i^2}\right) = -2.178\ 10^{-18}\ J\left(\frac{1}{5^2} - \frac{1}{1^2}\right) = 2.091 \times 10^{-18}\ J = E_{photon}$

$$\lambda = \frac{hc}{E} = \frac{6.6261\times10^{-34}\ J\ s\times2.9979\times10^{8}\ m/s}{2.091\times10^{-18}\ J} = 9.500 \times 10^{-8}\ m = 95.00\ nm$$

Because wavelength and energy are inversely related, visible light ($\lambda \approx 400 - 700$ nm) is not energetic enough to excite an electron in hydrogen from $n = 1$ to $n = 5$.

$$\Delta E = -2.178 \times 10^{-18}\ J\left(\frac{1}{6^2} - \frac{1}{2^2}\right) = 4.840 \times 10^{-19}\ J$$

$$\lambda = \frac{hc}{E} = \frac{6.6261\times10^{-34}\ J\ s\times2.9979\times10^{8}\ m/s}{4.840\times10^{-18}\ J} = 4.104 \times 10^{-7}\ m = 410.4\ nm$$

Visible light with $\lambda = 410.4$ nm will excite an electron from the $n = 2$ to the $n = 6$ energy level.

51. Ionization from $n = 1$ corresponds to the transition $n_i = 1 \rightarrow n_f = \infty$ where $E_\infty = 0$.

$$\Delta E = E_\infty - E_1 = -E_1 = 2.178 \times10^{-18}\left(\frac{1}{1^2}\right) = 2.178 \times 10^{-18}\ J = E_{photon}$$

$$\lambda = \frac{hc}{E} = \frac{6.6261\times10^{-34}\ J\ s\times2.9979\times10^{8}\ m/s}{2.178\times10^{-18}\ J} = 9.120 \times10^{-8}\ m = 91.20\ nm$$

To ionize from $n = 2$, $\Delta E = E_\infty - E_2 = -E_2 = 2.178 \times10^{-18}\left(\frac{1}{2^2}\right) = 5.445 \times10^{-19}\ J$

$$\lambda = \frac{6.6261\times10^{-34}\ J\ s\times2.9979\times10^{8}\ m/s}{5.445\times10^{-19}\ J} = 3.648 \times10^{-7}\ m = 364.8\ nm$$

53. $|\Delta E| = E_{photon} = h\nu = 6.662 \times10^{-34}\ J\ s \times 6.90 \times 10^{14}\ s^{-1} = 4.57 \times10^{-19}\ J$

$\Delta E = -4.57 \times10^{-19}\ J$ because we have an emission.

$$-4.57 \times 10^{-19} \text{ J} = E_n - E_5 = -2.178 \times 10^{-18} \text{ J} \left(\frac{1}{n^2} - \frac{1}{5^2}\right),$$

$$\frac{1}{n^2} - \frac{1}{25} = 0.210, \quad \frac{1}{n^2} = 0.250, \quad n^2 = 4, \ n = 2$$

The electronic transition is from $n = 5$ to $n = 2$.

Quantum Mechanics, Quantum Numbers, and Orbitals

55. a. $\Delta(mv) = m\Delta v = 9.11 \times 10^{-31} \text{ kg} \times 0.100 \text{ m/s} = \dfrac{9.11\times10^{-32} \text{ kg m}}{\text{s}}$

$$\Delta(mv) \bullet \Delta x \geq \frac{h}{4\pi}, \quad \Delta x = \frac{h}{4\pi\Delta(mv)} = \frac{6.626\times10^{-34} \text{ J s}}{4\times3.142\times(9.11\times10^{-32} \text{ kg m/s})}$$

$$= 5.79 \times 10^{-4} \text{ m}$$

b. $\Delta x = \dfrac{h}{4\pi\Delta(mv)} = \dfrac{6.626\times10^{-34} \text{ J s}}{4\times3.142\times0.145 \text{ kg} \times 0.100 \text{ m/s})} = 3.64 \times 10^{-33} \text{ m}$

c. The diameter of an H atom is roughly 1.0×10^{-8} cm. The uncertainty in position is much larger than the size of the atom.

d. The uncertainty is insignificant compared to the size of a baseball.

57. $n = 1, 2, 3, \ldots$; $\ell = 0, 1, 2, \ldots (n - 1)$; $m_\ell = -\ell \ldots -2, -1, 0, 1, 2, \ldots +\ell$

59. b. For $\ell = 3$, m_ℓ can range from -3 to +3; thus +4 is not allowed.

c. n cannot equal zero.

d. ℓ cannot be a negative number.

61. ψ^2 gives the probability of finding the electron at that point.

Polyelectronic Atoms

63. 5p: three orbitals; $3d_{z^2}$: one orbital; 4d: five orbitals

$n = 5$: $\ell = 0$ (1 orbital), $\ell = 1$ (3 orbitals), $\ell = 2$ (5 orbitals), $\ell = 3$ (7 orbitals), $\ell = 4$ (9 orbitals)

Total for $n = 5$ is 25 orbitals.

$n = 4$: $\ell = 0$ (1), $\ell = 1$ (3), $\ell = 2$ (5), $\ell = 3$ (7); Total for $n = 4$ is 16 orbitals.

65. a. $n = 4$: ℓ can be 0, 1, 2, or 3. Thus we have s (2 e^-), p (6 e^-), d (10 e^-) and f (14 e^-) orbitals present. Total number of electrons to fill these orbitals is 32.

b. $n = 5$, $m_\ell = +1$: For $n = 5$, $\ell = 0, 1, 2, 3, 4$. For $\ell = 1, 2, 3, 4$, all can have $m_\ell = +1$. Four distinct orbitals, thus 8 electrons.

c. $n = 5$, $m_s = +1/2$: For $n = 5$, $\ell = 0, 1, 2, 3, 4$. Number of orbitals = 1, 3, 5, 7, 9 for each value of ℓ, respectively. There are 25 orbitals with $n = 5$. They can hold 50 electrons and 25 of these electrons can have $m_s = +1/2$.

d. $n = 3$, $\ell = 2$: These quantum numbers define a set of 3d orbitals. There are 5 degenerate 3d orbitals which can hold a total of 10 electrons.

e. $n = 2$, $\ell = 1$: These define a set of 2p orbitals. There are 3 degenerate 2p orbitals which can hold a total of 6 electrons.

67. a. Na: $1s^22s^22p^63s^1$; Na has 1 unpaired electron.

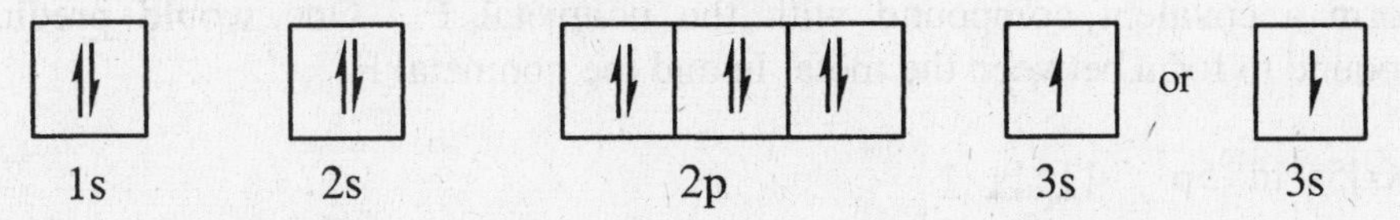

b. Co: $1s^22s^22p^63s^23p^64s^23d^7$; Co has 3 unpaired electrons.

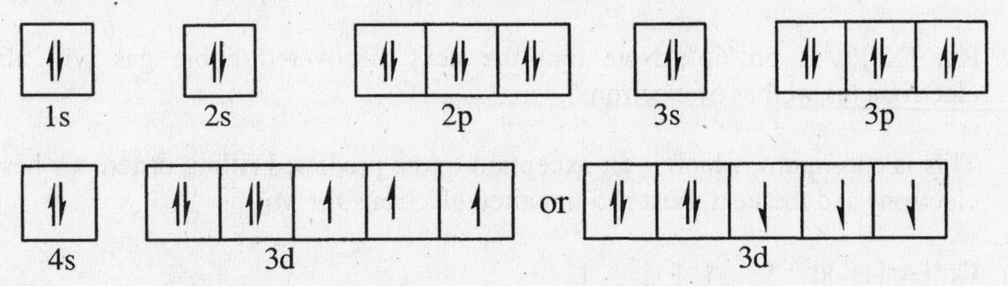

c. Kr: $1s^22s^22p^63s^23p^64s^23d^{10}4p^6$; Kr has 0 unpaired electrons.

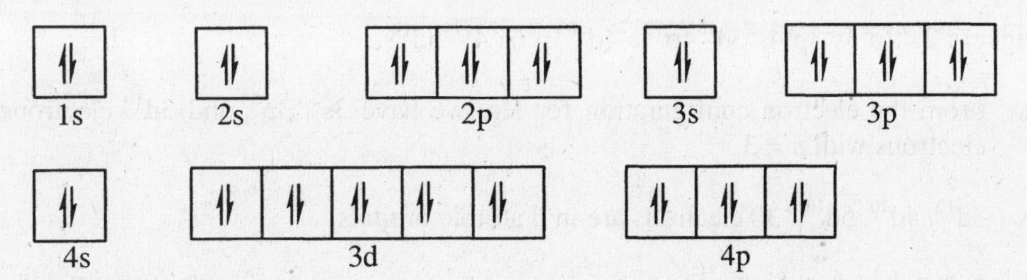

69. Si: $1s^22s^22p^63s^23p^2$ or $[Ne]3s^23p^2$; Ga: $1s^22s^22p^63s^23p^64s^23d^{10}4p^1$ or $[Ar]4s^23d^{10}4p^1$

As: $[Ar]4s^23d^{10}4p^3$; Ge: $[Ar]4s^23d^{10}4p^2$; Al: $[Ne]3s^23p^1$; Cd: $[Kr]5s^24d^{10}$

S: $[Ne]3s^23p^4$; Se: $[Ar]4s^23d^{10}4p^4$

71. The following are complete electron configurations. Noble gas shorthand notation could also be used.

Sc: $1s^22s^22p^63s^23p^64s^23d^1$; Fe: $1s^22s^22p^63s^23p^64s^23d^6$

P: $1s^22s^22p^63s^23p^3$; Cs: $1s^22s^22p^63s^23p^64s^23d^{10}4p^65s^24d^{10}5p^66s^1$

Eu: $1s^22s^22p^63s^23p^64s^23d^{10}4p^65s^24d^{10}5p^66s^24f^65d^1$*

Pt: $1s^22s^22p^63s^23p^64s^23d^{10}4p^65s^24d^{10}5p^66s^24f^{14}5d^8$*

Xe: $1s^22s^22p^63s^23p^64s^23d^{10}4p^65s^24d^{10}5p^6$; Br: $1s^22s^22p^63s^23p^64s^23d^{10}4p^5$

*Note: These electron configurations were predicted using only the periodic table. The actual electron configurations are: Eu: $[Xe]6s^24f^7$ and Pt: $[Xe]6s^14f^{14}5d^9$

73. a. Both In and I have one unpaired 5p electron, but only the nonmetal I would be expected to form a covalent compound with the nonmetal F. One would predict an ionic compound to form between the metal In and the nonmetal F.

I: $[Kr]5s^24d^{10}5p^5$ $\uparrow\downarrow$ $\uparrow\downarrow$ $\uparrow$
5p

b. From the periodic table, this will be element 120. Element 120: $[Rn]7s^25f^{14}6d^{10}7p^68s^2$

c. Rn: $[Xe]6s^24f^{14}5d^{10}6p^6$; Note that the next discovered noble gas will also have 4f electrons (as well as 5f electrons).

d. This is chromium, which is an exception to the predicted filling order. Cr has 6 unpaired electrons and the next most is 5 unpaired electrons for Mn.

Cr: $[Ar]4s^13d^5$ $\uparrow$ $\uparrow\uparrow\uparrow\uparrow\uparrow$
4s 3d

75. Hg: $1s^22s^22p^63s^23p^64s^23d^{10}4p^65s^24d^{10}5p^66s^24f^{14}5d^{10}$

a. From the electron configuration for Hg, we have $3s^2$, $3p^6$, and $3d^{10}$ electrons; 18 total electrons with $n = 3$.

b. $3d^{10}$, $4d^{10}$, $5d^{10}$; 30 electrons are in d atomic orbitals.

c. $2p^6$, $3p^6$, $4p^6$, $5p^6$; Each set of *np* orbitals contain one p_z atomic orbital. Because we have 4 sets of *np* orbitals and two electrons can occupy the p_z orbital, there are 4(2) = 8 electrons in p_z atomic orbitals.

d. All the electrons are paired in Hg, so one-half of the electrons are spin-up ($m_s = +1/2$) and the other half are spin-down ($m_s = -1/2$). 40 electrons have spin-up.

77. B: $1s^22s^22p^1$

	n	ℓ	m_ℓ	m_s
1s	1	0	0	+1/2
1s	1	0	0	- 1/2
2s	2	0	0	+1/2
2s	2	0	0	- 1/2
2p*	2	1	- 1	+1/2

*This is only one of several possibilities for the 2p electron. The 2p electron in B cold have $m_\ell = -1$, 0 or +1, and $m_s = +1/2$ or $-1/2$, for a total of six possibilities.

N: $1s^22s^22p^3$

	n	ℓ	m_ℓ	m_s	
1s	1	0	0	+1/2	
1s	1	0	0	- 1/2	
2s	2	0	0	+1/2	
2s	2	0	0	- 1/2	
2p	2	1	-1	+1/2	(Or all 2p electrons could have $m_s = -1/2$.)
2p	2	1	0	+1/2	
2p	2	1	+1	+1/2	

79. O: $1s^22s^22p_x^{\,2}2p_y^{\,2}$ ($\uparrow\downarrow$ $\uparrow\downarrow$ __); There are no unpaired electrons in this oxygen atom. This configuration would be an excited state, and in going to the more stable ground state ($\uparrow\downarrow$ $\uparrow$ $\uparrow$), energy would be released.

81. None of the s block elements have 2 unpaired electrons. In the p block, the elements with either ns^2np^2 or ns^2np^4 valence electron configurations have 2 unpaired electrons. For elements 1-36, these are elements C, Si, and Ge (with ns^2np^2), and element O, S, and Se (with ns^2np^4). For the d block, the elements with configurations nd^2 or nd^8 have two unpaired electrons. For elements 1-36, these are Ti ($3d^2$) and Ni ($3d^8$). A total of 8 elements from the first 36 elements have two unpaired electrons in the ground state.

83. We get the number of unpaired electrons by examining the incompletely filled subshells. The paramagnetic substances have unpaired electrons, and the ones with no unpaired electrons are not paramagnetic (they are called diamagnetic).

Li: $1s^22s^1$ $\underset{2s}{\uparrow}$; Paramagnetic with 1 unpaired electron.

N: $1s^22s^22p^3$ $\underset{2p}{\uparrow\ \uparrow\ \uparrow}$; Paramagnetic with 3 unpaired electrons.

Ni: $[Ar]4s^23d^8$ $\underset{3d}{\uparrow\downarrow\ \uparrow\downarrow\ \uparrow\downarrow\ \uparrow\ \uparrow}$; Paramagnetic with 2 unpaired electrons.

Te: $[Kr]5s^24d^{10}5p^4$ $\underset{5p}{\uparrow\downarrow\ \uparrow\ \uparrow}$; Paramagnetic with 2 unpaired electrons.

Ba: $[Xe]6s^2$ $\underset{6s}{\uparrow\downarrow}$; Not paramagnetic since no unpaired electrons.

Hg: $[Xe]6s^24f^{14}5d^{10}$ $\underset{5d}{\uparrow\downarrow\ \uparrow\downarrow\ \uparrow\downarrow\ \uparrow\downarrow\ \uparrow\downarrow}$; Not paramagnetic since no unpaired electrons.

The Periodic Table and Periodic Properties

85. Size (radii) decreases left to right across the periodic table, and size increases from top to bottom of the periodic table.

    a. S < Se < Te   b. Br < Ni < K   c. F < Si < Ba

87. The ionization energy trend is the opposite of the radii trend; ionization energy (IE), in general, increases left to right across the periodic table and decreases from top to bottom of the periodic table.

    a. Te < Se < S   b. K < Ni < Br   c. Ba < Si < F

89. a. He   b. Cl

    c. Element 117 is the next halogen to be discovered (under At), element 119 is the next alkali metal to be discovered (under Fr), and element 120 is the next alkaline earth metal to be discovered (under Ra). From the general radii trend, the halogen (element 117) will be the smallest.

    d. Si

    e. Na^+. This ion has the fewest electrons as compared to the other sodium species present. Na^+ has the smallest amount of electron-electron repulsions, which makes it the smallest ion with the largest ionization energy.

91. a. Sg: $[Rn]7s^25f^{14}6d^4$ b. W   c. SgO_3 or Sg_2O_3 and SgO_4^{2-} or $Sg_2O_7^{2-}$ (similar to Cr; Sg = 106)

93. As: $[Ar]4s^23d^{10}4p^3$; Se: $[Ar]4s^23d^{10}4p^4$; The general ionization energy trend predicts that Se should have a higher ionization energy than As. Se is an exception to the general ionization energy trend. There are extra electron-electron repulsions in Se because two electrons are in the same 4p orbital, resulting in a lower ionization energy for Se than predicted.

95. a. More favorable EA: C and Br; The electron affinity trend is very erratic. Both N and Ar have positive EA values (unfavorable) due to their electron configurations (see text for detailed explanation).

    b. Higher IE: N and Ar (follows the IE trend)

    c. Larger size: C and Br (follows the radii trend)

97. Al(−44), Si(−120), P(−74), S(−200.4), Cl(−348.7); Based on the increasing nuclear charge, we would expect the electron affinity (EA) values to become more exothermic as we go from left to right in the period. Phosphorus is out of line. The reaction for the EA of P is:

$$P(g) + e^- \rightarrow P^-(g)$$

$$[Ne]3s^23p^3 \qquad [Ne]3s^23p^4$$

The additional electron in P^- will have to go into an orbital that already has one electron. There will be greater repulsions between the paired electrons in P^-, causing the EA of P to be less favorable than predicted based solely on attractions to the nucleus.

99. The electron affinity trend is very erratic. In general, EA becomes more positive in going down a group and EA becomes more negative from left to right across a period (with many exceptions).

a. I < Br < F < Cl; Cl is most exothermic (F is an exception).

b. N < O < F; F is most exothermic.

101. a. $Se^{3+}(g) \rightarrow Se^{4+}(g) + e^-$ b. $S^-(g) + e^- \rightarrow S^{2-}(g)$

c. $Fe^{3+}(g) + e^- \rightarrow Fe^{2+}(g)$ d. $Mg(g) \rightarrow Mg^+(g) + e^-$

Alkali Metals

103. It should be potassium peroxide, K_2O_2; stable ionic compounds of potassium have K^+ ions, not K^{2+} ions.

105. $$\nu = \frac{c}{\lambda} = \frac{2.9979 \times 10^8 \text{ m/s}}{455.5 \times 10^{-9} \text{ m}} = 6.582 \times 10^{14} \text{ s}^{-1}$$

$$E = h\nu = 6.6261 \times 10^{-34} \text{ J s} \times 6.582 \times 10^{14} \text{ s}^{-1} = 4.361 \times 10^{-19} \text{ J}$$

107. Yes; the ionization energy general trend is to decrease down a group, and the atomic radius trend is to increase down a group. The data in Table 7.8 confirm both of these general trends.

109. a. $6 Li(s) + N_2(g) \rightarrow 2 Li_3N(s)$ b. $2 Rb(s) + S(s) \rightarrow Rb_2S(s)$

Additional Exercises

111. $$E = \frac{310 \text{ kJ}}{\text{mol}} \times \frac{1 \text{ mol}}{6.022 \times 10^{23}} = 5.15 \times 10^{-22} \text{ kJ} = 5.15 \times 10^{-19} \text{ J}$$

$$E = \frac{hc}{\lambda},\ \lambda = \frac{hc}{E} = \frac{6.626 \times 10^{-34} \text{ J s} \times 2.998 \times 10^8 \text{ m/s}}{5.15 \times 10^{-19} \text{ J}} = 3.86 \times 10^{-7} \text{ m} = 386 \text{ nm}$$

113. $60 \times 10^6 \text{ km} \times \dfrac{1000 \text{ m}}{\text{km}} \times \dfrac{1 \text{ s}}{3.00 \times 10^8 \text{ m}} = 200 \text{ s}$ (about 3 minutes)

115. $\Delta E = -R_H \left(\dfrac{1}{n_f^2} - \dfrac{1}{n_i^2} \right) = -2.178 \times 10^{-18} \text{ J} \left(\dfrac{1}{2^2} - \dfrac{1}{6^2} \right) = -4.840 \times 10^{-19} \text{ J}$

$$\lambda = \frac{hc}{|\Delta E|} = \frac{6.6261 \times 10^{-34} \text{ J s} \times 2.9979 \times 10^8 \text{ m/s}}{4.840 \times 10^{-19} \text{ J}} = 4.104 \times 10^{-7} \text{ m} \times \frac{100 \text{ cm}}{\text{m}}$$

$$= 4.104 \times 10^{-5} \text{ cm}$$

From the spectrum, $\lambda = 4.104 \times 10^{-5}$ cm is violet light, so the $n = 6$ to $n = 2$ visible spectrum line is violet.

117. a. True for H only. b. True for all atoms. c. True for all atoms.

119. When the p and d orbital functions are evaluated at various points in space, the results sometimes have positive values and sometimes have negative values. The term phase is often associated with the + and − signs. For example, a sine wave has alternating positive and negative phases. This is analogous to the positive and negative values (phases) in the p and d orbitals.

121. The general ionization energy trend is for ionization energy to increase going left to right across the periodic table. However, one of the exceptions to this trend occurs between groups 2A and 3A. Between these two groups, group 3A elements usually have a lower ionization energy than group 2A elements. Therefore, Al should have the lowest first ionization energy value, followed by Mg, with Si having the largest ionization energy. Looking at the values for the first ionization energy in the graph, the green plot is Al, the blue plot is Mg, and the red plot is Si.

Mg (the blue plot) is the element with the huge jump between I_2 and I_3. Mg has two valence electrons, so the third electron removed is an inner core electron. Inner core electrons are always much more difficult to remove compared to valence electrons since they are closer to the nucleus, on average, than the valence electrons.

123. Valence electrons are easier to remove than inner core electrons. The large difference in energy between I_2 and I_3 indicates that this element has two valence electrons. This element is most likely an alkaline earth metal since alkaline earth metal elements all have two valence electrons.

125. a.

$Na(g) \rightarrow Na^+(g) + e^-$ $\quad IE_1 = 495$ kJ

$Cl(g) + e^- \rightarrow Cl^-(g)$ $\quad EA = -348.7$ kJ

$Na(g) + Cl(g) \rightarrow Na^+(g) + Cl^-g)$ $\quad \Delta H = 146$ kJ

b. $Mg(g) \rightarrow Mg^{+}(g) + e^{-}$ $IE_1 = 735$ kJ

$F(g) + e^{-} \rightarrow F^{-}(g)$ $EA = -327.8$ kJ

$Mg(g) + F(g) \rightarrow Mg^{+}(g) + F^{-}(g)$ $\Delta H = 407$ kJ

c. $Mg^{+}(g) \rightarrow Mg^{2+}(g) + e^{-}$ $IE_2 = 1445$ kJ

$F(g) + e^{-} \rightarrow F^{-}(g)$ $EA = -327.8$ kJ

$Mg^{+}(g) + F(g) \rightarrow Mg^{2+}(g) + F^{-}(g)$ $\Delta H = 1117$ kJ

d. From parts b and c we get:

$Mg(g) + F(g) \rightarrow Mg^{+}(g) + F^{-}(g)$ $\Delta H = 407$ kJ

$Mg^{+}(g) + F(g) \rightarrow Mg^{2+}(g) + F^{-}(g)$ $\Delta H = 1117$ kJ

$Mg(g) + 2\ F(g) \rightarrow Mg^{2+}(g) + 2\ F^{-}(g)$ $\Delta H = 1524$ kJ

Challenge Problems

127. a. Because wavelength is inversely proportional to energy, the spectral line to the right of B (at a longer wavelength) represents the lowest possible energy transition; this is $n = 4$ to $n = 3$. The B line represents the next lowest energy transition, which is $n = 5$ to $n = 3$ and the A line corresponds to the $n = 6$ to $n = 3$ electronic transition.

b. This spectrum is for a one-electron ion, thus $E_n = -2.178 \times 10^{-18}$ J (Z^2/n^2). To determine ΔE and, in turn, the wavelength of spectral line A, we must determine Z, the atomic number of the one-electron species. Use spectral line B data to determine Z.

$$\Delta E_{5 \rightarrow 3} = -2.178 \times 10^{-18}\ \text{J}\ (Z)^2 \left(\frac{1}{3^2} - \frac{1}{5^2}\right) = -2.178 \times 10^{-18}\ \text{J} \left(\frac{Z^2}{3^2} - \frac{Z^2}{5^2}\right)$$

$$\Delta E_{5 \rightarrow 3} = -2.178 \times 10^{-18} \left(\frac{16\,Z^2}{9 \times 25}\right)$$

$$E = \frac{hc}{\lambda} = \frac{6.6261 \times 10^{-34}\ \text{J s} \times 2.9979 \times 10^{8}\ \text{m/s}}{142.5 \times 10^{-9}\ \text{m}} = 1.394 \times 10^{-18}\ \text{J}$$

Because an emission occurs, $\Delta E_{5 \rightarrow 3} = -1.394 \times 10^{-18}$ J.

$$\Delta E = -1.394 \times 10^{-18}\ \text{J} = -2.178 \times 10^{-18}\ \text{J} \left(\frac{16\,Z^2}{9 \times 25}\right),\quad Z^2 = 9.001,\ Z = 3;\ \text{The ion is } Li^{2+}.$$

Solving for the wavelength of line A:

$$\Delta E_{6 \rightarrow 3} = -2.178 \times 10^{-18}\ (3)^2 \left(\frac{1}{3^2} - \frac{1}{6^2}\right) = -1.634 \times 10^{-18}\ \text{J}$$

$$\lambda = \frac{hc}{|\Delta E|} = \frac{6.6261 \times 10^{-34}\ \text{J s} \times 2.9979 \times 10^{8}\ \text{m/s}}{1.634 \times 10^{-18}\ \text{J}} = 1.216 \times 10^{-7}\ \text{m} = 121.6\ \text{nm}$$

129. For $r = a_o$ and $\theta = 0°$ ($Z = 1$ for H):

$$\psi_{2p_z} = \frac{1}{4(2\pi)^{1/2}}\left(\frac{1}{5.29 \times 10^{-11}}\right)^{3/2} (1)\ e^{-1/2} \cos 0 = 1.57 \times 10^{14};\ \psi^2 = 2.46 \times 10^{28}$$

For $r = a_o$ and $\theta = 90°$, $\psi_{2p_z} = 0$ since $\cos 90° = 0$; $\psi^2 = 0$

There is no probability of finding an electron in the $2p_z$ orbital with $\theta = 0°$. As expected, the xy plane, which corresponds to $\theta = 0°$, is a node for the $2p_z$ atomic orbital.

131. a. 1st period: $p = 1,\ q = 1,\ r = 0,\ s = \pm 1/2$ (2 elements)

2nd period: $p = 2,\ q = 1,\ r = 0,\ s = \pm 1/2$ (2 elements)

3rd period: $p = 3,\ q = 1,\ r = 0,\ s = \pm 1/2$ (2 elements)

$p = 3,\ q = 3,\ r = -2,\ s = \pm 1/2$ (2 elements)

$p = 3,\ q = 3,\ r = 0,\ s = \pm 1/2$ (2 elements)

$p = 3,\ q = 3,\ r = +2,\ s = \pm 1/2$ (2 elements)

4th period: $p = 4$; q and r values are the same as with $p = 3$ (8 total elements)

1							2
3							4
5	6	7	8	9	10	11	12
13	14	15	16	17	18	19	20

b. Elements 2, 4, 12 and 20 all have filled shells and will be least reactive.

c. Draw similarities to the modern periodic table.

XY could be X^+Y^-, $X^{2+}Y^{2-}$ or $X^{3+}Y^{3-}$. Possible ions for each are:

X^+ could be elements 1, 3, 5 or 13; Y^- could be 11 or 19.

X^{2+} could be 6 or 14; Y^{2-} could be 10 or 18.

X^{3+} could be 7 or 15; Y^{3-} could be 9 or 17.

Note: X^{4+} and Y^{4-} ions probably won't form.

XY_2 will be $X^{2+}(Y^-)_2$; See above for possible ions.

X_2Y will be $(X^+)_2Y^{2-}$ See above for possible ions.

XY_3 will be $X^{3+}(Y^-)_3$; See above for possible ions.

X_2Y_3 will be $(X^{3+})_2(Y^{2-})_3$; See above for possible ions.

d. $p = 4,\ q = 3,\ r = -2,\ s = \pm 1/2$ (2)

$p = 4,\ q = 3,\ r = 0,\ s = \pm 1/2$ (2)

$p = 4,\ q = 3,\ r = +2,\ s = \pm 1/2$ (2)

A total of 6 electrons can have $p = 4$ and $q = 3$.

e. $p = 3,\ q = 0,\ r = 0$: This is not allowed; q must be odd. Zero electrons can have these quantum numbers.

f. $p = 6,\ q = 1,\ r = 0,\ s = \pm 1/2$ (2)

$p = 6,\ q = 3,\ r = -2, 0, +2;\ s = \pm 1/2$ (6)

$p = 6,\ q = 5,\ r = -4, -2, 0, +2, +4;\ s = \pm 1/2$ (10)

Eighteen electrons can have $p = 6$.

133. The ratios for Mg, Si, P, Cl, and Ar are about the same. However, the ratios for Na, Al, and S are higher. For Na, the second IE is extremely high because the electron is taken from $n = 2$ (the first electron is taken from $n = 3$). For Al, the first electron requires a bit less energy than expected due to the fact it is a 3p electron versus a 3s electron. For S, the first electron requires a bit less energy than expected due to electrons being paired in one of the p orbitals.

135. a. As we remove succeeding electrons, the electron being removed is closer to the nucleus, and there are fewer electrons left repelling it. The remaining electrons are more strongly attracted to the nucleus, and it takes more energy to remove these electrons; successive ionization energies should increase.

b. Al : $1s^22s^22p^63s^23p^1$; For I_4, we begin by removing an electron with $n = 2$. For I_3, we remove an electron with $n = 3$. In going from $n = 3$ to $n = 2$, there is a big jump in ionization energy because the $n = 2$ electrons (inner core electrons) are much closer to the nucleus on average than $n = 3$ electrons (valence electrons). Since the $n = 2$ electrons are closer to the nucleus, they are held more tightly and require a much larger amount of energy to remove them compared to the $n = 3$ electrons.

c. Al^{4+}; The electron affinity for Al^{4+} is ΔH for the reaction:

$$Al^{4+}(g) + e^- \rightarrow Al^{3+}(g) \quad \Delta H = -I_4 = -11{,}600 \text{ kJ/mol}$$

d. The greater the number of electrons, the greater the size.

Size trend: $Al^{4+} < Al^{3+} < Al^{2+} < Al^{+} < Al$

137. $$m = \frac{h}{\lambda v} = \frac{6.626 \times 10^{-34}\ \text{kg m}^2/\text{s}}{3.31 \times 10^{-15}\ \text{m} \times (0.0100 \times 2.998 \times 10^8\ \text{m/s})} = 6.68 \times 10^{-26}\ \text{kg/atom}$$

$$\frac{6.68 \times 10^{-26}\ \text{kg}}{\text{atom}} \times \frac{6.022 \times 10^{23}\ \text{atoms}}{1\ \text{mol}} \times \frac{1000\ \text{g}}{1\ \text{kg}} = 40.2\ \text{g/mol}$$

The element is calcium, Ca.

Integrated Problems

139. a. An atom of francium has 87 protons and 87 electrons. Francium is an alkali metal and forms stable 1+ cations in ionic compounds. This cation would have 86 electrons.

Therefore, the electron configurations will be:

Fr: $[Rn]7s^1$; Fr^+: $[Rn] = [Xe]6s^2 4f^{14} 5d^{10} 6p^6$

b. $$1.0\ \text{oz Fr} \times \frac{1\ \text{lb}}{16\ \text{oz}} \times \frac{1\ \text{kg}}{2.205\ \text{lb}} \times \frac{1000\ \text{g}}{1\ \text{kg}} \times \frac{1\ \text{mol Fr}}{223\ \text{g Fr}} \times \frac{6.02 \times 10^{23}\ \text{atoms}}{1\ \text{mol Fr}}$$

$$= 7.7 \times 10^{22}\ \text{atoms Fr}$$

c. ^{223}Fr is element 87, so it has 223 – 87 = 136 neutrons.

$$136\ \text{neutrons} \times \frac{1.67493 \times 10^{-27}\ \text{kg}}{1\ \text{neutron}} \times \frac{1000\ \text{g}}{1\ \text{kg}} = 2.27790 \times 10^{-22}\ \text{g neutrons}$$

CHAPTER EIGHT

BONDING: GENERAL CONCEPTS

Questions

13. Of the compounds listed, P_2O_5 is the only compound containing only covalent bonds. $(NH_4)_2SO_4$, $Ca_3(PO_4)_2$, K_2O, and KCl are all compounds composed of ions so they exhibit ionic bonding. The ions in $(NH_4)_2SO_4$ are NH_4^+ and SO_4^{2-}. Covalent bonds exist between the N and H atoms in NH_4^+ and between the S and O atoms in SO_4^{2-}. Therefore, $(NH_4)_2SO_4$ contains both ionic and covalent bonds. The same is true for $Ca_3(PO_4)_2$. The bonding is ionic between the Ca^{2+} and PO_4^{3-} ions and covalent between the P and O atoms in PO_4^{3-}. Therefore, $(NH_4)_2SO_4$ and $Ca_3(PO_4)_2$ are the compounds with both ionic and covalent bonds.

15. Electronegativity increases left to right across the periodic table and decreases from top to bottom. Hydrogen has an electronegativity value between B and C in the second row, and identical to P in the third row. Going further down the periodic table, H has an electronegativity value between As and Se (row 4) and identical to Te (row 5). It is important to know where hydrogen fits into the electronegativity trend, especially for rows 2 and 3. If you know where H fits into the trend, then you can predict bond dipole directions for nonmetals bonded to hydrogen.

17. For ions, concentrate on the number of protons and the number of electrons present. The species whose nucleus holds the electrons most tightly will be smallest. For example, anions are larger than the neutral atom. The anion has more electrons held by the same number of protons in the nucleus. These electrons will not be held as tightly, resulting in a bigger size for the anion as compared to the neutral atom. For isoelectronic ions, the same number of electrons are held by different numbers of protons in the various ions. The ion with the most protons holds the electrons tightest and is smallest in size.

19. Fossil fuels contain a lot of carbon and hydrogen atoms. Combustion of fossil fuels (reaction with O_2) produces CO_2 and H_2O. Both these compounds have very strong bonds. Because strong bonds are formed, combustion reactions are very exothermic.

21. CO_2, $4 + 2(6) = 16$ valence electrons

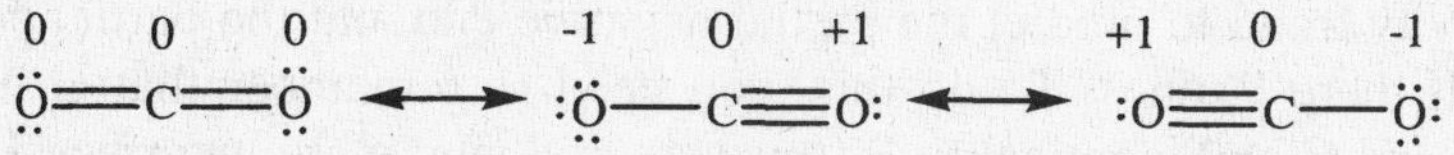

$$\overset{0}{\ddot{O}}=\overset{0}{C}=\overset{0}{\ddot{O}} \longleftrightarrow :\overset{-1}{\ddot{O}}-\overset{0}{C}\equiv\overset{+1}{O}: \longleftrightarrow :\overset{+1}{O}\equiv\overset{0}{C}-\overset{-1}{\ddot{O}}:$$

The formal charges are shown above the atoms in the three Lewis structures. The best Lewis structure for CO_2 from a formal charge standpoint is the first structure having each oxygen double bonded to carbon. This structure has a formal charge of zero on all atoms (which is preferred). The other two resonance structures have nonzero formal charges on the oxygens making them less reasonable. For CO_2, we usually ignore the last two resonance structures and think of the first structure as the true Lewis structure for CO_2.

Exercises

Chemical Bonds and Electronegativity

23. The general trend for electronegativity is:
    1) increase as we go from left to right across a period and
    2) decrease as we go down a group

    Using these trends, the expected orders are:

    a. C < N < O   b. Se < S < Cl   c. Sn < Ge < Si   d. Tl < Ge < S

25. The most polar bond will have the greatest difference in electronegativity between the two atoms. From positions in the periodic table, we would predict:

    a. Ge–F   b. P–Cl   c. S–F   d. Ti–Cl

27. The general trends in electronegativity used in Exercises 8.23 and 8.25 are only rules of thumb. In this exercise, we use experimental values of electronegativities and can begin to see several exceptions. The order of EN from Figure 8.3 is:

    a. C (2.5) < N (3.0) < O (3.5)   same as predicted

    b. Se (2.4) < S (2.5) < Cl (3.0)   same

    c. Si = Ge = Sn (1.8)   different

    d. Tl (1.8) = Ge (1.8) < S (2.5)   different

    Most polar bonds using actual EN values:

    a. Si–F and Ge–F have equal polarity (Ge–F predicted).

    b. P–Cl (same as predicted)

    c. S–F (same as predicted)   d. Ti–Cl (same as predicted)

29. Use the electronegativity trend to predict the partial negative end and the partial positive end of the bond dipole (if there is one). To do this, you need to remember that H has electronegativity between B and C and identical to P. Answers b, d, and e are incorrect. For d (Br_2), the bond between two Br atoms will be a pure covalent bond where there is equal sharing of the bonding electrons and no dipole moment. For b and e, the bond polarities are reversed. In Cl–I, the more electronegative Cl atom will be the partial negative end of the bond dipole with I having the partial positive end. In O–P, the more electronegative oxygen will be the partial negative end of the bond dipole with P having the partial positive end. In the following (on the next page), we used arrows to indicate the bond dipole. The arrow always points to the partial negative end of a bond dipole (which always is the most electronegative atom in the bond).

31. Electronegativity values increase from left to right across the periodic table. The order of electronegativities for the atoms from smallest to largest electronegativity will be H = P < C < N < O < F. The most polar bond will be F–H since it will have the largest difference in electronegativities, and the least polar bond will be P–H since it will have the smallest difference in electronegativities (ΔEN = 0). The order of the bonds in decreasing polarity will be F–H > O–H > N–H > C–H > P–H.

Ions and Ionic Compounds

33. Fr^+: $[Xe]6s^24f^{14}5d^{10}6p^6$ = [Rn]; Be^{2+}: $1s^2$ = [He]; P^{3-}: $[Ne]3s^23p^6$ = [Ar]

Cl^-: $[Ne]3s^23p^6$ = [Ar]; Se^{2-}: $[Ar]4s^23d^{10}4p^6$ = [Kr]

35. a. Sc^{3+}: [Ar] b. Te^{2-}: [Xe] c. Ce^{4+}: [Xe] and Ti^{4+}: [Ar] d. Ba^{2+}: [Xe]

All of these ions have the noble gas electron configuration shown in brackets.

37. There are many possible ions with 54 electrons. Some are: Sb^{3-}, Te^{2-}, I^-, Cs^+, Ba^{2+} and La^{3+}. In terms of size, the ion with the most protons will hold the electrons the tightest and will be the smallest. The largest ion will be the ion with the fewest protons. The size trend is:

$La^{3+} < Ba^{2+} < Cs^+ < I^- < Te^{2-} < Sb^{3-}$
smallest largest

39. a. $Cu > Cu^+ > Cu^{2+}$ b. $Pt^{2+} > Pd^{2+} > Ni^{2+}$ c. $O^{2-} > O^- > O$

d. $La^{3+} > Eu^{3+} > Gd^{3+} > Yb^{3+}$ e. $Te^{2-} > I^- > Cs^+ > Ba^{2+} > La^{3+}$

For answer a, as electrons are removed from an atom, size decreases. Answers b and d follow the radii trend. For answer c, as electrons are added to an atom, size increases. Answer e follows the trend for an isoelectronic series, i.e., the smallest ion has the most protons.

41. a. Al^{3+} and S^{2-} are the expected ions. The formula of the compound would be Al_2S_3 (aluminum sulfide).

b. K^+ and N^{3-}; K_3N, potassium nitride

c. Mg^{2+} and Cl^-; $MgCl_2$, magnesium chloride

d. Cs^+ and Br^-; CsBr, cesium bromide

43. Lattice energy is proportional to Q_1Q_2/r where Q is the charge of the ions and r is the distance between the ions. In general, charge effects on lattice energy are much greater than size effects.

a. NaCl; Na^+ is smaller than K^+.

b. LiF; F^- is smaller than Cl^-.

c. MgO; O^{2-} has a greater charge than OH^-.

d. $Fe(OH)_3$; Fe^{3+} has a greater charge than Fe^{2+}.

e. Na_2O; O^{2-} has a greater charge than Cl^-.

f. MgO; The ions are smaller in MgO.

45.

$K(s) \rightarrow K(g)$	$\Delta H =$ 64 kJ (sublimation)
$K(g) \rightarrow K^+(g) + e^-$	$\Delta H =$ 419 kJ (ionization energy)
$1/2\ Cl_2(g) \rightarrow Cl(g)$	$\Delta H =$ 239/2 kJ (bond energy)
$Cl(g) + e^- \rightarrow Cl^-(g)$	$\Delta H = -349$ kJ (electron affinity)
$K^+(g) + Cl^-(g) \rightarrow KCl(s)$	$\Delta H = -690.$ kJ (lattice energy)
$K(s) + 1/2\ Cl_2(g) \rightarrow KCl(s)$	$\Delta H_f^\circ = -437$ kJ/mol

47. From the data given, it takes less energy to produce $Mg^+(g) + O^-(g)$ than to produce $Mg^{2+}(g) + O^{2-}(g)$. However, the lattice energy for $Mg^{2+}O^{2-}$ will be much more exothermic than that for Mg^+O^- due to the greater charges in $Mg^{2+}O^{2-}$. The favorable lattice energy term dominates and $Mg^{2+}O^{2-}$ forms.

49. Use Figure 8.11 as a template for this problem.

$Li(s) \rightarrow Li(g)$	$\Delta H_{sub} = ?$
$Li(g) \rightarrow Li^+(g) + e^-$	$\Delta H = 520.$ kJ
$1/2\ I_2(g) \rightarrow I(g)$	$\Delta H = 151/2$ kJ
$I(g) + e^- \rightarrow I^-(g)$	$\Delta H = -295$ kJ
$Li^+(g) + I^-(g) \rightarrow LiI(s)$	$\Delta H = -753$ kJ
$Li(s) + 1/2\ I_2(g) \rightarrow LiI(s)$	$\Delta H = -272$ kJ

$\Delta H_{sub} + 520. + 151/2 - 295 - 753 = -272,\ \Delta H_{sub} = 181$ kJ

51. Ca^{2+} has a greater charge than Na^+, and Se^{2-} is smaller than Te^{2-}. The effect of charge on the lattice energy is greater than the effect of size. We expect the trend from most exothermic to least exothermic to be:

CaSe > CaTe > Na_2Se > Na_2Te

(−2862) (−2721) (−2130) (−2095) This is what we observe.

Bond Energies

53. a. H—H + Cl—Cl ⟶ 2 H—Cl

Bonds broken:

1 H – H (432 kJ/mol)
1 Cl – Cl (239 kJ/mol)

Bonds formed:

2 H – Cl (427 kJ/mol)

$\Delta H = \Sigma D_{broken} - \Sigma D_{formed}$, $\Delta H = 432\ kJ + 239\ kJ - 2(427)\ kJ = -183\ kJ$

b. N≡N + 3 H—H ⟶ 2 H—N(—H)—H

Bonds broken:

1 N ≡ N (941 kJ/mol)
3 H – H (432 kJ/mol)

Bonds formed:

6 N – H (391 kJ/mol)

$\Delta H = 941\ kJ + 3(432)\ kJ - 6(391)\ kJ = -109\ kJ$

55. H_3C—N≡C ⟶ H_3C—C≡N

Bonds broken: 1 C – N (305 kJ/mol)

Bonds formed: 1 C – C (347 kJ/mol)

$\Delta H = \Sigma D_{broken} - \Sigma D_{formed}$, $\Delta H = 305 - 347 = -42\ kJ$

Note: Sometimes some of the bonds remain the same between reactants and products. To save time, only break and form bonds that are involved in the reaction.

57. H_3C—CH_2—O—H + 3 O=O ⟶ 2 O=C=O + 3 H—O—H

Bonds broken:

5 C – H (413 kJ/mol)
1 C – C (347 kJ/mol)
1 C – O (358 kJ/mol)
1 O – H (467 kJ/mol)
3 O = O (495 kJ/mol)

Bonds formed:

2 × 2 C = O (799 kJ/mol)
3 × 2 O – H (467 kJ/mol)

$\Delta H = 5(413\text{ kJ}) + 347\text{ kJ} + 358\text{ kJ} + 467\text{ kJ} + 3(495\text{ kJ}) - [4(799\text{ kJ}) + 6(467\text{ kJ})]$

$= -1276\text{ kJ}$

59.

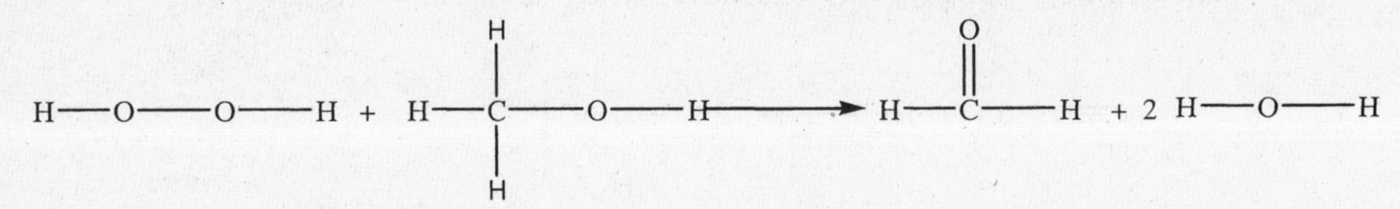

Bonds broken:
3 O–H (467 kJ/mol)
1 O–O (146 kJ/mol)
1 C–H (413 kJ/mol)
1 C–O (358 kJ/mol)

Bonds formed:
1 C=O (745 kJ/mol); This is not CO_2.
2 × 2 O–H (467 kJ/mol)

$\Delta H = 3(467\text{ kJ}) + 146\text{ kJ} + 413\text{ kJ} + 358\text{ kJ} - [745\text{ kJ} + 4(467\text{ kJ})] = -295\text{ kJ}$

61.

$H_2C{=}CH_2$ + F—F ⟶ $H{-}CHF{-}CHF{-}H$ $\quad \Delta H = -549\text{ kJ}$

Bonds broken:

1 C = C (614 kJ/mol)
1 F – F (154 kJ/mol)

Bonds formed:

1 C – C (347 kJ/mol)
2 C – F (D_{CF})

$\Delta H = -549\text{ kJ} = 614\text{ kJ} + 154\text{ kJ} - [347\text{ kJ} + 2\,D_{CF}],\ 2\,D_{CF} = 970.,\ D_{CF} = 485\text{ kJ/mol}$

63. a. $\Delta H° = 2\,\Delta H^{\circ}_{f,\,HCl} = 2\text{ mol }(-92\text{ kJ/mol}) = -184\text{ kJ}$ (= −183 kJ from bond energies)

b. $\Delta H° = 2\,\Delta H^{\circ}_{f,\,NH_3} = 2\text{ mol }(-46\text{ kJ/mol}) = -92\text{ kJ}$ (= −109 kJ from bond energies)

Comparing the values for each reaction, bond energies seem to give a reasonably good estimate of the enthalpy change for a reaction. The estimate is especially good for gas phase reactions.

65. a. Using SF_4 data: $SF_4(g) \rightarrow S(g) + 4\ F(g)$

$$\Delta H° = 4\ D_{SF} = 278.8 + 4\ (79.0) - (-775) = 1370.\ kJ$$

$$D_{SF} = \frac{1370.\ kJ}{4\ mol\ SF\ bonds} = 342.5\ kJ/mol$$

Using SF_6 data: $SF_6(g) \rightarrow S(g) + 6\ F(g)$

$$\Delta H° = 6\ D_{SF} = 278.8 + 6\ (79.0) - (-1209) = 1962\ kJ$$

$$D_{SF} = \frac{1962.\ kJ}{6\ mol} = 327.0\ kJ/mol$$

b. The S – F bond energy in the table is 327 kJ/mol. The value in the table was based on the S – F bond in SF_6.

c. S(g) and F(g) are not the most stable forms of the elements at 25°C. The most stable forms are $S_8(s)$ and $F_2(g)$; $\Delta H_f^\circ = 0$ for these two species.

Lewis Structures and Resonance

67. Drawing Lewis structures is mostly trial and error. However, the first two steps are always the same. These steps are 1) count the valence electrons available in the molecule/ion, and 2) attach all atoms to each other with single bonds (called the skeletal structure). Unless noted otherwise, the atom listed first is assumed to be the atom in the middle (called the central atom) and all other atoms in the formula are attached to this atom. The most notable exceptions to the rule are formulas which begin with H, e.g., H_2O, H_2CO, etc. Hydrogen can never be a central atom since this would require H to have more than two electrons. In these compounds, the atom listed second is assumed to be the central atom.

After counting valence electrons and drawing the skeletal structure, the rest is trial and error. We place the remaining electrons around the various atoms in an attempt to satisfy the octet rule (or duet rule for H). Keep in mind that practice makes perfect. After practicing you can (and will) become very adept at drawing Lewis structures.

a. HCN has 1 + 4 + 5 = 10 valence

b. PH_3 has 5 + 3(1) = 8 valence electrons.

Skeletal structure

Lewis structure

Skeletal structure

Lewis structure

c. $CHCl_3$ has 4 + 1 + 3(7) = 26 valence electrons.

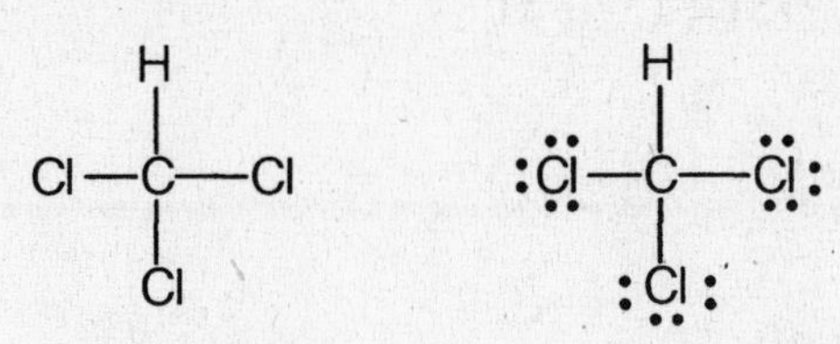

Skeletal structure

Lewis structure

d. NH_4^+ has 5 + 4(1) − 1 = 8 valence electrons.

Lewis structure

Note: Subtract valence electrons for positive charged ions.

e. H_2CO has 2(1) + 4 + 6 = 12 valence electrons.

f. SeF_2 has 6 + 2(7) = 20 valence electrons.

g. CO_2 has 4 + 2(6) = 16 valence electrons

h. O_2 has 2(6) = 12 valence electrons.

i. HBr has 1 + 7 = 8 valence electrons.

69. BeH_2, 2 + 2(1) = 4 valence electrons

BH_3, 3 + 3(1) = 6 valence electrons

71. PF_5, 5 +5(7) = 40 valence electrons

SF_4, 6 + 4(7) = 34 e^-

ClF_3, 7 + 3(7) = 28 e^-

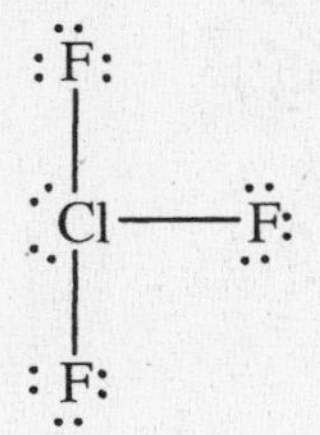

Br_3^-, 3(7) + 1 = 22 e^-

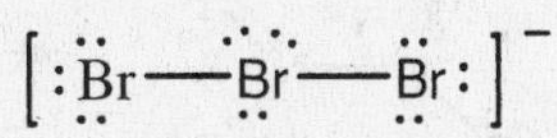

Row 3 and heavier nonmetals can have more than 8 electrons around them when they have to. Row 3 and heavier elements have empty d orbitals which are close in energy to valence s and p orbitals. These empty d orbitals can accept extra electrons.

For example, P in PF_5 has its five valence electrons in the 3s and 3p orbitals. These s and p orbitals have room for 3 more electrons, and if it has to, P can use the empty 3d orbitals for any electrons above 8.

73. a. NO_2^- has 5 + 2(6) + 1 = 18 valence electrons. The skeletal structure is: O–N–O

To get an octet about the nitrogen and only use 18 e^-, we must form a double bond to one of the oxygen atoms.

Since there is no reason to have the double bond to a particular oxygen atom, we can draw two resonance structures. Each Lewis structure uses the correct number of electrons and satisfies the octet rule, so each is a valid Lewis structure. Resonance structures occur when you have multiple bonds that can be in various positions. We say the actual structure is an average of these two resonance structures.

NO_3^- has 5 + 3(6) + 1 = 24 valence electrons. We can draw three resonance structures for NO_3^-, with the double bond rotating among the three oxygen atoms.

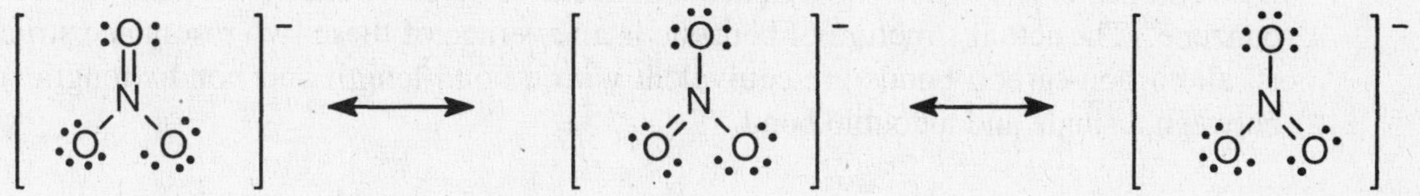

N_2O_4 has 2(5) + 4(6) = 34 valence electrons. We can draw four resonance structures for N_2O_4.

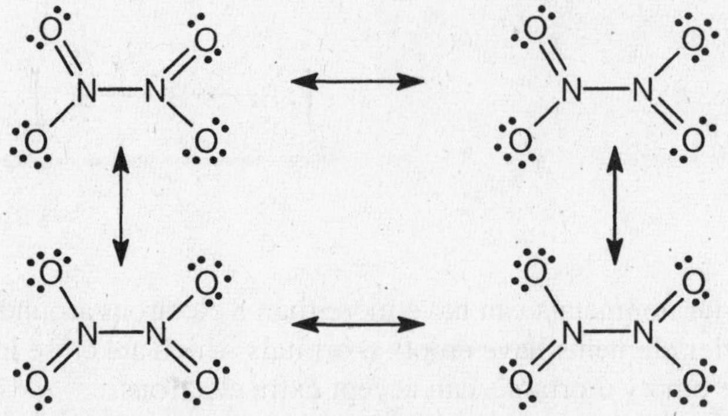

b. OCN^- has 6 + 4 + 5 + 1 = 16 valence electrons. We can draw three resonance structures for OCN^-.

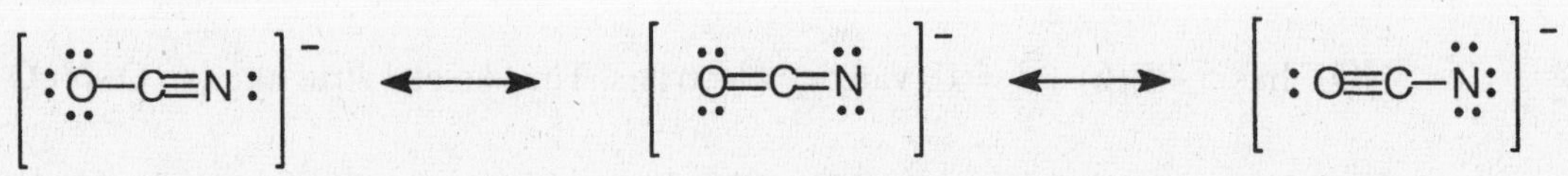

SCN⁻ has 6 + 4 + 5 + 1 = 16 valence electrons. Three resonance structures can be drawn.

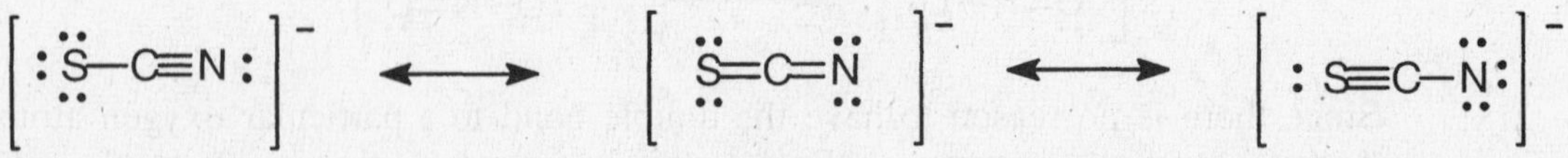

N_3^- has 3(5) + 1 = 16 valence electrons. As with OCN^- and SCN^-, three different resonance structures can be drawn.

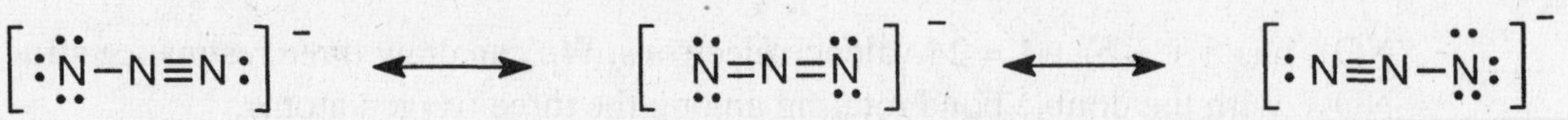

75. Benzene has 6(4) + 6(1) = 30 valence electrons. Two resonance structures can be drawn for benzene. The actual structure of benzene is an average of these two resonance structures, that is, all carbon-carbon bonds are equivalent with a bond length and bond strength somewhere between a single and a double bond.

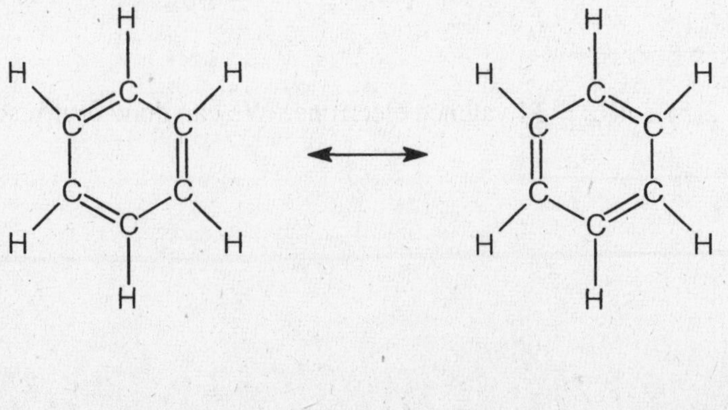

77. We will use a hexagon to represent the six-member carbon ring, and we will omit the 4 hydrogen atoms and the three lone pairs of electrons on each chlorine. If no resonance existed, we could draw 4 different molecules:

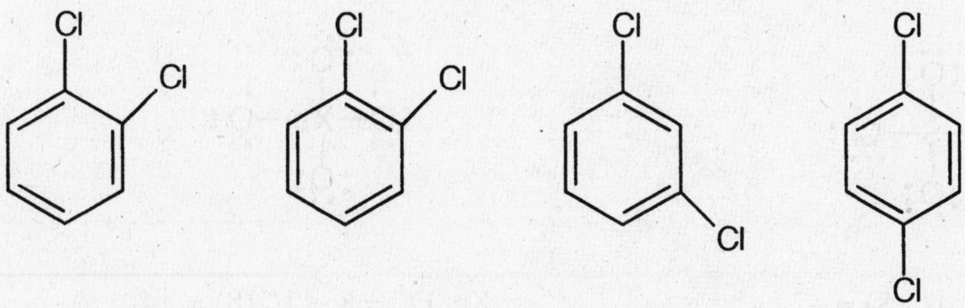

Cl Cl Cl Cl Cl Cl Cl Cl

If the double bonds in the benzene ring exhibit resonance, then we can draw only three different dichlorobenzenes. The circle in the hexagon represents the delocalization of the three double bonds in the benzene ring (see Exercise 8.75).

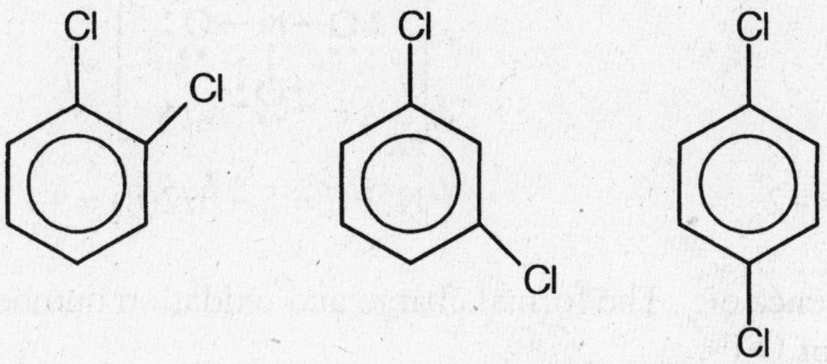

Cl Cl Cl Cl Cl Cl

With resonance, all carbon-carbon bonds are equivalent. We can't distinguish between a single and double bond between adjacent carbons that have a chlorine attached. That only 3 isomers are observed supports the concept of resonance.

79.

N_2 (10 e$^-$):	:N≡N:	Triple bond between N and N.
N_2F_4 (38 e$^-$):	:F—N—N—F: with :F: :F: on each N	Single bond between N and N.
N_2F_2 (24 e$^-$):	:F—N=N—F:	Double bond between N and N.

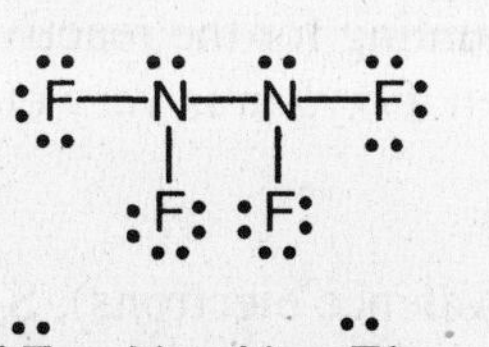

As the number of bonds increase between two atoms, bond strength increases and bond length decreases. From the Lewis structure, the shortest to longest N-N bonds are: $N_2 < N_2F_2 < N_2F_4$.

Formal Charge

81. See Exercise 8.68a for the Lewis structures of $POCl_3$, SO_4^{2-}, ClO_4^- and PO_4^{3-}. Formal charge = [number of valence electrons on free atom] - [number of lone pair electrons on atom + 1/2 (number of shared electrons of atom)].

a. $POCl_3$: P, FC = 5 - 1/2(8) = +1

b. SO_4^{2-}: S, FC = 6 - 1/2(8) = +2

c. ClO_4^-: Cl, FC = 7 - 1/2(8) = +3

d. PO_4^{3-}: P, FC = 5 - 1/2(8) = +1

e. SO_2Cl_2, 6 + 2(6) + 2(7) = 32 e^-

f. XeO_4, 8 + 4(6) = 32 e^-

S, FC = 6 - 1/2(8) = +2

Xe, FC = 8 - 1/2(8) = +4

g. ClO_3^-, 7 + 3(6) + 1 = 26 e^-

h. NO_4^{3-}, 5 + 4(6) + 3 = 32 e^-

Cl, FC = 7 - 2 - 1/2(6) = +2

N, FC = 5 - 1/2(8) = +1

83. O_2F_2 has 2(6) + 2(7) = 26 valence e^-. The formal charge and oxidation number of each atom is below the Lewis structure of O_2F_2.

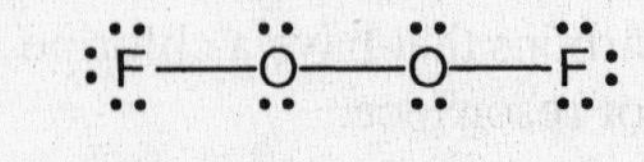

	F	O	O	F
Formal Charge	0	0	0	0
Oxid. Number	-1	+1	+1	-1

Oxidation numbers are more useful when accounting for the reactivity of O_2F_2. We are forced to assign +1 as the oxidation number for oxygen. Oxygen is very electronegative, and +1 is not a stable oxidation state for this element.

85. SCl, 6 + 7 = 13; the formula could be SCl (13 valence electrons), S_2Cl_2 (26 valence electrons), S_3Cl_3 (39 valence electrons), etc. For a formal charge of zero on S, we will need each sulfur in the Lewis structure to have two bonds to it and two lone pairs [FC = 6 – 4 – 1/2(4) = 0]. Cl will need one bond and three lone pairs for a formal charge of zero [FC = 7 – 6 – 1/2(2) = 0]. Since chlorine wants only one bond to it, it will not be a central atom here. With this in mind, only S_2Cl_2 can have a Lewis structure with a formal charge of zero on all atoms. The structure is:

:Cl—S—S—Cl:

Molecular Structure and Polarity

87. The first step always is to draw a valid Lewis structure when predicting molecular structure. When resonance is possible, only one of the possible resonance structures is necessary to predict the correct structure because all resonance structures give the same structure. The Lewis structures are in Exercises 8.67 and 8.73. The structures and bond angles for each follow.

8.67 a. HCN: linear, 180°

b. PH_3: trigonal pyramid, $< 109.5^\circ$

c. $CHCl_3$: tetrahedral, 109.5°

d. NH_4^+: tetrahedral, 109.5°

e. H_2CO: trigonal planar, 120°

f. SeF_2: V-shaped or bent, $< 109.5^\circ$

g. CO_2: linear, 180°

h and i. O_2 and HBr are both linear, but there is no bond angle in either.

Note: PH_3 and SeF_2 both have lone pairs of electrons on the central atom which result in bond angles that are something less than predicted from a tetrahedral arrangement (109.5°). However, we cannot predict the exact number. For the solutions manual, we will insert a less than sign to indicate this phenomenon. For bond angles equal to 120°, the lone pair phenomenon isn't as significant as compared to smaller bond angles. For these molecules, e.g., NO_2^-, we will insert an approximate sign in front of the 120° to note that there may be a slight distortion from the VSEPR predicted bond angle.

8.73 a. NO_2^-: V-shaped, $\approx 120^\circ$; NO_3^-: trigonal planar, 120°

N_2O_4: trigonal planar, 120° about both N atoms

b. OCN^-, SCN^- and N_3^- are all linear with 180° bond angles.

89. From the Lewis structures (see Exercise 8.71), Br_3^- would have a linear molecular structure, ClF_3 would have a T-shaped molecular structure and SF_4 would have a see-saw molecular structure. For example, consider ClF_3 (28 valence electrons):

```
 ..
:F:
 |
:Cl—F:
 |
:F:
 ..
```

The central Cl atom is surrounded by 5 electron pairs, which requires a trigonal bipyramid geometry. Since there are 3 bonded atoms and 2 lone pairs of electrons about Cl, we describe the molecular structure of ClF_3 as T-shaped with predicted bond angles of about 90°. The actual bond angles would be slightly less than 90° due to the stronger repulsive effect of the lone pair electrons as compared to the bonding electrons.

91. a. SeO_3, 6 + 3(6) = 24 e^-

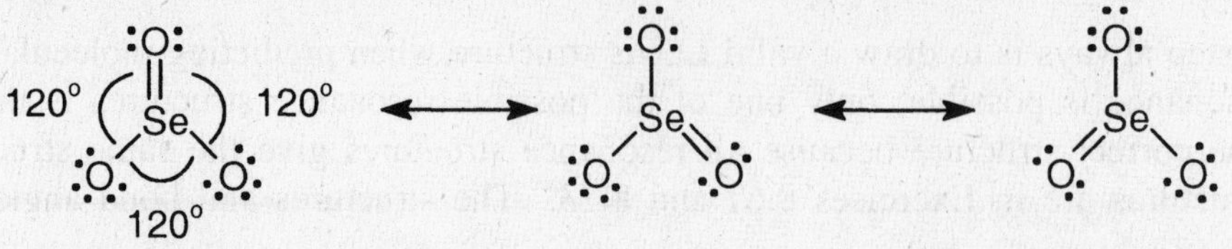

SeO₃ Lewis structures (three resonance forms), bond angles 120°, 120°, 120°

SeO_3 has a trigonal planar molecular structure with all bond angles equal to 120°. Note that any one of the resonance structures could be used to predict molecular structure and bond angles.

b. SeO_2, 6 + 2(6) = 18 e^-

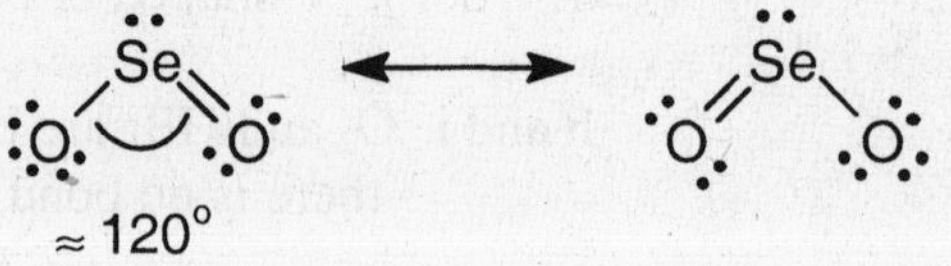

≈ 120°

SeO_2 has a V-shaped molecular structure. We would expect the bond angle to be approximately 120° as expected for trigonal planar geometry.

Note: Both of these structures have three effective pairs of electrons about the central atom. All of the structures are based on a trigonal planar geometry, but only SeO_3 is described as having a trigonal planar structure. Molecular structure always describes the relative positions of the atoms.

93. a. $XeCl_2$ has 8 + 2(7) = 22 valence electrons.

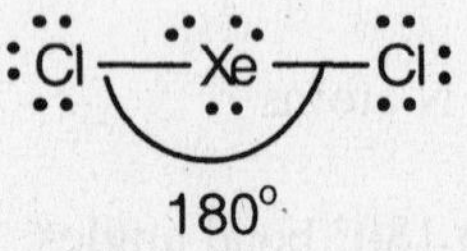

180°

There are 5 pairs of electrons about the central Xe atom. The structure will be based on a trigonal bipyramid geometry. The most stable arrangement of the atoms in $XeCl_2$ is a linear molecular structure with a 180° bond angle.

b. ICl_3 has 7 + 3(7) = 28 valence electrons.

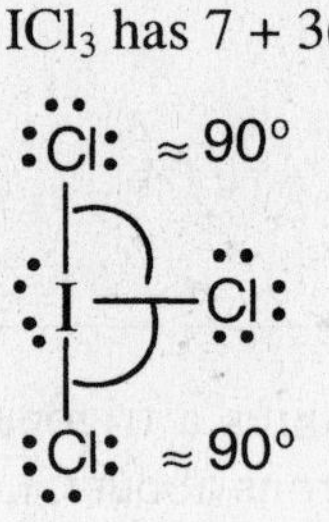

≈ 90°

≈ 90°

T-shaped; The ClICl angles are ≈ 90°. Since the lone pairs will take up more space, the ClICl bond angles will probably be slightly less than 90°.

c. TeF_4 has 6 + 4(7) = 34 valence electrons.

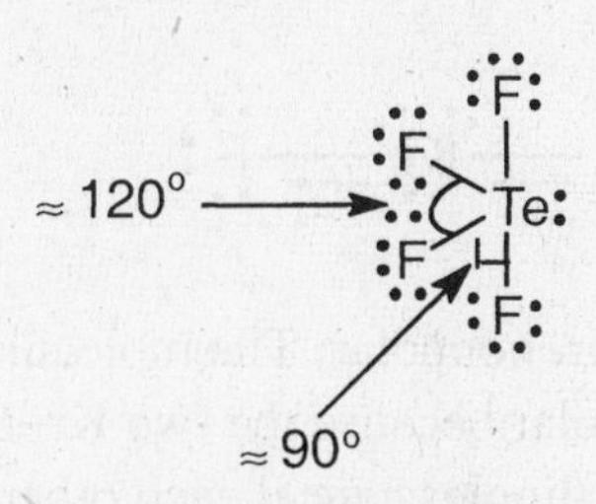

See-saw or teeter-totter or distorted tetrahedron

d. PCl_5 has 5 + 5(7) = 40 valence electrons.

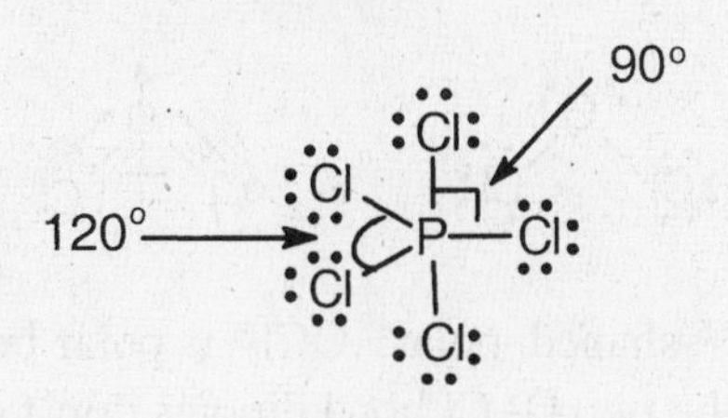

Trigonal bipyramid

All of the species in this exercise have 5 pairs of electrons around the central atom. All of the structures are based on a trigonal bipyramid geometry, but only in PCl_5 are all of the pairs bonding pairs. Thus, PCl_5 is the only one we describe as a trigonal bipyramid molecular structure. Still, we had to begin with the trigonal bipyramid geometry to get to the structures of the others.

95. SeO_3 and SeO_2 both have polar bonds but only SeO_2 has a dipole moment. The three bond dipoles from the three polar Se–O bonds in SeO_3 will all cancel when summed together. Hence, SeO_3 is nonpolar since the overall molecule has no resulting dipole moment. In SeO_2, the two Se–O bond dipoles do not cancel when summed together, hence SeO_2 has a dipole moment (is polar). Since O is more electronegative than Se, the negative end of the dipole moment is between the two O atoms, and the positive end is around the Se atom. The arrow in the following illustration represents the overall dipole moment in SeO_2. Note that to predict polarity for SeO_2, either of the two resonance structures can be used.

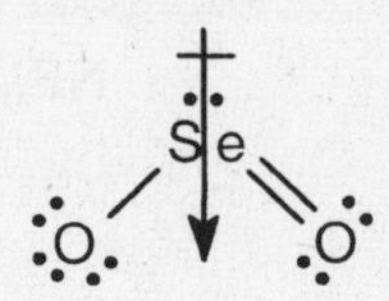

97. All have polar bonds, but only TeF_4 and ICl_3 have dipole moments. The bond dipoles from the five P–Cl bonds in PCl_5 cancel each other when summed together, so PCl_5 has no dipole moment. The bond dipoles in $XeCl_2$ also cancel:

:Cl—Xe—Cl:

Since the bond dipoles from the two Xe–Cl bonds are equal in magnitude but point in opposite directions, they cancel each other and $XeCl_2$ has no dipole moment (is nonpolar). For TeF_4 and ICl_3, the arrangement of these molecules is such that the individual bond dipoles do <u>not</u> all cancel, so each has an overall dipole moment (is polar).

99. Molecules which have an overall dipole moment are called polar molecules, and molecules which do not have an overall dipole moment are called nonpolar molecules.

a. OCl_2, $6 + 2(7) = 20\ e^-$

V-shaped, polar; OCl_2 is polar because the two O–Cl bond dipoles don't cancel each other. The resultant dipole moment is shown in the drawing.

KrF_2, $8 + 2(7) = 22\ e^-$

Linear, nonpolar; The molecule is nonpolar because the two Kr–F bond dipoles cancel each other.

BeH_2, $2 + 2(1) = 4\ e^-$

Linear, nonpolar; Be–H bond dipoles are equal and point in opposite directions. They cancel each other. BeH_2 is nonpolar.

SO_2, $6 + 2(6) = 18\ e^-$

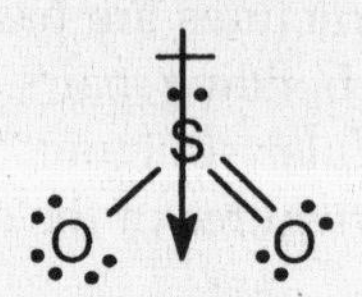

V-shaped, polar; The S–O bond dipoles do not cancel, so SO_2 is polar (has a dipole moment). Only one resonance structure is shown.

Note: All four species contain three atoms. They have different structures because the number of lone pairs of electrons around the central atom is different in each case. Polarity can only be predicted on an individual basis.

b. SO_3, $6 + 3(6) = 24\ e^-$

Trigonal planar, nonpolar; Bond dipoles cancel. Only one resonance structure is shown.

NF_3, $5 + 3(7) = 26\ e^-$

Trigonal pyramid, polar; Bond dipoles do not cancel.

IF_3 has $7 + 3(7) = 28$ valence electrons.

T-shaped, polar; bond dipoles do not cancel.

Note: Each molecule has the same number of atoms, but the structures are different because of differing numbers of lone pairs around each central atom.

c. CF_4, $4 + 4(7) = 32\ e^-$

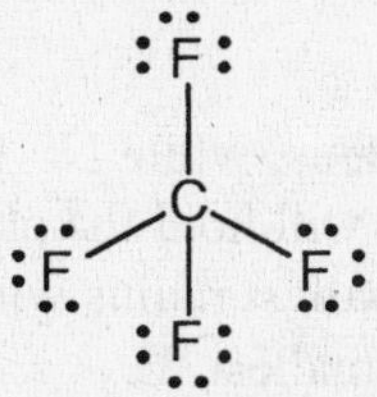

Tetrahedral, nonpolar;
Bond dipoles cancel.

SeF_4, $6 + 4(7) = 34\ e^-$

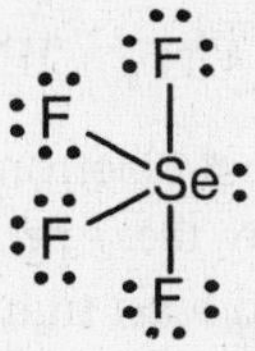

See-saw, polar;
Bond dipoles do not cancel.

KrF_4, $8 + 4(7) = 36$ valence electrons

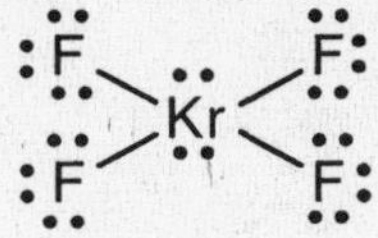

Square planar, nonpolar;
Bond dipoles cancel.

Again, each molecule has the same number of atoms, but a different structure because of differing numbers of lone pairs around the central atom.

d. IF_5, $7 + 5(7) = 42\ e^-$

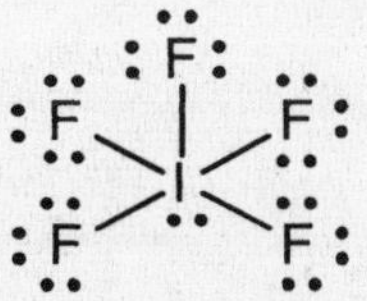

Square pyramid, polar;
Bond dipoles do not cancel.

AsF_5, $5 + 5(7) = 40\ e^-$

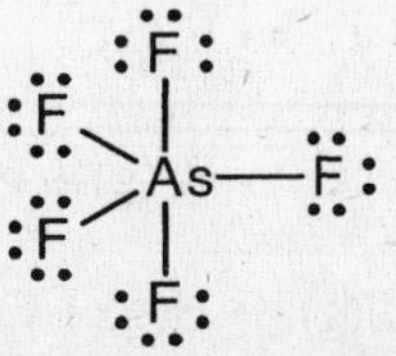

Trigonal bipyramid, nonpolar;
Bond dipoles cancel.

Yet again, the molecules have the same number of atoms, but different structures because of the presence of differing numbers of lone pairs.

101. EO_3^- is the formula of the ion. The Lewis structure has 26 valence electrons. Let x = number of valence electrons of element E.

$$26 = x + 3(6) + 1, \quad x = 7 \text{ valence electrons}$$

Element E is a halogen because halogens have 7 valence electrons. Some possible identities are F, Cl, Br and I. The EO_3^- ion has a trigonal pyramid molecular structure with bond angles $< 109.5^\circ$.

103. All these molecules have polar bonds that are symmetrically arranged about the central atoms. In each molecule, the individual bond dipoles cancel to give no net overall dipole moment, so they are all nonpolar.

Additional Exercises

105. a. Radius: $N^+ < N < N^-$; IE: $N^- < N < N^+$

N^+ has the fewest electrons held by the 7 protons in the nucleus, while N^- has the most electrons held by the 7 protons. The 7 protons in the nucleus will hold the electrons most tightly in N^+ and least tightly in N^-. Therefore, N^+ has the smallest radius with the largest ionization energy (IE) and N^- is the largest species with the smallest IE.

b. Radius: $Cl^+ < Cl < Se < Se^-$; IE: $Se^- < Se < Cl < Cl^+$

The general trends tell us that Cl has a smaller radius than Se and a larger IE than Se. Cl^+, with fewer electron-electron repulsions than Cl, will be smaller than Cl and have a larger IE. Se^-, with more electron-electron repulsions than Se, will be larger than Se and have a smaller IE.

c. Radius: $Sr^{2+} < Rb^+ < Br^-$; IE: $Br^- < Rb^+ < Sr^{2+}$

These ions are isoelectronic. The species with the most protons (Sr^{2+}) will hold the electrons most tightly and will have the smallest radius and largest IE. The ion with the fewest protons (Br^-) will hold the electrons least tightly and will have the largest radius and smallest IE.

107. a.

$$\begin{array}{ll} HF(g) \rightarrow H(g) + F(g) & \Delta H = 565\text{ kJ} \\ H(g) \rightarrow H^+(g) + e^- & \Delta H = 1312\text{ kJ} \\ F(g) + e^- \rightarrow F^-(g) & \Delta H = -327.8\text{ kJ} \\ \hline HF(g) \rightarrow H^+(g) + F^-(g) & \Delta H = 1549\text{ kJ} \end{array}$$

b.

$$\begin{array}{ll} Cl(g) \rightarrow H(g) + Cl(g) & \Delta H = 427\text{ kJ} \\ H(g) \rightarrow H^+(g) + e^- & \Delta H = 1312\text{ kJ} \\ l(g) + e^- \rightarrow Cl^-(g) & \Delta H = -348.7\text{ kJ} \\ \hline HCl(g) \rightarrow H^+(g) + Cl^-(g) & \Delta H = 1390.\text{ kJ} \end{array}$$

c.

$$\begin{array}{ll} HI(g) \rightarrow H(g) + I(g) & \Delta H = 295\text{ kJ} \\ H(g) \rightarrow H^+(g) + e^- & \Delta H = 1312\text{ kJ} \\ I(g) + e^- \rightarrow I^-(g) & \Delta H = -295.2\text{ kJ} \\ \hline HI(g) \rightarrow H^+(g) + I^-(g) & \Delta H = 1312\text{ kJ} \end{array}$$

d.

$$\begin{array}{ll} H_2O(g) \rightarrow OH(g) + H(g) & \Delta H = 467\text{ kJ} \\ H(g) \rightarrow H^+(g) + e^- & \Delta H = 1312\text{ kJ} \\ OH(g) + e^- \rightarrow OH^-(g) & \Delta H = -180.\text{ kJ} \\ \hline H_2O(g) \rightarrow H^+(g) + OH^-(g) & \Delta H = 1599\text{ kJ} \end{array}$$

109. The stable species are:

a. NaBr: In $NaBr_2$, the sodium ion would have a + 2 charge assuming each bromine has a − 1 charge. Sodium doesn't form stable Na^{2+} ionic compounds.

b. ClO_4^-: ClO_4 has 31 valence electrons so it is impossible to satisfy the octet rule for all atoms in ClO_4. The extra electron from the -1 charge in ClO_4^- allows for complete octets for all atoms.

c. XeO_4: We can't draw a Lewis structure that obeys the octet rule for SO_4 (30 electrons), unlike XeO_4 (32 electrons).

d. SeF_4: Both compounds require the central atom to expand its octet. O is too small and doesn't have low energy d orbitals to expand its octet (which is true for all row 2 elements).

111. a. $XeCl_4$, $8 + 4(7) = 36\ e^-$ $XeCl_2$, $8 + 2(7) = 22\ e^-$

:Cl Cl: Xe :Cl Cl: :Cl—Xe—Cl:

square planar, 90°, nonpolar linear, 180°, nonpolar

Both compounds have a central Xe atom that does not satisfy the octet rule. Both are nonpolar because the Xe–Cl bond dipoles are arranged in such a manner that they all cancel each other. The last item is that both have 180° bond angles. Although we haven't emphasized this, the bond angle between the Cl atoms on the diagonal in $XeCl_4$ are 180° apart from each other.

b. We didn't draw the Lewis structures, but all are polar covalent compounds. The bond dipoles do not cancel out each other when summed together. The reason the bond dipoles are not symmetrically arranged in these compounds is that they all have at least one lone pair of electrons on the central atom which disrupts the symmetry. Note that there are molecules that have lone pairs and are nonpolar, e.g., $XeCl_4$ and $XeCl_2$ in the previous problem. A lone pair on a central atom does not guarantee a polar molecule.

113. Yes, each structure has the same number of effective pairs around the central atom. (A multiple bond is counted as a single group of electrons.)

115. TeF_5^- has $6 + 5(7) + 1 = 42$ valence electrons.

[:F: :F F: Te :F F:] $^-$

The lone pair of electrons around Te exerts a stronger repulsion than the bonding pairs of electrons. This pushes the four square planar F's away from the lone pair and reduces the bond angles between the axial F atom and the square planar F atoms.

Challenge Problems

117. The reaction is:

$$1/2\ I_2(g) + 1/2\ Cl_2(g) \rightarrow ICl(g) \qquad \Delta H_f^\circ = ?$$

Using Hess's law:

$1/2\ I_2(s) \rightarrow 1/2\ I_2(g)$	$\Delta H = 1/2\ (62\ kJ)$	[Appendix 4]
$1/2\ I_2(g) \rightarrow I(g)$	$\Delta H = 1/2\ (149\ kJ)$	[Table 8.4]
$1/2\ Cl_2(g) \rightarrow Cl(g)$	$\Delta H = 1/2\ (239\ kJ)$	[Table 8.4]
$I(g) + Cl(g) \rightarrow ICl(g)$	$\Delta H = -208\ kJ$	[Table 8.4]
$1/2\ I_2(s) + 1/2\ Cl_2(g) \rightarrow ICl(g)$	$\Delta H = 17\ kJ$	

119. See Fig. 8.11 for data supporting MgO as an ionic compound. Note that the lattice energy is large enough to overcome all of the other processes (removing 2 electrons from Mg, etc.). The bond energy for O_2 (247 kJ/mol) and the electron affinity for O (737 kJ/mol) are the same when making CO. However, the energy needed to ionize carbon to form a C^{2+} ion must be too large. Fig. 7.30 shows that the first ionization energy for carbon is about 400 kJ/mol greater than the first IE for magnesium. If all other numbers were equal, the overall energy change would be down from ~ −600 kJ/mol to ~ −200 kJ/mol (see Fig. 8.11). It is not unreasonable to assume that the second ionization energy for carbon is more than 200 kJ/mol greater than the second ionization energy of magnesium. This would make ΔH_f° for CO a positive number (if it were ionic). One doesn't expect CO to be ionic as the energetics would be unfavorable.

121. As the halogen atoms get larger, it becomes more difficult to fit three halogen atoms around the small nitrogen atom, and the NX_3 molecule becomes less stable.

123. a. i. $C_6H_6N_{12}O_{12} \rightarrow 6\ CO + 6\ N_2 + 3\ H_2O + 3/2\ O_2$

The NO_2 groups are assumed to have one N–O single bond and one N=O double bond and each carbon atom has one C–H single bond. We must break and form all bonds.

Bonds broken:	Bonds formed:
3 C–C (347 kJ/mol)	6 C≡O (1072 kJ/mol)
6 C–H (413 kJ/mol)	6 N≡N (941 kJ/mol)
12 C–N (305 kJ/mol)	6 H–O (467 kJ/mol)
6 N–N (160. kJ/mol)	3/2 O=O (495 kJ/mol)
6 N–O (201 kJ/mol)	ΣD_{formed} = 15,623 kJ
6 N=O (607 kJ/mol)	
ΣD_{broken} = 12,987 kJ	

$$\Delta H = \Sigma D_{broken} - \Sigma D_{formed} = 12{,}987\ kJ - 15{,}623\ kJ = -2636\ kJ$$

ii. $C_6H_6N_{12}O_{12} \rightarrow 3\ CO + 3\ CO_2 + 6\ N_2 + 3\ H_2O$

Note: The bonds broken will be the same for all three reactions.

Bonds formed:

3 C≡O (1072 kJ/mol)
6 C=O (799 kJ/mol)
6 N≡N (941 kJ/mol)
6 H–O (467 kJ/mol)
ΣD_{formed} = 16,458 kJ

ΔH = 12,987 kJ − 16,458 kJ = −3471 kJ

iii. $C_6H_6N_{12}O_{12} \rightarrow 6\ CO_2 + 6\ N_2 + 3\ H_2$

Bonds formed:

12 C=O (799 kJ/mol)
6 N≡N (941 kJ/mol)
3 H–H (432 kJ/mol)
ΣD_{formed} = 16,530. kJ

ΔH = 12,987 kJ − 16,530. kJ = −3543 kJ

b. Reaction iii yields the most energy per mole of CL-20 so it will yield the most energy per kg.

$$\frac{-3543\ \text{kJ}}{\text{mol}} \times \frac{1\ \text{mol}}{438.23\ \text{g}} \times \frac{1000\ \text{g}}{\text{kg}} = -8085\ \text{kJ/kg}$$

125. PAN ($H_3C_2NO_5$) has 3(1) + 2(4) + 5 + 5(6) = 46 valence electrons.

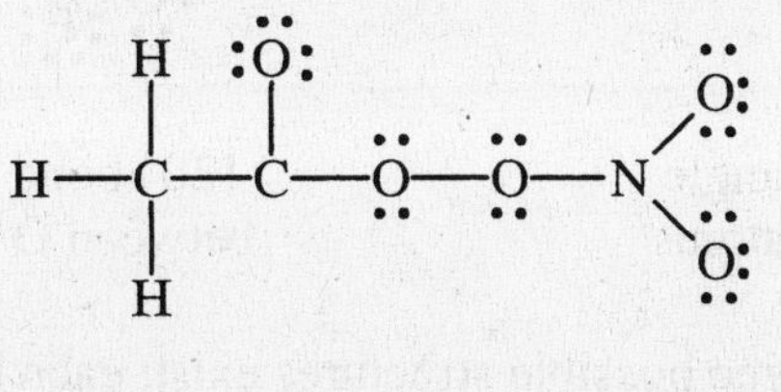

Skeletal structure with complete octets about oxygen atoms (46 electrons used).

This structure has used all 46 electrons, but there are only six electrons around one of the carbon atoms and the nitrogen atom. Two unshared pairs must become shared; we must form two double bonds.

(this form not important by formal charge arguments)

127. a. $BrFI_2$, 7 +7 + 2(7) = 28 e^-; Two possible structures exist; each has a T-shaped molecular structure.

90° bond angles between I atoms

180° bond angles between I atoms

b. XeO_2F_2, 8 + 2(6) + 2(7) = 34 e^-; Three possible structures exist; each has a see-saw molecular structure.

90° bond angle between O atoms

180° bond angle between O atoms

120° bond angle between O atoms

c. $TeF_2Cl_3^-$; 6 + 2(7) + 3(7) + 1 = 42 e^-; Three possible structures exist; each has a square pyramid molecular structure.

One F is 180° from lone pair.

Both F atoms are 90° from lone pair and 90° from each other.

Both F atoms are 90° from lone pair and 180° from each other.

129.

$$CH_3CH_2O\overset{*}{—}H + HO\overset{*}{—}\overset{\overset{\displaystyle O}{\|}}{C}CH_3 \longrightarrow CH_3CH_2O\overset{*}{—}\overset{\overset{\displaystyle O}{\|}}{C}CH_3 + H\overset{*}{—}OH$$

Bonds broken (*):
O–H
C–O

Bonds formed (*)
O–H
C–O

We make the same bonds that we have to break in order to convert reactants into products. Therefore, we would predict ΔH = 0 for this reaction using bond energies. This is probably not a great estimate for this reaction because this is not a gas phase reaction where bond energies work best.

Integrative Problems

131. Assuming 100.00 g of compound: $42.81 \text{ g F} = \frac{1 \text{ mol X}}{19.00 \text{ g F}} = 2.253 \text{ mol F}$

The number of moles of X in XF_5 is: $2.53 \text{ mol F} \times \frac{1 \text{ mol X}}{5 \text{ mol F}} = 0.4506 \text{ mol X}$

This number of moles of X has a mass of 57.19 g (= 100.00 g – 42.81 g). The molar mass of X is:

$\frac{57.19 \text{ g X}}{0.4506 \text{ mol X}} = 126.9 \text{ g/mol}$; This is element I.

IF_5, $7 + 5(7) = 42 \text{ e}^-$

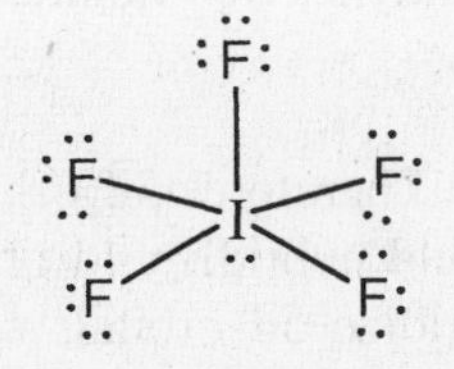

The molecular structure is square pyramid.

133. The elements are identified by their electron configurations:

$[Ar]4s^13d^5 = Cr$; $[Ne]3s^23p^3 = P$; $[Ar]4s^23d^{10}4p^3 = As$; $[Ne]3s^23p^5 = Cl$

Following the electronegativity trend, the order is Cr < As < P < Cl.

CHAPTER NINE

COVALENT BONDING: ORBITALS

Questions

7. In hybrid orbital theory, some or all of the valence atomic orbitals of the central atom in a molecule are mixed together to form hybrid orbitals; these hybrid orbitals point to where the bonded atoms and lone pairs are oriented. The sigma bonds are formed from the hybrid orbitals overlapping head to head with an appropriate orbital from the bonded atom. The π bonds in hybrid orbital theory are formed from unhybridized p atomic orbitals. The p orbitals overlap side to side to form the π bond where the π electrons occupy the space above and below a line joining the atoms (the internuclear axis). Assuming the z-axis is the internuclear axis, then the p_z atomic orbital will always be hybridized whether the hybridization is sp, sp^2, sp^3, dsp^3 or d^2sp^3. For sp hybridization, the p_x and p_y atomic orbitals are unhybridized; they are used to form two π bonds to the bonded atom(s). For sp^2 hybridization, either the p_x or p_y atomic orbital is hybridized (along with the s and p_z orbitals); the other p orbital is used to form a π bond to a bonded atom. For sp^3 hybridization, the s and all of the p orbitals are hybridized; no unhybridized p atomic orbitals are present, so no π bonds form with sp^3 hybridization. For dsp^3 and d^2sp^3 hybridization, we just mix in one or two d orbitals into the hybridization process. Which specific d orbitals are used is not important to our discussion.

9. We use d orbitals when we have to, i.e., we use d orbitals when the central atom on a molecule has more than eight electrons around it. The d orbitals are necessary to accommodate the electrons over eight. Row 2 elements never have more than eight electrons around them so they never hybridize d orbitals. We rationalize this by saying there are no d orbitals close in energy to the valence 2s and 2p orbitals (2d orbitals are forbidden energy levels). However, for row 3 and heavier elements, there are 3d, 4d, 5d, etc. orbitals which will be close in energy to the valence s and p orbitals. It is row 3 and heavier nonmetals that hybridize d orbitals when they have to.

   For phosphorus, the valence electrons are in 3s and 3p orbitals. Therefore, 3d orbitals are closest in energy and are available for hybridization. Arsenic would hybridize 4d orbitals to go with the valence 4s and 4p orbitals while iodine would hybridize 5d orbitals since the valence electrons are in $n = 5$.

11. Bonding and antibonding molecular orbitals are both solutions to the quantum mechanical treatment of the molecule. Bonding orbitals form when in-phase orbitals combine to give constructive interference. This results in enhanced electron probability located between the two nuclei. The end result is that a bonding MO is lower in energy than the atomic orbitals from which it is composed. Antibonding orbitals form when out-of-phase orbitals combine. The mismatched phases produce destructive interference leading to a node of electron probability between the two nuclei. With electron distribution pushed to the outside, the energy of an antibonding orbital is higher than the energy of the atomic orbitals from which it is composed.

13. The localized electron model does not deal effectively with molecules containing unpaired electrons. We can draw all of the possible structures for NO with its odd number of valence electrons, but still not have a good feel for whether the bond in NO is weaker or stronger than the bond in NO^-. MO theory can handle odd electron species without any modifications. In addition, hybrid orbital theory does not predict that NO^- is paramagnetic. The MO theory correctly makes this prediction.

Exercises

The Localized Electron Model and Hybrid Orbitals

15. H_2O has 2(1) + 6 = 8 valence electrons.

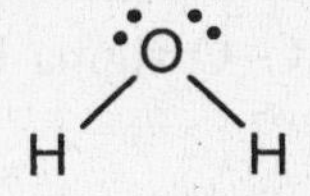

H_2O has a tetrahedral arrangement of the electron pairs about the O atom that requires sp^3 hybridization. Two of the four sp^3 hybrid orbitals are used to form bonds to the two hydrogen atoms and the other two sp^3 hybrid orbitals hold the two lone pairs of oxygen. The two O–H bonds are formed from overlap of the sp^3 hybrid orbitals on oxygen with the 1s atomic orbitals on the hydrogen atoms.

17. H_2CO has 2(1) + 4 + 6 = 12 valence electrons.

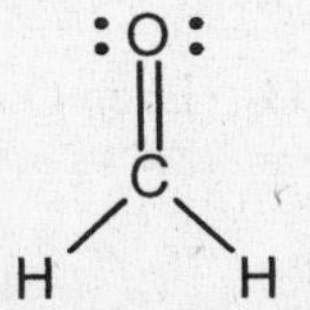

The central carbon atom has a trigonal planar arrangement of the electron pairs which requires sp^2 hybridization. The two C–H sigma bonds are formed from overlap of the sp^2 hybrid orbitals on carbon with the hydrogen 1s atomic orbitals. The double bond between carbon and oxygen consists of one σ and one π bond. The oxygen atom, like the carbon atom, also has a trigonal planar arrangement of the electrons which requires sp^2 hybridization. The σ bond in the double bond is formed from overlap of a carbon sp^2 hybrid orbital with an oxygen sp^2 hybrid orbital. The π bond in the double bond is formed from overlap of the unhybridized p atomic orbitals. Carbon and oxygen each have one unhybridized p atomic orbital which are parallel to each other. When two parallel p atomic orbitals overlap, a π bond results.

19. Ethane, C_2H_6, has 2(4) + 6(1) = 14 valence electrons.

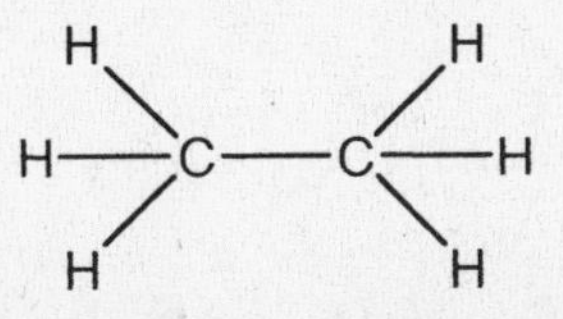

The carbon atoms are sp^3 hybridized. The six C–H sigma bonds are formed from overlap of the sp^3 hybrid orbitals on C with the 1s atomic orbitals from the hydrogen atoms. The carbon-carbon sigma bond is formed from overlap of an sp^3 hybrid orbital on each C atom.

Ethanol, C_2H_6O has $2(4) + 6(1) + 6 = 20\ e^-$

```
   H       H
    \     /   ..
H —— C —— C —— O:
    /     \     \
   H       H     H
```

The two C atoms and the O atom are sp^3 hybridized. All bonds are formed from overlap with these sp^3 hybrid orbitals. The C–H and O–H sigma bonds are formed from overlap of sp^3 hybrid orbitals with hydrogen 1s atomic orbitals. The C–C and C–O sigma bonds are formed from overlap of the sp^3 hybrid orbitals on each atom.

21. See Exercises 8.67 and 8.73 for the Lewis structures. To predict the hybridization, first determine the arrangement of electron pairs about each central atom using the VSEPR model; then utilize the information in Figure 9.24 of the text to deduce the hybridization required for that arrangement of electron pairs.

8.67

a. HCN; C is sp hybridized.

b. PH_3; P is sp^3 hybridized.

c. $CHCl_3$; C is sp^3 hybridized.

d. NH_4^+; N is sp^3 hybridized.

e. H_2CO; C is sp^2 hybridized.

f. SeF_2; Se is sp^3 hybridized.

g. CO_2; C is sp hybridized.

h. O_2; Each O atom is sp^2 hybridized.

i. HBr; Br is sp^3 hybridized.

8.73

a. The central N atom is sp^2 hybridized in NO_2^- and NO_3^-. In N_2O_4, both central N atoms are sp^2 hybridized.

b. In OCN^- and SCN^-, the central carbon atoms in each ion are sp hybridized and in N_3^-, the central N atom is also sp hybridized.

23. All exhibit dsp^3 hybridization. See Exercise 8.71 for the Lewis structures. All of these molecules/ions have a trigonal bipyramid arrangement of electron pairs about the central atom; all have central atoms with dsp^3 hybridization.

25. The molecules in Exercise 8.91 all have a trigonal planar arrangement of electron pairs about the central atom so all have central atoms with sp^2 hybridization. The molecules in Exercise 8.92 all have a tetrahedral arrangement of electron pairs about the central atom so all have central atoms with sp^3 hybridization. See Exercises 8.91 and 8.92 for the Lewis structures.

27. a.

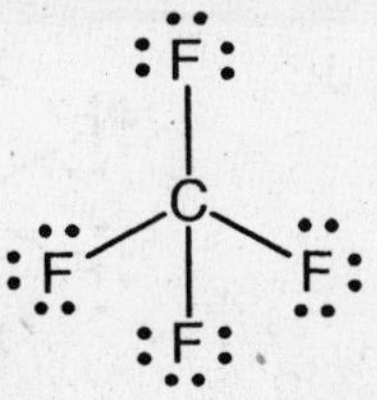

tetrahedral sp^3
109.5° nonpolar

b.

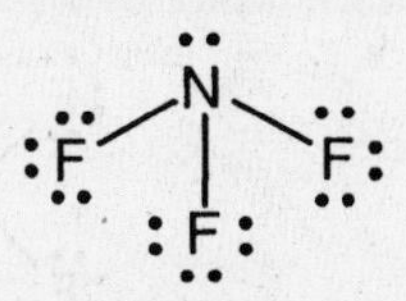

trigonal pyramid sp^3
< 109.5° polar

The angles in NF_3 should be slightly less than 109.5° because the lone pair requires more space than the bonding pairs.

c.

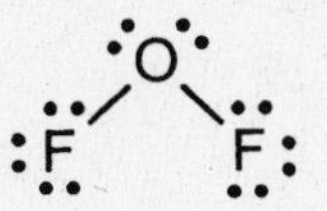

V-shaped sp^3
< 109.5° polar

d.

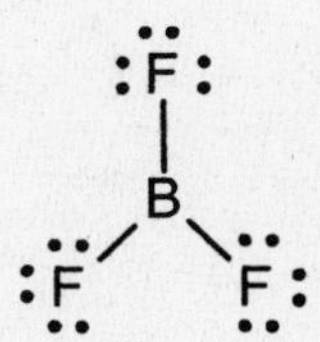

trigonal planar sp^2
120° nonpolar

e.

H—Be—H

linear sp
180° nonpolar

f.

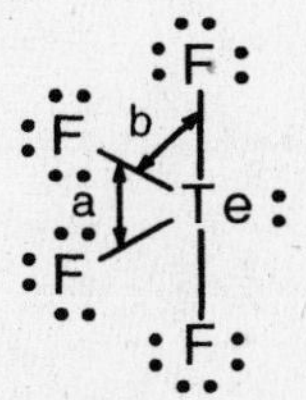

see-saw
a. ≈ 120°, b. ≈ 90°
dsp^3 polar

g.

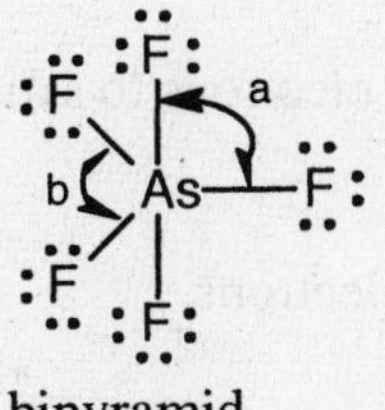

trigonal bipyramid dsp^3
a. 90°, b. 120° nonpolar

h.

F—Kr—F

linear dsp^3
120° nonpolar

i.

F F Kr 90° F F

square planar	d^2sp^3
90°	nonpolar

j.

F F F Se F F F

octahedral	d^2sp^3
90°	nonpolar

k.

F F F I F F

square pyramid	d^2sp^3
≈90°	polar

l.

F I—F F

T-shaped	dsp^3
≈ 90°	polar

29.

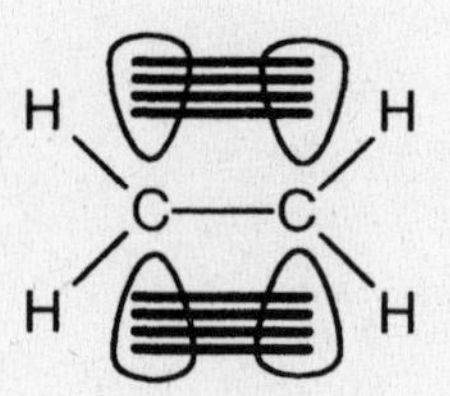

For the p-orbitals to properly line up to form the π bond, all six atoms are forced into the same plane. If the atoms were not in the same plane, the π bond could not form since the p-orbitals would no longer be parallel to each other.

31. To complete the Lewis structures, just add lone pairs of electrons to satisfy the octet rule for the atoms with fewer than eight electrons.

Biacetyl ($C_4H_6O_2$) has 4(4) + 6(1) + 2(6) = 34 valence electrons.

O sp2 H C O H—C sp3 C H H—C—H H

All CCO angles are 120°. The six atoms are not in the same plane because of free rotation about the carbon- carbon single (sigma) bonds. There are 11 σ and 2 π bonds in biacetyl.

Acetoin ($C_4H_8O_2$) has 4(4) + 8(1) + 2(6) = 36 valence electrons.

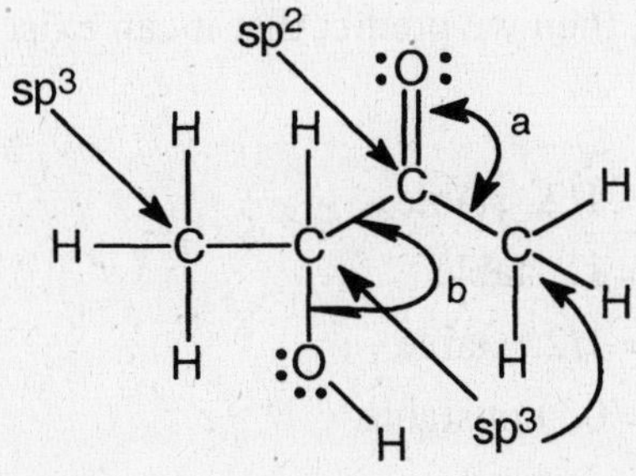

The carbon with the doubly-bonded O is sp^2 hybridized. The other 3 C atoms are sp^3 hybridized. Angle a = 120° and angle b = 109.5°. There are 13 σ and 1 π bonds in acetoin.

Note: All single bonds are σ bonds, all double bonds are one σ and one π bond, and all triple bonds are one σ and two π bonds.

33. To complete the Lewis structure, just add lone pairs of electrons to satisfy the octet rule for the atoms that have fewer than eight electrons.

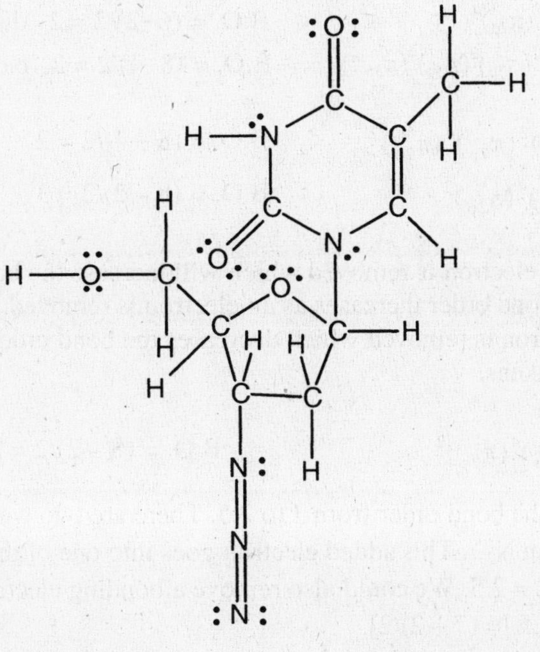

a. 6 b. 4 c. The center N in –N=N=N group

d. 33 σ e. 5 π bonds f. 180°

g. < 109.5° h. sp^3

The Molecular Orbital Model

35. If we calculate a non-zero bond order for a molecule, then we predict that it can exist (is stable).

a. H_2^+: $(\sigma_{1s})^1$ B.O. = (1−0)/2 = 1/2, stable

H_2: $(\sigma_{1s})^2$ B.O. = (2−0)/2 = 1, stable

H_2^-: $(\sigma_{1s})^2(\sigma_{1s}^*)^1$ B.O. = (2−1)/2 = 1/2, stable

H_2^{2-}: $(\sigma_{1s})^2(\sigma_{1s}^*)^2$ B.O. = (2−2)/2 = 0, not stable

b. He_2^{2+}: $(\sigma_{1s})^2$ B.O. = (2−0)/2 = 1, stable

He_2^+: $(\sigma_{1s})^2(\sigma_{1s}^*)^1$ B.O. = (2−1)/2 = 1/2, stable

He_2: $(\sigma_{1s})^2(\sigma_{1s}^*)^2$ B.O. = (2−2)/2 = 0, not stable

37. The electron configurations are:

a. Li_2: $(\sigma_{2s})^2$ B.O. = (2−0)/2 = 1, diamagnetic (0 unpaired e^-)

b. C_2: $(\sigma_{2s})^2(\sigma_{2s}^*)^2(\pi_{2p})^4$ B.O. = (6−2)/2 = 2, diamagnetic (0 unpaired e^-)

c. S_2: $(\sigma_{3s})^2(\sigma_{3s}^*)^2(\sigma_{3p})^2(\pi_{3p})^4(\pi_{3p}^*)^2$ B.O. = (8−4)/2 = 2, paramagnetic (2 unpaired e^-)

39. O_2: $(\sigma_{2s})^2(\sigma_{2s}^*)^2(\sigma_{2p})^2(\pi_{2p})^4(\pi_{2p}^*)^2$ B.O. = (8 − 4)/2 = 2

N_2: $(\sigma_{2s})^2(\sigma_{2s}^*)^2(\pi_{2p})^4(\sigma_{2p})^2$ B.O. = (8 − 2)/2 = 3

In O_2, an antibonding electron is removed which will increase the bond order to 2.5 [= (8–3)/2]. The bond order increases as an electron is removed, so the bond strengthens. In N_2, a bonding electron is removed which decreases the bond order to 2.5 = [(7 − 2)/2], so the bond strength weakens.

41. N_2: $(\sigma_{2s})^2(\sigma_{2s}^*)^2(\pi_{2p})^4(\pi_{2p})^2$ B.O. = (8 − 2)/2 = 3

We need to decrease the bond order from 3 to 2.5. There are two ways to do this. One is to add an electron to form N_2^- . This added electron goes into one of the π_{2p}^* orbitals giving a bond order = (8 − 3)/2 = 2.5. We could also remove a bonding electron to form N_2^+. The bond order for N_2^+ is also 2.5 [= (7 − 2)/2].

43. The electron configurations are (assuming the same orbital order as that for N_2):

a. CO: $(\sigma_{2s})^2(\sigma_{2s}^*)^2(\pi_{2p})^4(\sigma_{2p})^2$ B.O. = (8-2)/2 = 3, diamagnetic

b. CO^+: $(\sigma_{2s})^2(\sigma_{2s}^*)^2(\pi_{2p})^4(\sigma_{2p})^1$ B.O. = (7-2)/2 = 2.5, paramagnetic

c. CO^{2+}: $(\sigma_{2s})^2(\sigma_{2s}^*)^2(\pi_{2p})^4$ B.O. = (6-2)/2 = 2, diamagnetic

Since bond order is directly proportional to bond energy and inversely proportional to bond length, then:

shortest → longest bond length: $CO < CO^+ < CO^{2+}$

smallest → largest bond energy: $CO^{2+} < CO^+ < CO$

45. H_2: $(\sigma_{1s})^2$
B_2: $(\sigma_{2s})^2(\sigma_{2s}*)^2(\pi_{2p})^2$
C_2^{2-}: $(\sigma_{2s})^2(\sigma_{2s}*)^2(\pi_{2p})^4(\sigma_{2p})^2$
OF: $(\sigma_{2s})^2(\sigma_{2s}*)^2(\sigma_{2p})^2(\pi_{2p})^4(\pi_{2p}*)^3$

The bond strength will weaken if the electron removed comes from a bonding orbital. Of the molecules listed, H_2, B_2, and C_2^{2-} would be expected to have their bond strength weaken as an electron is removed. OF has the electron removed from an antibonding orbital, so its bond strength increases.

47. The two types of overlap that result in bond formation for p orbitals are in-phase side to side overlap (π bond) and in-phase head to head overlap (σ bond).

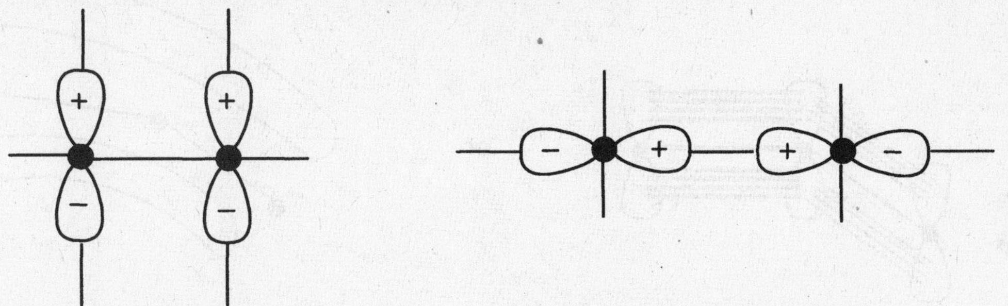

π_{2p} (in-phase; the signs match up) σ_{2p} (in-phase; the signs match up)

49. a. The electron density would be closer to F on the average. The F atom is more electronegative than the H atom, and the 2p orbital of F is lower in energy than the 1s orbital of H.

b. The bonding MO would have more fluorine 2p character since it is closer in energy to the fluorine 2p atomic orbital.

c. The antibonding MO would place more electron density closer to H and would have a greater contribution from the higher energy hydrogen 1s atomic orbital.

51. O_3 and NO_2^- are isoelectronic, so we only need consider one of them since the same bonding ideas apply to both. The Lewis structures for O_3 are:

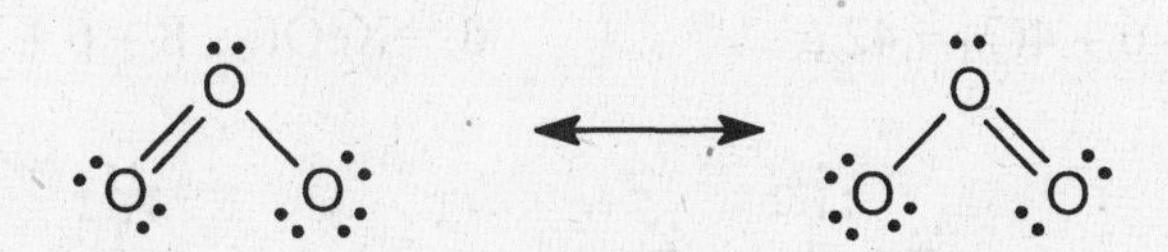

For each of the two resonance forms, the central O atom is sp^2 hybridized with one unhybridized p atomic orbital. The sp^2 hybrid orbitals are used to form the two sigma bonds to the central atom. The localized electron view of the π bond utilizes unhybridized p atomic orbitals. The π bond resonates between the two positions in the Lewis structures:

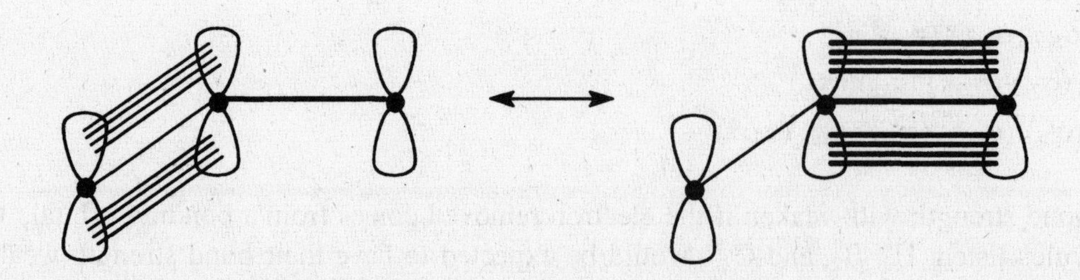

In the MO picture of the π bond, all three unhybridized p-orbitals overlap at the same time, resulting in π electrons that are delocalized over the entire surface of the molecule. This is represented as:

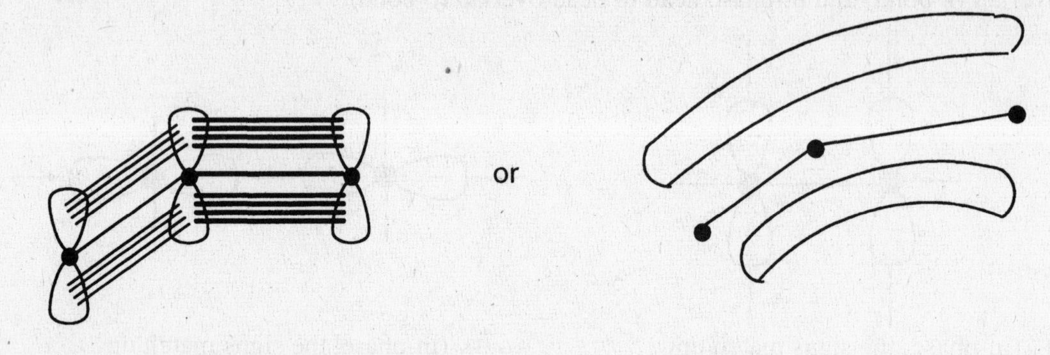

Additional Exercises

53. a. XeO_3, $8 + 3(6) = 26\ e^-$

trigonal pyramid; sp^3

b. XeO_4, $8 + 4(6) = 32\ e^-$

tetrahedral; sp^3

c. $XeOF_4$, $8 + 6 + 4(7) = 42\ e^-$

square pyramid; d^2sp^3

d. $XeOF_2$, $8 + 6 + 2(7) = 28\ e^-$

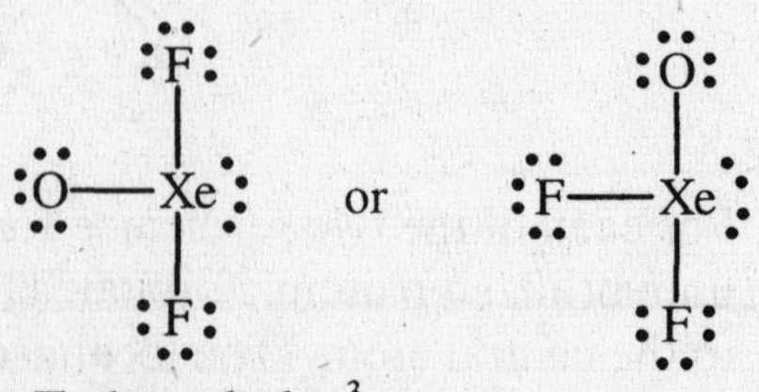

T-shaped; dsp^3

e. XeO_3F_2 has 8 + 3(6) + 2(7) = 40 valence electrons.

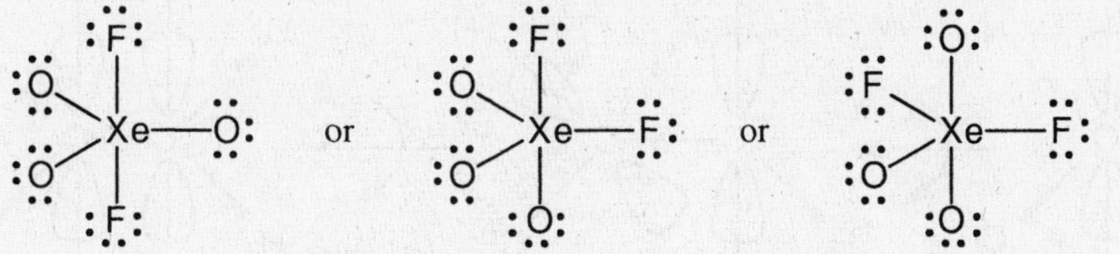

or or trigonal bipyramid; dsp^3

55. For carbon, nitrogen, and oxygen atoms to have formal charge values of zero, each C atom will form four bonds to other atoms and have no lone pairs of electrons, each N atom will form three bonds to other atoms and have one lone pair of electrons, and each O atom will form two bonds to other atoms and have two lone pairs of electrons. Following these bonding requirements gives the following two resonance structures for vitamin B_6:

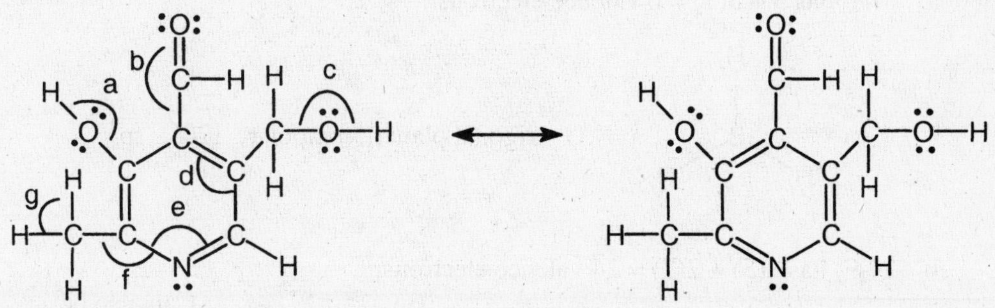

a. 21 σ bonds; 4 π bonds (The electrons in the 3 π bonds in the ring are delocalized.)

b. angles a, c, and g: ≈ 109.5°; angles b, d, e and f: ≈ 120°

c. 6 sp^2 carbons; the 5 carbon atoms in the ring are sp^2 hybridized, as is the carbon with the double bond to oxygen.

d. 4 sp^3 atoms; the 2 carbons which are not sp^2 hybridized are sp^3 hybridized, and the oxygens marked with angles a and c are sp^3 hybridized.

e. Yes, the π electrons in the ring are delocalized. The atoms in the ring are all sp^2 hybridized. This leaves a p orbital perpendicular to the plane of the ring from each atom. Overlap of all six of these p orbitals results in a π molecular orbital system where the electrons are delocalized above and below the plane of the ring (similar to benzene in Figure 9.48 of the text).

57.

In order to rotate about the double bond, the molecule must go through an intermediate stage where the π bond is broken while the sigma bond remains intact. Bond energies are 347 kJ/mol for C–C and 614 kJ/mol for C=C. If we take the single bond as the strength of the σ bond, then the strength of the π bond is (614 – 347 =) 267 kJ/mol. In theory, 267 kJ/mol must be supplied to rotate about a carbon-carbon double bond.

59. a. BH_3 has 3 + 3(1) = 6 valence electrons.

trigonal planar, nonpolar, 120°, sp^2

b. N_2F_2 has 2(5) + 2(7) = 24 valence electrons.

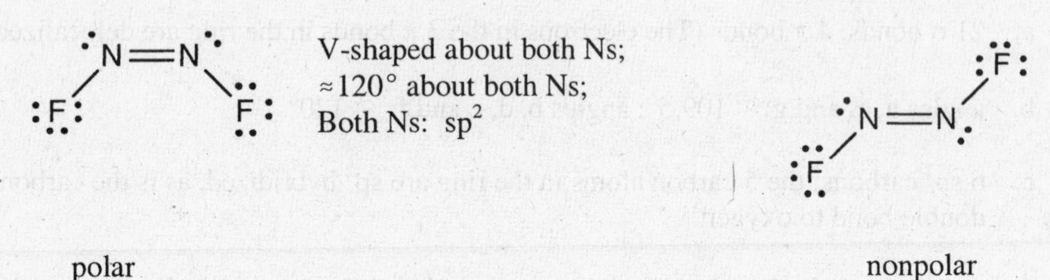

Can also be:

V-shaped about both Ns;
≈120° about both Ns;
Both Ns: sp^2

polar | nonpolar

These are distinctly different molecules.

c. C_4H_6 has 4(4) + 6(1) = 22 valence electrons.

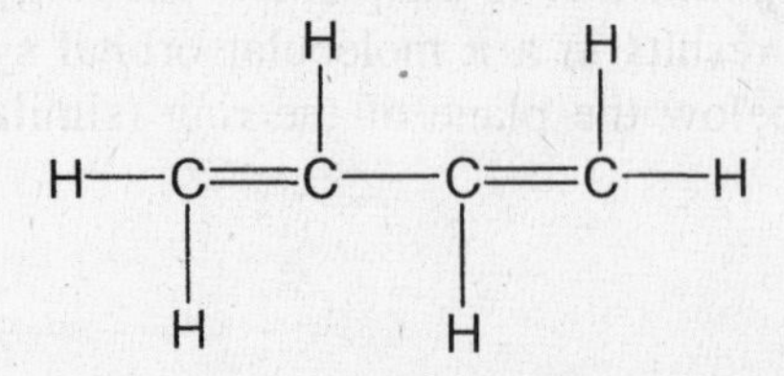

All Cs are trigonal planar with 120° bond angles and sp^2 hybridization. Because C and H have about equal electronegativities, the C–H bonds are essentially nonpolar so the molecule is nonpolar. All neutral compounds composed of only C and H atoms are nonpolar.

d. ICl_3 has 7 + 3(7) = 28 valence electrons.

T-shaped polar
a. ≈ 90°, dsp^3

61. a. The Lewis structures for NNO and NON are:

:N=N=O: ⟷ :N≡N—O: ⟷ :N—N≡O:

:N=O=N: ⟷ :N≡O—N: ⟷ :N—O≡N:

The NNO structure is correct. From the Lewis structures, we would predict both NNO and NON to be linear. However, we would predict NNO to be polar and NON to be nonpolar. Since experiments show N_2O to be polar, NNO is the correct structure.

b. Formal chare = number of valence electrons of atoms - [(number of lone pair electrons) + 1/2 (number of shared electrons)].

:N=N=O: ⟷ :N≡N—O: ⟷ :N—N≡O:

:N=N=O:			:N≡N—O:			:N—N≡O:		
-1	+1	0	0	+1	-1	-2	+1	+1

The formal charges for the atoms in the various resonance structures are below each atom. The central N is sp hybridized in all of the resonance structures. We can probably ignore the third resonance structure on the basis of the relatively large formal charges as compared to the first two resonance structures.

c. The sp hybrid orbitals on the center N overlap with atomic orbitals (or hybrid orbitals) on the other two atoms to form the two sigma bonds. The remaining two unhybridized p orbitals on the center N overlap with two p orbitals on the peripheral N to form the two π bonds.

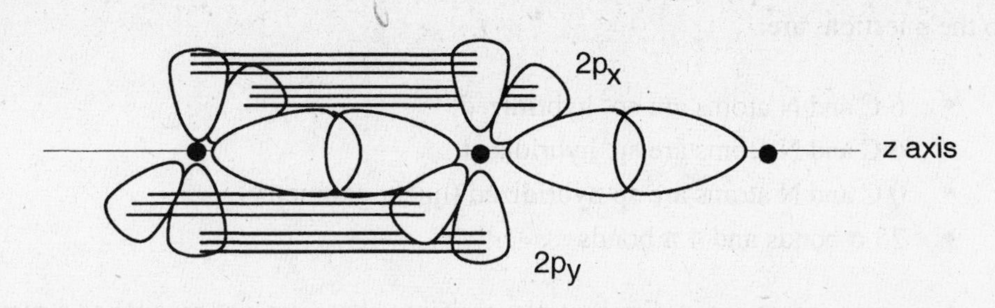

63. N_2 (ground state): $(\sigma_{2s})^2(\sigma_{2s}*)^2(\pi_{2p})^4(\sigma_{2p})^2$, B.O. = 3, diamagnetic (0 unpaired e^-)

N_2 (1st excited state): $(\sigma_{2s})^2(\sigma_{2s}*)^2(\pi_{2p})^4(\sigma_{2p})^1(\pi_{2p}*)^1$

B.O. = (7 − 3)/2 = 2, paramagnetic (2 unpaired e^-)

The first excited state of N_2 should have a weaker bond and should be paramagnetic.

65. F_2: $(\sigma_{2s})^2(\sigma_{2s}*)^2(\sigma_{2p})^2(\pi_{2p})^4(\pi_{2p}*)^4$; F_2 should have a lower ionization energy than F. The electron removed from F_2 is in a $\pi_{2p}*$ antibonding molecular orbital that is higher in energy than the 2p atomic orbitals from which the electron in atomic fluorine is removed. Since the electron removed from F_2 is higher in energy than the electron removed from F, it should be easier to remove an electron from F_2 than from F.

67. Side to side in-phase overlap of these d-orbitals would produce a π bonding molecular orbital. There would be no probability of finding an electron on the axis joining the two nuclei, which is characteristic of π MOs.

Challenge Problems

69.

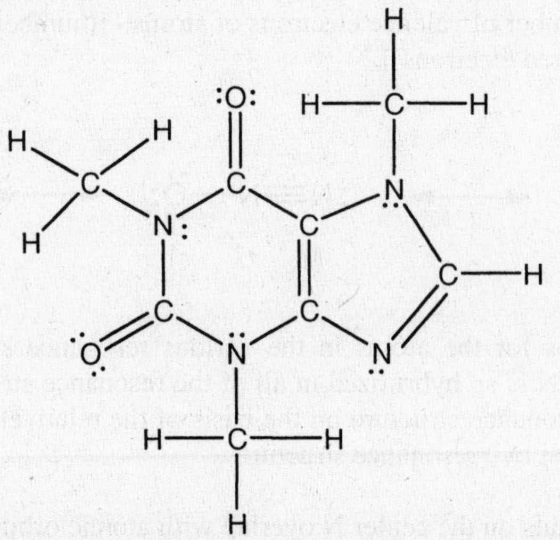

The three C atoms each bonded to three H atoms are sp^3 hybridized (tetrahedral geometry); the other five C atoms with trigonal planar geometry are sp^2 hybridized. The one N atom with the double bond is sp^2 hybridized and the other three N atoms are sp^3 hybridized. The answers to the questions are:

- 6 C and N atoms are sp^2 hybridized
- 6 C and N atoms are sp^3 hybridized
- 0 C and N atoms are sp hybridized (linear geometry)
- 25 σ bonds and 4 π bonds

71. a. No, some atoms are in different places. Thus, these are not resonance structures; they are different compounds.

b. For the first Lewis structure, all nitrogens are sp^3 hybridized and all carbons are sp^2 hybridized. In the second Lewis structure, all nitrogens and carbons are sp^2 hybridized.

c. For the reaction:

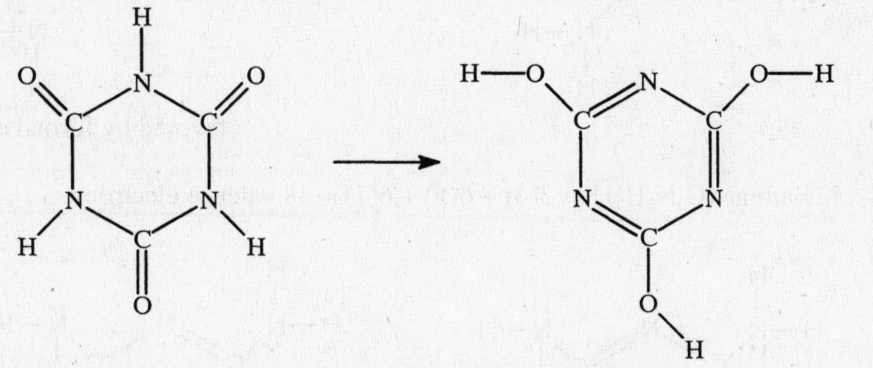

Bonds broken:

3 C=O (745 kJ/mol)

3 C–N (305 kJ/mol)

3 N–H (391 kJ/mol)

Bonds formed:

3 C=N (615 kJ/mol)

3 C–O (358 kJ/mol)

3 O–H (467 kJ/mol)

$\Delta H = 3(745) + 3(305) + 3(391) - [3(615) + 3(358) + 3(467)]$

$\Delta H = 4323 \text{ kJ} - 4320 \text{ kJ} = 3 \text{ kJ}$

The bonds are slightly stronger in the first structure with the carbon-oxygen double bonds since ΔH for the reaction is positive. However, the value of ΔH is so small that the best conclusion is that the bond strengths are comparable in the two structures.

73. a. NCN^{2-} has 5 + 4 + 5 + 2 = 16 valence electrons.

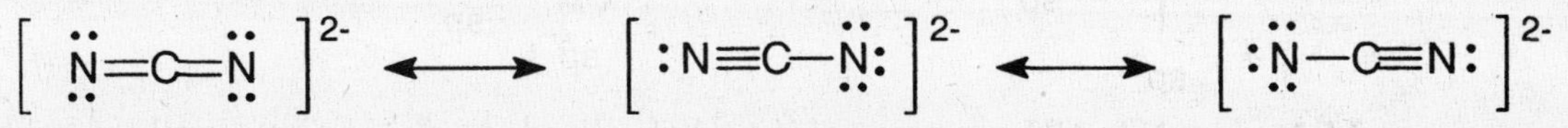

H_2NCN has 2(1) + 5 + 4 + 5 = 16 valence electrons.

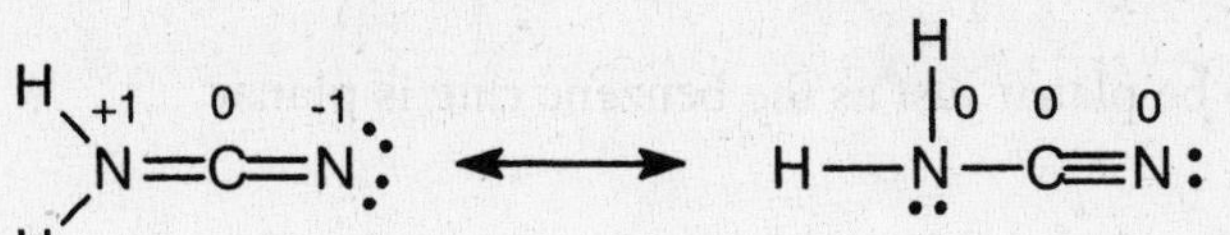

favored by formal charge

$NCNC(NH_2)_2$ has 5 + 4 + 5 + 4 + 2(5) + 4(1) = 32 valence electrons.

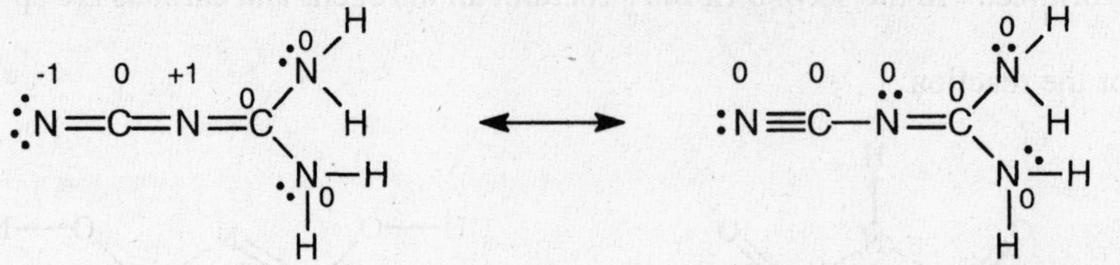

favored by formal charge

Melamine ($C_3N_6H_6$) has 3(4) + 6(5) + 6(1) = 48 valence electrons.

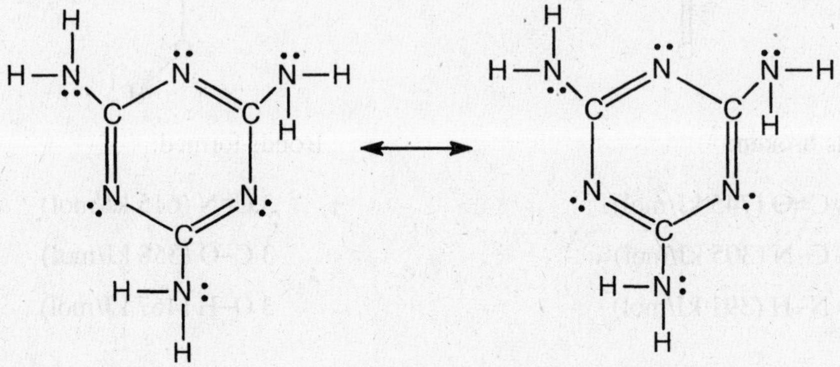

b. NCN^{2-}: C is sp hybridized. Each resonance structure predicts a different hybridization for the N atom. Depending on the resonance form, N can be sp, sp^2, or sp^3 hybridized. For the remaining compounds, we will give hybrids for the favored resonance structures as predicted from formal charge considerations.

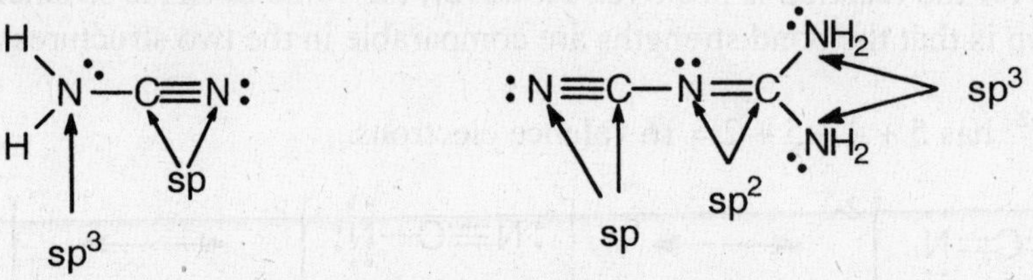

Melamine: N in NH_2 groups are all sp^3 hybridized. Atoms in ring are all sp^2 hybridized.

c. NCN^{2-}: 2 σ and 2 π bonds; H_2NCN: 4 σ and 2 π bonds; dicyandiamide: 9 σ and 3 π bonds; melamine: 15 σ and 3 π bonds

d. The π-system forces the ring to be planar just as the benzene ring is planar.

e. The structure:

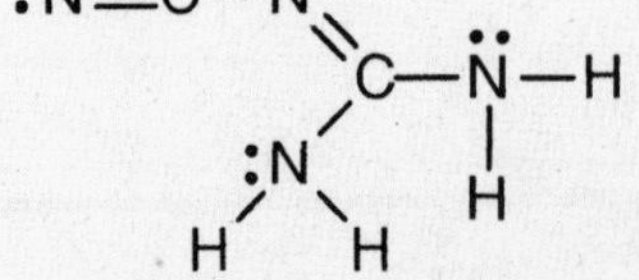

is the most important since it has three different CN bonds. This structure is also favored on the basis of formal charge.

75. a. $E = \frac{hc}{\lambda} = \frac{(6.626\times10^{-34}\ J\ s)(2.998\times10^{8}\ m/s)}{25\times10^{-9}\ m} = 7.9\times10^{-18}\ J$

$$7.9\times10^{-18}\ J \times \frac{6.022\times10^{23}}{mol} \times \frac{1\ kJ}{1000\ J} = 4800\ kJ/mol$$

Using ΔH values from the various reactions, 25 nm light has sufficient energy to ionize N_2 and N and to break the triple bond. Thus, N_2, N_2^+, N, and N^+ will all be present, assuming excess N_2.

b. To produce atomic nitrogen but no ions, the range of energies of the light must be from 941 kJ/mol to just below 1402 kJ/mol.

$$\frac{941\ kJ}{mol} \times \frac{mol}{6.022\times10^{23}} \times \frac{1000\ J}{kJ} = 1.56\times10^{-18}\ J/photon$$

$$\lambda = \frac{hc}{E} = \frac{(6.626\times10^{-34}\ J\ s)(2.998\times10^{8}\ m/s)}{1.56\times10^{-18}\ J} = 1.27\times10^{-7}\ m = 127\ nm$$

$$\frac{1402\ kJ}{mol} \times \frac{mol}{6.0221\times10^{23}} \times \frac{1000\ J}{kJ} = 2.328\times10^{-18}\ J/photon$$

$$\lambda = \frac{hc}{E} = \frac{(6.6261\times10^{-34}\ J\ s)(2.9979\times10^{8}\ m/s)}{2.328\times10^{-18}\ J} = 8.533\times10^{-8}\ m = 85.33\ nm$$

Light with wavelengths in the range of $85.33\ nm < \lambda \le 127\ nm$ will produce N but no ions.

c. N_2: $(\sigma_{2s})^2(\sigma_{2s}{*})^2(\pi_{2p})^4(\sigma_{2p})^2$; The electron removed from N_2 is in the σ_{2p} molecular orbital which is lower in energy than the 2p atomic orbital from which the electron in atomic nitrogen is removed. Since the electron removed from N_2 is lower in energy than the electron in N, the ionization energy of N_2 is greater than that for N.

77. O=N–Cl: The bond order of the NO bond in NOCl is 2 (a double bond).

NO: From molecular orbital theory, the bond order of this NO bond is 2.5.

Both reactions apparently involve only the breaking of the N–Cl bond. However, in the reaction ONCl → NO + Cl, some energy is released in forming the stronger NO bond, lowering the value of ΔH. Therefore, the apparent N–Cl bond energy is artificially low for this reaction. The first reaction involves only the breaking of the N–Cl bond.

79. a. The CO bond is polar with the negative end around the more electronegative oxygen atom. We would expect metal cations to be attracted to and bond to the oxygen end of CO on the basis of electronegativity.

b. :C≡O:

FC (carbon) = 4 − 2 − 1/2(6) = −1

FC (oxygen) = 6 − 2 − 1/2(6) = +1

From formal charge, we would expect metal cations to bond to the carbon (with the negative formal charge).

c. In molecular orbital theory, only orbitals with proper symmetry overlap to form bonding orbitals. The metals that form bonds to CO are usually transition metals, all of which have outer electrons in the d orbitals. The only molecular orbitals of CO that have proper symmetry to overlap with d orbitals are the π_{2p}* orbitals, whose shape is similar to the d orbitals (see Figure 9.34). Since the antibonding molecular orbitals have more carbon character (carbon is less electronegative than oxygen), one would expect the bond to form through carbon.

81. The electron configurations are:

N_2: $(\sigma_{2s})^2(\sigma_{2s}^*)^2(\pi_{2p})^4(\sigma_{2p})^2$

O_2: $(\sigma_{2s})^2(\sigma_{2s}^*)^2(\sigma_{2p})^2(\pi_{2p})^4(\pi_{2p}^*)^2$

N_2^{2-}: $(\sigma_{2s})^2(\sigma_{2s}^*)^2(\pi_{2p})^4(\sigma_{2p})^2(\pi_{2p}^*)^2$

N_2^-: $(\sigma_{2s})^2(\sigma_{2s}^*)^2(\pi_{2p})^4(\sigma_{2p})^2(\pi_{2p}^*)^1$

O_2^+: $(\sigma_{2s})^2(\sigma_{2s}^*)^2(\sigma_{2p})^2(\pi_{2p})^4(\pi_{2p}^*)^1$

Note: the ordering of the σ_{2p} and π_{2p} orbitals is not important to this question.

The species with the smallest ionization energy has the electron which is easiest to remove. From the MO electron configurations, O_2, N_2^{2-}, N_2^-, and O_2^+ all contain electrons in the same higher energy antibonding orbitals (π_{2p}^*), so they should have electrons that are easier to remove as compared to N_2 which has no π_{2p}^* electrons. To differentiate which has the easiest π_{2p}^* to remove, concentrate on the number of electrons in the orbitals attracted to the number of protons in the nucleus.

N_2^{2-} and N_2^- both have 14 protons in the two nuclei combined. Because N_2^{2-} has more electrons, one would expect N_2^{2-} to have more electron repulsions which translates into having an easier electron to remove. Between O_2 and O_2^+, the electron in O_2 should be easiest to remove. O_2 has one more electron than O_2^+, and one would expect the fewer electrons in O_2^+ to be better attracted to the nuclei (and harder to remove). Between N_2^{2-} and O_2, both have 16 electrons; the difference is the number of protons in the nucleus. Because N_2^{2-} has two fewer protons than O_2, one would expect the N_2^{2-} to have the easiest electron to remove which translates into the smallest ionization energy.

Integrative Problems

83. a. Li_2: $(\sigma_{2s})^2$ B.O. = (2 – 0)/2 = 1

B_2: $(\sigma_{2s})^2(\sigma_{2s}^*)^2(\pi_{2p})^2$ B.O. = (4 – 2)/2 = 1

Both have a bond order of 1.

b. B_2 has four more electrons than Li_2 so four electrons must be removed from B_2 to make it isoelectronic with Li_2. The isoelectronic ion is B_2^{4+}.

c. $1.5 \text{ kg } B_2 \times \frac{1000 \text{ g}}{1 \text{ kg}} \times \frac{1 \text{ mol } B_2}{21.62 \text{ g } B_2} \times \frac{6455 \text{ kJ}}{\text{mol } B_2} = 4.5 \times 10^5 \text{ kJ}$

85. Element X has 36 protons which identifies it as Kr. Element Y has one less electron than Y^-, so the electron configuration of Y is $1s^2 2s^2 2p^5$. This is F.

KrF_3^+, 8 + 3(7) – 1 = 28 e^-

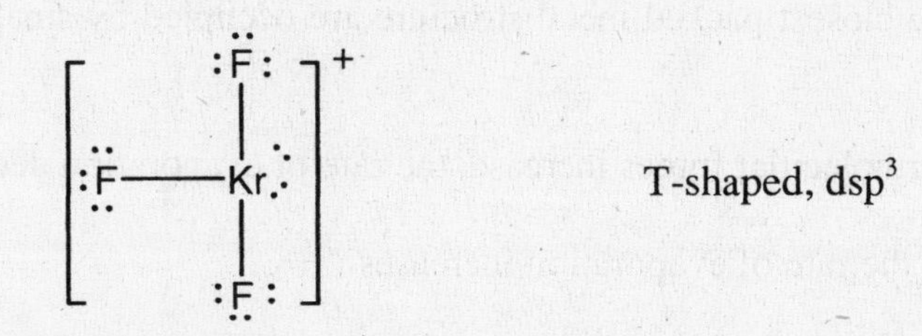

T-shaped, dsp^3

CHAPTER TEN

LIQUIDS AND SOLIDS

Questions

13. Atoms have an approximately spherical shape (on the average). It is impossible to pack spheres together without some empty space among the spheres.

15. Evaporation takes place when some molecules at the surface of a liquid have enough energy to break the intermolecular forces holding them in the liquid phase. When a liquid evaporates, the molecules that escape have high kinetic energies. The average kinetic energy of the remaining molecules is lower, thus, the temperature of the liquid is lower.

17. An alloy is a substance that contains a mixture of elements and has metallic properties. In a substitutional alloy, some of the host metal atoms are replaced by other metal atoms of similar size, e.g., brass, pewter, plumber's solder. An interstitial alloy is formed when some of the interstices (holes) in the closest packed metal structure are occupied by smaller atoms, e.g., carbon steels.

19. a. As the strength of the intermolecular forces increase, the rate of evaporation decreases.

    b. As temperature increases, the rate of evaporation increases.

    c. As surface area increases, the rate of evaporation increases.

21. Sublimation will occur allowing water to escape as $H_2O(g)$.

23. The strength of intermolecular forces determines relative boiling points. The types of intermolecular forces for covalent compounds are London dispersion forces, dipole forces, and hydrogen bonding. Because the three compounds are assumed to have similar molar mass and shape, the strength of the London dispersion forces will be about equal between the three compounds. One of the compounds will be nonpolar so it only has London dispersion forces. The other two compounds will be polar so they have additional dipole forces and will boil at a higher temperature than the nonpolar compound. One of the polar compounds will have an H covalently bonded to either N, O, or F. This gives rise to the strongest type of covalent intermolecular forces, hydrogen bonding. The compound which hydrogen bonds will have the highest boiling point while the polar compound with no hydrogen bonding will boil at a temperature in the middle of the other compounds.

25. a. Both CO_2 and H_2O are molecular solids. Both have an ordered array of the individual molecules, with the molecular units occupying the lattice points. A difference within each solid lattice is the strength of the intermolecular forces. CO_2 is nonpolar and only exhibits London dispersion forces. H_2O exhibits the relatively strong hydrogen bonding interactions. The differences in strength is evidenced by the solid phase changes that occur at 1 atm. $CO_2(s)$ sublimes at a relatively low temperature of $-78\,°C$. In sublimation,

all of the intermolecular forces are broken. However, $H_2O(s)$ doesn't have a phase change until 0°C, and in this phase change from ice to water, only a fraction of the intermolecular forces are broken. The higher temperature and the fact that only a portion of the intermolecular forces are broken are attributed to the strength of the intermolecular forces in $H_2O(s)$ as compared to $CO_2(s)$.

Related to the intermolecular forces are the relative densities of the solid and liquid phases for these two compounds. $CO_2(s)$ is denser than $CO_2(l)$ while $H_2O(s)$ is less dense than $H_2O(l)$. For $CO_2(s)$, the molecules pack together as close as possible, hence solids are usually more dense than the liquid phase. For H_2O, each molecule has two lone pairs and two bonded hydrogen atoms. Because of the equal number of lone pairs and O–H bonds, each H_2O molecule can form two hydrogen bonding interactions to other H_2O molecules. To keep this symmetric arrangement (which maximizes the hydrogen bonding interactions), the $H_2O(s)$ molecules occupy positions that create empty space in the lattice. This translates into a smaller density for $H_2O(s)$ as compared to $H_2O(l)$.

b. Both NaCl and CsCl are ionic compounds with the anions at the lattice points of the unit cells and the cations occupying the empty spaces created by anions (called holes). In NaCl, the Cl^- anions occupy the lattice points of a face-centered unit cell with the Na^+ cations occupying the octahedral holes. Octahedral holes are the empty spaces created by six Cl^- ions. CsCl has the Cl^- ions at the lattice points of a simple cubic unit cell with the Cs^+ cations occupying the middle of the cube.

27. The mathematical equation that relates the vapor pressure of a substance to temperature is:

$$\underset{y}{\ln P_{vap}} = \underset{m}{-\frac{\Delta H_{vap}}{R}}\underset{x}{\left(\frac{1}{T}\right)} \underset{+\,b}{+\,C}$$

As shown above, this equation is in the form of the straight line equation. If one plots $\ln P_{vap}$ vs. $1/T$, the slope of the straight line is $-\Delta H_{vap}/R$. Because ΔH_{vap} is always positive, the slope of the straight line will be negative.

Exercises

Intermolecular Forces and Physical Properties

29. Ionic compounds have ionic forces. Covalent compounds all have London Dispersion (LD) forces, while polar covalent compounds have dipole forces and/or hydrogen bonding forces. For H bonding forces, the covalent compound must have either a N–H, O–H or F–H bond in the molecule.

a. LD only b. dipole, LD c. H bonding, LD

d. ionic e. LD only (CH_4 in a nonpolar covalent compound.)

f. dipole, LD g. ionic

31. a. OCS; OCS is polar and has dipole-dipole forces in addition to London dispersion (LD) forces. All polar molecules have dipole forces. CO_2 is nonpolar and only has LD forces. To predict polarity, draw the Lewis structure and deduce whether the individual bond dipoles cancel.

b. SeO_2; Both SeO_2 and SO_2 are polar compounds, so they both have dipole forces as well as LD forces. However, SeO_2 is a larger molecule, so it would have stronger LD forces.

c. $H_2NCH_2CH_2NH_2$; More extensive hydrogen bonding is possible.

d. H_2CO; H_2CO is polar while CH_3CH_3 is nonpolar. H_2CO has dipole forces in addition to LD forces.

e. CH_3OH; CH_3OH can form relatively strong H bonding interactions, unlike H_2CO.

33. a. Neopentane is more compact than n-pentane. There is less surface area contact among neopentane molecules. This leads to weaker LD forces and a lower boiling point.

b. HF is capable of H bonding; HCl is not.

c. LiCl is ionic, and HCl is a molecular solid with only dipole forces and LD forces. Ionic forces are much stronger than the forces for molecular solids.

d. n-Hexane is a larger molecule, so it has stronger LD forces.

35. Boiling points and freezing points are assumed directly related to the strength of the intermolecular forces, while vapor pressure is inversely related to the strength of the intermolecular forces.

a. HBr; HBr is polar, while Kr and Cl_2 are nonpolar. HBr has dipole forces unlike Kr and Cl_2.

b. NaCl; Ionic forces are much stronger than molecular forces.

c. I_2; All are nonpolar, so the largest molecule (I_2) will have the strongest LD forces and the lowest vapor pressure.

d. N_2; Nonpolar and smallest, so has the weakest intermolecular forces.

e. CH_4; Smallest, nonpolar molecule so has the weakest LD forces.

f. HF; HF can form relatively strong H bonding interactions unlike the others.

g. $CH_3CH_2CH_2OH$; H bonding, unlike the others, so has strongest intermolecular forces.

Properties of Liquids

37. The attraction of H_2O for glass is stronger than the H_2O–H_2O attraction. The miniscus is concave to increase the area of contact between glass and H_2O. The Hg–Hg attraction is greater than the Hg–glass attraction. The miniscus is convex to minimize the Hg–glass contact.

39. The structure of H_2O_2 is H–O–O–H, which produces greater hydrogen bonding than water. Long chains of hydrogen–bonded H_2O_2 molecules then get tangled together.

Structures and Properties of Solids

41. $n\lambda = 2d \sin\theta,\ d = \dfrac{n\lambda}{2\sin\theta} = \dfrac{1\times 154\text{ pm}}{2\times\sin 14.22°} = 313\text{ pm} = 3.13\times 10^{-10}\text{ m}$

43. $\lambda = \dfrac{2d\sin\theta}{n} = \dfrac{2\times 1.36\times 10^{-10}\text{ m}\times\sin 15.0°}{1} = 7.04\times 10^{-11}\text{ m} = 0.704\text{ Å}$

45. A cubic closest packed structure has a face-centered cubic unit cell. In a face-centered cubic unit, there are:

$$8\text{ corners}\times\frac{1/8\text{ atom}}{\text{corner}} + 6\text{ faces}\times\frac{1/2\text{ atom}}{\text{face}} = 4\text{ atoms}$$

The atoms in a face-centered cubic unit cell touch along the face diagonal of the cubic unit cell. Using the Pythagorean formula where l = length of the face diagonal and r = radius of the atom:

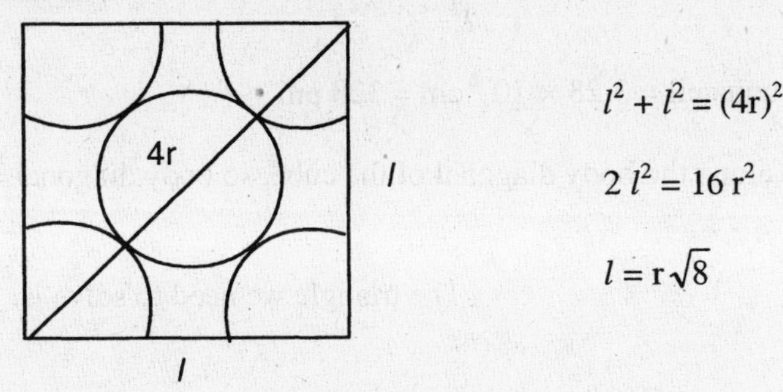

$$l^2 + l^2 = (4r)^2$$

$$2\,l^2 = 16\,r^2$$

$$l = r\sqrt{8}$$

$l = r\sqrt{8} = 197\times 10^{-12}\text{ m}\times\sqrt{8} = 5.57\times 10^{-10}\text{ m} = 5.57\times 10^{-8}\text{ cm}$

Volume of a unit cell $= l^3 = (5.57\times 10^{-8}\text{ cm})^3 = 1.73\times 10^{-22}\text{ cm}^3$

Mass of a unit cell $= 4\text{ Ca atoms}\times\dfrac{1\text{ mol Ca}}{6.022\times 10^{23}\text{ atoms}}\times\dfrac{40.08\text{ g Ca}}{\text{mol Ca}} = 2.662\times 10^{-22}\text{ g Ca}$

$\text{density} = \dfrac{\text{mass}}{\text{volume}} = \dfrac{2.662\times 10^{-22}\text{ g}}{1.73\times 10^{-22}\text{ cm}^3} = 1.54\text{ g/cm}^3$

47. The unit cell for cubic closest packing is the face-centered unit cell. The volume of a unit cell is:

$$V = l^3 = (492 \times 10^{-10}\ \text{cm})^3 = 1.19 \times 10^{-22}\ \text{cm}^3$$

There are 4 Pb atoms in the unit cell, as is the case for all face-centered cubic unit cells. The mass of atoms in a unit cell is:

$$\text{mass} = 4\ \text{Pb atoms} \times \frac{1\ \text{mol Pb}}{6.022\times10^{23}\ \text{atoms}} \times \frac{207.2\ \text{g Pb}}{\text{mol Pb}} = 1.38 \times 10^{-21}\ \text{g}$$

$$\text{density} = \frac{\text{mass}}{\text{volume}} = \frac{1.38\times10^{-21}\ \text{g}}{1.19\times10^{-22}\ \text{cm}^3} = 11.6\ \text{g/cm}^3$$

From Exercise 45, the relationship between the cube edge length, l, and the radius of an atom in a face-centered unit cell is: $l = r\sqrt{8}$.

$$r = \frac{l}{\sqrt{8}} = \frac{492\ \text{pm}}{\sqrt{8}} = 174\ \text{pm} = 1.74 \times 10^{-10}\ \text{m}$$

49. For a body-centered unit cell: $8\ \text{corners} \times \frac{1/8\ \text{Ti}}{\text{corner}} + \text{Ti at body center} = 2\ \text{Ti atoms}$

All body-centered unit cells have 2 atoms per unit cell. For a unit cell:

$$\text{density} = 4.50\ \text{g/cm}^3 = \frac{2\ \text{atoms Ti} \times \frac{1\ \text{mol Ti}}{6.022\times10^{23}\ \text{atoms}} \times \frac{47.88\ \text{g Ti}}{\text{mol Ti}}}{l^3}$$

Solving: l = edge length of unit cell = 3.28×10^{-8} cm = 328 pm

Assume Ti atoms just touch along the body diagonal of the cube, so body diagonal = 4 × radius of atoms = 4r.

The triangle we need to solve is:

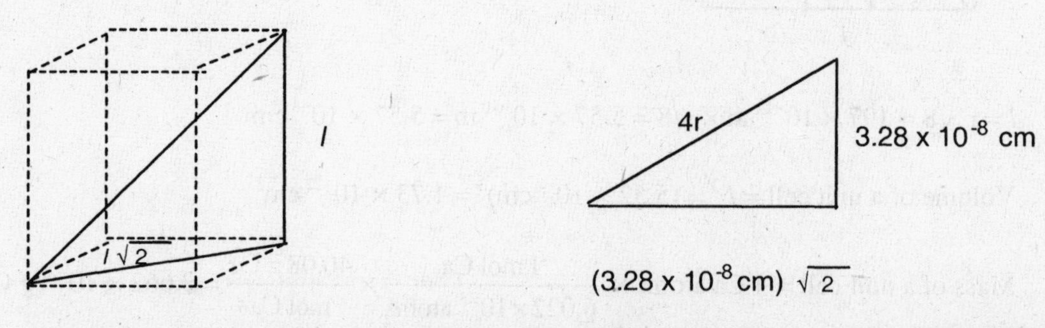

$$(4r)^2 = (3.28 \times 10^{-8}\ \text{cm})^2 + [(3.28 \times 10^{-8}\ \text{cm})\ \sqrt{2}\ ]^2,\quad r = 1.42 \times 10^{-8}\ \text{cm} = 142\ \text{pm}$$

For a body-centered unit cell (bcc), the radius of the atom is related to the cube edge length by: $4r = l\sqrt{3}$ or $l = 4r/\sqrt{3}$.

51. If a face-centered cubic structure, then 4 atoms/unit cell and from Exercise 10.45:

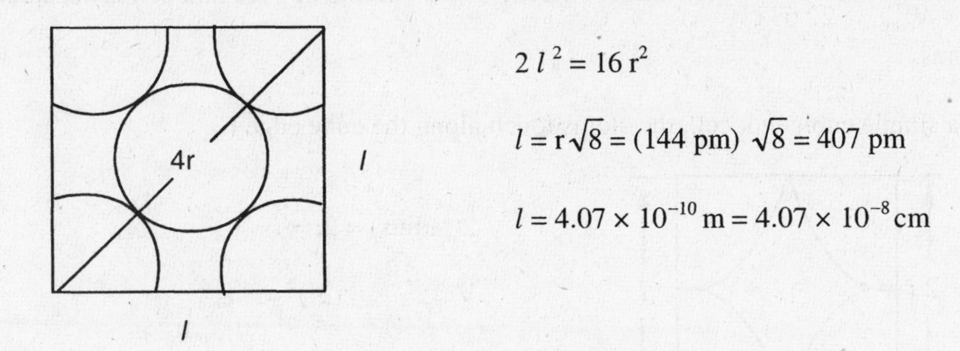

$$2\,l^2 = 16\,r^2$$

$$l = r\sqrt{8} = (144\text{ pm})\ \sqrt{8} = 407\text{ pm}$$

$$l = 4.07 \times 10^{-10}\text{ m} = 4.07 \times 10^{-8}\text{ cm}$$

$$\text{density} = \frac{4\text{ atoms Au} \times \dfrac{1\text{ mol Au}}{6.022\times10^{23}\text{ atoms}} \times \dfrac{197.0\text{ g Au}}{\text{mol Au}}}{(4.07\times10^{-8}\text{ cm})^3} = 19.4\text{ g/cm}^3$$

If a body-centered cubic structure, then 2 atoms/unit cell and from Exercise 10.49:

$$16\,r^2 = l^2 + 2\,l^2$$

$$l = 4r/\sqrt{3} = 333\text{ pm} = 3.33 \times 10^{-10}\text{ m}$$

$$l = 3.33 \times 10^{-8}\text{ cm}$$

$$\text{density} = \frac{2\text{ atoms Au} \times \dfrac{1\text{ mol Au}}{6.022\times10^{23}\text{ atoms}} \times \dfrac{197.0\text{ g Au}}{\text{mol Au}}}{(3.33\times10^{-8}\text{ cm})^3} = 17.7\text{ g/cm}^3$$

The measured density is consistent with a face-centered cubic unit cell.

53. In a face-centered unit cell (ccp structure), the atoms touch along the face diagonal:

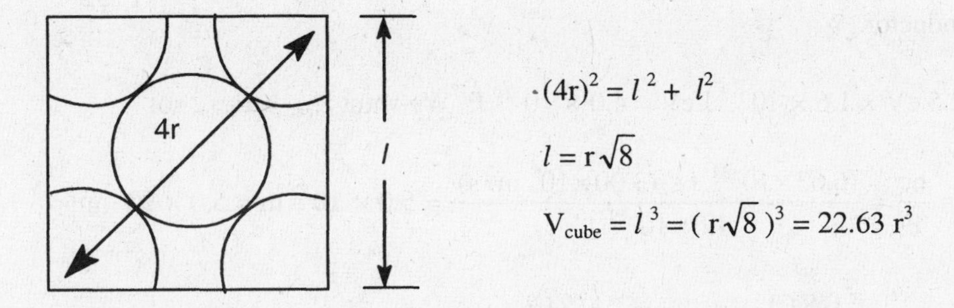

$$(4r)^2 = l^2 + l^2$$

$$l = r\sqrt{8}$$

$$V_{cube} = l^3 = (r\sqrt{8})^3 = 22.63\,r^3$$

There are four atoms in a face-centered cubic cell (see Exercise 10.45). Each atom has a volume of $4/3\ \pi r^3$.

$$V_{atoms} = 4 \times \frac{4}{3}\pi r^3 = 16.76\,r^3$$

So, $\frac{V_{atoms}}{V_{cube}} = \frac{16.76\ r^3}{22.63\ r^3} = 0.7406$ or 74.06% of the volume of each unit cell is occupied by atoms.

In a simple cubic unit cell, the atoms touch along the cube edge (l):

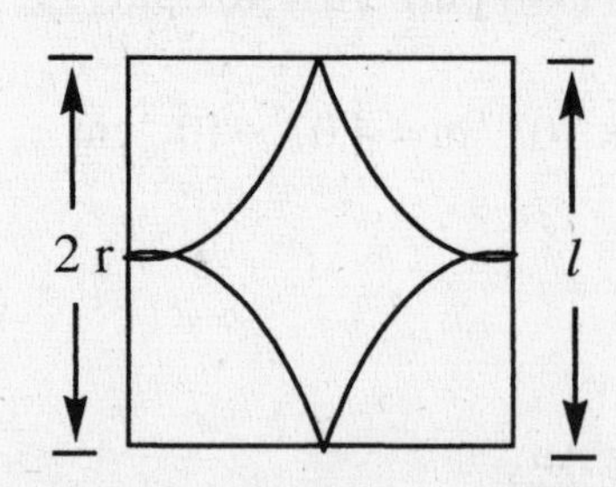

$2(\text{radius}) = 2r = l$

$V_{cube} = l^3 = (2r)^3 = 8\ r^3$

There is one atom per simple cubic cell (8 corner atoms × 1/8 atom per corner = 1 atom/unit cell). Each atom has an assumed volume of 4/3 πr^3 = volume of a sphere.

$$V_{atom} = \frac{4}{3}\pi r^3 = 4.189\ r^3$$

So, $\frac{V_{atom}}{V_{cube}} = \frac{4.189\ r^3}{8\ r^3} = = 0.5236$ or 52.36% of the volume of each unit cell is occupied by atoms.

A cubic closest packed structure packs the atoms much more efficiently than a simple cubic structure.

55. Doping silicon with phosphorus produces an n-type semiconductor. The phosphorus adds electrons at energies near the conduction band of silicon. Electrons do not need as much energy to move from filled to unfilled energy levels so conduction increases. Doping silicon with gallium produces a p-type semiconductor. Because gallium has fewer valence electrons than silicon, holes (unfilled energy levels) at energies in the previously filled molecular orbitals are created, which induces greater electron movement (greater conductivity).

57. In has fewer valence electrons than Se, thus, Se doped with In would be a p-type semiconductor.

59. $E_{gap} = 2.5\ \text{eV} \times 1.6 \times 10^{-19}\ \text{J/eV} = 4.0 \times 10^{-19}\ \text{J}$; We want $E_{gap} = E_{light}$, so:

$$\lambda = \frac{hc}{E} = \frac{(6.63\times10^{-34}\ \text{J s})(3.00\times10^{8}\ \text{m/s})}{4.0\times10^{-19}\ \text{J}} = 5.0 \times 10^{-7}\ \text{m} = 5.0 \times 10^{2}\ \text{nm}$$

61. a. $8\ \text{corners} \times \frac{1/8\ \text{Cl}}{\text{corner}} + 6\ \text{faces} \times \frac{1/2\ \text{Cl}}{\text{face}} = 4\ \text{Cl ions}$

$12\ \text{edges} \times \frac{1/4\ \text{Na}}{\text{edge}} + 1\ \text{Na at body center} = 4\ \text{Na ions}$; NaCl is the formula.

b. 1 Cs ion at body center; 8 corners $\times \dfrac{1/8\text{ Cl}}{\text{corner}}$ = 1 Cl ion; CsCl is the formula.

c. There are 4 Zn ions inside the cube.

$$8\text{ corners} \times \frac{1/8\text{ S}}{\text{corner}} + 6\text{ faces} \times \frac{1/2\text{ S}}{\text{face}} = 4\text{ S ions};\ \ ZnS\text{ is the formula.}$$

d. $8\text{ corners} \times \dfrac{1/8\text{ Ti}}{\text{corner}} + 1\text{ Ti at body center} = 2\text{ Ti ions}$

$$4\text{ faces} \times \frac{1/2\text{ O}}{\text{face}} + 2\text{ O inside cube} = 4\text{ O ions};\ \ TiO_2\text{ is the formula.}$$

63. There is one octahedral hole per closest packed anion in a closest packed structure. If half of the octahedral holes are filled, there is a 2:1 ratio of fluoride ions to cobalt ions in the crystal. The formula is CoF_2.

65. In a cubic closest packed array of anions, there are twice the number of tetrahedral holes as anions present and an equal number of octahedral holes as anions present. A cubic closest packed array of sulfide ions will have 4 S^{2-} ions, 8 tetrahedral holes, and 4 octahedral holes. In this structure we have 1/8(8) = 1 Zn^{2+} ion and 1/2(4) = 2 Al^{3+} ions present along with the 4 S^{2-} ions. The formula is $ZnAl_2S_4$.

67. 8 F^- ions at corners × 1/8 F^-/corner = 1 F^- ion per unit cell; Because there is one cubic hole per cubic unit cell, there is a 2:1 ratio of F^- ions to metal ions in the crystal if only ½ of the body centers are filled with the metal ions. The formula is MF_2 where M^{2+} is the metal ion.

69. From Fig. 10.37, MgO has the NaCl structure containing 4 Mg^{2+} ions and 4 O^{2-} ions per face-centered unit cell.

$$4\text{ MgO formula units} \times \frac{1\text{ mol MgO}}{6.022\times10^{23}\text{ atoms}} \times \frac{40.31\text{ g MgO}}{1\text{ mol MgO}} = 2.678\times10^{-22}\text{ g MgO}$$

$$\text{Volume of unit cell} = 2.678\times10^{-22}\text{ g MgO} \times \frac{1\text{ cm}^3}{3.58\text{ g}} = 7.48\times10^{-23}\text{ cm}^3$$

Volume of unit cell = l^3, l = cube edge length; $l = (7.48\times10^{-23}\text{ cm}^3)^{1/3} = 4.21\times10^{-8}$ cm

For a face-centered unit cell, the O^{2-} ions touch along the face diagonal:

$$\sqrt{2}\,l = 4r_{O^{2-}},\ \ r_{O^{2-}} = \frac{\sqrt{2}\times4.21\times10^{-8}\text{ cm}}{4} = 1.49\times10^{-8}\text{ cm}$$

The cube edge length goes through two radii of the O^{2-} anions and the diameter of the Mg^{2+} cation. So:

$$l = 2r_{O^{2-}} + 2r_{Mg^{2+}},\ \ 4.21\times10^{-8}\text{ cm} = 2(1.49\times10^{-8}\text{ cm}) + 2r_{Mg^{2+}},\ \ r_{Mg^{2+}} = 6.15\times10^{-9}\text{ cm}$$

71. a. CO_2: molecular b. SiO_2: network c. Si: atomic, network

d. CH_4: molecular e. Ru: atomic, metallic f. I_2: molecular

g. KBr: ionic h. H_2O: molecular i. NaOH: ionic

j. U: atomic, metallic k. $CaCO_3$: ionic l. PH_3: molecular

73. a. The unit cell consists of Ni at the cube corners and Ti at the body center, or Ti at the cube corners and Ni at the body center.

b. 8 × 1/8 = 1 atom from corners + 1 atom at body center; Empirical formula = NiTi

c. Both have a coordination number of 8 (both are surrounded by 8 atoms).

75. Structure 1

$$8 \text{ corners} \times \frac{1/8 \text{ Ca}}{\text{corner}} = 1 \text{ Ca atom}$$

$$6 \text{ faces} \times \frac{1/2 \text{ O}}{\text{face}} = 3 \text{ O atoms}$$

1 Ti at body center. Formula = $CaTiO_3$

Structure 2

$$8 \text{ corners} \times \frac{1/8 \text{ Ti}}{\text{corner}} = 1 \text{ Ti atom}$$

$$12 \text{ edges} \times \frac{1/4 \text{ O}}{\text{corner}} = 3 \text{ O atoms}$$

1 Ca at body center. Formula = $CaTiO_3$

In the extended lattice of both structures, each Ti atom is surrounded by six O atoms.

77. a. Y: 1 Y in center; Ba: 2 Ba in center

$$\text{Cu: } 8 \text{ corners} \times \frac{1/8 \text{ Cu}}{\text{corner}} = 1 \text{ Cu}, \ 8 \text{ edges} \times \frac{1/4 \text{ Cu}}{\text{edge}} = 2 \text{ Cu}, \ \text{total} = 3 \text{ Cu atoms}$$

$$\text{O: } 20 \text{ edges} \times \frac{1/4 \text{ O}}{\text{edge}} = 5 \text{ oxygen}, \ 8 \text{ faces} \times \frac{1/2 \text{ O}}{\text{face}} = 4 \text{ oxygen}, \ \text{total} = 9 \text{ O atoms}$$

Formula: $YBa_2Cu_3O_9$

b. The structure of this superconductor material follows the second perovskite structure described in Exercise 10.75. The $YBa_2Cu_3O_9$ structure is three of these cubic perovskite unit cells stacked on top of each other. The oxygen atoms are in the same places, Cu takes the place of Ti, two of the calcium atoms are replaced by two barium atoms, and one Ca is replaced by Y.

c. Y, Ba, and Cu are the same. Some oxygen atoms are missing.

$$12 \text{ edges} \times \frac{1/4 \text{ O}}{\text{edge}} = 3 \text{ O}, \ 8 \text{ faces} \times \frac{1/2 \text{ O}}{\text{face}} = 4 \text{ O}, \ \text{total} = 7 \text{ O atoms}$$

Superconductor formula is $YBa_2Cu_3O_7$.

Phase Changes and Phase Diagrams

79. If we graph ln P_{vap} vs 1/T, the slope of the resulting straight line will be $-\Delta H_{vap}/R$.

P_{vap}	ln P_{vap}	T (Li)	1/T	T (Mg)	1/T
1 torr	0	1023 K	9.775×10^{-4} K^{-1}	893 K	11.2×10^{-4} K^{-1}
10.	2.3	1163	8.598×10^{-4}	1013	9.872×10^{-4}
100.	4.61	1353	7.391×10^{-4}	1173	8.525×10^{-4}
400.	5.99	1513	6.609×10^{-4}	1313	7.616×10^{-4}
760.	6.63	1583	6.317×10^{-4}	1383	7.231×10^{-4}

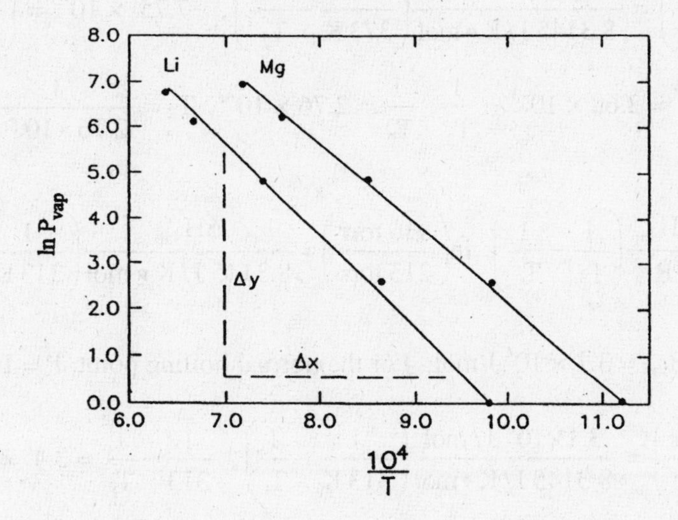

For Li:

We get the slope by taking two points (*x*, *y*) that are on the line we draw. For a line:

$$\text{slope} = \frac{\Delta y}{\Delta x} = \frac{y_2 - y_1}{x_2 - x_1}$$

or we can determine the straight line equation using a calculator. The general straight line equation is y = mx + b where m = slope and b = y-intercept.

The equation of the Li line is: $\ln P_{vap} = -1.90 \times 10^4(1/T) + 18.6$, slope = -1.90×10^4 K

Slope = $-\Delta H_{vap}/R$, $\Delta H_{vap} = -\text{slope} \times R = 1.90 \times 10^4 \text{ K} \times 8.3145$ J/K•mol

$\Delta H_{vap} = 1.58 \times 10^5$ J/mol = 158 kJ/mol

For Mg:

The equation of the line is: $\ln P_{vap} = -1.67 \times 10^4(1/T) + 18.7$, slope = -1.67×10^4 K

ΔH_{vap} = -slope × R = 1.67×10^4 K × 8.3145 J/K•mol,

$\Delta H_{vap} = 1.39 \times 10^5$ J/mol = 139 kJ/mol

The bonding is stronger in Li since ΔH_{vap} is larger for Li.

81. At 100.°C (373 K), the vapor pressure of H_2O is 1.00 atm = 760. torr.
For water, ΔH_{vap} = 40.7 kJ/mol.

$$\ln\left(\frac{P_1}{P_2}\right) = \frac{\Delta H_{vap}}{R}\left(\frac{1}{T_2} - \frac{1}{T_1}\right) \text{ or } \ln\left(\frac{P_2}{P_1}\right) = \frac{\Delta H_{vap}}{R}\left(\frac{1}{T_1} - \frac{1}{T_2}\right)$$

$$\ln\left(\frac{520.\text{ torr}}{760.\text{ torr}}\right) = \frac{40.7 \times 10^3 \text{ J/mol}}{8.3145 \text{ J/K}\bullet\text{mol}}\left(\frac{1}{373\text{ K}} - \frac{1}{T_2}\right),\ -7.75 \times 10^{-5} = \left(\frac{1}{373\text{ K}} - \frac{1}{T_2}\right)$$

$$-7.75 \times 10^{-5} = 2.68 \times 10^{-3} - \frac{1}{T_2},\ \frac{1}{T_2} = 2.76 \times 10^{-3},\ T_2 = \frac{1}{2.76 \times 10 - 3} = 362 \text{ K or } 89°\text{C}$$

83. $$\ln\left(\frac{P_1}{P_2}\right) = \frac{\Delta H_{vap}}{R}\left(\frac{1}{T_2} - \frac{1}{T_1}\right),\ \ln\left(\frac{836\text{ torr}}{213\text{ torr}}\right) = \frac{\Delta H_{vap}}{8.3145 \text{ J/K}\bullet\text{mol}}\left(\frac{1}{313\text{ K}} - \frac{1}{353\text{ K}}\right)$$

Solving: $\Delta H_{vap} = 3.1 \times 10^4$ J/mol; For the normal boiling point, P = 1.00 atm = 760. torr.

$$\ln\left(\frac{760.\text{ torr}}{213\text{ torr}}\right) = \frac{3.1 \times 10^4 \text{ J/mol}}{8.3145 \text{ J/K}\bullet\text{mol}}\left(\frac{1}{313\text{ K}} - \frac{1}{T_1}\right),\ \frac{1}{313} - \frac{1}{T_1} = 3.4 \times 10^{-4}$$

T_1 = 350. K = 77°C; The normal boiling point of CCl_4 is 77°C.

85.

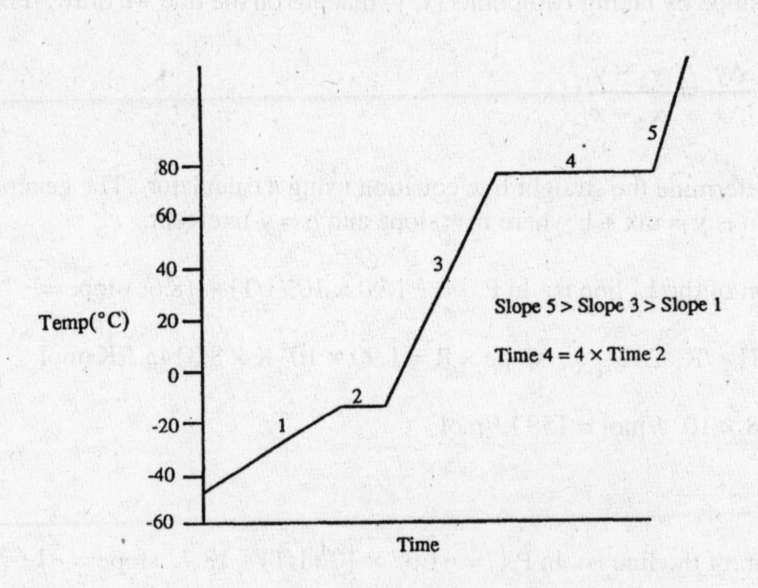

87. $H_2O(s, -20.°C) \rightarrow H_2O(s, 0°C)$, $\Delta T = 20.°C$

$$q_1 = s_{ice} \times m \times \Delta T = \frac{2.03\ J}{g\ °C} \times 5.00 \times 10^2\ g \times 20.°C = 2.0 \times 10^4\ J = 20.\ kJ$$

$H_2O(s, 0°C) \rightarrow H_2O(l, 0°C)$, $q_2 = 5.00 \times 10^2\ g\ H_2O \times \frac{1\ mol}{18.02\ g} \times \frac{6.02\ kJ}{mol} = 167\ kJ$

$H_2O(l, 0°C) \rightarrow H_2O(l, 100.°C)$, $q_3 = \frac{4.2\ J}{g\ °C} \times 5.00 \times 10^2\ g \times 100.°C = 2.1 \times 10^5\ J = 210\ kJ$

$H_2O(l, 100.°C) \rightarrow H_2O(g, 100.°C)$, $q_4 = 5.00 \times 10^2\ g \times \frac{1\ mol}{18.02\ g} \times \frac{40.7\ kJ}{mol} = 1130\ kJ$

$H_2O(g, 100.°C) \rightarrow H_2O(g, 250.°C)$, $q_5 = \frac{2.0\ J}{g\ °C} \times 5.00 \times 10^2\ g \times 150.°C = 1.5 \times 10^5\ J$

$= 150\ kJ$

$$q_{total} = q_1 + q_2 + q_3 + q_4 + q_5 = 20. + 167 + 210 + 1130 + 150 = 1680\ kJ$$

89. Total mass H_2O = 18 cubes $\times \frac{30.0\ g}{cube} = 540.\ g$; $540.\ g\ H_2O \times \frac{1\ mol\ H_2O}{18.02\ g} = 30.0\ mol\ H_2O$

Heat removed to produce ice at -5.0°C:

$$\frac{4.18\ J}{g\ °C} \times 540.\ g \times 22.0\ °C + \frac{6.02 \times 10^3\ J}{mol} \times 30.0\ mol + \frac{2.03\ J}{g\ °C} \times 540.\ g \times 5.0\ °C$$

$$4.97 \times 10^4\ J + 1.81 \times 10^5\ J + 5.5 \times 10^3\ J = 2.36 \times 10^5\ J$$

$2.36 \times 10^5\ J \times \frac{1\ g\ CF_2Cl_2}{158\ J} = 1.49 \times 10^3\ g\ CF_2Cl_2$ must be vaporized.

91. A: solid B: liquid C: vapor

D: solid + vapor E: solid + liquid + vapor

F: liquid + vapor G: liquid + vapor H: vapor

triple point: E critical point: G

normal freezing point: temperature at which solid - liquid line is at 1.0 atm (see following plot).

normal boiling point: temperature at which liquid - vapor line is at 1.0 atm (see following plot).

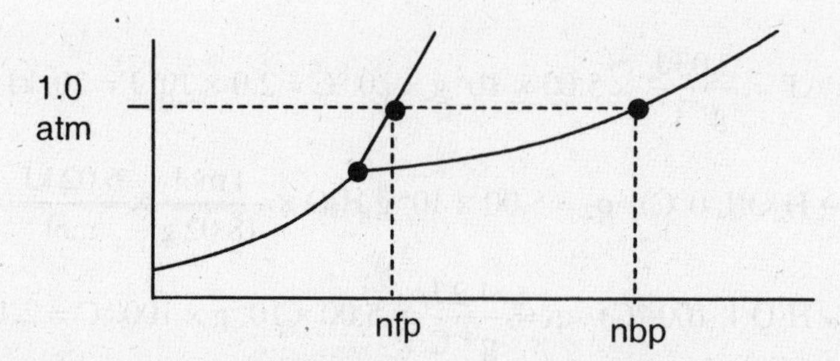

Since the solid-liquid line has a positive slope, the solid phase is denser than the liquid phase.

93. a. two

b. Higher pressure triple point: graphite, diamond and liquid; Lower pressure triple point: graphite, liquid and vapor

c. It is converted to diamond (the more dense solid form).

d. Diamond is more dense, which is why graphite can be converted to diamond by applying pressure.

95. Because the density of the liquid phase is greater than the density of the solid phase, the slope of the solid-liquid boundary line is negative (as in H_2O). With a negative slope, the melting points increase with a decrease in pressure so the normal melting point of X should be greater than 225°C.

Additional Exercises

97. Chalk is composed of the ionic compound calcium carbonate ($CaCO_3$). The electrostatic forces in ionic compounds are much stronger than the intermolecular forces in covalent compounds. Therefore, $CaCO_3$ should have a much higher boiling point than the covalent compounds found in motor oil and in H_2O. Motor oil is composed of nonpolar C–C and C–H bonds. The intermolecular forces in motor oil are therefore London dispersion forces. We generally consider these forces to be weak. However, with compounds that have large molar masses, these London dispersion forces add up significantly and can overtake the relatively strong hydrogen bonding interactions in water.

99. At any temperature, the plot tells us that substance A has a higher vapor pressure than substance B, with substance C having the lowest vapor pressure. Therefore, the substance with the weakest intermolecular forces is A, and the substance with the strongest intermolecular forces is C.

NH_3 can form hydrogen bonding interactions while the others cannot. Substance C is NH_3. The other two are nonpolar compounds with only London dispersion forces. Since CH_4 is smaller than SiH_4, CH_4 will have weaker LD forces and is substance A. Therefore, substance B is SiH_4.

101. If TiO_2 conducts electricity as a liquid, then it is an ionic solid; if not, then TiO_2 is a network solid.

103. B_2H_6: This compound contains only nonmetals so it is probably a molecular solid with covalent bonding. The low boiling point confirms this.

SiO_2: This is the empirical formula for quartz, which is a network solid.

CsI: This is a metal bonded to a nonmetal, which generally form ionic solids. The electrical conductivity in aqueous solution confirms this.

W: Tungsten is a metallic solid as the conductivity data confirms.

105. $1.00 \text{ lb} \times \frac{454 \text{ g}}{\text{lb}} = 454 \text{ g } H_2O$; A change of 1.00°F is equal to a change of $\frac{5}{9}$ °C.

The amount of heat in J in 1 Btu is: $\frac{4.18 \text{ J}}{\text{g °C}} \times 454 \text{ g} \times \frac{5}{9} \text{ °C} = 1.05 \times 10^3 \text{ J} = 1.05 \text{ kJ}$

It takes 40.7 kJ to vaporize 1 mol H_2O (ΔH_{vap}). Combining these:

$$\frac{1.00 \times 10^4 \text{ Bu}}{\text{hr}} \times \frac{1.05 \text{ kJ}}{\text{Btu}} \times \frac{1 \text{ mol } H_2O}{40.7 \text{ kJ}} = 258 \text{ mol/hr}$$

or: $$\frac{258 \text{ mol}}{\text{hr}} \times \frac{18.02 \text{ g } H_2O}{\text{mol}} = 4650 \text{ g/hr} = 4.65 \text{ kg/hr}$$

Challenge Problems

107. $\Delta H = q_p = 30.79 \text{ kJ}$; $\Delta E = q_p + w$, $w = -P\Delta V$

$$w = -P\Delta V = -1.00 \text{ atm } (28.90 \text{ L}) = -28.9 \text{ L atm} \times \frac{101.3 \text{ J}}{\text{L atm}} = -2930 \text{ J}$$

$$\Delta E = 30.79 \text{ kJ} + (-2.93 \text{ kJ}) = 27.86 \text{ kJ}$$

109. A single hydrogen bond in H_2O has a strength of 21 kJ/mol. Each H_2O molecule forms two H bonds. Thus, it should take 42 kJ/mol of energy to break all of the H bonds in water. Consider the phase transitions:

$$\text{solid} \xrightarrow{6.0 \text{ kJ}} \text{liquid} \xrightarrow{40.7 \text{ kJ}} \text{vapor} \qquad \Delta H_{sub} = \Delta H_{fus} + \Delta H_{vap}$$

It takes a total of 46.7 kJ/mol to convert solid H_2O to vapor (ΔH_{sub}). This would be the amount of energy necessary to disrupt all of the intermolecular forces in ice. Thus, $(42 \div 46.7) \times 100 = 90\%$ of the attraction in ice can be attributed to H bonding.

111. $NaCl$, $MgCl_2$, NaF, MgF_2 AlF_3 all have very high melting points indicative of strong intermolecular forces. They are all ionic solids. $SiCl_4$, SiF_4, F_2, Cl_2, PF_5 and SF_6 are nonpolar covalent molecules. Only LD forces are present. PCl_3 and SCl_2 are polar molecules. LD forces and dipole forces are present. In these 8 molecular substances, the intermolecular forces are weak and the melting points low. $AlCl_3$ doesn't seem to fit in as well. From the melting point, there are much stronger forces present than in the nonmetal halides, but they aren't as strong as we would expect for an ionic solid. $AlCl_3$ illustrates a gradual transition from ionic to covalent bonding, from an ionic solid to discrete molecules.

113. Out of 100.00 g: $28.31\text{ g O} \times \frac{1\text{ mol}}{16.00\text{ g}} = 1.769\text{ mol O}$; $71.69\text{ g Ti} \times \frac{1\text{ mol}}{47.88\text{ g}} = 1.497\text{ mol Ti}$

$\frac{1.769}{1.497} = 1.182$; $\frac{1.497}{1.769} = 0.8462$; The formula is $TiO_{1.182}$ or $Ti_{0.8462}O$.

For $Ti_{0.8462}O$, let $x = Ti^{2+}$ per mol O^{2-} and $y = Ti^{3+}$ per mol O^{2-}. Setting up two equations and solving:

$$x + y = 0.8462 \text{ (mass balance) and } 2x + 3y = 2 \text{ (charge balance)}; \quad 2x + 3(0.8462 - x) = 2$$

$$x = 0.539 \text{ mol Ti}^{2+}/\text{mol O}^{2-} \text{ and } y = 0.307 \text{ mol Ti}^{3+}/\text{mol O}^{2-}$$

$\frac{0.539}{0.8462} \times 100 = 63.7\%$ of the titanium ions are Ti^{2+} and 36.3% are Ti^{3+} (a 1.75:1 ion ratio).

115. $$\frac{\text{density}_{Mn}}{\text{density}_{Cu}} = \frac{\text{mass}_{Mn} \times \text{volume}_{Cu}}{\text{volume}_{Mn} \times \text{mass}_{Cu}} = \frac{\text{mass}_{Mn}}{\text{mass}_{Cu}} \times \frac{\text{volume}_{Cu}}{\text{volume}_{Mn}}$$

The type of cubic cell formed is not important; only that Cu and Mn crystallize in the same type of cubic unit cell is important. Each cubic unit cell has a specific relationship between the cube edge length, l, and the radius, r. In all cases $l \propto r$. Therefore, $V \propto l^3 \propto r^3$. For the mass ratio, we can use the molar masses of Mn and Cu since each unit cell must contain the same number of Mn and Cu atoms. Solving:

$$\frac{\text{density}_{Mn}}{\text{density}_{Cu}} = \frac{\text{mass}_{Mn}}{\text{mass}_{Cu}} \times \frac{\text{volume}_{Cu}}{\text{volume}_{Mn}} = \frac{54.94\text{ g/mol}}{63.55\text{ g/mol}} \times \frac{(r_{Cu})^3}{(1.056\, r_{Cu})^3}$$

$$\frac{\text{density}_{Mn}}{\text{density}_{Cu}} = 0.8645 \times \left(\frac{1}{1.056}\right)^3 = 0.7341$$

$$\text{density}_{Mn} = 0.7341 \times \text{density}_{Cu} = 0.7341 \times 8.96\text{ g/cm}^3 = 6.58\text{ g/cm}^3$$

117.

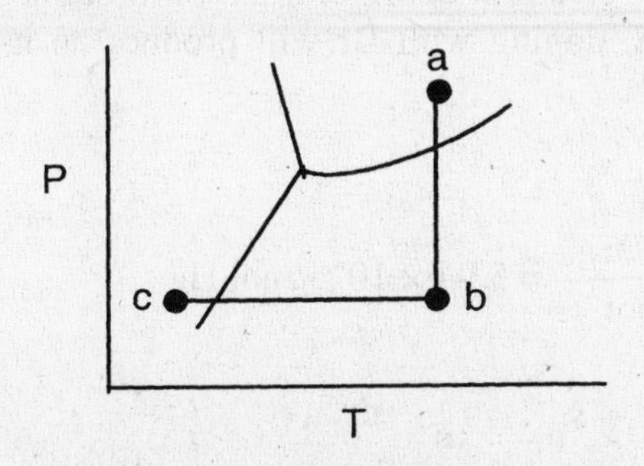

As P is lowered, we go from a to b on the phase diagram. The water boils. The boiling of water is endothermic and the water is cooled (b →c), forming some ice. If the pump is left on, the ice will sublime until none is left. This is the basis of freeze drying.

119. For a cube: (body diagonal)^2 = (face diagonal)^2 + (cube edge length)^2

In a simple cubic structure, the atoms touch on cube edge so the cube edge = 2 r where r = radius of sphere.

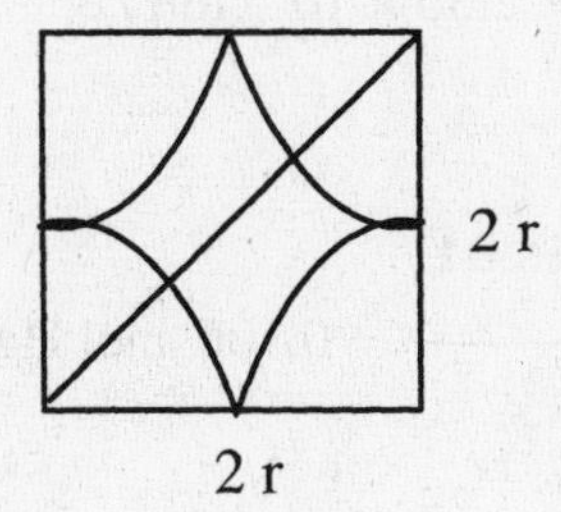

face diagonal = $\sqrt{(2\,r)^2 + (2\,r)^2} = \sqrt{4\,r^2 + 4\,r^2} = r\sqrt{8} = 2\sqrt{2}\ r$

body diagonal = $\sqrt{(2\sqrt{2}\ r)^2 + (2\,r)^2} = \sqrt{12\,r^2} = 2\sqrt{3}\ r$

The diameter of the hole = body diagonal – 2 radius of atoms at corners.

diameter = $2\sqrt{3}\ r - 2\,r$; Thus, the radius of the hole is: $\dfrac{2\sqrt{3}\,r - 2\,r}{2} = \left(\dfrac{2\sqrt{3} - 2}{2}\right) r$

The volume of the hole is: $\dfrac{4}{3}\pi\left[\left(\dfrac{2\sqrt{3} - 2}{2}\right) r\right]^3$

Integrative Problems

121. molar mass XY = $\dfrac{19.0\text{ g}}{0.132\text{ mol}}$ = 144 g/mol

X: [Kr] $5s^2 4d^{10}$; This is cadmium, Cd.

molar mass Y = 144 – 112.4 = 32 g/mol; Y is sulfur, S.

The semiconductor is CdS. The dopant has the electron configuration of bromine, Br. Because Br has one more valence electron than S, doping with Br will produce an n-type semiconductor.

123. $\ln\left(\frac{P_1}{P_2}\right) = \frac{\Delta H_{vap}}{R}\left(\frac{1}{T_2} - \frac{1}{T_1}\right)$; $\Delta H_{vap} = \frac{296\ J}{g} \times \frac{200.6\ g}{mol} = 5.94 \times 10^4$ J/mol Hg

$$\ln\left(\frac{2.56\times10^{-3}\ \text{torr}}{P_2}\right) = \frac{5.94\times10^4\ \text{J/mol}}{8.3145\times10^4\ \text{J/K} \bullet \text{mol}}\left(\frac{1}{573\ \text{K}} - \frac{1}{298.2\ \text{K}}\right)$$

$$\ln\left(\frac{2.56\times10^{-3}\ \text{torr}}{P_2}\right) = -11.5,\quad P_2 = 2.56 \times 10^{-3}\ \text{torr}/e^{-11.5} = 253\ \text{torr}$$

$$n = \frac{PV}{RT} = \frac{\left(253\ \text{torr} \times \frac{1\ \text{atm}}{760\ \text{torr}}\right) \times 15.0\ \text{L}}{\frac{0.08206\ \text{L atm}}{\text{K mol}} \times 573\ \text{K}} = 0.106\ \text{mol Hg}$$

$$0.106\ \text{mol Hg} \times \frac{6.022\times10^{23}\ \text{atoms Hg}}{\text{mol Hg}} = 6.38 \times 10^{22}\ \text{atoms Hg}$$

CHAPTER ELEVEN

PROPERTIES OF SOLUTIONS

Solution Review

9. $$\frac{585\text{ g } C_3H_7OH \times \frac{1\text{ mol } C_3H_7OH}{60.09\text{ g } C_3H_7OH}}{1.00\text{ L}} = 9.74\ M$$

11. $$\text{mol } Na_2CO_3 = 0.0700\text{ L} \times \frac{3.0\text{ mol } Na_2CO_3}{\text{L}} = 0.21\text{ mol } Na_2CO_3$$

$$Na_2CO_3(s) \rightarrow 2\ Na^+(aq) + CO_3^{2-}(aq);\ \ \text{mol } Na^+ = 2(0.21) = 0.42\text{ mol}$$

$$\text{mol } NaHCO_3 = 0.0300\text{ L} \times \frac{1.0\text{ mol } NaHCO_3}{\text{L}} = 0.030\text{ mol } NaHCO_3$$

$$NaHCO_3(s) \rightarrow Na^+(aq) + HCO_3^-(aq);\ \ \text{mol } Na^+ = 0.030\text{ mol}$$

$$M_{Na^+} = \frac{\text{total mol } Na^+}{\text{total volume}} = \frac{0.42\text{ mol} + 0.030\text{ mol}}{0.0700\text{ L} + 0.030\text{ L}} = \frac{0.45\text{ mol}}{0.1000\text{ L}} = 4.5\ M\ Na^+$$

Questions

13. As the temperature increases, the gas molecules will have a greater average kinetic energy. A greater fraction of the gas molecules in solution will have kinetic energy greater than the attractive forces between the gas molecules and the solvent molecules. More gas molecules will escape to the vapor phase and the solubility of the gas will decrease.

15 Because the solute is volatile, both the water and solute will transfer back and forth between the two beakers. The volume in each beaker will become constant when the concentrations of solute in the beakers are equal to each other. Because the solute is less volatile than water, one would expect there to be a larger net transfer of water molecules into the right beaker than the net transfer of solute molecules into the left beaker. This results in a larger solution volume in the right beaker when equilibrium is reached, i.e., when the solute concentration is identical in each beaker.

17. No, the solution is not ideal. For an ideal solution, the strength of intermolecular forces in solution is the same as in pure solute and pure solvent. This results in $\Delta H_{soln} = 0$ for an ideal solution. ΔH_{soln} for methanol/water is not zero. Because $\Delta H_{soln} < 0$ (heat is released), this solution shows a negative deviation from Raoult's law.

19. Normality is the number of equivalents per liter of solution. For an acid or a base, an equivalent is the mass of acid or base that can furnish 1 mol of protons (if an acid) or accept 1 mol of protons (if a base). A proton is an H^+ ion. Molarity is defined as the moles of solute per liter of solution. When the number of equivalents equals the number of moles of solute, then normality = molarity. This is true for acids which only have one acidic proton in them and for bases that accept only one proton per formula unit. Examples of acids where equivalents = moles solute are HCl, HNO_3, HF, and $HC_2H_3O_2$. Examples of bases where equivalents = moles solute are NaOH, KOH, and NH_3. When equivalents ≠ moles solute, then normality ≠ molarity. This is true for acids that donate more than one proton (H_2SO_4, H_3PO_4, H_2CO_3, etc.) and for bases that react with more than one proton per formula unit [$Ca(OH)_2$, $Ba(OH)_2$, $Sr(OH)_2$, etc.].

21. Only statement b is true. A substance freezes when the vapor pressure of the liquid and solid are the same. When a solute is added to water, the vapor pressure of the solution at 0°C is less than the vapor pressure of the solid and the net result is for any ice present to convert to liquid in order to try to equalize the vapor pressures (which never can occur at 0°C). A lower temperature is needed to equalize the vapor pressure of water and ice, hence the freezing point is depressed.

For statement a, the vapor pressure of a solution is directly related to the mole fraction of solvent (not solute) by Raoult's law. For statement c, colligative properties depend on the number of solute particles present and not on the identity of the solute. For statement d, the boiling point of water is increased because the sugar solute decreases the vapor pressure of the water; a higher temperature is required for the vapor pressure of the solution to equal the external pressure so boiling can occur.

23. Isotonic solutions are those which have identical osmotic pressures. Crenation and hemolysis refer to phenomena that occur when red blood cells are bathed in solutions having a mismatch in osmotic pressures inside and outside the cell. When red blood cells are in a solution having a higher osmotic pressure than that of the cells, the cells shrivel as there is a net transfer of water out of the cells. This is called crenation. Hemolysis occurs when the red blood cells are bathed in a solution having lower osmotic pressure than that inside the cell. Here, the cells rupture as there is a net transfer of water to into the red blood cells.

Exercises

Concentration of Solutions

25. Because the density of water is 1.00 g/mL, 100.0 mL of water has a mass of 100. g.

$$\text{density} = \frac{\text{mass}}{\text{volume}} = \frac{10.0\text{ g } H_3PO_4 + 100.\text{ g } H_2O}{104\text{ mL}} = 1.06\text{ g/mL} = 1.06\text{ g/cm}^3$$

$$\text{mol } H_3PO_4 = 10.0 \text{ g} \times \frac{1 \text{ mol}}{97.99 \text{ g}} = 0.102 \text{ mol } H_3PO_4$$

$$\text{mol } H_2O = 100. \text{ g} \times \frac{1 \text{ mol}}{18.02 \text{ g}} = 5.55 \text{ mol } H_2O$$

$$\text{mole fraction of } H_3PO_4 = \frac{0.102 \text{ mol } H_3PO_4}{(0.102 + 5.55) \text{ mol}} = 0.0180$$

$$\chi_{H_2O} = 1.000 - 0.0180 = 0.9820$$

$$\text{molarity} = \frac{0.102 \text{ mol } H_3PO_4}{0.104 \text{ L}} = 0.981 \text{ mol/L}$$

$$\text{molality} = \frac{0.102 \text{ mol } H_3PO_4}{0.100 \text{ kg}} = 1.02 \text{ mol/kg}$$

27. Hydrochloric acid:

$$\text{molarity} = \frac{38 \text{ g HCl}}{100. \text{ g soln}} \times \frac{1.19 \text{ g soln}}{\text{cm}^3 \text{ soln}} \times \frac{1000 \text{ cm}^3}{\text{L}} \times \frac{1 \text{ mol HCl}}{36.46 \text{ g}} = 12 \text{ mol/L}$$

$$\text{molality} = \frac{38 \text{ g HCl}}{62 \text{ g solvent}} \times \frac{1000 \text{ g}}{\text{kg}} \times \frac{1 \text{ mol HCl}}{36.46 \text{ g}} = 17 \text{ mol/kg}$$

$$38 \text{ g HCl} \times \frac{1 \text{ mol}}{36.46 \text{ g}} = 1.0 \text{ mol HCl}; \;\; 62 \text{ g } H_2O \times \frac{1 \text{ mol}}{18.02 \text{ g}} = 3.4 \text{ mol } H_2O$$

$$\text{mole fraction of HCl} = \chi_{HCl} = \frac{1.0}{3.4 + 1.0} = 0.23$$

Nitric acid:

$$\frac{70. \text{ g } HNO_3}{100. \text{ g soln}} \times \frac{1.42 \text{ g soln}}{\text{cm}^3 \text{ soln}} \times \frac{1000 \text{ cm}^3}{\text{L}} \times \frac{1 \text{ mol } HNO_3}{63.02 \text{ g}} = 16 \text{ mol/L}$$

$$\frac{70. \text{ g } HNO_3}{30. \text{ g solvent}} \times \frac{1000 \text{ g}}{\text{kg}} \times \frac{1 \text{ mol } HNO_3}{63.02 \text{ g}} = 37 \text{ mol/kg}$$

$$70. \text{ g } HNO_3 \times \frac{1 \text{ mol}}{63.02 \text{ g}} = 1.1 \text{ mol } HNO_3; \;\; 30. \text{ g } H_2O \times \frac{1 \text{ mol}}{18.02 \text{ g}} = 1.7 \text{ mol } H_2O$$

$$\chi_{HNO_3} = \frac{1.1}{1.7 + 1.1} = 0.39$$

Sulfuric acid:

$$\frac{95\text{ g H}_2\text{SO}_4}{100.\text{ g soln}} \times \frac{1.84\text{ g soln}}{\text{cm}^3\text{ soln}} \times \frac{1000\text{ cm}^3}{\text{L}} \times \frac{1\text{ mol H}_2\text{SO}_4}{98.09\text{ g H}_2\text{SO}_4} = 18\text{ mol/L}$$

$$\frac{95\text{ g H}_2\text{SO}_4}{5\text{ g H}_2\text{O}} \times \frac{1000\text{ g}}{\text{kg}} \times \frac{1\text{ mol}}{98.09\text{ g}} = 194\text{ mol/kg} \approx 200\text{ mol/kg}$$

$$95\text{ g H}_2\text{SO}_4 \times \frac{1\text{ mol}}{98.09\text{ g}} = 0.97\text{ mol H}_2\text{SO}_4;\ \ 5\text{ g H}_2\text{O} \times \frac{1\text{ mol}}{18.02\text{ g}} = 0.3\text{ mol H}_2\text{O}$$

$$\chi_{\text{H}_2\text{SO}_4} = \frac{0.97}{0.97 + 0.3} = 0.76$$

Acetic Acid:

$$\frac{99\text{ g HC}_2\text{H}_3\text{O}_2}{100.\text{ g soln}} \times \frac{1.05\text{ g soln}}{\text{cm}^3\text{ soln}} \times \frac{1000\text{ cm}^3}{\text{L}} \times \frac{1\text{ mol}}{60.05\text{ g H}} = 17\text{ mol/L}$$

$$\frac{99\text{ g HC}_2\text{H}_3\text{O}_2}{1\text{ g H}_2\text{O}} \times \frac{1000\text{ g}}{\text{kg}} \times \frac{1\text{ mol}}{60.05\text{ g}} = 1600\text{ mol/kg} \approx 2000\text{ mol/kg}$$

$$99\text{ g HC}_2\text{H}_3\text{O}_2 \times \frac{1\text{ mol}}{60.05\text{ g}} = 1.6\text{ mol HC}_2\text{H}_3\text{O}_2;\ 1\text{ g H}_2\text{O} \times \frac{1\text{ mol}}{18.02\text{ g}} = 0.06\text{ mol H}_2\text{O}$$

$$\chi_{\text{HC}_2\text{H}_3\text{O}_2} = \frac{1.6}{1.6 + 0.06} = 0.96$$

Ammonia:

$$\frac{28\text{ g NH}_3}{100.\text{ g soln}} \times \frac{0.90\text{ g}}{\text{cm}^3} \times \frac{1000\text{ cm}^3}{\text{L}} \times \frac{1\text{ mol}}{17.03\text{ g}} = 15\text{ mol/L}$$

$$\frac{28\text{ g NH}_3}{72\text{ g H}_2\text{O}} \times \frac{1000\text{ g}}{\text{kg}} \times \frac{1\text{ mol}}{17.03\text{ g}} = 23\text{ mol/kg}$$

$$28\text{ g NH}_3 \times \frac{1\text{ mol}}{17.03\text{ g}} = 1.6\text{ mol NH}_3;\ \ 72\text{ g H}_2\text{O} \times \frac{1\text{ mol}}{18.02\text{ g}} = 4.0\text{ mol H}_2\text{O}$$

$$\chi_{\text{NH}_3} = \frac{1.6}{4.0 + 1.6} = 0.29$$

29. $$25\text{ mL C}_5\text{H}_{12} \times \frac{0.63\text{ g}}{\text{mL}} = 16\text{ g C}_5\text{H}_{12};\ 25\text{ mL} \times \frac{0.63\text{ g}}{\text{mL}} \times \frac{1\text{ mol}}{72.15\text{ g}} = 0.22\text{ mol C}_5\text{H}_{12}$$

$$45\text{ mL C}_6\text{H}_{14} \times \frac{0.66\text{ g}}{\text{mL}} = 30.\text{ g C}_6\text{H}_{14};\ 45\text{ mL} \times \frac{0.66\text{ g}}{\text{mL}} \times \frac{1\text{ mol}}{86.17\text{ g}} = 0.34\text{ mol C}_6\text{H}_{14}$$

$$\text{mass \% pentane} = \frac{\text{mass pentane}}{\text{total mass}} \times 100 = \frac{16\text{ g}}{16\text{ g} + 30.\text{ g}} = 35\%$$

$$\chi_{\text{pentane}} = \frac{\text{mol pentane}}{\text{total mol}} = \frac{0.22\text{ mol}}{0.22\text{ mol} + 0.34\text{ mol}} = 0.39$$

$$\text{molality} = \frac{\text{mol pentane}}{\text{kg hexane}} = \frac{0.22\text{ mol}}{0.030\text{ kg}} = = 7.3\text{ mol/kg}$$

$$\text{molarity} = \frac{\text{mol pentane}}{\text{L solution}} = \frac{0.22\text{ mol}}{25\text{ mL} + 45\text{ mL}} \times \frac{1000\text{ mL}}{1\text{ L}} = 3.1\text{ mol/L}$$

31. If we have 1.00 L of solution:

$$1.37\text{ mol citric acid} \times \frac{192.12\text{ g}}{\text{mol}} = 263\text{ g citric acid } (H_3C_6H_5O_7)$$

$$1.00 \times 10^3\text{ mL solution} \times \frac{1.10\text{ g}}{\text{mL}} = 1.10 \times 10^3\text{ g solution}$$

$$\text{mass \% of citric acid} = \frac{263\text{ g}}{1.10 \times 10^3\text{ g}} \times 100 = 23.9\%$$

In 1.00 L of solution, we have 263 g citric acid and $(1.10 \times 10^3 - 263) = 840$ g of H_2O.

$$\text{molality} = \frac{1.37\text{ mol citric acid}}{0.84\text{ kg } H_2O} = 1.6\text{ mol/kg}$$

$$840\text{ g } H_2O \times \frac{1\text{ mol}}{18.02\text{ g}} = 47\text{ mol } H_2O;\ \chi_{\text{citric acid}} = \frac{1.37}{47 + 1.37} = 0.028$$

Since citric acid is a triprotic acid, the number of protons citric acid can provide is three times the molarity. Therefore, normality = 3 × molarity:

normality = $3 \times 1.37\ M = 4.11\ N$

Energetics of Solutions and Solubility

33. Using Hess's law:

$NaI(s) \rightarrow Na^+(g) + I^-(g)$	$\Delta H = -\Delta H_{LE} = -(-686\text{ kJ/mol})$
$Na^+(g) + I^-(g) \rightarrow Na^+(aq) + I^-(aq)$	$\Delta H = \Delta H_{hyd} = -694\text{ kJ/mol}$
$NaI(s) \rightarrow Na^+(aq) + I^-(aq)$	$\Delta H_{soln} = -8\text{ kJ/mol}$

ΔH_{soln} refers to the heat released or gained when a solute dissolves in a solvent. Here, an ionic compound dissolves in water.

35. Both $Al(OH)_3$ and NaOH are ionic compounds. Since the lattice energy is proportional to the charge of the ions, the lattice energy of aluminum hydroxide is greater than that of sodium hydroxide. The attraction of water molecules for Al^{3+} and OH^- cannot overcome the larger lattice energy and $Al(OH)_3$ is insoluble. For NaOH, the favorable hydration energy is large enough to overcome the smaller lattice energy and NaOH is soluble.

37. Water is a polar solvent and dissolves polar solutes and ionic solutes. Carbon tetrachloride (CCl_4) is a nonpolar solvent and dissolves nonpolar solutes (like dissolves like). To predict the polarity of the following molecules, draw the correct Lewis structure and then determine if the individual bond dipoles cancel or not. If the bond dipoles are arranged in such a manner that they cancel each other out, then the molecule is nonpolar. If the bond dipoles do not cancel each other out, then the molecule is polar.

a. KrF_2, $8 + 2(7) = 22\ e^-$

nonpolar; soluble in CCl_4

b. SF_2, $6 + 2(7) = 20\ e^-$

polar; soluble in H_2O

c. SO_2, $6 + 2(6) = 18\ e^-$

+ 1 more

polar; soluble in H_2O

d. CO_2, $4 + 2(6) = 16\ e^-$

nonpolar; soluble in CCl_4

e. MgF_2 is an ionic compound so it is soluble in water.

f. CH_2O, $4 + 2(1) + 6 = 12\ e^-$

polar; soluble in H_2O

g. C_2H_4, $2(4) + 4(1) = 12\ e^-$

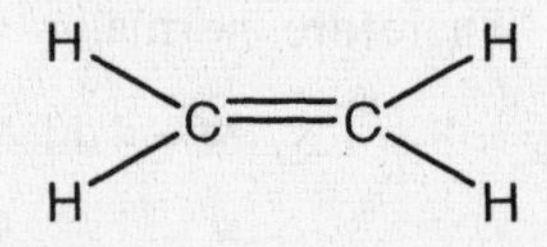

nonpolar (like all compounds made up of only carbon and hydrogen); soluble in CCl_4

39. Water is a polar molecule capable of hydrogen bonding. Polar molecules, especially molecules capable of hydrogen bonding, and ions are all attracted to water. For covalent compounds, as polarity increases, the attraction to water increases. For ionic compounds, as the charge of the ions increases and/or the size of the ions decreases, the attraction to water increases.

a. CH_3CH_2OH; CH_3CH_2OH is polar while $CH_3CH_2CH_3$ is nonpolar.

b. $CHCl_3$; $CHCl_3$ is polar while CCl_4 is nonpolar.

c. CH_3CH_2OH; CH_3CH_2OH is much more polar than $CH_3(CH_2)_{14}CH_2OH$.

41. As the length of the hydrocarbon chain increases, the solubility decreases. The –OH end of the alcohols can hydrogen bond with water. The hydrocarbon chain, however, is basically nonpolar and interacts poorly with water. As the hydrocarbon chain gets longer, a greater portion of the molecule cannot interact with the water molecules and the solubility decreases, i.e., the effect of the –OH group decreases as the alcohols get larger.

43. $C = kP, \quad \frac{8.21 \times 10^{-4}\ \text{mol}}{\text{L}} = k \times 0.790\ \text{atm}, \quad k = 1.04 \times 10^{-3}\ \text{mol/L•atm}$

$$C = kP, \ C = \frac{1.04 \times 10^{-4}\ \text{mol}}{\text{L atm}} \times 1.10\ \text{atm} = 1.14 \times 10^{-3}\ \text{mol/L}$$

Vapor Pressures of Solutions

45. $\text{mol } C_3H_8O_3 = 164\ \text{g} \times \frac{1\ \text{mol}}{92.09\ \text{g}} = 1.78\ \text{mol } C_3H_8O_3$

$$\text{mol } H_2O = 338\ \text{mL} \times \frac{0.992\ \text{g}}{\text{mL}} \times \frac{1\ \text{mol}}{18.02\ \text{g}} = 18.6\ \text{mol } H_2O$$

$$P_{H_2O} = \chi_{H_2O} P^{\circ}_{H_2O} = \frac{18.6\ \text{mol}}{(1.78 + 18.6)\ \text{mol}} \times 54.74\ \text{torr} = 0.913 \times 54.74\ \text{torr} = 50.0\ \text{torr}$$

47. $P_B = \chi_B P^{\circ}_B, \ \chi_B = P_B / P^{\circ}_B = 0.900\ \text{atm}/0.930\ \text{atm} = 0.968$

$$0.968 = \frac{\text{mol benzene}}{\text{total mol}}; \ \text{mol benzene} = 78.11\ \text{g } C_6H_6 \times \frac{1\ \text{mol}}{78.11\ \text{g}} = 1.000\ \text{mol}$$

Let x = mol solute, then: $\chi_B = 0.968 = \frac{1.000\ \text{mol}}{1.000 + x}$, $0.968 + 0.968\,x = 1.000$, $x = 0.033$ mol

$$\text{molar mass} = \frac{10.0\ \text{g}}{0.033\ \text{mol}} = 303\ \text{g/mol} \approx 3.0 \times 10^2\ \text{g/mol}$$

49. a. $25\ \text{mL } C_5H_{12} \times \frac{0.63\ \text{g}}{\text{mL}} \times \frac{1\ \text{mol}}{72.15\ \text{g}} = 0.22\ \text{mol } C_5H_{12}$

$$45\ \text{mL } C_6H_{14} \times \frac{0.66\ \text{g}}{\text{mL}} \times \frac{1\ \text{mol}}{86.17\ \text{g}} = 0.34\ \text{mol } C_6H_{14}; \ \text{total mol} = 0.22 + 0.34 = 0.56\ \text{mol}$$

$$\chi^L_{pen} = \frac{\text{mol pentane in solution}}{\text{total mol in solution}} = \frac{0.22\ \text{mol}}{0.56\ \text{mol}} = 0.39, \ \chi^L_{hex} = 1.00 - 0.39 = 0.61$$

$$P_{pen} = \chi^L_{pen} P^o_{pen} = 0.39(511 \text{ torr}) = 2.0 \times 10^2 \text{ torr}; \quad P_{hex} = 0.61(150. \text{ torr}) = 92 \text{ torr}$$

$$P_{total} = P_{pen} + P_{hex} = 2.0 \times 10^2 + 92 = 292 \text{ torr} = 290 \text{ torr}$$

b. From Chapter 5 on gases, the partial pressure of a gas is proportional to the number of moles of gas present. For the vapor phase:

$$\chi^V_{pen} = \frac{\text{mol pentane in vapor}}{\text{total mol vapor}} = \frac{P_{pen}}{P_{total}} = \frac{2.0 \times 10^2 \text{ torr}}{290 \text{ torr}} = 0.69$$

Note: In the Solutions Guide, we have added V or L to the mole fraction symbol to emphasize the value for which we are solving. If the L or V is omitted, then the liquid phase is assumed.

51. $P_{total} = P_{meth} + P_{prop}$, $174 \text{ torr} = \chi^L_{meth}(303 \text{ torr}) + \chi^L_{prop}(44.6 \text{ torr})$; $\chi^L_{prop} = 1.000 - \chi^L_{meth}$

$$174 = 303\,\chi^L_{meth} + (1.000 - \chi^L_{meth})\,44.6, \quad \frac{129}{258} = \chi^L_{meth} = 0.500; \quad \chi^L_{prop} = 1.000 - 0.500 = 0.500$$

53. Compared to H_2O, solution d (methanol/water) will have the highest vapor pressure because methanol is more volatile than water. Both solution b (glucose/water) and solution c (NaCl/water) will have a lower vapor pressure than water by Raoult's law. NaCl dissolves to give Na^+ ions and Cl^- ions; glucose is a nonelectrolyte. Since there are more solute particles in solution c, the vapor pressure of solution c will be the lowest.

55. $50.0 \text{ g } CH_3COCH_3 \times \frac{1 \text{ mol}}{58.08 \text{ g}} = 0.861 \text{ mol acetone}$

$$50.0 \text{ g } CH_3OH \times \frac{1 \text{ mol}}{32.04 \text{ g}} = 1.56 \text{ mol methanol}$$

$$\chi^L_{acetone} = \frac{0.861}{0.861 + 1.56} = 0.356; \quad \chi^L_{methanol} = 1.000 - \chi^L_{acetone} = 0.644$$

$$P_{total} = P_{methanol} + P_{acetone} = 0.644(143 \text{ torr}) + 0.356(271 \text{ torr}) = 92.1 \text{ torr} + 96.5 \text{ torr} = 188.6 \text{ torr}$$

Because partial pressures are proportional to the moles of gas present, then in the vapor phase:

$$\chi^V_{acetone} = \frac{P_{acetone}}{P_{total}} = \frac{96.5 \text{ torr}}{188.6 \text{ torr}} = 0.512; \quad \chi^V_{methanol} = 1.000 - 0.512 = 0.488$$

The actual vapor pressure of the solution (161 torr) is less than the calculated pressure assuming ideal behavior (188.6 torr). Therefore, the solution exhibits negative deviations from Raoult's law. This occurs when the solute-solvent interactions are stronger than in pure solute and pure solvent.

Colligative Properties

57. $\text{molality} = m = \dfrac{\text{mol solute}}{\text{kg solvent}} = \dfrac{27.0\text{ g } N_2H_4CO}{150.0\text{ g } H_2O} \times \dfrac{1000\text{ g}}{\text{kg}} \times \dfrac{1\text{ mol } N_2H_4CO}{60.06\text{ g } N_2H_4CO} = 3.00\text{ molal}$

$\Delta T_b = K_b m = \dfrac{0.51\,°C}{\text{molal}} \times 3.00\text{ molal} = 1.5°C$

The boiling point is raised from 100.0°C to 101.5°C (assuming P = 1 atm).

59. $\Delta T_f = K_f m,\ \Delta T_f = 1.50°C = \dfrac{1.86\,°C}{\text{molal}} \times m,\ m = 0.806\text{ mol/kg}$

$0.200\text{ kg } H_2O \times \dfrac{0.806\text{ mol } C_3H_8O_3}{\text{kg } H_2O} \times \dfrac{92.09\text{ g } C_3H_8O_3}{\text{mol } C_3H_8O_3} = 14.8\text{ g } C_3H_8O_3$

61. $\text{molality} = m = \dfrac{50.0\text{ g } C_2H_6O_2}{50.0\text{ g } H_2O} \times \dfrac{1000\text{ g}}{\text{kg}} \times \dfrac{1\text{ mol}}{62.07\text{ g}} = 16.1\text{ mol/kg}$

$\Delta T_f = K_f m = 1.86°C/\text{molal} \times 16.1\text{ molal} = 29.9°C;\ T_f = 0.0°C - 29.9°C = -29.9°C$

$\Delta T_b = K_b m = 0.51°C/\text{molal} \times 16.1\text{ molal} = 8.2°C;\ T_b = 100.0°C + 8.2°C = 108.2°C$

63. $\Delta T_f = K_f m,\ m = \dfrac{\Delta T_f}{K_f} = \dfrac{0.300°C}{5.12\,°C\text{ kg/mol}} = \dfrac{5.86 \times 10^{-2}\text{ mol thyroxine}}{\text{kg benzene}}$

The mol of thyroxine present is:

$0.0100\text{ kg benzene} \times \dfrac{5.86 \times 10^{-2}\text{ mol thyroxine}}{\text{kg benzene}} = 5.86 \times 10^{-4}\text{ mol thyroxine}$

From the problem, 0.455 g thyroxine were used; this must contain 5.86×10^{-4} mol thyroxine. The molar mass of the thyroxine is:

$\text{molar mass} = \dfrac{0.455\text{ g}}{5.86 \times 10^{-4}\text{ mol}} = 776\text{ g/mol}$

65. a. $M = \dfrac{1.0\text{ g protein}}{\text{L}} \times \dfrac{1\text{ mol}}{9.0 \times 10^{-4}\text{ g}} = 1.1 \times 10^{-5}\text{ mol/L};\ \pi = MRT$

At 298 K: $\pi = \dfrac{1.1 \times 10^{-5}\text{ mol}}{\text{L}} \times \dfrac{0.08206\text{ L atm}}{\text{mol K}} \times 298\text{ K} \times \dfrac{760\text{ torr}}{\text{atm}},\ \pi = 0.20\text{ torr}$

Because d = 1.0 g/cm^3, 1.0 L solution has a mass of 1.0 kg. Because only 1.0 g of protein is present per liter of solution, 1.0 kg of H_2O is present and molality equals molarity.

$$\Delta T_f = K_f m = \frac{1.86°C}{molal} \times 1.1 \times 10^{-5} \text{ molal} = 2.0 \times 10^{-5} \text{ °C}$$

b. Osmotic pressure is better for determining the molar mass of large molecules. A temperature change of 10^{-5} °C is very difficult to measure. A change in height of a column of mercury by 0.2 mm (0.2 torr) is not as hard to measure precisely.

67. $\pi = MRT,\ M = \dfrac{\pi}{RT} = \dfrac{8.00 \text{ atm}}{\dfrac{0.08206 \text{ L atm}}{\text{mol K}} \times 298 \text{ K}} = 0.327 \text{ mol/L}$

Properties of Electrolyte Solutions

69. $Na_3PO_4(s) \rightarrow 3\ Na^+(aq) + PO_4^{3-}(aq)$, i = 4.0; $CaBr_2(s) \rightarrow Ca^{2+}(aq) + 2\ Br^-(aq)$, i = 3.0

$KCl(s) \rightarrow K^+(aq) + Cl^-(aq)$, i = 2.0.

The effective particle concentrations of the solutions are:

4.0(0.010 molal) = 0.040 molal for Na_3PO_4 solution; 3.0(0.020 molal) = 0.060 molal for $CaBr_2$ solution; 2.0(0.020 molal) = 0.040 molal for KCl solution; slightly greater than 0.020 molal for HF solution since HF only partially dissociates in water (it is a weak acid).

a. The 0.010 *m* Na_3PO_4 solution and the 0.020 *m* KCl solution both have effective particle concentrations of 0.040 *m* (assuming complete dissociation), so both of these solutions should have the same boiling point as the 0.040 *m* $C_6H_{12}O_6$ solution (a nonelectrolyte).

b. $P = \chi P°$; As the solute concentration decreases, the solvent's vapor pressure increases since χ increases. Therefore, the 0.020 *m* HF solution will have the highest vapor pressure since it has the smallest effective particle concentration.

c. $\Delta T = K_f m$; The 0.020 *m* $CaBr_2$ solution has the largest effective particle concentration so it will have the largest freezing point depression (largest ΔT).

71. a. $MgCl_2(s) \rightarrow Mg^{2+}(aq) + 2\ Cl^-(aq)$, i = 3.0 mol ions/mol solute

$\Delta T_f = iK_f m = 3.0 \times 1.86°C/\text{molal} \times 0.050 \text{ molal} = 0.28°C$; $T_f = -0.28°C$ (Assuming water freezes at 0.00°C.)

$\Delta T_b = iK_b m = 3.0 \times 0.51°C/\text{molal} \times 0.050 \text{ molal} = 0.077°C$; $T_b = 100.077°C$ (Assuming water boils at 100.000°C.)

b. $FeCl_3(s) \rightarrow Fe^{3+}(aq) + 3\ Cl^-(aq)$, i = 4.0 mol ions/mol solute

$\Delta T_f = iK_f m = 4.0 \times 1.86°C/\text{molal} \times 0.050 \text{ molal} = 0.37°C$; $T_f = -0.37°C$

$\Delta T_b = iK_b m = 4.0 \times 0.51°C/\text{molal} \times 0.050 \text{ molal} = 0.10°C$; $T_b = 100.10°C$

73. $\Delta T_f = iK_f m,\ i = \frac{\Delta T_f}{K_f m} = \frac{0.110°C}{1.86°C/molal \times 0.0225\ molal} = 2.63$ for 0.0225 m $CaCl_2$

$i = \frac{0.440}{1.86 \times 0.0910} = 2.60$ for 0.0910 m $CaCl_2$; $i = \frac{1.330}{1.86 \times 0.278} = 2.57$ for 0.278 m $CaCl_2$

$i_{ave} = (2.63 + 2.60 + 2.57)/3 = 2.60$

Note that i is less than the ideal value of 3.0 for $CaCl_2$. This is due to ion pairing in solution. Also note that as molality increases, i decreases. More ion pairing occurs as the solute concentration increases.

75. a. $T_C = 5(T_F - 32)/9 = 5(-29 - 32)/9 = -34°C$; Assuming the solubility of $CaCl_2$ is temperature independent, the molality of a saturated $CaCl_2$ solution is:

$$\frac{74.5\ g\ CaCl_2}{100.0\ g} \times \frac{1000\ g}{kg} \times \frac{1\ mol\ CaCl_2}{110.98\ g\ CaCl_2} = \frac{6.71\ mol\ CaCl_2}{kg\ H_2O}$$

$\Delta T_f = iK_f m = 3.00 \times 1.86°C\ kg/mol \times 6.71\ mol/kg = 37.4°C$

Assuming i = 3.00, a saturated solution of $CaCl_2$ can lower the freezing point of water to −37.4°C. Assuming these conditions, a saturated $CaCl_2$ solution should melt ice at −34°C (−29°F).

b. From Exercise 11.73, $i_{ave} = 2.60$; $\Delta T_f = iK_f m = 2.60 \times 1.86 \times 6.71 = 32.4°C$

$T_f = -32.4°C$

Assuming i = 2.60, a saturated $CaCl_2$ solution will not melt ice at −34°C(−29°F).

Additional Exercises

77 a. $NH_4NO_3(s) \rightarrow NH_4^+(aq) + NO_3^-(aq)\quad \Delta H_{soln} = ?$

Heat gain by dissolution process = heat loss by solution; We will keep all quantities positive in order to avoid sign errors. Since the temperature of the water decreased, the dissolution of NH_4NO_3 is endothermic (ΔH is positive). Mass of solution = 1.60 + 75.0 = 76.6 g.

$$\text{heat loss by solution} = \frac{4.18\ J}{g\ °C} \times 76.6\ g \times (25.00°C - 23.34°C) = 532\ J$$

$$\Delta H_{soln} = \frac{532\ J}{1.60\ g\ NH_4NO_3} \times \frac{80.05\ g\ NH_4NO_3}{mol\ NH_4NO_3} = 2.66 \times 10^4\ J/mol = 26.6\ kJ/mol$$

b. We will use Hess's law to solve for the lattice energy. The lattice energy equation is:

$$NH_4^+(g) + NO_3^-(g) \rightarrow NH_4NO_3(s) \qquad \Delta H = \text{lattice energy}$$

$$NH_4^+(g) + NO_3^-(g) \rightarrow NH_4^+(aq) + NO_3^-(aq) \qquad \Delta H = \Delta H_{hyd} = -630.\text{ kJ/mol}$$

$$NH_4^+(aq) + NO_3^-(aq) \rightarrow NH_4NO_3(s) \qquad \Delta H = -\Delta H_{soln} = -26.6\text{ kJ/mol}$$

$$NH_4^+(g) + NO_3^-(g) \rightarrow NH_4NO_3(s) \qquad \Delta H = \Delta H_{hyd} - \Delta H_{soln} = -657\text{ kJ/mol}$$

79. a. Water boils when the vapor pressure equals the pressure above the water. In an open pan, $P_{atm} \approx 1.0$ atm. In a pressure cooker, $P_{inside} > 1.0$ atm, and water boils at a higher temperature. The higher the cooking temperature, the faster the cooking time.

b. Salt dissolves in water forming a solution with a melting point lower than that of pure water ($\Delta T_f = K_f m$). This happens in water on the surface of ice. If it is not too cold, the ice melts. This won't work if the ambient temperature is lower than the depressed freezing point of the salt solution.

c. When water freezes from a solution, it freezes as pure water, leaving behind a more concentrated salt solution. Therefore, the melt of frozen sea ice is pure water.

d. On the CO_2 phase diagram in chapter 10, the triple point is above 1 atm, so $CO_2(g)$ is the stable phase at 1 atm and room temperature. $CO_2(l)$ can't exist at normal atmospheric pressures. Therefore, dry ice sublimes instead of boils. In a fire extinguisher, P > 1 atm and $CO_2(l)$ can exist. When CO_2 is released from the fire extinguisher, $CO_2(g)$ forms as predicted from the phase diagram.

e. Adding a solute to a solvent increases the boiling point and decreases the freezing point of the solvent. Thus, the solvent is a liquid over a wider range of temperatures when a solute is dissolved.

81. Because partial pressures are proportional to the moles of gas present, then $\chi^V_{CS_2} = P_{CS_2}/P_{tot}$.

$$P_{CS_2} = \chi^V_{CS_2} P_{tot} = 0.855\ (263\text{ torr}) = 225\text{ torr}$$

$$P_{CS_2} = \chi^L_{CS_2} P^o_{CS_2},\quad \chi^L_{CS_2} = \frac{P_{CS_2}}{P^o_{CS_2}} = \frac{225\text{ torr}}{375\text{ torr}} = 0.600$$

83. Out of 100.00 g, there are:

$$31.57\text{ g C} \times \frac{1\text{ mol C}}{12.01\text{ g}} = 2.629\text{ mol C};\quad \frac{2.629}{2.629} = 1.000$$

$$5.30\text{ g H} \times \frac{1\text{ mol H}}{1.008\text{ g}} = 5.26\text{ mol H};\quad \frac{5.26}{2.629} = 2.00$$

$$63.13\text{ g O} \times \frac{1\text{ mol O}}{16.00\text{ g}} = 3.946\text{ mol O};\quad \frac{3.946}{2.629} = 1.501$$

empirical formula: $C_2H_4O_3$; Use the freezing point data to determine the molar mass.

$$m = \frac{\Delta T_f}{K_f} = \frac{5.20°C}{1.86\ °C/\text{molal}} = 2.80 \text{ molal}$$

$$\text{mol solute} = 0.0250 \text{ kg} \times \frac{2.80 \text{ mol solute}}{\text{kg}} = 0.0700 \text{ mol solute}$$

$$\text{molar mass} = \frac{10.56 \text{ g}}{0.0700 \text{ mol}} = 151 \text{ g/mol}$$

The empirical formula mass of $C_2H_4O_3$ = 76.05 g/mol. Since the molar mass is about twice the empirical mass, the molecular formula is $C_4H_8O_6$, which has a molar mass of 152.10 g/mol.

Note: We use the experimental molar mass to determine the molecular formula. Knowing this, we calculate the molar mass precisely from the molecular formula using the atomic masses in the periodic table.

85. If ideal, NaCl dissociates completely and i = 2.00. $\Delta T_f = iK_fm$; Assuming water freezes at 0.00°C:

$$1.28°C = 2 \times 1.86°C \text{ kg/mol} \times m, \quad m = 0.344 \text{ mol NaCl/kg } H_2O$$

Assume an amount of solution which contains 1.00 kg of water (solvent).

$$0.344 \text{ mol NaCl} \times \frac{58.44 \text{ g}}{\text{mol}} = 20.1 \text{ g NaCl}; \quad \text{mass \% NaCl} = \frac{20.1 \text{ g}}{1.00\times 10^3 \text{ g} + 20.1 \text{ g}} \times 100$$

$$= 1.97\%$$

87. $\Delta T = K_fm, \quad m = \frac{\Delta T}{K_f} = \frac{2.79°C}{1.86°C/\text{molal}} = 1.50 \text{ molal}$

a. $\Delta T = K_bm, \quad \Delta T = (0.51°C/\text{molal})(1.50 \text{ molal}) = 0.77°C, \quad T_b = 100.77°C$

b. $P_{water} = \chi_{water}P^o_{water}, \quad \chi_{water} = \frac{\text{mol } H_2O}{\text{mol } H_2O + \text{mol solute}}$

Assuming 1.00 kg of water, we have 1.50 mol solute and:

$$\text{mol } H_2O = 1.00 \times 10^3 \text{ g } H_2O \times \frac{1 \text{ mol } H_2O}{18.02 \text{ g } H_2O} = 55.5 \text{ mol } H_2O$$

$$\chi_{water} = \frac{55.5 \text{ mol}}{1.50 + 55.5} = 0.974; \quad P_{water} = (0.974)(23.76 \text{ mm Hg}) = 23.1 \text{ mm Hg}$$

c. We assumed ideal behavior in solution formation and assumed i = 1 (no ions form).

Challenge Problems

89. For 30.% A by moles in the vapor, $30. = \frac{P_A}{P_A + P_B} \times 100$:

$$0.30 = \frac{\chi_A x}{\chi_A x + \chi_B y},\quad 0.30 = \frac{\chi_A x}{\chi_A x + (1.00 - \chi_A)y}$$

$$\chi_A x = 0.30(\chi_A x) + 0.30\, y - 0.30\, \chi_A y,\ \chi_A x - 0.30\, \chi_A x + 0.30\, \chi_A y = 0.30\, y$$

$$\chi_A(x - 0.30\, x + 0.30\, y) = 0.30\, y,\ \chi_A = \frac{0.30\, y}{0.70x + 0.30y};\ \chi_B = 1.00 - \chi_A$$

Similarly, if vapor above is 50.% A: $\chi_A = \frac{y}{x+y};\ \chi_B = 1.00 - \frac{y}{x+y}$

If vapor above is 80%A: $\chi_A = \frac{0.80\, y}{0.20x + 0.80y};\ \chi_B = 1.00 - \chi_A$

If the liquid solution is 30.%A by moles, $\chi_A = 0.30$.

Thus, $\chi_A^V = \frac{P_A}{P_A + P_B} = \frac{0.30\, x}{0.30x + 0.70y}$ and $\chi_B^V = 1.00 - \frac{0.30\, x}{0.30x + 0.70y}$

If solution is 50.%A: $\chi_A^V = \frac{x}{x+y}$ and $\chi_B^V = 1.00 - \chi_A^V$

If solution is 80.%A: $\chi_A^V = \frac{0.80\, x}{0.80x + 0.20y}$ and $\chi_B^V = 1.00 - \chi_A^V$

91. $m = \frac{\Delta T}{K_f} = \frac{0.426°C}{1.86°C/molal} = 0.229$ molal

Let x = mol NaCl in mixture and y = mol $C_{12}H_{22}O_{11}$.

$NaCl(aq) \rightarrow Na^+(aq) + Cl^-(aq)$; In solution, NaCl exists as separate Na^+ and Cl^- ions.

$$0.229 \text{ mol} = \text{mol } Na^+ + \text{mol } Cl^- + \text{mol } C_{12}H_{22}O_{11} = x + x + y = 2x + y$$

The molar mass of NaCl is 58.44 g/mol and the molar mass of $C_{12}H_{22}O_{11}$ = 342.3 g/mol. Setting up another equation for the mass of the mixture:

$$20.0 \text{ g} = x(58.44) + y(342.3)$$

Substituting: $20.0 = 58.44\, x + (0.229 - 2x)\, 342.3$

Solving: x = mol NaCl = 0.0932 mol and

$$y = \text{mol } C_{12}H_{22}O_{11} = 0.229 - 2(0.0932) = 0.043 \text{ mol}$$

mass NaCl = 0.0932 mol × 58.44 g/mol = 5.45 g NaCl

$$\text{mass\% NaCl} = \frac{5.45\text{ g}}{20.0\text{ g}} \times 100 = 27.3\%,\ \text{mass \% } C_{12}H_{22}O_{11} = 100.0 - 27.3 = 72.7\%$$

$$\chi_{sucrose} = \frac{0.043\text{ mol}}{(0.043 + 0.0932)\text{ mol}} = 0.32$$

93. $\chi_{pen}^{V} = 0.15 = \frac{P_{pen}}{P_{total}}$; $P_{pen} = \chi_{pen}^{L} P_{pen}^{o} = \chi_{pen}^{L}$ (511 torr); $P_{total} = P_{pen} + P_{hex} = \chi_{pen}^{L}$ (511) + χ_{hex}^{L} (150.)

Since $\chi_{hex}^{L} = 1.000 - \chi_{pen}^{L}$, then: $P_{total} = \chi_{pen}^{L}$ (511 torr) + (1.000− χ_{pen}^{L})(150.) = 150. + $361\chi_{pen}^{L}$

$$\chi_{pen}^{V} = \frac{P_{pen}}{P_{total}},\ 0.15 = \frac{\chi_{pen}^{L}(511)}{150. + 361\chi_{pen}^{L}},\ 0.15\,(150. + 361\chi_{pen}^{L}) = 511\chi_{pen}^{L}$$

$$23 + 54\chi_{pen}^{L} = 511\chi_{pen}^{L},\ \chi_{pen}^{L} = \frac{23}{457} = 0.050$$

95. $\Delta T_f = 5.51°C - 2.81°C = 2.70°C;\ m = \frac{\Delta T_f}{K_f} = \frac{2.70°C}{5.12\ °C/\text{molal}} = 0.527\text{ molal}$

Let x = mass of naphthalene (molar mass = 128.2 g/mol). Then, 1.60 − x = mass of anthracene (molar mass = 178.2 g/mol).

$$\frac{x}{128.2} = \text{moles napthalene and } \frac{1.60 - x}{178.2} = \text{moles anthracene}$$

$$\frac{0.527\text{ mol solute}}{\text{kg solvent}} = \frac{\frac{x}{128.2} + \frac{1.60 - x}{178.2}}{0.0200\text{ kg solvent}},\ 1.05 \times 10^{-2} = \frac{178.2\,x + 1.60(128.2) - 128.2\,x}{128.2(178.2)}$$

$50.0\,x + 205 = 240.,\ 50.0\,x = 35,\ x = 0.70$ g naphthalene

So mixture is: $\frac{0.70\text{ g}}{1.60\text{ g}} \times 100 = 44\%$ naphthalene by mass and 56% anthracene by mass

97. $HCO_2H \rightarrow H^+ + HCO_2^-$; Only 4.2% of HCO_2H ionizes. The amount of H^+ or HCO_2^- produced is:

$0.042 \times 0.10\ M = 0.0042\ M$

The amount of HCO_2H remaining in solution after ionization is:

$0.10\ M - 0.0042\ M = 0.10\ M$

The total molarity of species present = $M_{HCO_2H} + M_{H^+} + M_{HCO_2^-}$

$= 0.10 + 0.0042 + 0.0042 = 0.11\ M$

Assuming 0.11 M = 0.11 molal and assuming ample significant figures in the freezing point and boiling point of water at P = 1 atm:

$\Delta T = K_f m = 1.86°C/\text{molal} \times 0.11\ \text{molal} = 0.20°C$; freezing point = −0.20°C

$\Delta T = K_b m = 0.51°C/\text{molal} \times 0.11\ \text{molal} = 0.056°C$; boiling point = 100.056°C

99. a. Assuming $MgCO_3(s)$ does not dissociate, the solute concentration in water is:

$$\frac{560\ \mu g\ MgCO_3(s)}{mL} = \frac{560\ mg}{L} = \frac{560 \times 10^{-3}\ g}{L} \times \frac{1\ mol\ MgCO_3}{84.32\ g} = 6.6 \times 10^{-3}\ mol\ MgCO_3/L$$

An applied pressure of 8.0 atm will purify water up to a solute concentration of:

$$M = \frac{\pi}{RT} = \frac{8.0\ atm}{\frac{0.08206\ L\ atm}{mol\ K} \times 300.\ K} = \frac{0.32\ mol}{L}$$

When the concentration of $MgCO_3(s)$ reaches 0.32 mol/L, the reverse osmosis unit can no longer purify the water. Let V = volume (L) of water remaining after purifying 45 L of H_2O. When V + 45 L of water have been processed, the moles of solute particles will equal:

$$6.6 \times 10^{-3}\ mol/L \times (45\ L + V) = 0.32\ mol/L \times V$$

Solving: $0.30 = (0.32 - 0.0066) \times V$, $V = 0.96$ L

The minimum total volume of water that must be processed is 45 L + 0.96 L = 46 L.

Note: If $MgCO_3$ does dissociate into Mg^{2+} and CO_3^{2-} ions, the solute concentration will increase to 1.3×10^{-2} M and at least 47 L of water must be processed.

b. No; A reverse osmosis system that applies 8.0 atm can only purify water with a solute concentration less than 0.32 mol/L. Salt water has a solute concentration of 2(0.60 M) = 1.20 M ions. The solute concentration of salt water is much too high for this reverse osmosis unit to work.

Integrative Problems

101. $\Delta T = imK_f$, $i = \dfrac{\Delta T}{mK_f} = \dfrac{2.79^\circ C}{\dfrac{0.250\text{ mol}}{0.500\text{ kg}} \times \dfrac{1.86^\circ C\text{ kg}}{\text{mol}}} = 3.00$

We have 3 ions in solutions and we have twice as many anions as cations. Therefore, the formula of Q is MCl_2. Assuming 100.00 g compound:

$$38.68\text{ g Cl} \times \frac{1\text{ mol Cl}}{35.45\text{ g}} = 1.091\text{ mol Cl}$$

$$\text{mol M} = 1.091\text{ mol Cl} \times \frac{1\text{ mol M}}{2\text{ mol Cl}} = 0.5455\text{ mol M}$$

$$\text{molar mass of M} = \frac{61.32\text{ g M}}{0.5455\text{ mol M}} = 112.4\text{ g/mol; M is Cd so Q} = CdCl_2.$$

CHAPTER TWELVE

CHEMICAL KINETICS

Questions

9. In a unimolecular reaction, a single reactant molecule decomposes to products. In a bimolecular reaction, two molecules collide to give products. The probability of the simultaneous collision of three molecules with enough energy and proper orientation is very small, making termolecular steps very unlikely.

11. All of these choices would affect the rate of the reaction, but only b and c affect the rate by affecting the value of the rate constant k. The value of the rate constant is dependent on temperature. The value of the rate constant also depends on the activation energy. A catalyst will change the value of k because the activation energy changes. Increasing the concentration (partial pressure) of either H_2 or NO does not affect the value of k, but it does increase the rate of the reaction because both concentrations appear in the rate law.

13. The average rate decreases with time because the reverse reaction occurs more frequently as the concentration of products increase. Initially, with no products present, the rate of the forward reaction is at its fastest; but as time goes on, the rate gets slower and slower since products are converting back into reactants. The instantaneous rate will also decrease with time. The only rate that is constant is the initial rate. This is the instantaneous rate taken at $t \approx 0$. At this time, the amount of products is insignificant and the rate of the reaction only depends on the rate of the forward reaction.

15. $$\frac{\text{rate}_2}{\text{rate}_1} = \frac{k[A]_2^x}{k[A]_1^x} = \left(\frac{[A]_2}{[A]_1}\right)^x$$

The rate doubles as the concentration quadruples:

$$2 = (4)^x, \quad x = 1/2$$

The order is 1/2 (the square root of the concentration of reactant).

For a reactant that has an order of −1 and the reactant concentration is doubled:

$$\frac{\text{rate}_2}{\text{rate}_1} = (2)^{-1} = \frac{1}{2}$$

The rate will decrease by a factor of 1/2 when the reactant concentration is doubled for a −1 order reaction.

17. Two reasons are:

a. The collision must involve enough energy to produce the reaction, i.e., the collision energy must be equal to or exceed the activation energy.

b. The relative orientation of the reactants when they collide must allow formation of any new bonds necessary to produce products.

Exercises

Reaction Rates

19. The coefficients in the balanced reaction relate the rate of disappearance of reactants to the rate of production of products. From the balanced reaction, the rate of production of P_4 will be 1/4 the rate of disappearance of PH_3, and the rate of production of H_2 will be 6/4 the rate of disappearance of PH_3. By convention, all rates are given as positive values.

$$\text{Rate} = \frac{-\Delta[PH_3]}{\Delta t} = \frac{(0.0048\ \text{mol}/2.0\ \text{L})}{\text{s}} = 2.4 \times 10^{-3}\ \text{mol/L•s}$$

$$\frac{\Delta[P_4]}{\Delta t} = -\frac{1}{4}\frac{\Delta[PH_3]}{\Delta t} = 2.4 \times 10^{-3}/4 = 6.0 \times 10^{-4}\ \text{mol/L•s}$$

$$\frac{\Delta[H_2]}{\Delta t} = -\frac{6}{4}\frac{\Delta[PH_3]}{\Delta t} = 6(2.4 \times 10^{-3})/4 = 3.6 \times 10^{-3}\ \text{mol/L•s}$$

21. a. $$\text{average rate} = \frac{-\Delta[H_2O_2]}{\Delta t} = \frac{-(0.500\ M - 1.000\ M)}{(2.16 \times 10^4\ \text{s} - 0)} = 2.31 \times 10^{-5}\ \text{mol/L•s}$$

From the coefficients in the balanced equation:

$$\frac{\Delta[O_2]}{\Delta t} = -\frac{1}{2}\frac{\Delta[H_2O_2]}{\Delta t} = 1.16 \times 10^{-5}\ \text{mol/L•s}$$

b. $$\frac{-\Delta[H_2O_2]}{\Delta t} = \frac{-(0.250 - 0.500)\ M}{(4.32 \times 10^4 - 2.16 \times 10^4)\ \text{s}} = 1.16 \times 10^{-5}\ \text{mol/L•s}$$

$$\frac{\Delta[O_2]}{\Delta t} = 1/2\,(1.16 \times 10^{-5}) = 5.80 \times 10^{-6}\ \text{mol/L•s}$$

Notice that as time goes on in a reaction, the average rate decreases.

23. a. The units for rate are always mol/L•s. b. Rate = k; k must have units of mol/L•s.

c. Rate = k[A], $\frac{mol}{L\,s} = k\left(\frac{mol}{L}\right)$

k must have units of s^{-1}.

d. Rate = $k[A]^2$, $\frac{mol}{L\,s} = k\left(\frac{mol}{L}\right)^2$

k must have units of L/mol•s.

e. $L^2/mol^2 \bullet s$

Rate Laws from Experimental Data: Initial Rates Method

25. a. In the first two experiments, [NO] is held constant and $[Cl_2]$ is doubled. The rate also doubled. Thus, the reaction is first order with respect to Cl_2. Or mathematically: Rate = $k[NO]^x[Cl_2]^y$

$$\frac{0.36}{0.18} = \frac{k(0.10)^x(0.20)^y}{k(0.10)^x(0.10)^y} = \frac{(0.20)^y}{(0.10)^y},\ 2.0 = 2.0^y,\ y = 1$$

We can get the dependence on NO from the second and third experiments. Here, as the NO concentration doubles (Cl_2 concentration is constant), the rate increases by a factor of four. Thus, the reaction is second order with respect to NO. Or mathematically:

$$\frac{1.45}{0.36} = \frac{k(0.20)^x(0.20)}{k(0.10)^x(0.20)} = \frac{(0.20)^x}{(0.10)^x},\ 4.0 = 2.0^x,\ x = 2;\ \text{So, Rate} = k[NO]^2[Cl_2]$$

Try to examine experiments where only one concentration changes at a time. The more variables that change, the harder it is to determine the orders. Also, these types of problems can usually be solved by inspection. In general, we will solve using a mathematical approach, but keep in mind you probably can solve for the orders by simple inspection of the data.

b. The rate constant k can be determined from the experiments. From experiment 1:

$$\frac{0.18\ mol}{L\ min} = k\left(\frac{0.10\ mol}{L}\right)^2\left(\frac{0.10\ mol}{L}\right),\ k = 180\ L^2/mol^2 \bullet min$$

From the other experiments:

$k = 180\ L^2/mol^2 \bullet min$ (2nd exp.); $k = 180\ L^2/mol^2 \bullet min$ (3rd exp.)

The average rate constant is $k_{mean} = 1.8 \times 10^2\ L^2/mol^2 \bullet min$.

27. a. Rate = $k[NOCl]^n$; Using experiments two and three:

$$\frac{2.66 \times 10^4}{6.64 \times 10^3} = k\frac{(2.0 \times 10^{16})^n}{(1.0 \times 10^{16})^n},\ 4.01 = 2.0^n,\ n = 2;\ \text{Rate} = k[NOCl]^2$$

b. $$\frac{5.98 \times 10^4\ molecules}{cm^3\ s} = k\left(\frac{3.0 \times 10^{16}\ molecules}{cm^3}\right)^2,\ k = 6.6 \times 10^{-29}\ cm^3/molecules \bullet s$$

The other three experiments give (6.7, 6.6 and 6.6) × 10^{-29} cm^3/molecules•s, respectively.

The mean value for k is 6.6 × 10^{-29} cm^3/molecules•s.

c. $$\frac{6.6\times10^{-29}\ \text{cm}^3}{\text{molecules s}}\times\frac{1\ \text{L}}{1000\ \text{cm}^3}\times\frac{6.022\times10^{23}\ \text{molecules}}{\text{mol}}=\frac{4.0\times10^{-8}\ \text{L}}{\text{mol s}}$$

29. a. Rate = $k[Hb]^x[CO]^y$; Comparing the first two experiments, [CO] is unchanged, [Hb] doubles, and the rate doubles. Therefore, x = 1 and the reaction is first order in Hb. Comparing the second and third experiments, [Hb] is unchanged, [CO] triples. and the rate triples. Therefore, y = 1 and the reaction is first order in CO.

b. Rate = k[Hb][CO]

c. From the first experiment:

$$0.619\ \mu\text{mol/L•s} = k\ (2.21\ \mu\text{mol/L})(1.00\ \mu\text{mol/L}),\ \ k = 0.280\ \text{L}/\mu\text{mol•s}$$

The second and third experiments give similar k values, so k_{mean} = 0.280 L/μmol•s.

d. $$\text{Rate} = k[\text{Hb}][\text{CO}] = \frac{0.280\ \text{L}}{\mu\text{mol s}}\times\frac{3.36\ \mu\text{mol}}{\text{L}}\times\frac{2.40\ \mu\text{mol}}{\text{L}} = 2.26\ \mu\text{mol/L•s}$$

Integrated Rate Laws

31. The first assumption to make is that the reaction is first order because first-order reactions are most common. For a first-order reaction, a graph of ln $[H_2O_2]$ vs time will yield a straight line. If this plot is not linear, then the reaction is not first order and we make another assumption. The data and plot for the first-order assumption follows.

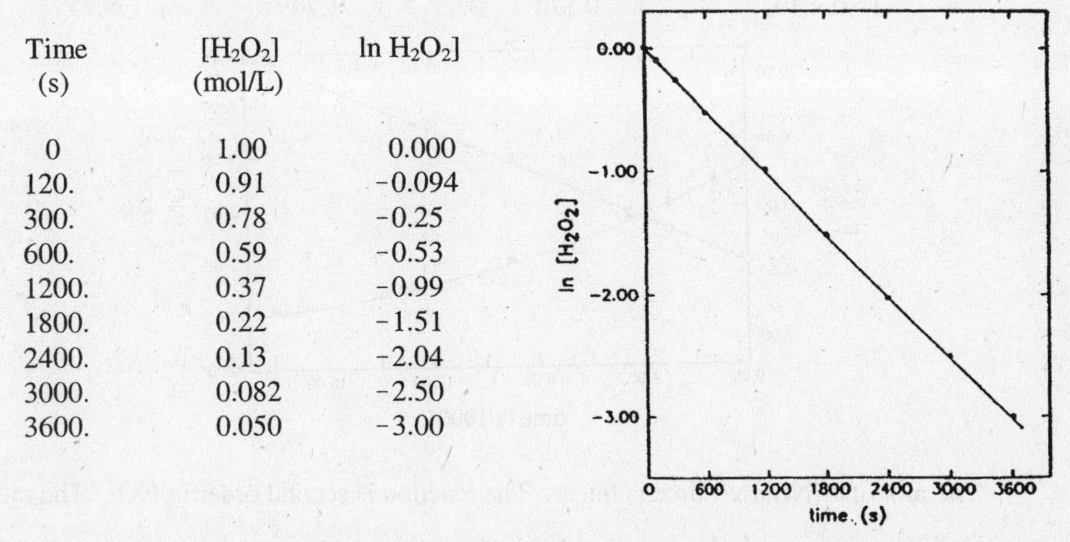

Time (s)	$[H_2O_2]$ (mol/L)	ln $H_2O_2]$
0	1.00	0.000
120.	0.91	−0.094
300.	0.78	−0.25
600.	0.59	−0.53
1200.	0.37	−0.99
1800.	0.22	−1.51
2400.	0.13	−2.04
3000.	0.082	−2.50
3600.	0.050	−3.00

Note: We carried extra significant figures in some of the ln values in order to reduce round-off error. For the plots, we will do this most of the time when the ln function is involved.

The plot of ln $[H_2O_2]$ vs. time is linear. Thus, the reaction is first order. The rate law and integrated rate law are: Rate = $k[H_2O_2]$ and ln $[H_2O_2] = -kt + \ln [H_2O_2]_o$.

We determine the rate constant k by determining the slope of the ln $[H_2O_2]$ vs time plot (slope = $-k$). Using two points on the curve gives:

$$\text{slope} = -k = \frac{\Delta y}{\Delta x} = \frac{0-(3.00)}{0-3600.} = -8.3 \times 10^{-4}\ s^{-1},\quad k = 8.3 \times 10^{-4}\ s^{-1}$$

To determine $[H_2O_2]$ at 4000. s, use the integrated rate law where at t = 0, $[H_2O_2]_o = 1.00\ M$.

$$\ln [H_2O_2] = -kt + \ln [H_2O_2]_o \text{ or } \ln\left(\frac{[H_2O_2]}{[H_2O_2]_o}\right) = -kt$$

$$\ln\left(\frac{[H_2O_2]}{1.00}\right) = -8.3 \times 10^{-4}\ s^{-1} \times 4000.\ s,\ \ln [H_2O_2] = -3.3,\ [H_2O_2] = e^{-3.3} = 0.037\ M$$

33. Assume the reaction is first order and see if the plot of ln $[NO_2]$ vs. time is linear. If this isn't linear, try the second-order plot of $1/[NO_2]$ vs. time because second-order reactions are the next most common after first-order reactions. The data and plots follow.

Time (s)	$[NO_2]$ (M)	ln $[NO_2]$	$1/[NO_2]$ (M^{-1})
0	0.500	-0.693	2.00
1.20×10^3	0.444	-0.812	2.25
3.00×10^3	0.381	-0.965	2.62
4.50×10^3	0.340	-1.079	2.94
9.00×10^3	0.250	-1.386	4.00
1.80×10^4	0.174	-1.749	5.75

The plot of $1/[NO_2]$ vs. time is linear. The reaction is second order in NO_2. The rate law and integrated rate law are: Rate = $k[NO_2]^2$ and $\frac{1}{[NO_2]} = kt + \frac{1}{[NO_2]_o}$.

The slope of the plot $1/[NO_2]$ vs. t gives the value of k. Using a couple of points on the plot:

$$\text{slope} = k = \frac{\Delta y}{\Delta x} = \frac{(5.75 - 2.00)\ M^{-1}}{(1.80\times10^4 - 0)\ s} = 2.08 \times 10^{-4}\ \text{L/mol•s}$$

To determine $[NO_2]$ at 2.70×10^4 s, use the integrated rate law where $1/[NO_2]_o = 1/0.500\ M = 2.00\ M^{-1}$.

$$\frac{1}{[NO_2]} = kt + \frac{1}{[NO_2]_o},\quad \frac{1}{[NO_2]} = \frac{2.08\times10^{-4}\ L}{mol\ s} \times 2.70 \times 10^4\ s + 2.00\ M^{-1}$$

$$\frac{1}{[NO_2]} = 7.62,\quad [NO_2] = 0.131\ M$$

35. a. Because the $[C_2H_5OH]$ vs. time plot was linear, the reaction is zero order in C_2H_5OH. The slope of the $[C_2H_5OH]$ vs. time plot equals -k. Therefore, the rate law, the integrated rate law and the rate constant value are: Rate = $k[C_2H_5OH]^0 = k$; $[C_2H_5OH] = -kt + [C_2H_5OH]_o$; $k = 4.00 \times 10^{-5}$ mol/L•s.

b. The half-life expression for a zero-order reaction is: $t_{1/2} = [A]_o/2k$.

$$t_{1/2} = \frac{[C_2H_5OH]_o}{2k} = \frac{1.25\times10^{-2}\ \text{mol/L}}{2\times4.00\times10^{-5}\ \text{mol/L•s}} = 156\ s$$

Note: we could have used the integrated rate law to solve for $t_{1/2}$ where $[C_2H_5OH] = (1.25 \times 10^{-2}/2)$ mol/L.

c. $[C_2H_5OH] = -kt + [C_2H_5OH]_o$, $0\ \text{mol/L} = -(4.00 \times10^{-5}\ \text{mol/L•s})\ t + 1.25 \times 10^{-2}\ \text{mol/L}$

$$t = \frac{1.25\times10^{-2}\ \text{mol/L}}{4.00\times10^{-5}\ \text{mol/L•s}} = 313\ s$$

37. The first assumption to make is that the reaction is first order. For a first-order reaction, a graph of $\ln [C_4H_6]$ vs. t should yield a straight line. If this isn't linear, then try the second-order plot of $1/[C_4H_6]$ vs. t. The data and the plots follow.

Time	195	604	1246	2180	6210 s
$[C_4H_6]$	1.6×10^{-2}	1.5×10^{-2}	1.3×10^{-2}	1.1×10^{-2}	$0.68 \times 10^{-2}\ M$
$\ln [C_4H_6]$	-4.14	-4.20	-4.34	-4.51	-4.99
$1/[C_4H_6]$	62.5	66.7	76.9	90.9	$147\ M^{-1}$

Note: To reduce round-off error, we carried extra sig. figs. in the data points.

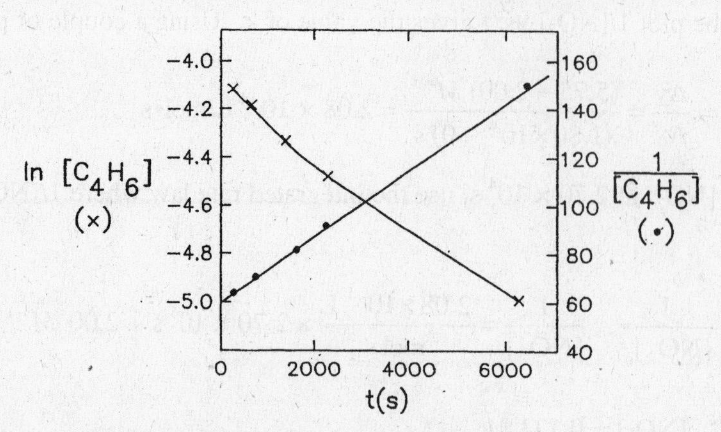

The ln plot is not linear, so the reaction is not first order. Since the second-order plot of $1/[C_4H_6]$ vs. t is linear, we can conclude that the reaction is second order in butadiene. The rate law is:

$$\text{Rate} = k[C_4H_6]^2$$

For a second order reaction, the integrated rate law is: $\dfrac{1}{[C_4H_6]} = kt + \dfrac{1}{[C_4H_6]_o}$

The slope of the straight line equals the value of the rate constant. Using the points on the line at 1000. and 6000. s:

$$k = \text{slope} = \frac{144\ \text{L/mol} - 73\ \text{L/mol}}{6000.\ \text{s} - 1000.\ \text{s}} = 1.4 \times 10^{-2}\ \text{L/mol}\bullet\text{s}$$

39. Because the 1/[A] vs. time plot is linear with a positive slope, the reaction is second order with respect to A. The y-intercept in the plot will equal $1/[A]_o$. Extending the plot, the y-intercept will be about 10, so $1/10 = 0.1\ M = [A]_o$.

41. a. $[A] = -kt + [A]_o$, $[A] = -(5.0 \times 10^{-2}\ \text{mol/L}\bullet\text{s})\ t + 1.0 \times 10^{-3}\ \text{mol/L}$

b. The half-life expression for a zero-order reaction is: $t_{1/2} = \dfrac{[A]_o}{2k}$

$$t_{1/2} = \frac{1.0\times10^{-3}\ \text{mol/L}}{2\times5.0\times10^{-2}\ \text{mol/L}\bullet\text{s}} = 1.0 \times 10^{-2}\ \text{s}$$

c. $[A] = -5.0 \times 10^{-2}\ \text{mol/L}\bullet\text{s} \times 5.0 \times 10^{-3}\ \text{s} + 1.0 \times 10^{-3}\ \text{mol/L} = 7.5 \times 10^{-4}\ \text{mol/L}$

Because $7.5 \times 10^{-4}\ M$ A remains, $2.5 \times 10^{-4}\ M$ A reacted, which means that $2.5 \times 10^{-4}\ M$ B has been produced.

43. a. When a reaction is 75.0% complete (25.0% of reactant remains), this represents two half lives (100% → 50%→ 25%). The first-order half-life expression is: $t_{1/2} = (\ln 2)/k$. Because there is no concentration dependence for a first order half-life: 320. s = two half-lives, $t_{1/2} = 320./2 = 160.$ s. This is both the first half-life, the second half-life, etc.

b. $t_{1/2} = \frac{\ln 2}{k}, \quad k = \frac{\ln 2}{t_{1/2}} = \frac{\ln 2}{160.\text{ s}} = 4.33 \times 10^{-3}\text{ s}^{-1}$

At 90.0% complete, 10.0% of the original amount of the reactant remains, so $[A] = 0.100[A]_0$.

$$\ln\left(\frac{[A]}{[A]_0}\right) = -kt, \quad \ln\frac{0.100[A]_0}{[A]_0} = -(4.33 \times 10^{-3}\text{ s}^{-1})t, \quad t = \frac{\ln 0.100}{-4.33\times10^{-3}\text{ s}^{-1}} = 532\text{ s}$$

45. Comparing experiments 1 and 2, as the concentration of AB is doubled, the initial rate increases by a factor of 4. The reaction is second order in AB.

Rate = $k[AB]^2$, $3.20 \times 10^{-3}\text{ mol/L}\bullet\text{s} = k_1(0.200\ M)^2$

$k = 8.00 \times 10^{-2}\text{ L/mol}\bullet\text{s} = k_{mean}$

For a second order reaction:

$$t_{1/2} = \frac{1}{k[AB]_0} = \frac{1}{8.00\times10^{-2}\text{ L/mol}\bullet\text{s} \times 1.00\text{ mol/L}} = 12.5\text{ s}$$

47. Successive half-lives double as concentration is decreased by one-half. This is consistent with second-order reactions so assume the reaction is second order in A.

$$t_{1/2} = \frac{1}{k[A]_0}, \quad k = \frac{1}{t_{1/2}[A]_0} = \frac{1}{10.0\text{ min}(0.10\ M)} = 1.0\text{ L/mol}\bullet\text{min}$$

a. $\frac{1}{[A]} = kt + \frac{1}{[A]_0} = \frac{1.0\text{ L}}{\text{mol min}} \times 80.0\text{ min} + \frac{1}{0.10\ M} = 90.\ M^{-1}, \quad [A] = 1.1 \times 10^{-2}\ M$

b. 30.0 min = 2 half-lives, so 25% of original A is remaining.

$[A] = 0.25(0.10\ M) = 0.025\ M$

Reaction Mechanisms

49. For elementary reactions, the rate law can be written using the coefficients in the balanced equation to determine orders.

a. Rate = $k[CH_3NC]$ b. Rate = $k[O_3][NO]$

c. Rate = $k[O_3]$ d. Rate = $k[O_3][O]$

51. A mechanism consists of a series of elementary reactions where the rate law for each step can be determined using the coefficients in the balanced equations. For a plausible mechanism, the rate law derived from a mechanism must agree with the rate law determined from

experiment. To derive the rate law from the mechanism, the rate of the reaction is assumed to equal the rate of the slowest step in the mechanism.

Because step 1 is the rate-determining step, the rate law for this mechanism is: Rate = $[C_4H_9Br]$. To get the overall reaction, we sum all the individual steps of the mechanism.

Summing all steps gives:

$$C_4H_9Br \rightarrow C_4H_9^+ + Br^-$$
$$C_4H_9^+ + H_2O \rightarrow C_4H_9OH_2^+$$
$$C_4H_9OH_2^+ + H_2O \rightarrow C_4H_9OH + H_3O^+$$

$$C_4H_9Br + 2\ H_2O \rightarrow C_4H_9OH + Br^- + H_3O^+$$

Intermediates in a mechanism are species that are neither reactants nor products, but that are formed and consumed during the reaction sequence. The intermediates for this mechanism are $C_4H_9^+$ and $C_4H_9OH_2^+$.

Temperature Dependence of Rate Constants and the Collision Model

53. In the following plot, R = reactants, P = products, E_a = activation energy and RC = reaction coordinate which is the same as reaction progress. Note for this reaction that ΔE is positive since the products are at a higher energy than the reactants.

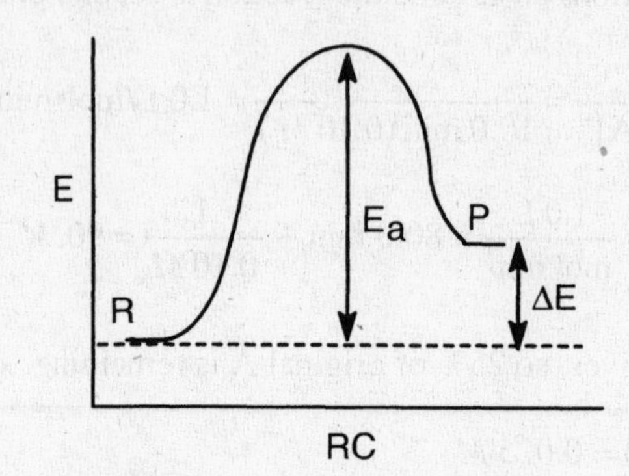

55.

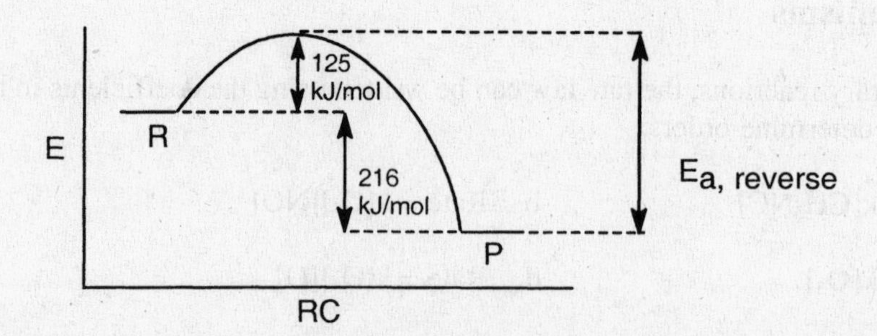

The activation energy for the reverse reaction is:

$$E_{a,\ reverse} = 216\ kJ/mol + 125\ kJ/mol = 341\ kJ/mol$$

57. The Arrhenius equation is: $k = A \exp(-E_a/RT)$ or in logarithmic form, $\ln k = -E_a/RT + \ln A$. Hence, a graph of ln k vs. 1/T should yield a straight line with a slope equal to $-E_a/R$ since the logarithmic form of the Arrhenius equation is in the form of a straight line equation, $y = mx + b$. Note: We carried extra significant figures in the following ln k values in order to reduce round off error.

T (K)	1/T (K^{-1})	k (s^{-1})	ln k
338	2.96×10^{-3}	4.9×10^{-3}	-5.32
318	3.14×10^{-3}	5.0×10^{-4}	-7.60
298	3.36×10^{-3}	3.5×10^{-5}	-10.26

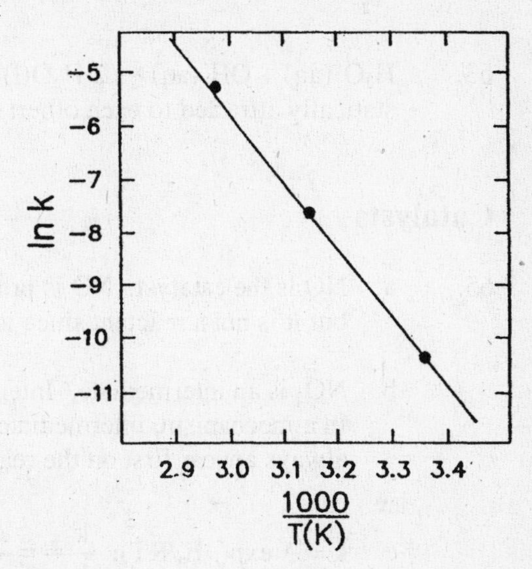

$$\text{Slope} = \frac{-10.76 - (-5.85)}{(3.40 \times 10^{-3} - 3.00 \times 10^{-3})\ K^{-1}} = -1.2 \times 10^4\ K = -E_a/R$$

$$E_a = -\text{slope} \times R = 1.2 \times 10^4\ K \times \frac{8.3145\ J}{K\ mol},\quad E_a = 1.0 \times 10^5\ J/mol = 1.0 \times 10^2\ kJ/mol$$

59. $k = A \exp(-E_a/RT)$ or $\ln k = \frac{-E_a}{RT} + \ln A$ (the Arrhenius equation)

For two conditions: $\ln\left(\frac{k_2}{k_1}\right) = \frac{E_a}{R}\left(\frac{1}{T_1} - \frac{1}{T_2}\right)$ (Assuming A is temperature independent.)

Let $k_1 = 3.52 \times 10^{-7}$ L/mol•s, $T_1 = 555$ K; $k_2 = ?$, $T_2 = 645$ K; $E_a = 186 \times 10^3$ J/mol

$$\ln\left(\frac{k_2}{3.52 \times 10^{-7}}\right) = \frac{1.86 \times 10^5\ J/mol}{8.3145\ J/mol \bullet K}\left(\frac{1}{555\ K} - \frac{1}{645\ K}\right) = 5.6$$

$$\frac{k_2}{3.52 \times 10^{-7}} = e^{5.6} = 270,\quad k_2 = 270(3.52 \times 10^{-7}) = 9.5 \times 10^{-5}\ L/mol\bullet s$$

61. $\ln\left(\frac{k_2}{k_1}\right) = \frac{E_a}{R}\left(\frac{1}{T_1} - \frac{1}{T_2}\right)$; $\frac{k_2}{k_1} = 7.00$, $T_1 = 295$ K, $E_a = 54.0 \times 10^3$ J/mol

$$\ln(7.00) = \frac{5.0\times10^3\ \text{J/mol}}{8.3145\ \text{J/mol}\bullet\text{K}}\left(\frac{1}{295\ \text{K}} - \frac{1}{T_2}\right),\quad \frac{1}{295\ \text{K}} - \frac{1}{T_2} = 3.00 \times 10^{-4}$$

$$\frac{1}{T_2} = 3.09\times10^{-3},\ T_2 = 324\ \text{K} = 51°\text{C}$$

63. $H_3O^+(aq) + OH^-(aq) \rightarrow 2\ H_2O(l)$ should have the faster rate. H_3O^+ and OH^- will be electrostatically attracted to each other; Ce^{4+} and Hg_2^{2+} will repel each other (so E_a is much larger).

Catalysts

65. a. NO is the catalyst. NO is present in the first step of the mechanism on the reactant side, but it is not a reactant since it is regenerated in the second step.

b. NO_2 is an intermediate. Intermediates also never appear in the overall balanced equation. In a mechanism, intermediates always appear first on the product side while catalysts always appear first on the reactant side.

c. $k = A\ \exp(-E_a/RT)$; $\dfrac{k_{cat}}{k_{un}} = \dfrac{A\exp[-E_a(cat)/RT]}{A\exp[-E(un)/RT]} = \exp\left(\dfrac{E_a(un) - E_a(cat)}{RT}\right)$

$$\frac{k_{cat}}{k_{un}} = \exp\left(\frac{2100\ \text{J/mol}}{8.3145\ \text{J/mol}\bullet\text{K}\times298\ \text{K}}\right) = e^{0.85} = 2.3$$

The catalyzed reaction is 2.3 times faster than the uncatalyzed reaction at 25°C.

67. The reaction at the surface of the catalyst is assumed to follow the steps:

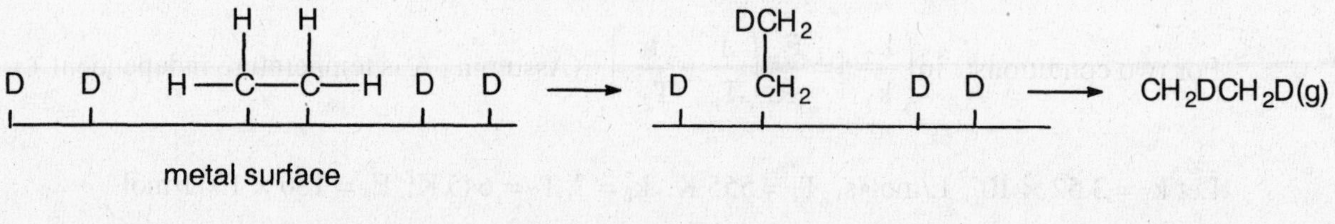

D D H—C(H)—C(H)—H D D ⟶ D CH2(DCH2) D D ⟶ $CH_2DCH_2D(g)$

metal surface

Thus, $CH_2D–CH_2D$ should be the product. If the mechanism is possible, then the reaction must be:

$$C_2H_4 + D_2 \rightarrow CH_2DCH_2D$$

If we got this product, then we could conclude that this is a possible mechanism. If we got some other product, e.g., CH_3CHD_2, then we would conclude that the mechanism is wrong. Even though this mechanism correctly predicts the products of the reaction, we cannot say conclusively that this is the correct mechanism; we might be able to conceive of other mechanisms that would give the same products as our proposed one.

69. Assuming the catalyzed and uncatalyzed reactions have the same form and orders and because concentrations are assumed equal, the rates will be equal when the k values are equal.

$k = A\ exp(-E_a/RT)$, $k_{cat} = k_{un}$ when $E_{a,cat}/RT_{cat} = E_{a,un}/RT_{un}$

$$\frac{4.20\times10^4\ J/mol}{8.3145\ J/mol\bullet K\times293\ K} = \frac{7.00\times10^4\ J/mol}{8.3145\ J/mol\bullet K\times T_{un}},\ T_{un} = 488\ K = 215°C$$

Additional Exercises

71. Rate = $k[NO]^x[O_2]^y$; comparing the first two experiments, $[O_2]$ is unchanged, [NO] is tripled, and the rate increases by a factor of nine. Therefore, the reaction is second order in NO (3^2 = 9). The order of O_2 is more difficult to determine. Comparing the second and third experiments;

$$\frac{3.13\times10^{17}}{1.80\times10^{17}} = \frac{k(2.50\times10^{18})^2(2.50\times10^{18})^y}{k(3.00\times10^{18})^2(1.00\times10^{18})^y},\ 1.74 = 0.694\ (2.50)^y,\ 2.51 = 2.50^y,\ y = 1$$

Rate = $k[NO]^2[O_2]$; From experiment 1:

2.00×10^{16} molecules/cm^3•s = k (1.00×10^{18} molecules/cm^3)^2 (1.00×10^{18} molecules/cm^3)

$k = 2.00\times10^{-38}$ cm^6/molecules^2•s = k_{mean}

$$\text{Rate} = \frac{2.00\times10^{-38}\ cm^6}{molecules^2\bullet s}\times\left(\frac{6.21\times10^{18}\ molecules}{cm^3}\right)^2\times\frac{7.36\times10^{18}\ molecules}{cm^3}$$

$= 5.68\times10^{18}$ molecules/cm^3•s

73. From 338 K data, a plot of $\ln[N_2O_5]$ vs. t is linear and the slope = -4.86×10^{-3} (plot not included). This tells us the reaction is first order in N_2O_5 with $k = 4.86\times10^{-3}$ at 338 K.

From 318 K data, the slope of $\ln[N_2O_5]$ vs t plot is equal to -4.98×10^{-4}, so $k = 4.98\times10^{-4}$ at 318 K. We now have two values of k at two temperatures, so we can solve for E_a.

$$\ln\left(\frac{k_2}{k_1}\right) = \frac{E_a}{R}\left(\frac{1}{T_1}-\frac{1}{T_2}\right),\ \ln\left(\frac{4.86\times10^{-3}}{4.98\times10^{-4}}\right) = \frac{E_a}{8.3145\ J/mol\bullet K}\left(\frac{1}{318\ K}-\frac{1}{338\ K}\right)$$

$E_a = 1.0\times10^5$ J/mol = 1.0×10^2 kJ/mol

75. At high [S], the enzyme is completely saturated with substrate. Once the enzyme is completely saturated, the rate of decomposition of ES can no longer increase, and the overall rate remains constant.

77. a. Because $[A]_o << [B]_o$ or $[C]_o$, the B and C concentrations remain constant at 1.00 *M* for this experiment. So: rate = $k[A]^2[B][C] = k'[A]^2$ where $k' = k[B][C]$

For this pseudo-second-order reaction:

$$\frac{1}{[A]} = k't + \frac{1}{[A]_0}, \quad \frac{1}{3.26\times10^{-5}\ M} = k'(3.00\ \text{min}) + \frac{1}{1.00\times10^{-4}\ M}$$

k′ = 6890 L/mol•min = 115 L/mol•s

$$k' = k[B][C],\ k = \frac{k'}{[B][C]} = \frac{115\ \text{L/mol}\bullet\text{s}}{(1.00\ M)(1.00\ M)} = 115\ \text{L}^3/\text{mol}^3\bullet\text{s}$$

b. For this pseudo-second-order reaction:

$$\text{rate} = k'[A]^2,\ t_{1/2} = \frac{1}{k'[A]_0} = \frac{1}{115\ \text{L/mol}\bullet\text{s}\ (1.00\times10^{-4}\ \text{mol/L})} = 87.0\ \text{s}$$

c. $$\frac{1}{[A]} = k't + \frac{1}{[A]_0} = 115\ \text{L/mol}\bullet\text{s} \times 600.\ \text{s} + \frac{1}{1.00\times10^{-4}\ \text{mol/L}} = 7.90\times10^4\ \text{L/mol},$$

$$[A] = \frac{1}{7.90\times10^{-4}\ \text{L/mol}} = 1.27\times10^{-5}\ \text{mol/L}$$

From the stoichiometry in the balanced reaction, 1 mol of B reacts with every 3 mol of A.

amount A reacted = $1.00 \times 10^{-4}\ M - 1.27 \times 10^{-5}\ M = 8.7 \times 10^{-5}\ M$

amount B reacted = 8.7×10^{-5} mol/L × 1 mol B / 3 mol A = $2.9 \times 10^{-5}\ M$

[B] = $1.00\ M - 2.9 \times 10^{-5}\ M = 1.00\ M$

As we mentioned in part a, the concentration of B (and C) remains constant since the A concentration is so small.

Challenge Problems

79. Rate = $k[I^-]^x[OCl^-]^y[OH^-]^z$; Comparing the first and second experiments:

$$\frac{18.7\times10^{-3}}{9.4\times10^{-3}} = \frac{k(0.0026)^x(0.012)^y(0.10)^z}{k(0.0013)^x(0.012)^y(0.10)^z},\ 2.0 = 2.0^x,\ x = 1$$

Comparing the first and third experiments:

$$\frac{9.4\times10^{-3}}{4.7\times10^{-3}} = \frac{k(0.0013)(0.012)^y(0.10)^z}{k(0.0013)^x(0.0060)^y(0.10)^z},\ 2.0 = 2.0^y,\ y = 1$$

Comparing the first and sixth experiments:

$$\frac{4.8\times10^{-3}}{9.4\times10^{-3}} = \frac{k(0.0013)(0.012)(0.20)^z}{k(0.0013)(0.012)(0.10)^z},\ 1/2 = 2.0^z,\ z = -1$$

$$\text{Rate} = \frac{k[I^-][OCl^-]}{[OH^-]}$$; The presence of OH^- decreases the rate of the reaction.

For the first experiment:

$$\frac{9.4\times10^{-3}\ \text{mol}}{\text{L s}} = k\frac{(0.0013\ \text{mol/L})(0.012\ \text{mol/L})}{(0.10\ \text{mol/L})},\ k = 60.3\ s^{-1} = 60.\ s^{-1}$$

For all experiments, $k_{mean} = 60.\ s^{-1}$.

81. a. We check for first-order dependence by graphing ln [concentration] vs. time for each set of data. The rate dependence on NO is determined from the first set of data because the ozone concentration is relatively large compared to the NO concentration, so $[O_3]$ is effectively constant.

Time (ms)	[NO] (molecules/cm^3)	ln [NO]
0	6.0×10^8	20.21
100.	5.0×10^8	20.03
500.	2.4×10^8	19.30
700.	1.7×10^8	18.95
1000.	9.9×10^7	18.41

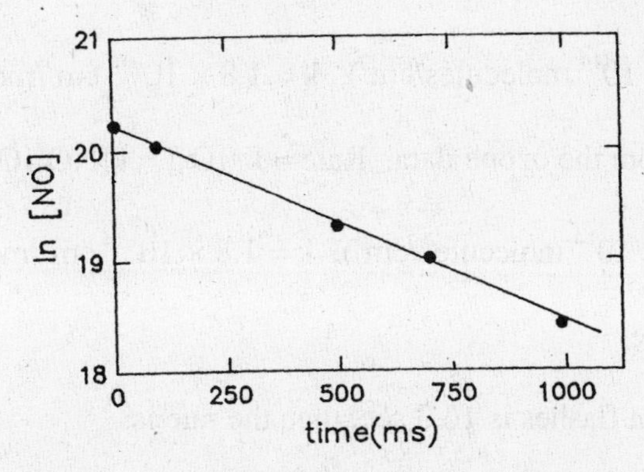

Because ln [NO] vs. t is linear, the reaction is first order with respect to NO.

We follow the same procedure for ozone using the second set of data. The data and plot are:

Time (ms)	$[O_3]$ (molecules/cm^3)	ln $[O_3]$
0	1.0×10^{10}	23.03
50.	8.4×10^9	22.85
100.	7.0×10^9	22.67
200.	4.9×10^9	22.31
300.	3.4×10^9	21.95

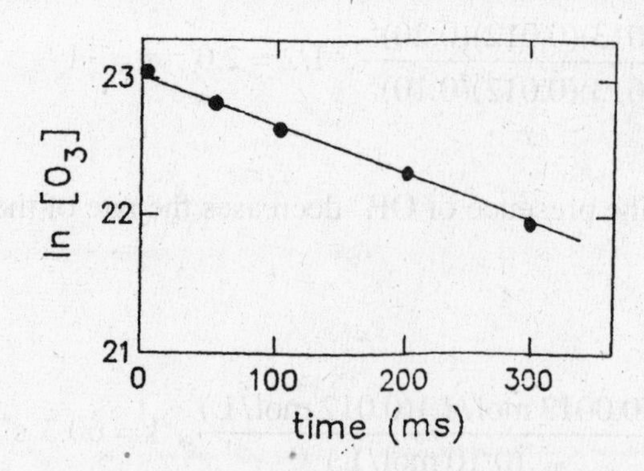

The plot of ln $[O_3]$ vs. t is linear. Hence, the reaction is first order with respect to ozone.

b. Rate = k[NO]$[O_3]$ is the overall rate law.

c. For NO experiment, Rate = k′[NO] and k′ = −(slope from graph of ln [NO] vs. t).

$$k' = \text{-slope} = -\frac{18.41 - 20.21}{(1000. - 0)\times 10^{-3}\ \text{s}} = 1.8\ \text{s}^{-1}$$

For ozone experiment, Rate = k″$[O_3]$ and k″ = −(slope from ln $[O_3]$ vs. t plot).

$$k'' = -\text{slope} = -\frac{(21.95 - 23.03)}{(300. - 0)\times 10^{-3}\ \text{s}} = 3.6\ \text{s}^{-1}$$

d. From NO experiment, Rate = k[NO]$[O_3]$ = k′[NO] where k′ = k$[O_3]$.

$k' = 1.8\ s^{-1} = k(1.0 \times 10^{14}$ molecules/cm^3), $k = 1.8 \times 10^{-14}$ cm^3/molecules•s

We can check this from the ozone data. Rate = k″$[O_3]$ = k[NO]$[O_3]$ where k″ = k[NO].

$k'' = 3.6\ s^{-1} = k(2.0 \times 10^{14}$ molecules/cm^3), $k = 1.8 \times 10^{-14}$ cm^3/molecules•s

Both values of k agree.

83. a. If the interval between flashes is 16.3 sec, then the rate is:

1 flash/16.3 s = $6.13 \times 10^{-2}\ s^{-1}$ = k

Interval	k	T
16.3 s	$6.13 \times 10^{-2}\ s^{-1}$	21.0°C (294.2 K)
13.0 s	$7.69 \times 10^{-2}\ s^{-1}$	27.8°C (301.0 K)

$$\ln\left(\frac{k_2}{k_1}\right) = \frac{E_a}{R}\left(\frac{1}{T_1} - \frac{1}{T_2}\right)$$; Solving: $E_a = 2.5 \times 10^4$ J/mol = 25 kJ/mol

b. $\ln\left(\frac{k}{6.13\times10^{-2}}\right) = \frac{2.5\times10^4 \text{ J/mol}}{8.3145 \text{ J/mol}\bullet\text{K}}\left(\frac{1}{294.2 \text{ K}} - \frac{1}{303.2 \text{ K}}\right) = 0.30$

$k = e^{0.30} \times 6.13\times10^{-2} = 8.3\times10^{-2} \text{ s}^{-1}$; Interval = 1/k = 12 seconds

c.

T	Interval	54-2(Intervals)
21.0 °C	16.3 s	21 °C
27.8 °C	13.0 s	28 °C
30.0 °C	12 s	30. °C

This rule of thumb gives excellent agreement to two significant figures.

85. a. [B] >> [A] so that [B] can be considered constant over the experiments. (This gives us a pseudo-order rate law equation.)

b. Note in each data set that successive half-lives double (in expt. 1 the first half-life is 40. sec, the second is 80. sec; in expt. 2 the first half-life is 20. sec, the second is 40. sec). Thus, the reaction is second order in [A] since $t_{1/2}$ for second order reactions is inversely proportional to concentration. Between expt. 1 and expt. 2, we double [B] and the reaction rate doubles, thus it is first order in [B]. The overall rate law equation is rate = $k[A]^2[B]$.

Using $t_{1/2} = \frac{1}{k[A]_0}$, we get $k = \frac{1}{(40.\text{ s})(10.0\times10^{-2}\text{ mol/L})} = 0.25$ L/mol•s. But this is actually k′ where rate = $k'[A]^2$ and $k' = k[B]$.

$$k = \frac{k'}{[B]} = \frac{0.25 \text{ L/mol}\bullet\text{s}}{5.0 \text{ mol/L}} = 0.050 \text{ L}^2/\text{mol}^2\bullet\text{s}$$

87. Rate = $k[A]^x[B]^y[C]^z$; During the course of experiment 1, [A] and [C] are essentially constant, and Rate = $k'[B]^y$ where $k' = k[A]_0^x[C]_0^z$.

[B] (*M*)	time (s)	ln[B]	1/[B] (M^{-1})
1.0×10^{-3}	0	-6.91	1.0×10^3
2.7×10^{-4}	1.0×10^5	-8.22	3.7×10^3
1.6×10^{-4}	2.0×10^5	-8.74	6.3×10^3
1.1×10^{-4}	3.0×10^5	-9.12	9.1×10^3
8.5×10^{-5}	4.0×10^5	-9.37	12×10^3
6.9×10^{-5}	5.0×10^5	-9.58	14×10^3
5.8×10^{-5}	6.0×10^5	-9.76	17×10^3

A plot of 1/[B] vs. t is linear (plot not included). So the reaction is second order in B and the integrated rate equation from the slope and y-intercept of the straight line in the plot is:

$1/[B] = (2.7\times10^{-2} \text{ L/mol}\bullet\text{s})\,t + 1.0\times10^3$ L/mol; $k' = 2.7\times10^{-2}$ L/mol•s

For experiment 2, [B] and [C] are essentially constant and Rate = $k''[A]^x$ where $k'' = k[B]_0^y[C]_0^z = k[B]_0^2[C]_0^z$.

[A] (M)	time (s)	ln[A]	1/[A] (M^{-1})
1.0×10^{-2}	0	-4.61	1.0×10^2
8.9×10^{-3}	1.0	-4.72	110
7.1×10^{-3}	3.0	-4.95	140
5.5×10^{-3}	5.0	-5.20	180
3.8×10^{-3}	8.0	-5.57	260
2.9×10^{-3}	10.0	-5.84	340
2.0×10^{-3}	13.0	-6.21	5.0×10^2

A plot of ln[A] vs. t is linear. So the reaction is first order in A and the integrated rate law from the slope and y-intercept of the straight line in the plot is:

$$\ln[A] = -(0.123 \text{ s}^{-1})\,t - 4.61; \quad k'' = 0.123 \text{ s}^{-1}$$

Note: We will carry an extra significant figure in k''.

Experiment 3: [A] and [B] are constant; Rate = $k'''[C]^z$

The plot of [C] vs. t is linear. Thus, z = 0.

The overall rate law is: Rate = $k[A][B]^2$

From Experiment 1 (to determine k):

$$k' = 2.7 \times 10^{-2} \text{ L/mol} \bullet \text{s} = k[A]_0^x[C]_0^z = k[A]_o = k(2.0\ M),\ k = 1.4 \times 10^{-2} \text{ L}^2/\text{mol}^2 \bullet \text{s}$$

From Experiment 2: $k'' = 0.123 \text{ s}^{-1} = k[B]_0^2,\ k = \dfrac{0.123 \text{ s}^{-1}}{(3.0\ M)^2} = 1.4 \times 10^{-2} \text{ L}^2/\text{mol}^2 \bullet \text{s}$

Thus, Rate = $k[A][B]^2$ and $k = 1.4 \times 10^{-2}$ L^2/mol^2•s.

Integrative Problems

89. $8.75 \text{ h} \times \dfrac{3600 \text{ s}}{\text{h}} = 3.15 \times 10^4 \text{ s};\ k = \dfrac{\ln 2}{t_{1/2}} = \dfrac{\ln 2}{3.15 \times 10^4 \text{ s}} = 2.20 \times 10^{-5} \text{ s}^{-1}$

The partial pressure of a gas is directly related to the concentration in mol/L. So instead of using mol/L as the concentration units in the integrated first order rate law, we can use partial pressures of SO_2Cl_2.

$$\ln\left(\frac{P}{P_0}\right) = -kt,\ \ln\left(\frac{P}{791 \text{ torr}}\right) = -(2.20 \times 10^{-5} \text{ s}^{-1})(12.5 \text{ h} \times \frac{3600 \text{ s}}{\text{h}})$$

$$P_{SO_2Cl_2} = 294 \text{ torr} \times \frac{1 \text{ atm}}{760 \text{ torr}} = 0.387 \text{ atm}$$

$$n = \frac{PV}{RT} = \frac{0.387 \text{ atm} \times 1.25 \text{ L}}{\frac{0.08206 \text{ L atm}}{\text{mol K}} \times 593 \text{ K}} = 9.94 \times 10^{-3} \text{ mol } SO_2Cl_2$$

$$9.94 \times 10^{-3} \text{ mol} \times \frac{6.022 \times 10^{23} \text{ molecules}}{\text{mol}} = 5.99 \times 10^{21} \text{ molecules } SO_2Cl_2$$

91. $\ln\left(\frac{k_2}{k_1}\right) = \frac{E_a}{R}\left(\frac{1}{T_1} - \frac{1}{T_2}\right)$; $\ln\left(\frac{1.7 \times 10^{-2} \text{ s}^{-1}}{7.2 \times 10^{-4} \text{ s}^{-1}}\right) = \frac{E_a}{8.3145 \text{ J/K} \bullet \text{mol}}\left(\frac{1}{660. \text{ K}} - \frac{1}{720. \text{ K}}\right)$

$E_a = 2.1 \times 10^5$ J/mol

For k at 325°C (598 K):

$$\ln\left(\frac{1.7 \times 10^{-2} \text{ s}^{-1}}{k}\right) = \frac{2.1 \times 10^5 \text{ J/mol}}{8.3145 \text{ J/K} \bullet \text{mol}}\left(\frac{1}{598 \text{ K}} - \frac{1}{720. \text{ K}}\right), \; k = 1.3 \times 10^{-5} \text{ s}^{-1}$$

For three half-lives, we go from 100% → 50% → 25% → 12.5%. After three half-lives, 12.5% of the original amount of C_2H_5I remains. Partial pressures are directly related to gas concentrations in mol/L:

$$P_{C_2H_5I} = 894 \text{ torr} \times 0.125 = 112 \text{ torr after 3 half-lives}$$

CHAPTER THIRTEEN

CHEMICAL EQUILIBRIUM

Questions

9. No, equilibrium is a dynamic process. Both reactions:

   $H_2O + CO \rightarrow H_2 + CO_2$ and $H_2 + CO_2 \rightarrow H_2O + CO$

   are occurring, but at equal rates. Thus, ^{14}C atoms will be distributed between CO and CO_2.

11. $H_2O(g) + CO(g) \rightleftharpoons H_2(g) + CO_2(g) \quad K = \dfrac{[H_2][CO_2]}{[H_2O][CO]} = 2.0$

   K is a unitless number since there is an equal number of moles of product gases as compared to moles of reactant gases in the balanced equation. Therefore, we can use units of molecules per liter instead of moles per liter to determine K.

   We need to start somewhere, so let's assume 3 molecules of CO react. If 3 molecules of CO react, then 3 molecules of H_2O must react, and 3 molecules each of H_2 and CO_2 are formed. We would have 6 − 3 = 3 molecules CO, 8 − 3 = 5 molecules H_2O, 0 + 3 = 3 molecules H_2, and 0 + 3 = 3 molecules CO_2 present. This will be an equilibrium mixture if K = 2.0:

$$K = \frac{\left(\frac{3 \text{ molecules } H_2}{L}\right)\left(\frac{3 \text{ molecules } CO_2}{L}\right)}{\left(\frac{5 \text{ molecules } H_2O}{L}\right)\left(\frac{3 \text{ molecules } CO}{L}\right)} = \frac{3}{5}$$

   Because this mixture does not give a value of K = 2.0, this is not an equilibrium mixture. Let's try 4 molecules of CO reacting to reach equilibrium.

   molecules CO remaining = 6 − 4 = 2 molecules CO;
   molecules H_2O remaining = 8 − 4 = 4 molecules H_2O;
   molecules H_2 present = 0 + 4 = 4 molecules H_2;
   molecules CO_2 present = 0 + 4 = 4 molecules CO_2

$$K = \frac{\left(\frac{4 \text{ molecules } H_2}{L}\right)\left(\frac{4 \text{ molecules } CO_2}{L}\right)}{\left(\frac{4 \text{ molecules } H_2O}{L}\right)\left(\frac{2 \text{ molecules } CO}{L}\right)} = 2.0$$

   Since K = 2.0 for this reaction mixture, we are at equilibrium.

13. K and K_p are equilibrium constants as determined by the law of mass action. For K, concentration units of mol/L are used, and for K_p, partial pressures in units of atm are used (generally). Q is called the reaction quotient. Q has the exact same form as K or K_p, but instead of equilibrium concentrations, initial concentrations are used to calculate the Q value. The use of Q is when it is compared to the K value. When Q = K (or when $Q_p = K_p$), the reaction is at equilibrium. When $Q \neq K$, the reaction is not at equilibrium and one can deduce the net change that must occur for the system to get to equilibrium.

15. We always try to make good assumptions that simplify the math. In some problems, we can set-up the problem so that the net change, x, that must occur to reach equilibrium is a small number. This comes in handy when you have expressions like 0.12 – x or 0.727 + 2x, etc. When x is small, we can assume that it makes little difference when subtracted from or added to some relatively big number. When this is the case, $0.12 - x \approx 0.12$ and $0.727 + 2x \approx 0.727$, etc. If the assumption holds by the 5% rule, the assumption is assumed valid. The 5% rule refers to x (or 2x or 3x, etc.) that is assumed small compared to some number. If x (or 2x or 3x, etc.) is less than 5% of the number the assumption was made against, then the assumption will be assumed valid. If the 5% rule fails to work, one can use a math procedure called the method of successive approximations to solve the quadratic or cubic equation. Of course, one could always solve the quadratic or cubic equation exactly. This is generally a last resort (and is usually not necessary).

The Equilibrium Constant

17. a. $K = \dfrac{[NO]^2}{[N_2][O_2]}$ b. $K = \dfrac{[NO_2]^2}{[N_2O_4]}$

c. $K = \dfrac{[SiCl_4][H_2]^2}{[SiH_4][Cl_2]^2}$ d. $K = \dfrac{[PCl_3]^2[Br_2]^3}{[PBr_3]^2[Cl_2]^3}$

19. $K = 1.3 \times 10^{-2} = \dfrac{[NH_3]^2}{[N_2][H_2]^3}$ for $N_2(g) + 3\,H_2(g) \rightleftharpoons 2\,NH_3(g)$

When a reaction is reversed, then $K_{new} = 1/K_{original}$. When a reaction is multiplied through by a value of n, then $K_{new} = (K_{original})^n$.

a. $1/2\,N_2(g) + 3/2\,H_2(g) \rightleftharpoons NH_3(g)$ $K' = \dfrac{[NH_3]^2}{[N_2]^{1/2}[H_2]^{3/2}} = K^{1/2} = (1.3 \times 10^{-2})^{1/2} = 0.11$

b. $2\,NH_3(g) \rightleftharpoons N_2(g) + 3\,H_2(g)$ $K'' = \dfrac{[N_2][H_2]^3}{[NH_3]^2} = \dfrac{1}{K} = \dfrac{1}{1.3 \times 10^{-2}} = 77$

c. $NH_3(g) \rightleftharpoons 1/2\,N_2(g) + 3/2\,H_2(g)$ $K''' = \dfrac{[N_2]^{1/2}[H_2]^{3/2}}{[NH_3]} = \left(\dfrac{1}{K}\right)^{1/2} = \left(\dfrac{1}{1.3 \times 10^{-2}}\right)^{1/2}$

$= 8.8$

d. $2\,N_2(g) + 6\,H_2(g) \rightleftharpoons 4\,NH_3(g)$ $K = \dfrac{[NH_3]^4}{[N_2]^2[H_2]^6} = (K)^2 = (1.3 \times 10^{-2})^2 = 1.7 \times 10^{-4}$

21. $2\,NO(g) + 2\,H_2(g) \rightleftharpoons N_2(g) + 2\,H_2O(g)$ $K = \dfrac{[N_2][H_2O]^2}{[NO]^2[H_2]^2}$

$$K = \frac{(5.3\times10^{-2})(2.9\times10^{-3})^2}{(8.1\times10^{-3})^2\,(4.1\times10^{-5})^2} = 4.0 \times 10^6$$

23. $[NO] = \dfrac{4.5\times10^{-3}\text{ mol}}{3.0\text{ L}} = 1.5 \times 10^{-3}\,M$; $[Cl_2] = \dfrac{2.4\text{ mol}}{3.0\text{ L}} = 0.80\,M$

$[NOCl] = \dfrac{1.0\text{ mol}}{3.0\text{ L}} = 0.33\,M$; $K = \dfrac{[NO]^2[Cl_2]}{[NOCl]^2} = \dfrac{(1.5\times10^{-3})^2(0.80)}{(0.33)^2} = 1.7 \times 10^{-5}$

25. $K_p = \dfrac{P_{NO}^2 \times P_{O_2}}{P_{NO_2}^2} = \dfrac{(6.5\times10^{-5})^2(4.5\times10^{-5})}{(0.55)^2} = 6.3 \times 10^{-13}$

27. $K_p = K(RT)^{\Delta n}$ where Δn = sum of gaseous product coefficients − sum of gaseous reactant coefficients. For this reaction, $\Delta n = 3 - 1 = 2$.

$$K = \frac{[CO][H_2]^2}{[CH_3OH]} = \frac{(0.24)(1.1)^2}{(0.15)} = 1.9$$

$K_p = K(RT)^2 = 1.9\,(0.08206\text{ L/atm/K}\bullet\text{ mol} \times 600.\text{ K})^2 = 4.6 \times 10^3$

29. Solids and liquids do not appear in equilibrium expressions. Only gases and dissolved solutes appear in equilibrium expressions.

a. $K = \dfrac{[H_2O]}{[NH_3]^2[CO_2]}$; $K_p = \dfrac{P_{H_2O}}{P_{NH_3}^2 \times P_{CO_2}}$

b. $K = [N_2][Br_2]^3$; $K_p = P_{N_2} \times P_{Br_2}^3$

c. $K = [O_2]^3$; $K_p = P_{O_2}^3$

d. $K = \dfrac{[H_2O]}{[H_2]}$; $K_p = \dfrac{P_{H_2O}}{P_{H_2}}$

31. Because solids do not appear in the equilibrium constant expression, $K = 1/[O_2]^3$.

$[O_2] = \dfrac{1.0\times10^{-3}\text{ mol}}{2.0\text{ L}}$; $K = \dfrac{1}{[O_2]^3} = \dfrac{1}{\left(\dfrac{1.0\times10^{-3}}{2.0}\right)^3} = \dfrac{1}{(5.0\times10^{-4})^3} = 8.0 \times 10^9$

Equilibrium Calculations

33. $2\ NO(g) \rightleftharpoons N_2(g) + O_2(g) \quad K = \frac{[N_2][O_2]}{[NO]^2} = 2.4 \times 10^3$

Use the reaction quotient Q to determine which way the reaction shifts to reach equilibrium. For the reaction quotient, initial concentrations given in a problem are used to calculate the value for Q. If Q < K, then the reaction shifts right to reach equilibrium. If Q > K, then the reaction shifts left to reach equilibrium. If Q = K, then the reaction does not shift in either direction because the reaction is at equilibrium.

a. $[N_2] = \frac{2.0\ \text{mol}}{1.0\ \text{L}} = 2.0\ M;\ [O_2] = \frac{2.6\ \text{mol}}{1.0\ \text{L}} = 2.6\ M;\ [NO] = \frac{0.024\ \text{mol}}{1.0\ \text{L}} = 0.024\ M$

$$Q = \frac{[N_2]_o[O_2]_o}{[NO]_o^2} = \frac{(2.0)(2.6)}{(0.024)^2} = 9.0 \times 10^3$$

Q > K so the reaction shifts left to produce more reactants in order to reach equilibrium.

b. $[N_2] = \frac{0.62\ \text{mol}}{2.0\ \text{L}} = 0.31\ M;\ [O_2] = \frac{4.0\ \text{mol}}{2.0\ \text{L}} = 2.0\ M;\ [NO] = \frac{0.032\ \text{mol}}{2.0\ \text{L}} = 0.016\ M$

$$Q = \frac{(0.31)(2.0)}{(0.016)^2} = 2.4 \times 10^3 = K;\ \text{at equilibrium}$$

c. $[N_2] = \frac{2.4\ \text{mol}}{3.0\ \text{L}} = 0.80\ M;\ [O_2] = \frac{1.7\ \text{mol}}{3.0\ \text{L}} = 0.57\ M;\ [NO] = \frac{0.060\ \text{mol}}{3.0\ \text{L}} = 0.020\ M$

$$Q = \frac{(0.80)(0.57)}{(0.020)^2} = 1.1 \times 10^3 < K;\ \text{Reaction shifts right to reach equilibrium.}$$

35. $CaCO_3(s) \rightleftharpoons CaO(s) + CO_2(g) \quad K_p = P_{CO_2} = 1.04$

a. $Q = P_{CO_2}$; We only need the partial pressure of CO_2 to determine Q since solids do not appear in equilibrium expressions (or Q expressions). At this temperature all CO_2 will be in the gas phase. Q = 2.55 so $Q > K_p$; Reaction will shift to the left to reach equilibrium; the mass of CaO will decrease.

b. $Q = 1.04 = K_p$ so the reaction is at equilibrium; mass of CaO will not change.

c. $Q = 1.04 = K_p$ so the reaction is at equilibrium; mass of CaO will not change.

d. $Q = 0.211 < K_p$; The reaction will shift to the right to reach equilibrium; the mass of CaO will increase.

37. $K = \frac{[H_2]^2[O_2]}{[H_2O]^2}$, $2.4\times10^{-3} = \frac{(1.9\times10^{-2})^2[O_2]}{(0.11)^2}$, $[O_2] = 0.080\ M$

39. $SO_2(g) + NO_2(g) \rightleftharpoons SO_3(g) + NO(g)$ $K = \frac{[SO_3][NO]}{[SO_2][NO_2]}$

To determine K, we must calculate the equilibrium concentrations. The initial concentrations are:

$$[SO_3]_o = [NO]_o = 0;\ \ [SO_2]_o = [NO_2]_o = \frac{2.00\ \text{mol}}{1.00\ \text{L}} = 2.00\ M$$

Next, we determine the change required to reach equilibrium. At equilibrium, [NO] = 1.30 mol/1.00 L = 1.30 *M*. Since there was zero NO present initially, 1.30 *M* of SO_2 and 1.30 *M* NO_2 must have reacted to produce 1.30 *M* NO as well as 1.30 *M* SO_3, all required by the balanced reaction. The equilibrium concentration for each substance is the sum of the initial concentration plus the change in concentration necessary to reach equilibrium. The equilibrium concentrations are:

$$[SO_3] = [NO] = 0 + 1.30\ M = 1.30\ M;\ \ [SO_2] = [NO_2] = 2.00\ M - 1.30\ M = 0.70\ M$$

We now use these equilibrium concentrations to calculate K:

$$K = \frac{[SO_3][NO]}{[SO_2][NO_2]} = \frac{(1.30)(1.30)}{(0.70)(0.70)} = 3.4$$

41. When solving equilibrium problems, a common method to summarize all the information in the problem is to set up a table. We call this table the ICE table since it summarizes initial concentrations, changes that must occur to reach equilibrium and equilibrium concentrations (the sum of the initial and change columns). For the change column, we will generally use the variable *x*, which will be defined as the amount of reactant (or product) that must react to reach equilibrium. In this problem, the reaction must shift right to reach equilibrium because there are no products present initially. Therefore, *x* is defined as the amount (in units of mol/L) of reactant SO_3 that reacts to reach equilibrium; we use the coefficients in the balanced equation to relate the net change in SO_3 to the net change in SO_2 and O_2. The general ICE table for this problem is:

	$2\ SO_3(g)$	$\rightleftharpoons$	$2\ SO_2(g)$	+	$O_2(g)$	$K = \frac{[SO_2]^2[O_2]}{[SO_3]^2}$
Initial	12.0 mol/3.0 L		0		0	
	Let *x* mol/L of SO_3 react to reach equilibrium					
Change	$-x$	$\rightarrow$	$+x$		$+x/2$	
Equil.	$4.0 - x$		x		$x/2$	

From the problem, we are told that the equilibrium SO_2 concentration is 3.0 mol/3.0 L = 1.0 *M* ($[SO_2]_e$ = 1.0 *M*). From the ICE table, $[SO_2]_e = x$ so $x = 1.0$. Solving for the other equilibrium concentrations: $[SO_3]_e = 4.0 - x = 4.0 - 1.0 = 3.0\ M$; $[O_2] = x/2 = 1.0/2 = 0.50\ M$.

$$K = \frac{[SO_2]^2[O_2]}{[SO_3]^2} = \frac{(1.0)^2(0.50)}{(3.0)^2} = 0.056$$

Alternate Method: Fractions in the change column can be avoided (if you want) by defining x differently. If we were to let $2x$ mol/L of SO_3 react to reach equilibrium, then the ICE table is:

	$2\ SO_3(g)$	$\rightleftharpoons$	$2\ SO_2(g)$	+	$O_2(g)$	$K = \frac{[SO_2]^2[O_2]}{[SO_3]^2}$
Initial	4.0 M		0		0	
	Let $2x$ mol/L of SO_3 react to reach equilibrium					
Change	$-2x$	$\rightarrow$	$+2x$		$+x$	
Equil.	$4.0 - 2x$		$2x$		x	

Solving: $2x = [SO_2]_e = 1.0\ M$, $x = 0.50\ M$; $[SO_3]_e = 4.0 - 2(0.50) = 3.0\ M$; $[O_2]_e = x = 0.50\ M$

These are exactly the same equilibrium concentrations as solved for previously, thus K will be the same (as it must be). The moral of the story is to define x in a manner that is most comfortable for you. Your final answer is independent of how you define x initially.

43.

	$3\ H_2(g)$	+	$N_2(g)$	$\rightleftharpoons$	$2\ NH_3(g)$
Initial	$[H_2]_o$		$[N_2]_o$		0
	x mol/L of N_2 reacts to reach equilibrium				
Change	$-3x$		$-x$	$\rightarrow$	$+2x$
Equil	$[H_2]_o - 3x$		$[N_2]_o - x$		$2x$

From the problem:

$$[NH_3]_e = 4.0\ M = 2x,\ x = 2.0\ M;\ [H_2]_e = 5.0\ M = [H_2]_o - 3x;\ [N_2]_e = 8.0\ M = [N_2]_o - x$$

$$5.0\ M = [H_2]_o - 3(2.0\ M),\ [H_2]_o = 11.0\ M;\ 8.0\ M = [N_2]_o - 2.0\ M,\ [N_2]_o = 10.0\ M$$

45. Q = 1.00, which is less than K. Reaction shifts to the right to reach equilibrium. Summarizing the equilibrium problem in a table:

	$SO_2(g)$	+	$NO_2(g)$	$\rightleftharpoons$	$SO_3(g)$	+	$NO(g)$	K = 3.75
Initial	0.800 M		0.800 M		0.800 M		0.800 M	
	x mol/L of SO_2 reacts to reach equilibrium							
Change	$-x$		$-x$	$\rightarrow$	$+x$		$+x$	
Equil.	$0.800 - x$		$0.800 - x$		$0.800 + x$		$0.800 + x$	

Plug the equilibrium concentrations into the equilibrium constant expression:

$$K = \frac{[SO_3][NO]}{[SO_2][NO_2]} = 3.75 = \frac{(0.800 + x)^2}{(0.800 - x)^2}$$; Take the square root of both sides and solve for x:

$$\frac{0.800 + x}{0.800 - x} = 1.94,\ 0.800 + x = 1.55 - 1.94\,x,\ 2.94\,x = 0.75,\ x = 0.26\ M$$

The equilibrium concentrations are:

$[SO_3] = [NO] = 0.800 + x = 0.800 + 0.26 = 1.06\ M$; $[SO_2] = [NO_2] = 0.800 - x = 0.54\ M$

47. Because only reactants are present initially, the reaction must proceed to the right to reach equilibrium. Summarizing the problem in a table:

	$N_2(g)$	+	$O_2(g)$	$\rightleftharpoons$	$2\ NO(g)$	$K_p = 0.050$
Initial	0.80 atm		0.20 atm		0	
	x atm of N_2 reacts to reach equilibrium					
Change	$-x$		$-x$	$\rightarrow$	$+2x$	
Equil.	$0.80 - x$		$0.20 - x$		$2x$	

$$K_p = 0.050 = \frac{P_{NO}^2}{P_{N_2} \times P_{O_2}} = \frac{(2x)^2}{(0.80 - x)(0.20 - x)},\ 0.050(0.16 - 1.00\,x + x^2) = 4\,x^2$$

$$4\,x^2 = 8.0 \times 10^{-3} - 0.050\,x + 0.050\,x^2,\ 3.95\,x^2 + 0.050\,x - 8.0 \times 10^{-3} = 0$$

Solving using the quadratic formula (see Appendix 1.4 of the text):

$$x = \frac{-b \pm (b^2 - 4ac)^{1/2}}{2a} = \frac{-0.050 \pm [(0.050)^2 - 4(3.95)(-8.0 \times 10^{-3})]^{1/2}}{2(3.95)}$$

$x = 3.9 \times 10^{-2}$ atm or $x = -5.2 \times 10^{-2}$ atm; Only $x = 3.9 \times 10^{-2}$ atm makes sense (x cannot be negative), so the equilibrium NO partial pressure is:

$$P_{NO} = 2x = 2(3.9 \times 10^{-2}\ \text{atm}) = 7.8 \times 10^{-2}\ \text{atm}$$

49.

	$2\ SO_2(g)$	+	$O_2(g)$	$\rightleftharpoons$	$2\ SO_3(g)$	$K_p = 0.25$
Initial	0.50 atm		0.50 atm		0	
	$2x$ atm of SO_2 reacts to reach equilibrium					
Change	$-2x$		$-x$	$\rightarrow$	$+2x$	
Equil.	$0.50 - 2x$		$0.50 - x$		$2x$	

$$K_p = 0.25 = \frac{P_{SO_3}^2}{P_{SO_2}^2 \times P_{O_2}} = \frac{(2x)^2}{(0.50 - 2x)^2(0.50 - x)}$$

This will give a cubic equation. Graphing calculators can be used to solve this expression. If you don't have a graphing calculator, an alternative method for solving a cubic equation is to use the method of successive approximations (see Appendix 1.4 of the text). The first step is to guess a value for x. Because the value of K is small (K < 1), then not much of the forward reaction will occur to reach equilibrium. This tells us that x is small. Lets guess that $x = 0.050$ atm. Now we take this estimated value for x and substitute it into the equation everywhere that x appears except for one. For equilibrium problems, we will substitute the estimated value for x into the denominator, then solve for the numerator value of x. We continue this process until the estimated value of x and the calculated value of x converge on the same number. This is the same answer we would get if we were to solve the cubic equation exactly. Applying the method of successive approximations and carrying extra significant figures:

$$\frac{4x^2}{[0.50-2(0.050)]^2[0.50-(0.050)]} = \frac{4x^2}{(0.40)^2(0.45)} = 0.25,\ x = 0.067$$

$$\frac{4x^2}{[0.50-2(0.067)]^2[0.50-(0.067)]} = \frac{4x^2}{(0.366)^2(0.433)} = 0.25,\ x = 0.060$$

$$\frac{4x^2}{(0.38)^2(0.44)},\ 0.25,\ x = 0.063;\quad \frac{4x^2}{(0.374)^2(0.437)} = 0.25,\ x = 0.062$$

The next trial gives the same value for $x = 0.062$ atm. We are done except for determining the equilibrium concentrations. They are:

$$P_{SO_2} = 0.50 - 2x = 0.50 - 2(0.062) = 0.376 = 0.38 \text{ atm}$$

$$P_{O_2} = 0.50 - x = 0.438 = 0.44 \text{ atm};\ P_{SO_3} = 2x = 0.124 = 0.12 \text{ atm}$$

51. a. The reaction must proceed to products to reach equilibrium since only reactants are present initially. Summarizing the problem in a table:

	$2\ NOCl(g)$	$\rightleftharpoons$	$2\ NO(g)$	+	$Cl_2(g)$	$K = 1.6 \times 10^{-5}$
Initial	$\frac{2.0 \text{ mol}}{2.0 \text{ L}}$		0		0	
	$2x$ mol/L of NOCl reacts to reach equilibrium					
Change	$-2x$	$\rightarrow$	$+2x$		$+x$	
Equil.	$1.0 - 2x$		$2x$		x	

$$K = 1.6 \times 10^{-5} = \frac{[NO]^2[Cl_2]}{[NOCl]^2} = \frac{(2x)^2(x)}{(1.0-2x)^2}$$

If we assume that $1.0 - 2x \approx 1.0$ (from the small size of K, we know that not a lot of products are present at equilibrium so x is small), then:

$$1.6 \times 10^{-5} = \frac{4x^3}{1.0^2},\quad x = 1.6 \times 10^{-2};$$ Now we must check the assumption.

$$1.0 - 2x = 1.0 - 2(0.016) = 0.97 = 1.0 \text{ (to proper significant figures)}$$

Our error is about 3%, i.e., $2x$ is 3.2% of 1.0 M. Generally, if the error we introduce by making simplifying assumptions is less than 5%, we go no further and the assumption is said to be valid. We call this the 5% rule. Solving for the equilibrium concentrations:

$$[NO] = 2x = 0.032\ M;\ [Cl_2] = x = 0.016\ M;\ [NOCl] = 1.0 - 2x = 0.97\ M \approx 1.0\ M$$

Note: If we were to solve this cubic equation exactly (a long and tedious process), we would get x = 0.016. This is the exact same answer we determined by making a simplifying assumption. We saved time and energy. Whenever K is a very small value, always make the assumption that x is small. If the assumption introduces an error of less than 5%, then the answer you calculated making the assumption will be considered the correct answer.

b.

	$2\ NOCl(g)$	$\rightleftharpoons$	$2\ NO(g)$	+	$Cl_2(g)$
Initial	1.0 M		1.0 M		0
	$2x$ mol/L of NOCl reacts to reach equilibrium				
Change	$-2x$	$\rightarrow$	$+2x$		$+x$
Equil.	$1.0 - 2x$		$1.0 + 2x$		x

$$1.6 \times 10^{-5} = \frac{(1.0 + 2x)^2(x)}{(1.0 - 2x)^2} = \frac{(1.0)^2(x)}{(1.0)^2} \quad \text{(assuming } 2x << 1.0\text{)}$$

$x = 1.6 \times 10^{-5}$; Assumptions are great ($2x$ is 3.2×10^{-3} % of 1.0).

$[Cl_2] = 1.6 \times 10^{-5}\ M$ and $[NOCl] = [NO] = 1.0\ M$

c.

	$2\ NOCl(g)$	$\rightleftharpoons$	$2\ NO(g)$	+	$Cl_2(g)$
Initial	2.0 M		0		1.0 M
	$2x$ mol/L of NOCl reacts to reach equilibrium				
Change	$-2x$	$\rightarrow$	$+2x$		$+x$
Equil.	$2.0 - 2x$		$2x$		$1.0 + x$

$$1.6 \times 10^{-5} = \frac{(2x)^2(1.0 + x)}{(2.0 - 2x)^2} = \frac{4x^2}{4.0} \quad \text{(assuming } x << 1.0\text{)}$$

Solving: $x = 4.0 \times 10^{-3}$; Assumptions good (x is 0.4% of 1.0 and $2x$ is 0.4% of 2.0).

$[Cl_2] = 1.0 + x = 1.0\ M$; $[NO] = 2(4.0 \times 10^{-3}) = 8.0 \times 10^{-3}\ M$; $[NOCl] = 2.0\ M$

53. $2\ CO_2(g) \rightleftharpoons 2\ CO(g) + O_2(g) \quad K = \frac{[CO]^2[O_2]}{[CO_2]^2} = 2.0 \times 10^{-6}$

	$2\ CO_2(g)$		$2\ CO(g)$	$O_2(g)$
Initial	2.0 mol/5.0 L		0	0
	$2x$ mol/L of CO_2 reacts to reach equilibrium			
Change	$-2x$	→	$+2x$	$+x$
Equil.	$0.40 - 2x$		$2x$	x

$$K = 2.0 \times 10^{-6} = \frac{[CO]^2[O_2]}{[CO_2]^2} = \frac{(2x)^2(x)}{(0.40 - 2x)^2};\ \text{Assuming } 2x \ll 0.40:$$

$$2.0 \times 10^{-6} \approx \frac{4x^3}{(0.40)^2},\ 2.0 \times 10^{-6} = \frac{4x^3}{0.16},\ x = 4.3 \times 10^{-3}\ M$$

Checking assumption: $\frac{2(4.3 \times 10^{-3})}{0.40} \times 100 = 2.2\%$; Assumption valid by the 5% rule.

$[CO_2] = 0.40 - 2x = 0.40 - 2(4.3 \times 10^{-3}) = 0.39\ M$

$[CO] = 2x = 2(4.3 \times 10^{-3}) = 8.6 \times 10^{-3}\ M$; $[O_2] = x = 4.3 \times 10^{-3}\ M$

55. This is a typical equilibrium problem except that the reaction contains a solid. Whenever solids and liquids are present, we basically ignore them in the equilibrium problem.

$$NH_4OCONH_2(s) \rightleftharpoons 2\ NH_3(g) + CO_2(g) \qquad K_p = 2.9 \times 10^{-3}$$

	$NH_4OCONH_2(s)$		$2\ NH_3(g)$	$CO_2(g)$
Initial	–		0	0
	Some NH_4OCONH_2 decomposes to produce $2x$ atm of NH_3 and x atm of CO_2.			
Change	–	→	$+2x$	$+x$
Equil.	–		$2x$	x

$$K_p = 2.9 \times 10^{-3} = P_{NH_3}^2 \times P_{CO_2} = (2x)^2(x) = 4x^3$$

$$x = \left(\frac{2.9 \times 10^{-3}}{4}\right)^{1/3} = 9.0 \times 10^{-2}\ \text{atm};\ P_{NH_3} = 2x = 0.18\ \text{atm};\ P_{CO_2} = x = 9.0 \times 10^{-2}\ \text{atm}$$

$$P_{total} = P_{NH_3} + P_{CO_2} = 0.18\ \text{atm} + 0.090\ \text{atm} = 0.27\ \text{atm}$$

Le Chatelier's Principle

57. a. No effect; Adding more of a pure solid or pure liquid has no effect on the equilibrium position.

b. Shifts left; HF(g) will be removed by reaction with the glass. As HF(g) is removed, the reaction will shift left to produce more HF(g).

c. Shifts right; As $H_2O(g)$ is removed, the reaction will shift right to produce more $H_2O(g)$.

59. a. right b. right c. no effect; He(g) is neither a reactant nor a product.

d. left; The reaction is exothermic; heat is a product:

$$CO(g) + H_2O(g) \rightarrow H_2(g) + CO_2(g) + \text{Heat}$$

Increasing T will add heat. The equilibrium shifts to the left to use up the added heat.

e. no effect; There are equal numbers of gas molecules on both sides of the reaction, so a change in volume has no effect on the equilibrium position.

61. a. left b. right c. left

d. no effect (reactant and product concentrations are unchanged)

e. no effect; Since there are equal numbers of product and reactant gas molecules, a change in volume has no effect on this equilibrium position.

f. right; A decrease in temperature will shift the equilibrium to the right since heat is a product in this reaction (as is true in all exothermic reactions).

63. An endothermic reaction, where heat is a reactant, will shift right to products with an increase in temperature. The amount of $NH_3(g)$ will increase as the reaction shifts right so the smell of ammonia will increase.

Additional Exercises

65.

$$O(g) + NO(g) \rightleftharpoons NO_2(g) \qquad K = 1/6.8 \times 10^{-49} = 1.5 \times 10^{48}$$

$$NO_2(g) + O_2(g) \rightleftharpoons NO(g) + O_3(g) \qquad K = 1/5.8 \times 10^{-34} = 1.7 \times 10^{33}$$

$$O_2(g) + O(g) \rightleftharpoons O_3(g) \qquad K = (1.5 \times 10^{48})(1.7 \times 10^{33}) = 2.6 \times 10^{81}$$

67. a.

	$2\ AsH_3(g)$	$\rightleftharpoons$ $2\ As(s)$	+ $3\ H_2(g)$
Initial	392.0 torr	-	0
Equil.	$392.0 - 2x$	-	$3x$

Using Dalton's Law of Partial Pressure:

$$P_{total} = 488.0 \text{ torr} = P_{AsH_3} + P_{H_2} = 392.0 - 2x + 3x, \ x = 96.0 \text{ torr}$$

$$P_{H_2} = 3x = 3(96.0) = 288 \text{ torr} \times \frac{1 \text{ atm}}{760 \text{ torr}} = 0.379 \text{ atm}$$

b. $P_{AsH_3} = 392.0 - 2(96.0) = 200.0 \text{ torr} \times \frac{1 \text{ atm}}{760 \text{ torr}} = 0.2632 \text{ atm}$

$$K_p = \frac{(P_{H_2})^3}{(P_{AsH_3})^2} = \frac{(0.379)^3}{(0.2632)^2} = 0.786$$

69. a. $P_{Cl_5} = \frac{nRT}{V} = \frac{\frac{2.450 \text{ g PCl}_5}{208.22 \text{ g/mol}} \times \frac{0.08206 \text{ L atm}}{\text{mol K}} \times 600.\text{ K}}{0.500 \text{ L}} = 1.16 \text{ atm}$

b. $PCl_5(g) \rightleftharpoons PCl_3(g) + Cl_2(g) \qquad K_p = \frac{P_{PCl_3} \times P_{Cl_2}}{P_{PCl_5}} = 11.5$

	$PCl_5(g)$		$PCl_3(g)$	$Cl_2(g)$
Initial	1.16 atm		0	0
	x atm of PCl_5 reacts to reach equilibrium			
Change	$-x$	$\rightarrow$	$+x$	$+x$
Equil.	$1.16 - x$		x	x

$K_p = \frac{x^2}{1.16 - x} = 11.5$, $x^2 + 11.5x - 13.3 = 0$; Using the quadratic formula: $x = 1.06$ atm

$P_{PCl_5} = 1.16 - 1.06 = 0.10 \text{ atm}$

c. $P_{PCl_3} = P_{Cl_2} = 1.06$ atm; $P_{PCl_5} = 0.10$ atm

$P_{tot} = P_{PCl_5} + P_{PCl_3} + P_{Cl_2} = 0.10 + 1.06 + 1.06 = 2.22 \text{ atm}$

d. Percent dissociation $= \frac{x}{1.16} \times 100 = \frac{1.06}{1.16} \times 100 = 91.4\%$

71. $K = \frac{[HF]^2}{[H_2][F_2]} = \frac{(0.400)^2}{(0.0500)(0.0100)} = 320.$; 0.200 mol F_2/5.00 L = 0.0400 M F_2 added

From LeChatelier's principle, added F_2 causes the reaction to shift right to reestablish equilibrium.

	$H_2(g)$	+	$F_2(g)$	$\rightleftharpoons$	$2\ HF(g)$
Initial	0.0500 M		0.0500 M		0.400 M
	x mol/L of F_2 reacts to reach equilibrium				
Change	$-x$		$-x$	$\rightarrow$	$+2x$
Equil.	$0.0500 - x$		$0.0500 - x$		$0.400 + 2x$

$K = 320. = \frac{(0.400 + 2x)^2}{(0.500 - x)^2}$; Taking the square root of the equation:

$$17.9 = \frac{0.400 + 2x}{0.500 - x},\ 0.895 - 17.9\,x = 0.400 + 2x,\ 19.9\,x = 0.495,\ x = 0.0249\ \text{mol/L}$$

$[HF] = 0.400 + 2(0.0249) = 0.450\ M$; $[H_2] = [F_2] = 0.0500 - 0.0249 = 0.0251\ M$

73. $H^+ + OH^- \rightarrow H_2O$; Sodium hydroxide (NaOH) will react with the H^+ on the product side of the reaction. This effectively removes H^+ from the equilibrium, which will shift the reaction to the right to produce more H^+ and CrO_4^{2-}. Because more CrO_4^{2-} is produced, the solution turns yellow.

75. $$PCl_5(g) \rightleftharpoons PCl_3(g) + Cl_2(g) \quad K = \frac{[PCl_3][Cl_2]}{[PCl_5]} = 4.5 \times 10^{-3}$$

At equilibrium, $[PCl_5] = 2[PCl_3]$.

$$4.5 \times 10^{-3} = \frac{[PCl_3][Cl_2]}{2[PCl_3]},\ [Cl_2] = 2(4.5 \times 10^{-3}) = 9.0 \times 10^{-3}\ M$$

77. $$PCl_5(g) \rightleftharpoons PCl_3(g) + Cl_2(g) \qquad K_p = P_{PCl_3} \times P_{Cl_2} / P_{PCl_5}$$

	$PCl_5(g)$		$PCl_3(g)$	$Cl_2(g)$	
Initial	P_0		0	0	P_0 = initial PCl_5 pressure
Change	$-x$	$\rightarrow$	$+x$	$+x$	
Equil.	$P_0 - x$		x	x	

$P_{total} = P_0 - x + x + x = P_0 + x = 358.7$ torr

$$P_0 = \frac{n_{PCl_5}RT}{V} = \frac{\frac{2.4156\ g}{208.22\ g/mol} \times \frac{0.08206\ L\ atm}{K\ mol} \times 523.2\ K}{2.000\ L} = 0.2490\ atm = 189.2\ torr$$

$x = P_{total} - P_0 = 358.7 - 189.2 = 169.5$ torr

$P_{PCl_3} = P_{Cl_2} = 169.5$ torr $= 0.2230$ atm

$P_{PCl_5} = 189.2 - 169.5 = 19.7$ torr $= 0.0259$ atm

$$K_p = \frac{(0.2230)^2}{0.0259} = 1.92$$

Challenge Problems

79. There is a little trick we can use to solve this problem in order to avoid solving a cubic equation. Because K for this reaction is very small, the dominant reaction is the reverse reaction. We will let the products react to completion by the reverse reaction, then we will solve the forward equilibrium problem to determine the equilibrium concentrations. Summarizing these steps in a table:

	$2\ NOCl(g)$	$\rightleftharpoons$	$2\ NO(g)$	+	$Cl_2(g)$	$K = 1.6 \times 10^{-5}$
Before	0		2.0 *M*		1.0 *M*	
	Let 1.0 mol/L Cl_2 react completely.					(K is small, reactants dominate.)
Change	+2.0	←	−2.0		−1.0	React completely
After	2.0		0		0	New initial conditions
	$2x$ mol/L of NOCl reacts to reach equilibrium					
Change	$-2x$	→	$+2x$		$+x$	
Equil.	$2.0 - 2x$		$2x$		x	

$$K = 1.6 \times 10^{-5} = \frac{(2x)^2(x)}{(2.0-2x)^2} \approx \frac{4x^3}{2.0^2} \quad \text{(assuming } 2.0 - 2x \approx 2.0\text{)}$$

$x^3 = 1.6 \times 10^{-5}$, $x = 2.5 \times 10^{-2}$; Assumption good by the 5% rule ($2x$ is 2.5% of 2.0).

[NOCl] = 2.0 − 0.050 = 1.95 *M* = 2.0 *M*; [NO] = 0.050 *M*; [Cl_2] = 0.025 *M*

Note: If we do not break this problem into two parts (a stoichiometric part and an equilibrium part), we are faced with solving a cubic equation. The set-up would be:

	$2\ NOCl(g)$	$\rightleftharpoons$	$2\ NO(g)$	+	$Cl_2(g)$
Initial	0		2.0 *M*		1.0 *M*
Change	+2.0	←	$-2y$		$-y$
Equil.	$2y$		$2.0 - 2y$		$1.0 - y$

$$1.6 \times 10^{-5} = \frac{(2.0-2y)^2(1.0-y)}{(2y)^2}$$ If we say that y is small to simplify the problem, then:

$$1.6 \times 10^{-5} = \frac{2.0^2}{4y^2};$$ We get $y = 250$. This is impossible!

To solve this equation, we cannot make any simplifying assumptions; we have to solve a cubic equation. If you don't have a graphing calculator, this is difficult. Alternatively, we can use some chemical common sense and solve the problem as illustrated above.

81.

	$N_2(g)$	+	$3\ H_2(g)$	$\rightleftharpoons$	$2\ NH_3(g)$	
Initial	0		0		P_0	P_0 = initial pressure of NH_3 in atm
	$2x$ atm of NH_3 reacts to reach equilibrium					
Change	$+x$		$+3x$	←	$-2x$	
Equil.	x		$3x$		$P_0 - 2x$	

From problem, $P_0 - 2x = \frac{P_0}{2.00}$, so $P_0 = 4.00\ x$

$$K_p = \frac{(4.00\,x - 2x)^2}{(x)(3x)^3} = \frac{(2.00\,x)^2}{(x)(3x)^3} = \frac{4.00\,x^2}{27x^4} = \frac{4.00}{27x^2} = 5.3 \times 10^5,\ x = 5.3 \times 10^{-4}\text{ atm}$$

$P_0 = 4.00\,x = 4.00 \times (5.3 \times 10^{-4})$ atm $= 2.1 \times 10^{-3}$ atm

83. $N_2O_4(g) \rightleftharpoons 2\ NO_2(g)$ $\quad K_p = \frac{P^2_{NO_2}}{P_{N_2O_4}} = \frac{(1.20)^2}{0.34} = 4.2$

Doubling the volume decreases each partial pressure by a factor of 2 (P = nRT/V).

$P_{NO_2} = 0.600$ atm and $P_{N_2O_4} = 0.17$ atm are the new partial pressures.

$Q = \frac{(0.600)^2}{0.17} = 2.1$, so Q < K; Equilibrium will shift to the right.

	$N_2O_4(g)$	$\rightleftharpoons$	$2\ NO_2(g)$
Initial	0.17 atm		0.600 atm
	x atm of N_2O_4 reacts to reach equilibrium		
Change	$-x$	$\rightarrow$	$+2x$
Equil.	$0.17 - x$		$0.600 + 2x$

$K_p = 4.2 = \frac{(0.600 + 2x)^2}{(0.17 - x)}$, $4x^2 + 6.6\,x - 0.354 = 0$ (carrying extra sig. figs.)

Solving using the quadratic formula: $x = 0.052$

$P_{NO_2} = 0.600 + 2(0.052) = 0.704$ atm; $P_{N_2O_4} = 0.17 - 0.052 = 0.12$ atm

85.

	$SO_3(g)$	$\rightleftharpoons$	$SO_2(g)$	+	$1/2\ O_2(g)$	
Initial	P_0		0		0	P_0 = initial pressure of SO_3
Change	$-x$	$\rightarrow$	$+x$		$+x/2$	
Equil.	$P_0 - x$		x		$x/2$	

The average molar mass of the mixture is:

$$\text{average molar mass} = \frac{dRT}{P} = \frac{(1.60\text{ g/L})(0.08206\text{ L atm/mol} \bullet \text{K})(873\text{ K})}{1.80\text{ atm}} = 63.7\text{ g/mol}$$

The average molar mass is determined by:

$$\text{average molar mass} = \frac{n_{SO_3}(80.07\text{ g/mol}) + n_{SO_2}(64.07\text{ g/mol}) + n_{O_2}(32.00\text{ g/mol})}{n_{total}}$$

Because χ_A = mol fraction of component A = $n_A/n_{total} = P_A/P_{total}$:

$$63.7 \text{ g/mol} = \frac{P_{SO_3}(80.07) + P_{SO_2}(64.07) + P_{O_2}(32.00)}{P_{total}}$$

$P_{total} = P_0 - x + x + x/2 = P_0 + x/2 = 1.80$ atm, $P_0 = 1.80 - x/2$

$$63.7 = \frac{(P_0 - x)(80.07) + x(64.07) + \frac{x}{2}(32.00)}{1.80}$$

$$63.7 = \frac{(1.80 - 3/2\,x)(80.07) + x(64.07) + \frac{x}{2}(32.00)}{1.80}$$

$115 = 144 - 120.1\,x + 64.07\,x + 16.00\,x$, $40.0\,x = 29$, $x = 0.73$ atm

$P_{SO_3} = P_0 - x = 1.80 - 3/2\,x = 0.71$ atm; $P_{SO_2} = 0.73$ atm; $P_{O_2} = x/2 = 0.37$ atm

$$K_p = \frac{P_{SO_2} \times P_{O_2}^{1/2}}{P_{SO_3}} = \frac{(0.73)(0.37)^{1/2}}{(0.71)} = 0.63$$

87. $O_2 \rightleftharpoons 2\,O$ Assuming exactly 100 O_2 molecules

	O_2	$2\,O$
Initial	100	0
Change	−83	+166
Equil.	17	166

Thus: $\chi_O = \frac{166}{183} = 0.9071$ and $\chi_{O_2} = 0.0929$

$P_A = \chi_A P_{TOTAL}$; Because $P_{TOTAL} = 1.000$ atm, $P_{O_2} = 0.0929$ atm and $P_O = 0.9071$ atm.

$$K_p = \frac{(0.9071)^2}{0.0929} = 8.857$$

$O_2 \rightleftharpoons 2\,O$

	O_2		$2\,O$
Initial	x		0
Change	$-y$	→	$+2y$
Equil.	$x - y$		$2y$

$$\frac{(2y)^2}{x - y} = 8.857; \quad \frac{y}{x} \times 100 = 95.0$$

Solving: x = 0.123 atm and y = 0.117 atm; $P_{total} = 0.240$ atm

89. $2\ NOBr(g) \rightleftharpoons 2\ NO(g) + Br_2(g)$

Initial	P_0	0	0	P_0 = initial pressure of NOBr
Equil.	$P_0 - 2x$	$2x$	x	Note: $P_{NO} = 2\,P_{Br_2}$

$P_{total} = 0.0515\text{ atm} = (P_0 - 2x) + (2x) + (x) = P_0 + x;\quad 0.0515\text{ atm} = P_{NOBr} + 3\,P_{Br_2}$

$$d = \frac{P \times (\text{molar mass})}{RT} = 0.1861\text{ g/L} = \frac{P_{NOBr}(109.9) + 2\,P_{Br_2}(30.01) + P_{Br_2}(159.8)}{0.08206 \times 298}$$

$4.55 = 109.9\,P_{NOBr} + 219.8\,P_{Br_2}$

Solving using simultaneous equations:

$$\begin{aligned} 0.0515 &= P_{NOBr} + 3\,P_{Br_2} \\ -0.0414 &= -P_{NOBr} - 2.000\,P_{Br_2} \\ \hline 0.0101 &= P_{Br_2} \end{aligned}$$

$P_{Br_2} = 1.01 \times 10^{-2}$ atm; $P_{NO} = 2\,P_{Br_2} = 2.02 \times 10^{-2}$ atm

$P_{NOBr} = 0.0515 - 3(1.01 \times 10^{-2}) = 2.12 \times 10^{-2}$ atm

$$K_p = \frac{P_{Br_2} \times P_{NO}^2}{P_{NOBr}^2} = \frac{(1.01\times10^{-2})(2.02\times10^{-2})^2}{(2.12\times10^{-2})^2} = 9.17 \times 10^{-3}$$

Integrative Problems

91. $NH_3(g) + H_2S(g) \rightleftharpoons NH_4HS(s) \qquad K = 400. = \dfrac{1}{[NH_3][H_2S]}$

Initial $\quad \dfrac{2.00\text{ mol}}{5.00\text{ L}} \quad \dfrac{2.00\text{ mol}}{5.00\text{ L}} \quad -$

x mol/L of NH_3 reacts to reach equilibrium

Change	$-x$	$-x$	$-$
Equil.	$0.400 - x$	$0.400 - x$	$-$

$$K = 400. = \frac{1}{(0.400 - x)(0.400 - x)},\quad 0.400 - x = \left(\frac{1}{400.}\right)^{1/2} = 0.0500$$

$x = 0.350\ M$; mol $NH_4HS(s)$ produced $= 5.00\ L \times \frac{0.350\ mol\ NH_3}{L} \times \frac{1\ mol\ NH_4HS}{mol\ NH_3}$
$= 1.75$ mol

Total mol $NH_4HS(s) = 2.00$ mol initially + 1.75 mol produced = 3.75 mol total

$$3.75\ mol\ NH_4HS \times \frac{51.12\ g\ NH_4HS}{mol\ NH_4HS} = 192\ g\ NH_4HS$$

$[H_2S]_e = 0.400\ M - x = 0.400\ M - 0.350\ M = 0.050\ M\ H_2S$

$$P_{H_2S} = \frac{n_{H_2S}RT}{V} = \frac{n_{H_2S}}{V} \times RT = \frac{0.050\ mol}{L} \times \frac{0.08206\ L\ atm}{K\ mol} \times 308\ K = 1.3\ atm$$

93. Assuming 100.00 g naphthalene:

$$93.71\ C \times \frac{1\ mol\ C}{12.01\ g} = 7.803\ mol\ C$$

$$6.29\ g\ H \times \frac{1\ mol\ H}{1.008\ g} = 6.240\ mol\ H; \quad \frac{7.803}{6.240} = 1.25$$

empirical formula $= (C_{1.25}H) \times 4 = C_5H_4$; molar mass $= \frac{32.8\ g}{0.256\ mol} = 128$ g/mol

Because the empirical mass (64.08 g/mol) is one-half of 128, the molecular formula is $C_{10}H_8$.

	$C_{10}H_8(s)$	$\rightleftharpoons$	$C_{10}H_8(g)$	$K = 4.29 \times 10^{-6} = [C_{10}H_8]$
Initial	–		0	
Equil.	–		x	

$K = 4.29 \times 10^{-6} = [C_{10}H_8] = x$

mol $C_{10}H_8$ sublimed $= 5.00\ L \times 4.29 \times 10^{-6}$ mol/L $= 2.15 \times 10^{-5}$ mol $C_{10}H_8$ sublimed

mol $C_{10}H_8$ initially $= 3.00\ g \times \frac{1\ mol\ C_{10}H_8}{128.16\ g} = 2.34 \times 10^{-2}$ mol $C_{10}H_8$ initially

$$\%\ C_{10}H_8\ sublimed = \frac{2.15 \times 10^{-5}\ mol}{2.34 \times 10^{-2}\ mol} \times 100 = 9.19 \times 10^{-2}\ \%$$

CHAPTER FOURTEEN

ACIDS AND BASES

Questions

17. 10.78 (4 S.F.); 6.78 (3 S.F.); 0.78 (2 S.F.); A pH value is a logarithm. The numbers to the left of the decimal point identify the power of ten to which $[H^+]$ is expressed in scientific notation, e.g., 10^{-11}, 10^{-7}, 10^{-1}. The number of decimal places in a pH value identifies the number of significant figures in $[H^+]$. In all three pH values, the $[H^+]$ should be expressed only to two significant figures because these pH values have only two decimal places.

19. a. These are strong acids like HCl, HBr, HI, HNO_3, H_2SO_4 or $HClO_4$.

    b. These are salts of the conjugate acids of the bases in Table 14.3. These conjugate acids are all weak acids. NH_4Cl, $CH_3NH_3NO_3$, and $C_2H_5NH_3Br$ are three examples. Note that the anions used to form these salts (Cl^-, NO_3^-, and Br^-) are conjugate bases of strong acids; this is because they have no acidic or basic properties in water (with the exception of HSO_4^-, which has weak acid properties).

    c. These are strong bases like LiOH, NaOH, KOH, RbOH, CsOH, $Ca(OH)_2$, $Sr(OH)_2$, and $Ba(OH)_2$.

    d. These are salts of the conjugate bases of the neutrally charged weak acids in Table 14.2. The conjugate bases of weak acids are weak bases themselves. Three examples are $NaClO_2$, $KC_2H_3O_2$, and CaF_2. The cations used to form these salts are Li^+, Na^+, K^+, Rb, Cs^+, Ca^{2+}, Sr^{2+}, and Ba^{2+} since these cations have no acidic or basic properties in water. Notice that these are the cations of the strong bases you should memorize.

    e. There are two ways to make a neutral salt. The easiest way is to combine a conjugate base of a strong acid (except for HSO_4^-) with one of the cations from a strong base. These ions have no acidic/basic properties in water so salts of these ions are neutral. Three examples are NaCl, KNO_3, and SrI_2. Another type of strong electrolyte that can produce neutral solutions are salts that contain an ion with weak acid properties combined with an ion of opposite charge having weak base properties. If the K_a for the weak acid ion is equal to the K_b for the weak base ion, then the salt will produce a neutral solution. The most common example of this type of salt is ammonium acetate, $NH_4C_2H_3O_2$. For this salt, K_a for NH_4^+ = K_b for $C_2H_3O_2^-$ = 5.6×10^{-10}. This salt at any concentration produces a neutral solution.

21. a. $H_2O(l) + H_2O(l) \rightleftharpoons H_3O^+(aq) + OH^-(aq)$ or

    $H_2O(l) \rightleftharpoons H^+(aq) + OH^-(aq) \quad K = K_w = [H^+][OH^-]$

b. $HF(aq) + H_2O(l) \rightleftharpoons F^-(aq) + H_3O^+(aq)$ or

$HF(aq) \rightleftharpoons H+(aq) + F^-(aq) \quad K = K_a = \dfrac{[H^+][F^-]}{[HF]}$

c. $C_5H_5N(aq) + H_2O(l) \rightleftharpoons C_5H_5NH^+(aq) + OH^-(aq) \quad K = K_b = \dfrac{[C_5H_5NH^+][OH^-]}{[C_5H_5N]}$

23. a. This expression holds true for solutions of strong acids having a concentration greater than 1.0×10^{-6} *M*. 0.10 *M* HCl, 7.8 *M* HNO_3, and 3.6×10^{-4} *M* $HClO_4$ are examples where this expression holds true.

b. This expression holds true for solutions of weak acids where the two normal assumptions hold. The two assumptions are that water does not contribute enough H^+ to solution to matter and that the acid is less than 5% dissociated in water (from the assumption that x is small compared to some number). This expression will generally hold true for solutions of weak acids having a K_a value less than 1×10^{-4}, as long as there is a significant amount of weak acid present. Three example solutions are 1.5 *M* $HC_2H_3O_2$, 0.10 *M* HOCl, and 0.72 *M* HCN.

c. This expression holds true for strong bases that donate 2 OH^- ions per formula unit. As long as the concentration of the base is above 5×10^{-7} *M*, this expression will hold true. Three examples are 5.0×10^{-3} *M* $Ca(OH)_2$, 2.1×10^{-4} *M* $Sr(OH)_2$, and 9.1×10^{-5} *M* $Ba(OH)_2$.

d. This expression holds true for solutions of weak bases where the two normal assumptions hold. The assumptions are that the OH^- contribution from water is negligible and that and that the base is less than 5% ionized in water (for the 5% rule to hold). For the 5% rule to hold, you generally need bases with $K_b < 1 \times 10^{-4}$ and concentrations of weak base greater than 0.10 *M*. Three examples are 0.10 *M* NH_3, 0.54 *M* $C_6H_5NH_2$, and 1.1 *M* C_5H_5N.

25. One reason HF is a weak acid is that the H–F bond is unusually strong and is difficult to break. This contributes significantly to the reluctance of the HF molecules to dissociate in water.

Exercises

Nature of Acids and Bases

27. a. $HClO_4(aq) + H_2O(l) \rightarrow H_3O^+(aq) + ClO_4^-(aq)$. Only the forward reaction is indicated since $HClO_4$ is a strong acid and is basically 100% dissociated in water. For acids, the dissociation reaction is commonly written without water as a reactant. The common abbreviation for this reaction is: $HClO_4(aq) \rightarrow H^+(aq) + ClO_4^-(aq)$. This reaction is also called the K_a reaction as the equilibrium constant for this reaction is called K_a.

b. Propanoic acid is a weak acid, so it is only partially dissociated in water. The dissociation reaction is: $CH_3CH_2CO_2H(aq) + H_2O(l) \rightleftharpoons H_3O^+(aq) + CH_3CH_2CO_2^-(aq)$ or $CH_3CH_2CO_2H(aq) \rightleftharpoons H^+(aq) + CH_3CH_2CO^-(aq)$.

c. NH_4^+ is a weak acid. Similar to propanoic acid, the dissociation reaction is:

$$NH_4^+(aq) + H_2O(l) \rightleftharpoons H_3O^+(aq) + NH_3(aq) \text{ or } NH_4^+(aq) \rightleftharpoons H^+(aq) + NH_3(aq)$$

29. An acid is a proton (H^+) donor and a base is a proton acceptor. A conjugate acid-base pair differs by only a proton (H^+).

	Acid	Base	Conjugate Base of Acid	Conjugate Acid of Base
a.	H_2CO_3	H_2O	HCO_3^-	H_3O^+
b.	$C_5H_5NH^+$	H_2O	C_5H_5N	H_3O^+
c.	$C_5H_5NH^+$	HCO_3^-	C_5H_5N	H_2CO_3

31. Strong acids have a $K_a >> 1$ and weak acids have $K_a < 1$. Table 14.2 in the text lists some K_a values for weak acids. K_a values for strong acids are hard to determine so they are not listed in the text. However, there are only a few common strong acids so if you memorize the strong acids, then all other acids will be weak acids. The strong acids to memorize are HCl, HBr, HI, HNO_3, $HClO_4$ and H_2SO_4.

a. $HClO_4$ is a strong acid.

b. HOCl is a weak acid ($K_a = 3.5 \times 10^{-8}$).

c. H_2SO_4 is a strong acid.

d. H_2SO_3 is a weak diprotic acid with K_{a1} and K_{a2} values less than one.

33. The K_a value is directly related to acid strength. As K_a increases, acid strength increases. For water, use K_w when comparing the acid strength of water to other species. The K_a values are:

$HClO_4$: strong acid ($K_a >> 1$); $HClO_2$: $K_a = 1.2 \times 10^{-2}$

NH_4^+: $K_a = 5.6 \times 10^{-10}$; H_2O: $K_a = K_w = 1.0 \times 10^{-14}$

From the K_a values, the ordering is: $HClO_4 > HClO_2 > NH_4^+ > H_2O$.

35. a. HCl is a strong acid and water is a very weak acid with $K_a = K_w = 1.0 \times 10^{-14}$. HCl is a much stronger acid than H_2O.

b. H_2O, $K_a = K_w = 1.0 \times 10^{-14}$; HNO_2, $K_a = 4.0 \times 10^{-4}$; HNO_2 is a stronger acid than H_2O because K_a for $HNO_2 > K_w$ for H_2O.

c. HOC_6H_5, $K_a = 1.6 \times 10^{-10}$; HCN, $K_a = 6.2 \times 10^{-10}$; HCN is a stronger acid than HOC_6H_5 because K_a for HCN > K_a for HOC_6H_5.

Autoionization of Water and the pH Scale

37. At 25°C, the relationship: $[H^+]\,[OH^-] = K_w = 1.0 \times 10^{-14}$ always holds for aqueous solutions. When $[H^+]$ is greater than 1.0×10^{-7} *M*, the solution is acidic; when $[H^+]$ is less than 1.0×10^{-7} *M*, the solution is basic; when $[H^+] = 1.0 \times 10^{-7}$ *M*, the solution is neutral. In terms of $[OH^-]$, an acidic solution has $[OH^-] < 1.0 \times 10^{-7}$ *M*, a basic solution has $[OH^-] > 1.0 \times 10^{-7}$ *M*, and a neutral solution has $[OH^-] = 1.0 \times 10^{-7}$ *M*.

a. $[OH^-] = \dfrac{K_w}{[H^+]} = \dfrac{1.0\times10^{-14}}{1.0\times10^{-7}} = 1.0 \times 10^{-7}\ M$; The solution is neutral.

b. $[OH^-] = \dfrac{1.0\times10^{-14}}{8.3\times10^{-16}} = 12\ M$; The solution is basic.

c. $[OH^-] = \dfrac{1.0\times10^{-14}}{12} = 8.3 \times 10^{-16}\ M$; The solution is acidic.

d. $[OH^-] = \dfrac{1.0\times10^{-14}}{5.4\times10^{-5}} = 1.9 \times 10^{-10}\ M$; The solution is acidic.

39. a. Because the value of the equilibrium constant increases as the temperature increases, the reaction is endothermic. In endothermic reactions, heat is a reactant so an increase in temperature (heat) shifts the reaction to produce more products and increases K in the process.

b. $H_2O(l) \rightleftharpoons H^+(aq) + OH^-(aq)$ $K_w = 5.47 \times 10^{-14} = [H^+][OH^-]$ at 50.°C

In pure water $[H^+] = [OH^-]$, so $5.47 \times 10^{-14} = [H^+]^2$, $[H^+] = 2.34 \times 10^{-7}\ M = [OH^-]$

41. $pH = -\log [H^+]$; $pOH = -\log [OH^-]$; At 25°C, pH + pOH = 14.00; For Exercise 13.37:

a. $pH = -\log [H^+] = -\log (1.0 \times 10^{-7}) = 7.00$; $pOH = 14.00 - pH = 14.00 - 7.00 = 7.00$

b. $pH = -\log (8.3 \times 10^{-16}) = 15.08$; $pOH = 14.00 - 15.08 = -1.08$

c. $pH = -\log (12) = -1.08$; $pOH = 14.00 - (-1.08) = 15.08$

d. $pH = -\log (5.4 \times 10^{-5}) = 4.27$; $pOH = 14.00 - 4.27 = 9.73$

Note that pH is less than zero when $[H^+]$ is greater than 1.0 *M* (an extremely acidic solution). For Exercise 13.38:

a. $pOH = -\log [OH^-] = -\log (1.5) = -0.18$; $pH = 14.00 - pOH = 14.00 - (-0.18) = 14.18$

b. $pOH = -\log (3.6 \times 10^{-15}) = 14.44$; $pH = 14.00 - 14.44 = -0.44$

c. $pOH = -\log(1.0 \times 10^{-7}) = 7.00$; $pH = 14.00 - 7.00 = 7.00$

d. $pOH = -\log(7.3 \times 10^{-4}) = 3.14$; $pH = 14.00 - 3.14 = 10.86$

Note that pH is greater than 14.00 when $[OH^-]$ is greater than 1.0 *M* (an extremely basic solution).

43. a. $pOH = 14.00 - 6.88 = 7.12$; $[H^+] = 10^{-6.88} = 1.3 \times 10^{-7}\ M$

$[OH^-] = 10^{-7.12} = 7.6 \times 10^{-8}\ M$; acidic

b. $[H^+] = \frac{1.0 \times 10^{-14}}{8.4 \times 10^{-14}} = 0.12\ M$; $pH = -\log(0.12) = 0.92$

$pOH = 14.00 - 0.92 = 13.08$; acidic

c. $pH = 14.00 - 3.11 = 10.89$; $[H^+] = 10^{-10.89} = 1.3 \times 10^{-11}\ M$

$[OH^-] = 10^{-3.11} = 7.8 \times 10^{-4}\ M$; basic

d. $pH = -\log(1.0 \times 10^{-7}) = 7.00$; $pOH = 14.00 - 7.00 = 7.00$

$[OH^-] = 10^{-7.00} = 1.0 \times 10^{-7}\ M$; neutral

45. $pOH = 14.0 - pH = 14.0 - 2.1 = 11.9$; $[H^+] = 10^{-pH} = 10^{-2.1} = 8 \times 10^{-3}\ M$ (1 sig fig)

$[OH^-] = \frac{K_w}{[H^+]} = \frac{1.0 \times 10^{-14}}{8 \times 10^{-3}} = 1 \times 10^{-12}\ M$ or $[OH^-] = 10^{-pOH} = 10^{-11.9} = 1 \times 10^{-12}\ M$

The sample of gastric juice is acidic since the pH is less than 7.00 at 25°C.

Solutions of Acids

47. All the acids in this problem are strong acids that are always assumed to completely dissociate in water. The general dissociation reaction for a strong acid is: $HA(aq) \rightarrow H^+(aq) + A^-(aq)$ where A^- is the conjugate base of the strong acid HA. For 0.250 *M* solutions of these strong acids, 0.250 *M* H^+ and 0.250 *M* A^- are present when the acids completely dissociate. The amount of H^+ donated from water will be insignificant in this problem since H_2O is a very weak acid.

a. Major species present after dissociation = H^+, ClO_4^- and H_2O;

$pH = -\log[H^+] = -\log(0.250) = 0.602$

b. Major species = H^+, NO_3^- and H_2O; $pH = 0.602$

49. Both are strong acids.

$0.0500 \text{ L} \times 0.050 \text{ mol/L} = 2.5 \times 10^{-3} \text{ mol HCl} = 2.5 \times 10^{-3} \text{ mol H}^+ + 2.5 \times 10^{-3} \text{ mol Cl}^-$

$0.1500 \text{ L} \times 0.10 \text{ mol/L} = 1.5 \times 10^{-2} \text{ mol HNO}_3 = 1.5 \times 10^{-2} \text{ mol H}^+ + 1.5 \times 10^{-2} \text{ mol NO}_3^-$

$$[\text{H}^+] = \frac{(2.5\times10^{-3} + 1.5\times10^{-2}) \text{ mol}}{0.2000 \text{ L}} = 0.088\ M; \ [\text{OH}^-] = \frac{K_w}{[\text{H}^+]} = 1.1 \times 10^{-13}\ M$$

$$[\text{Cl}^-] = \frac{2.5\times10^{-3} \text{ mol}}{0.2000 \text{ L}} = 0.013\ M; \ [\text{NO}_3^-] = \frac{1.5\times10^{-2} \text{ mol}}{0.2000 \text{ L}} = 0.075\ M$$

51. $[\text{H}^+] = 10^{-1.50} = 3.16 \times 10^{-2}\ M$ (carryiung one extra sig fig); $M_1V_1 = M_2V_2$

$$V_1 = \frac{M_2V_2}{M_1} = \frac{3.16\times10^{-2} \text{ mol/L}\times1.6 \text{ L}}{12 \text{ mol/L}} = 4.2 \times 10^{-3} \text{ L}$$

To 4.2 mL of 12 M HCl, add enough water to make 1600 mL of solution. The resulting solution will have $[\text{H}^+] = 3.2 \times 10^{-2}\ M$ and pH = 1.50.

53. a. HNO_2 ($K_a = 4.0 \times 10^{-4}$) and H_2O ($K_a = K_w = 1.0 \times 10^{-14}$) are the major species. HNO_2 is a much stronger acid than H_2O so it is the major source of H^+. However, HNO_2 is a weak acid ($K_a < 1$) so it only partially dissociates in water. We must solve an equilibrium problem to determine $[H^+]$. In the Solutions Guide, we will summarize the initial, change and equilibrium concentrations into one table called the ICE table. Solving the weak acid problem:

	HNO_2	$\rightleftharpoons$	H^+	+	NO_2^-
Initial	0.250 M		~0		0
	x mol/L HNO_2 dissociates to reach equilibrium				
Change	$-x$	$\rightarrow$	$+x$		$+x$
Equil.	$0.250 - x$		x		x

$$K_a = \frac{[H^+][NO_2^-]}{[HNO_2]} = 4.0 \times 10^{-4} = \frac{x^2}{0.250 - x}; \text{ If we assume } x << 0.250, \text{ then:}$$

$$4.0 \times 10^{-4} \approx \frac{x^2}{0.250}, \quad x = \sqrt{4.0\times10^{-4}\,(0.250)} = 0.010\ M$$

We must check the assumption: $\frac{x}{0.250}\times100 = \frac{0.010}{0.250} \times 100 = 4.0\%$

All the assumptions are good. The H^+ contribution from water ($10^{-7}\ M$) is negligible, and x is small compared to 0.250 (percent error = 4.0%). If the percent error is less than 5% for an assumption, we will consider it a valid assumption (called the 5% rule). Finishing the problem: $x = 0.010\ M = [H^+]$; pH = −log(0.010) = 2.00

b. CH_3CO_2H ($K_a = 1.8 \times 10^{-5}$) and H_2O ($K_a = K_w = 1.0 \times 10^{-14}$) are the major species. CH_3CO_2H is the major source of H^+. Solving the weak acid problem:

	CH_3CO_2H	$\rightleftharpoons$	H^+	+	$CH_3CO_2^-$
Initial	0.250 M		~0		0
	x mol/L CH_3CO_2H dissociates to reach equilibrium				
Change	$-x$	$\rightarrow$	$+x$		$+x$
Equil.	$0.250 - x$		x		x

$$K_a = \frac{[H^+][CH_3CO_2^-]}{[CH_3CO_2H]} = 1.8 \times 10^{-5} = \frac{x^2}{0.250 - x} \approx \frac{x^2}{0.250} \quad \text{(assuming } x << 0.250\text{)}$$

$x = 2.1 \times 10^{-3}$ M; Checking assumption: $\frac{2.1 \times 10^{-3}}{0.250} \times 100 = 0.84\%$. Assumptions good.

$[H^+] = x = 2.1 \times 10^{-3}$ M; pH = $-\log(2.1 \times 10^{-3}) = 2.68$

55. $$[CH_3COOH]_0 = [HC_2H_3O_2]_0 = \frac{0.0560 \text{ g } HC_2H_3O_2 \times \frac{1 \text{ mol } HC_2H_3O_2}{60.05 \text{ g}}}{0.05000 \text{ L}} = 1.87 \times 10^{-2}\ M$$

	$HC_2H_3O_2$	$\rightleftharpoons$	H^+	+	$C_2H_3O_2^-$	$K_a = 1.8 \times 10^{-5}$
Initial	0.0187 M		~0		0	
	x mol/L $HC_2H_3O_2$ dissociates to reach equilibrium					
Change	$-x$	$\rightarrow$	$+x$		$+x$	
Equil.	$0.0187 - x$		x		x	

$$K_a = 1.8 \times 10^{-5} = \frac{[H^+][C_2H_3O_2^-]}{[HC_2H_3O_2]} = \frac{x^2}{0.0187 - x} \approx \frac{x^2}{0.0187}$$

$x = [H^+] = 5.8 \times 10^{-4}$ M; pH = 3.24 Assumptions good (x is 3.1% of 0.0187).

$[H^+] = [C_2H_3O_2^-] = [CH_3COO^-] = 5.8 \times 10^{-4}$ M; $[CH_3COOH] = 0.0187 - 5.8 \times 10^{-4}$ $= 0.0181$ M

57. This is a weak acid in water. Solving the weak acid problem:

	HF	$\rightleftharpoons$	H^+	+	F^-	$K_a = 7.2 \times 10^{-4}$
Initial	0.020 M		~0		0	
	x mol/L HF dissociates to reach equilibrium					
Change	$-x$	$\rightarrow$	$+x$		$+x$	
Equil.	$0.020 - x$		x		x	

$$K_a = 7.2 \times 10^{-4} = \frac{[H^+][F^-]}{[HF]} = \frac{x^2}{0.020 - x} \approx \frac{x^2}{0.020} \quad \text{(assuming } x << 0.020\text{)}$$

$x = [H^+] = 3.8 \times 10^{-3}\ M$; Check assumptions: $\frac{x}{0.020} \times 100 = \frac{3.8 \times 10^{-3}}{0.020} = 19\%$

The assumption $x << 0.020$ is not good (x is more than 5% of 0.020). We must solve $x^2/(0.020 - x) = 7.2 \times 10^{-4}$ exactly by using either the quadratic formula or by the method of successive approximations (see Appendix 1.4 of text). Using successive approximations, we let 0.016 M be a new approximation for [HF]. That is, in the denominator, try $x = 0.0038$ (the value of x we calculated making the normal assumption), so $0.020 - 0.0038 = 0.016$, then solve for a new value of x in the numerator.

$$\frac{x^2}{0.020 - x} \approx \frac{x^2}{0.016} = 7.2 \times 10^{-4}, \quad x = 3.4 \times 10^{-3}$$

We use this new value of x to further refine our estimate of [HF], i.e., $0.020 - x = 0.020 - 0.0034 = 0.0166$ (carry extra significant figure).

$$\frac{x^2}{0.020 - x} \approx \frac{x^2}{0.0166} = 7.2 \times 10^{-4}, \quad x = 3.5 \times 10^{-3}$$

We repeat until we get an answer that repeats itself. This would be the same answer we would get solving exactly using the quadratic equation. In this case it is: $x = 3.5 \times 10^{-3}$

So: $[H^+] = [F^-] = x = 3.5 \times 10^{-3}\ M$; $[OH^-] = K_w/[H^+] = 2.9 \times 10^{-12}\ M$

$[HF] = 0.020 - x = 0.020 - 0.0035 = 0.017\ M$; pH = 2.46

Note: When the 5% assumption fails, use whichever method you are most comfortable with to solve exactly. The method of successive approximations is probably fastest when the percent error is less than ~25% (unless you have a calculator that can solve quadratic equations).

59. Major species: $HC_2H_2ClO_2$ ($K_a = 1.35 \times 10^{-3}$) and H_2O; Major source of H^+: $HC_2H_2ClO_2$

$$HC_2H_2ClO_2 \rightleftharpoons H^+ + C_2H_2ClO_2^-$$

	$HC_2H_2ClO_2$	$\rightleftharpoons$	H^+	$C_2H_2ClO_2^-$
Initial	0.10 M		~0	0
	x mol/L $HC_2H_2ClO_2$ dissociates to reach equilibrium			
Change	$-x$	$\rightarrow$	$+x$	$+x$
Equil.	$0.10 - x$		x	x

$$K_a = 1.35 \times 10^{-3} = \frac{x^2}{0.10 - x} \approx \frac{x^2}{0.10}, \quad x = 1.2 \times 10^{-2}\ M$$

Checking the assumptions finds that x is 12% of 0.10 which fails the 5% rule. We must solve $1.35 \times 10^{-3} = x^2/(0.10 - x)$ exactly using either the method of successive approximations or the quadratic equation. Using either method gives $x = [H^+] = 1.1 \times 10^{-2}\ M$. $pH = -\log [H^+] = -\log(1.1 \times 10^{-2}) = 1.96$.

61. a. HCl is a strong acid. It will produce 0.10 *M* H^+. HOCl is a weak acid. Let's consider the equilibrium:

	HOCl	⇌	H^+	+	OCl^-	$K_a = 3.5 \times 10^{-8}$
Initial	0.10 *M*		0.10 *M*		0	
	x mol/L HOCl dissociates to reach equilibrium					
Change	$-x$	→	$+x$		$+x$	
Equil.	$0.10 - x$		$0.10 + x$		x	

$$K_a = 3.5 \times 10^{-8} = \frac{[H^+][OCl^-]}{[HOCl]} = \frac{(0.10 + x)(x)}{0.10 - x} \approx x,\ x = 3.5 \times 10^{-8}\ M$$

Assumptions are great (x is 3.5×10^{-5}% of 0.10). We are really assuming that HCl is the only important source of H^+, which it is. The $[H^+]$ contribution from HOCl, x, is negligible. Therefore, $[H^+] = 0.10\ M$; pH = 1.00

b. HNO_3 is a strong acid, giving an initial concentration of H^+ equal to 0.050 *M*. Consider the equilibrium:

	$HC_2H_3O_2$	⇌	H^+	+	$C_2H_3O_2^-$	$K_a = 1.8 \times 10^{-5}$
Initial	0.50 *M*		0.050 *M*		0	
	x mol/L $HC_2H_3O_2$ dissociates to reach equilibrium					
Change	$-x$	→	$+x$		$+x$	
Equil.	$0.50 - x$		$0.050 + x$		x	

$$K_a = 1.8 \times 10^{-5} = \frac{[H^+][C_2H_3O_2^-]}{[HC_2H_3O_2]} = \frac{(0.050 + x)x}{0.50 - x} \approx \frac{0.050\,x}{0.50}$$

$x = 1.8 \times 10^{-4}$; Assumptions are good (well within the 5% rule).

$[H^+] = 0.050 + x = 0.050\ M$ and pH = 1.30

63. In all parts of this problem, acetic acid ($HC_2H_3O_2$) is the best weak acid present. We must solve a weak acid problem.

a.

	$HC_2H_3O_2$	⇌	H^+	+	$C_2H_3O_2^-$
Initial	0.50 *M*		~0		0
	x mol/L $HC_2H_3O_2$ dissociates to reach equilibrium				
Change	$-x$	→	$+x$		$+x$
Equil.	$0.50 - x$		x		x

$$K_a = 1.8 \times 10^{-5} = \frac{[H^+][C_2H_3O_2^-]}{[HC_2H_3O_2]} = \frac{x^2}{0.50 - x} \approx \frac{x^2}{0.50}$$

$x = [H^+] = [C_2H_3O_2^-] = 3.0 \times 10^{-3}\ M$ Assumptions good.

$$\text{Percent dissociation} = \frac{[H^+]}{[HC_2H_3O_2]_0} \times 100 = \frac{3.0\times10^{-3}}{0.50} \times 100 = 0.60\%$$

b. The setups for solutions b and c are similar to solution a except the final equation is slightly different, reflecting the new concentration of $HC_2H_3O_2$.

$$K_a = 1.8 \times 10^{-5} = \frac{x^2}{0.050 - x} \approx \frac{x^2}{0.050}$$

$x = [H^+] = [C_2H_3O_2^-] = 9.5 \times 10^{-4}\ M$ Assumptions good.

$$\%\ \text{dissociation} = \frac{9.5\times10^{-4}}{0.050} \times 100 = 1.9\%$$

c. $$K_a = 1.8 \times 10^{-5} = \frac{x^2}{0.0050 - x} \approx \frac{x^2}{0.0050}$$

$x = [H^+] = [C_2H_3O_2^-] = 3.0 \times 10^{-4}\ M$; Check assumptions.

Assumption that x is negligible is borderline (6.0% error). We should solve exactly. Using the method of successive approximations (see Appendix 1.4 of text):

$$1.8 \times 10^{-5} = \frac{x^2}{0.0050 - 3.0\times10^{-4}} = \frac{x^2}{0.0047},\quad x = 2.9 \times 10^{-4}$$

Next trial also gives $x = 2.9 \times 10^{-4}$.

$$\%\ \text{dissociation} = \frac{2.9\times10^{-4}}{5.0\times10^{-3}} \times 100 = 5.8\%$$

d. As we dilute a solution, all concentrations decrease. Dilution will shift the equilibrium to the side with the greater number of particles. For example, suppose we double the volume of an equilibrium mixture of a weak acid by adding water, then:

$$Q = \frac{\left(\frac{[H^+]_{eq}}{2}\right)\left(\frac{[X^-]_{eq}}{2}\right)}{\left(\frac{[HX]_{eq}}{2}\right)} = \frac{1}{2}K_a$$

$Q < K_a$, so the equilibrium shifts to the right or towards a greater percent dissociation.

e. $[H^+]$ depends on the initial concentration of weak acid and on how much weak acid dissociates. For solutions a-c the initial concentration of acid decreases more rapidly than the percent dissociation increases. Thus, $[H^+]$ decreases.

65. Let HX symbolize the weak acid. Setup the problem like a typical weak acid equilibrium problem.

	HX	⇌	H^+	+	X^-
Initial	0.15 *M*		~0		0
	x mol/L HX dissociates to reach equilibrium				
Change	$-x$	→	$+x$		$+x$
Equil.	$0.15 - x$		x		x

If the acid is 3.0% dissociated, then $x = [H^+]$ is 3.0% of 0.15: $x = 0.030 \times (0.15\ M) = 4.5 \times 10^{-3}\ M$. Now that we know the value of *x*, we can solve for K_a.

$$K_a = \frac{[H^+][X^-]}{[HX]} = \frac{x^2}{0.15 - x} = \frac{(4.5\times10^{-3})^2}{0.15 - 4.5\times10^{-3}} = 1.4 \times 10^{-4}$$

67. Setup the problem using the K_a equilibrium reaction for HOCN.

	HOCN	⇌	H^+	+	OCN^-
Initial	0.0100 *M*		~0		0
	x mol/L HOCN dissociates to reach equilibrium				
Change	$-x$	→	$+x$		$+x$
Equil.	$0.0100 - x$		x		x

$$K_a = \frac{[H^+][OCN^-]}{[HOCN]} = \frac{x^2}{0.0100 - x}; \text{ pH} = 2.77: \ x = [H^+] = 10^{-pH} = 10^{-2.77} = 1.7 \times 10^{-3}\ M$$

$$K_a = \frac{(1.7\times10^{-3})^2}{0.0100 - 1.7\times10^{-3}} = 3.5 \times 10^{-4}$$

69. Major species: HCOOH and H_2O; Major source of H^+: HCOOH

	HCOOH	⇌	H^+	+	$HCOO^-$	
Initial	C		~0		0	where $C = [HCOOH]_o$
	x mol/L HCOOH dissociates to reach equilibrium					
Change	$-x$	→	$+x$		$+x$	
Equil.	$C - x$		x		x	

$$K_a = 1.8 \times 10^{-4} = \frac{[H^+][HCOO^-]}{[HCOOH]} = \frac{x^2}{C - x} \text{ where } x = [H^+]$$

$$1.8 \times 10^{-4} = \frac{[H^+]^2}{C - [H^+]}; \text{ pH} = 2.70, \text{ so: } [H^+] = 10^{-2.70} = 2.0 \times 10^{-3}\ M$$

$$1.8 \times 10^{-4} = \frac{(2.0\times10^{-3})^2}{C - (2.0\times10^{-3})}, \ C - (2.0\times10^{-3}) = \frac{4.0\times10^{-6}}{1.8\times10^{-4}}, \ C = 2.4 \times 10^{-2}\ M$$

A 0.024 *M* formic acid solution will have pH = 2.70.

Solutions of Bases

71. a. $NH_3(aq) + H_2O(l) \rightleftharpoons NH_4^+(aq) + OH^-(aq)$ $K_b = \frac{[NH_4^+][OH^-]}{[NH_3]}$

b. $C_5H_5N(aq) + H_2O(l) \rightleftharpoons C_5H_5NH^+(aq) + OH^-(aq)$ $K_b = \frac{[C_5H_5NH^+][OH^-]}{[C_5H_5N]}$

73. NO_3^-: $K_b << K_w$ since HNO_3 is a strong acid. All conjugate bases of strong acids have no base strength. H_2O: $K_b = K_w = 1.0 \times 10^{-14}$; NH_3: $K_b = 1.8 \times 10^{-5}$; C_5H_5N: $K_b = 1.7 \times 10^{-9}$

$NH_3 > C_5H_5N > H_2O > NO_3^-$ (As K_b increases, base strength increases.)

75. a. $C_6H_5NH_2$ b. $C_6H_5NH_2$ c. OH^- d. CH_3NH_2

The base with the largest K_b value is the strongest base ($K_{b,\, C_6H_5NH_2} = 3.8 \times 10^{-10}$, $K_{b,\, CH_3NH_2} = 4.4\ 10^{-4}$. OH^- is the strongest base possible in water.

77. $NaOH(aq) \rightarrow Na^+(aq) + OH^-(aq)$; NaOH is a strong base which completely dissociates into Na^+ and OH^-. The initial concentration of NaOH will equal the concentration of OH^- donated by NaOH.

a. $[OH^-] = 0.10\ M$; $pOH = -\log[OH^-] = -\log(0.10) = 1.00$

$pH = 14.00 - pOH = 14.00 - 1.00 = 13.00$

Note that H_2O is also present, but the amount of OH^- produced by H_2O will be insignificant compared to the 0.10 *M* OH^- produced from the NaOH.

b. The $[OH^-]$ concentration donated by the NaOH is $1.0 \times 10^{-10}\ M$. Water by itself donates $1.0 \times 10^{-7}\ M$. In this problem, water is the major OH^- contributor and $[OH^-] = 1.0 \times 10^{-7}\ M$.

$pOH = -\log(1.0 \times 10^{-7}) = 7.00$; $pH = 14.00 - 7.00 = 7.00$

c. $[OH^-] = 2.0\ M$; $pOH = -\log(2.0) = -0.30$; $pH = 14.00 - (-0.30) = 14.30$

79. a. Major species: K^+, OH^-, H_2O (KOH is a strong base.)

$[OH^-] = 0.015\ M$, $pOH = -\log(0.015) = 1.82$; $pH = 14.00 - pOH = 12.18$

b. Major species: Ba^{2+}, OH^-, H_2O; $Ba(OH)_2(aq) \rightarrow Ba^{2+}(aq) + 2\ OH^-(aq)$; Since each mol of the strong base $Ba(OH)_2$ dissolves in water to produce two mol OH^-, then $[OH^-] = 2(0.015\ M) = 0.030\ M$.

$pOH = -\log(0.030) = 1.52$; $pH = 14.00 - 1.52 = 12.48$

81. $pOH = 14.00 - 11.56 = 2.44$; $[OH^-] = [KOH] = 10^{-2.44} = 3.6 \times 10^{-3}\ M$

$$0.8000\ L \times \frac{3.6 \times 10^{-3}\ mol\ KOH}{L} \times \frac{56.11\ g\ KOH}{mol\ KOH} = 0.16\ g\ KOH$$

83. NH_3 is a weak base with $K_b = 1.8 \times 10^{-5}$. The major species present will be NH_3 and H_2O ($K_b = K_w = 1.0 \times 10^{-14}$). Since NH_3 has a much larger K_b value compared to H_2O, NH_3 is the stronger base present and will be the major producer of OH^-. To determine the amount of OH^- produced from NH_3, we must perform an equilibrium calculation.

$$NH_3(aq) + H_2O(l) \rightleftharpoons NH_4^+(aq) + OH^-(aq)$$

	$NH_3(aq)$	$\rightleftharpoons$	$NH_4^+(aq)$	$OH^-(aq)$
Initial	0.150 M		0	~0
	x mol/L NH_3 reacts with H_2O to reach equilibrium			
Change	$-x$	$\rightarrow$	$+x$	$+x$
Equil.	$0.150 - x$		x	x

$$K_b = 1.8 \times 10^{-5} = \frac{[NH_4^+][OH^-]}{[NH_3]} = \frac{x^2}{0.150 - x} \approx \frac{x^2}{0.150} \quad \text{(assuming } x << 0.150\text{)}$$

$x = [OH^-] = 1.6 \times 10^{-3}$ M; Check assumptions: x is 1.1% of 0.150 so the assumption $0.150 - x \approx 0.150$ is valid by the 5% rule. Also, the contribution of OH^- from water will be insignificant (which will usually be the case). Finishing the problem, pOH = $-\log [OH^-] = -\log (1.6 \times 10^{-3}\ M)$ = 2.80; pH = 14.00 − pOH = 14.00 − 2.80 = 11.20.

85. These are solutions of weak bases in water. We must solve the equilibrium weak base problem.

a. $$(C_2H_5)_3N + H_2O \rightleftharpoons (C_2H_5)_3NH^+ + OH^- \qquad K_b = 4.0 \times 10^{-4}$$

	$(C_2H_5)_3N$		$(C_2H_5)_3NH^+$	OH^-
Initial	0.20 M		0	~0
	x mol/L of $(C_2H_5)_3N$ reacts with H_2O to reach equilibrium			
Change	$-x$	$\rightarrow$	$+x$	$+x$
Equil.	$0.20 - x$		x	x

$$K_b = 4.0 \times 10^{-4} = \frac{[(C_2H_5)_3NH^+][OH^-]}{[(C_2H_5)_3N]} = \frac{x^2}{0.20 - x} \approx \frac{x^2}{0.20}, \quad x = [OH^-] = 8.9 \times 10^{-3}\ M$$

Assumptions good (x is 4.5% of 0.20). $[OH^-] = 8.9 \times 10^{-3}\ M$

$$[H^+] = \frac{K_w}{[OH^-]} = \frac{1.0 \times 10^{-14}}{8.9 \times 10^{-3}} = 1.1 \times 10^{-12}\ M; \ \text{pH} = 11.96$$

b. $$HONH_2 + H_2O \rightleftharpoons HONH_3^+ + OH^- \qquad K_b = 1.1 \times 10^{-8}$$

	$HONH_2$	$HONH_3^+$	OH^-
Initial	0.20 M	0	~0
Equil.	$0.20 - x$	x	x

$$K_b = 1.1 \times 10^{-8} = \frac{x^2}{0.20 - x} \approx \frac{x^2}{0.20}, \quad x = [OH^-] = 4.7 \times 10^{-5}\ M; \ \text{Assumptions good.}$$

$[H^+] = 2.1 \times 10^{-10}\ M$; pH = 9.68

87. This is a solution of a weak base in water. We must solve the weak base equilibrium problem.

$$C_2H_5NH_2 + H_2O \rightleftharpoons C_2H_5NH_3^+ + OH^- \quad K_b = 5.6 \times 10^{-4}$$

	$C_2H_5NH_2$		$C_2H_5NH_3^+$	OH^-
Initial	0.20 M		0	~0
	x mol/L $C_2H_5NH_2$ reacts with H_2O to reach equilibrium			
Change	$-x$	→	$+x$	$+x$
Equil.	$0.20 - x$		x	x

$$K_b = \frac{[C_2H_5NH_3^+][OH^-]}{[C_2H_5NH_2]} = \frac{x^2}{0.20 - x} \approx \frac{x^2}{0.20} \quad \text{(assuming } x << 0.20\text{)}$$

$$x = 1.1 \times 10^{-2}; \text{ Checking assumption: } \frac{1.1 \times 10^{-2}}{0.20} \times 100 = 5.5\%$$

Assumption fails the 5% rule. We must solve exactly using either the quadratic equation or the method of successive approximations (see Appendix 1.4 of the text). Using successive approximations and carrying extra significant figures:

$$\frac{x^2}{0.20 - 0.011} = \frac{x^2}{0.189} = 5.6 \times 10^{-4}, \; x = 1.0 \times 10^{-2}\, M \text{ (consistent answer)}$$

$$x = [OH^-] = 1.0 \times 10^{-2}\, M; \; [H^+] = \frac{K_w}{[OH^-]} = \frac{1.0 \times 10^{-14}}{1.0 \times 10^{-2}} = 1.0 \times 10^{-12}\, M; \; pH = 12.00$$

89. To solve for percent ionization, just solve the weak base equilibrium problem.

a. $$NH_3 + H_2O \rightleftharpoons NH_4^+ + OH^- \quad K_b = 1.8 \times 10^{-5}$$

	NH_3	NH_4^+	OH^-
Initial	0.10 M	0	~0
Equil.	$0.10 - x$	x	x

$$K_b = 1.8 \times 10^{-5} = \frac{x^2}{0.10 - x} \approx \frac{x^2}{0.10}, \; x = [OH^-] = 1.3 \times 10^{-3}\, M; \text{ Assumptions good.}$$

$$\text{Percent ionization} = \frac{[OH^-]}{[NH_3]_0} \times 100 = \frac{1.3 \times 10^{-3}\, M}{0.10\, M} \times 100 = 1.3\%$$

b. $$NH_3 + H_2O \rightleftharpoons NH_4^+ + OH^-$$

	NH_3	NH_4^+	OH^-
Initial	0.010 M	0	~0
Equil.	$0.010 - x$	x	x

$$1.8 \times 10^{-5} = \frac{x^2}{0.010 - x} \approx \frac{x^2}{0.010}, \; x = [OH^-] = 4.2 \times 10^{-4}\, M; \text{ Assumptions good.}$$

$$\text{Percent ionization} = \frac{4.2 \times 10^{-4}}{0.010} \times 100 = 4.2\%$$

Note: For the same base, the percent ionization increases as the initial concentration of base decreases.

91. Let cod = codeine, $C_{18}H_{21}NO_3$; using the K_b reaction to solve:

	cod + H_2O	$\rightleftharpoons$	$codH^+$	+	OH^-
Initial	1.7×10^{-3} *M*		0		~0
	x mol/L codeine reacts with H_2O to reach equilibrium				
Change	$-x$	$\rightarrow$	$+x$		$+x$
Equil.	$1.7 \times 10^{-3} - x$		x		x

$$K_b = \frac{x^2}{1.7\times10^{-3} - x}; \quad pH = 9.59; \quad so: pOH = 14.00 - 9.59 = 4.41.$$

$$[OH^-] = x = 10^{-4.41} = 3.9 \times 10^{-5} M; \quad K_b = \frac{(3.9\times10^{-5})^2}{1.7\times10^{-3} - 3.9\times10^{-5}} = 9.2 \times 10^{-7}$$

Polyprotic Acids

93. $H_2SO_3(aq) \rightleftharpoons HSO_3^-(aq) + H^+(aq) \qquad K_{a_1} = \frac{[HSO_3^-][H^+]}{[H_2SO_3]}$

$HSO_3^-(aq) \rightleftharpoons SO_3^{2-}(aq) + H^+(aq) \qquad K_{a_2} = \frac{[SO_3^{2-}][H^+]}{[HSO_3^-]}$

95. In both these polyprotic acid problems, the dominate equilibrium is the K_{a_1} reaction. The amount of H^+ produced from the subsequent K_a reactions will be minimal since they are all have much smaller K_a values.

a.

	H_3PO_4	$\rightleftharpoons$	H^+	+	$H_2PO_4^-$	$K_{a_1} = 7.5 \times 10^{-3}$
Initial	0.10 *M*		~0		0	
	x mol/L H_3PO_4 dissociates to reach equilibrium					
Change	$-x$	$\rightarrow$	$+x$		$+x$	
Equil.	$0.10 - x$		x		x	

$$K_{a_1} = 7.5 \times 10^{-3} = \frac{[H^+][H_2PO_4^-]}{[H_3PO_4]} = \frac{x^2}{0.10 - x} \approx \frac{x^2}{0.10}, \quad x = 2.7 \times 10^{-2}$$

Assumption is bad (*x* is 27% of 0.10). Using successive approximations:

$$\frac{x^2}{0.10 - 0.027} = 7.5\times10^{-3}, \ x = 2.3 \times 10^{-2}; \quad \frac{x^2}{0.10} = 7.5 \times 10^{-3}, \ x = 2.4 \times 10^{-2}$$

(consistent answer)

$x = [H^+] = 2.4 \times 10^{-2}$ *M*; $pH = -\log(2.4 \times 10^{-2}) = 1.62$

b.

	H_2CO_3	$\rightleftharpoons$	H^+	+	HCO_3^-	$K_{a_1} = 4.3 \times 10^{-7}$
Initial	0.10 *M*		~0		0	
Equil.	$0.10 - x$		x		x	

$$K_{a_1} = 4.3 \times 10^{-7} = \frac{[H^+][HCO_3^-]}{[H_2CO_3]} = \frac{x^2}{0.10 - x} \approx \frac{x^2}{0.10}$$

$x = [H^+] = 2.1 \times 10^{-4}$ *M*; pH = 3.68; Assumptions good.

97. The dominant H^+ producer is the strong acid H_2SO_4. A 2.0 *M* H_2SO_4 solution produces 2.0 *M* HSO_4^- and 2.0 *M* H^+. However, HSO_4^- is a weak acid which could also add H^+ to the solution.

	HSO_4^-	$\rightleftharpoons$	H^+	+	SO_4^{2-}
Initial	2.0 *M*		2.0 *M*		0
	x mol/L HSO_4^- dissociates to reach equilibrium				
Change	$-x$	$\rightarrow$	$+x$		$+x$
Equil.	$2.0 - x$		$2.0 + x$		x

$$K_{a_2} = 1.2 \times 10^{-2} = \frac{[H^+][SO_4^{2-}]}{[HSO_4^-]} = \frac{(2.0 + x)(x)}{2.0 - x} \approx \frac{2.0\,(x)}{2.0},\ x = 1.2 \times 10^{-2}$$

Because *x* is 0.60% of 2.0, the assumption is valid by the 5% rule. The amount of additional H^+ from HSO_4^- is 1.2×10^{-2}. The total amount of H^+ present is:

$[H^+] = 2.0 + 1.2 \times 10^{-2} = 2.0$ *M*; pH = −log (2.0) = −0.30

Note: In this problem, H^+ from HSO_4^- could have been ignored. However, this is not always the case, especially in more dilute solutions of H_2SO_4.

Acid-Base Properties of Salts

99. One difficult aspect of acid-base chemistry is recognizing what types of species are present in solution, i.e., whether a species is a strong acid, strong base, weak acid, weak base or a neutral species. Below are some ideas and generalizations to keep in mind that will help in recognizing types of species present.

 a. Memorize the following strong acids: HCl, HBr, HI, HNO_3, $HClO_4$ and H_2SO_4
 b. Memorize the following strong bases: LiOH, NaOH, KOH, RbOH, $Ca(OH)_2$, $Sr(OH)_2$ and $Ba(OH)_2$
 c. All weak acids have a K_a value less than 1 but greater than K_w. Some weak acids are in Table 14.2 of the text. All weak bases have a K_b value less than 1 but greater than K_w. Some weak bases are in Table 14.3 of the text.
 d. All conjugate bases of weak acids are weak bases, i.e., all have a K_b value less than 1 but greater than K_w. Some examples of these are the conjugate bases of the weak acids in Table 14.2 of the text.
 e. All conjugate acids of weak bases are weak acids, i.e., all have a K_a value less than 1 but greater than K_w. Some examples of these are the conjugate acids of the weak bases in Table 14.3 of the text.
 f. Alkali metal ions (Li^+, Na^+, K^+, Rb^+, Cs^+) and heavier alkaline earth metal ions (Ca^{2+}, Sr^{2+}, Ba^{2+}) have no acidic or basic properties in water.
 g. All conjugate bases of strong acids (Cl^-, Br^-, I^-, NO_3^-, ClO_4^-, HSO_4^-) have no basic properties in water ($K_b << K_w$) and only HSO_4^- has any acidic properties in water.

Let's apply these ideas to this problem to see what type of species are present. The letters in parenthesis is/are the generalization(s) above which identifies the species.

KOH: strong base (b)
KCl: neutral; K^+ and Cl^- have no acidic/basic properties (f and g).
KCN: CN^- is a weak base, $K_b = 1.0 \times 10^{-14}/6.2 \times 10^{-10} = 1.6 \times 10^{-5}$ (c and d). Ignore K^+ (f).
NH_4Cl: NH_4^+ is a weak acid, $K_a = 5.6 \times 10^{-10}$ (c and e). Ignore Cl^- (g).
HCl: strong acid (a)

The most acidic solution will be the strong acid followed by the weak acid. The most basic solution will be the strong base followed by the weak base. The KCl solution will be between the acidic and basic solutions at pH = 7.00.

Most acidic → most basic: HCl > NH_4Cl > KCl > KCN > KOH

101. From the K_a values, acetic acid is a stronger acid than hypochlorous acid. Conversely, the conjugate base of acetic acid, $C_2H_3O_2^-$, will be a weaker base than the conjugate base of hypochlorous acid, OCl^-. Thus, the hypochlorite ion, OCl^-, is a stronger base than the acetate ion, $C_2H_3O_2^-$. In general, the stronger the acid, the weaker the conjugate base. This statement comes from the relationship $K_w = K_a \times K_b$, which holds for all conjugate acid-base pairs.

103. $NaN_3 \rightarrow Na^+ + N_3^-$; Azide, N_3^-, is a weak base since it is the conjugate base of a weak acid. All conjugate bases of weak acids are weak bases ($K_w < K_b < 1$). Ignore Na^+.

$$N_3^- + H_2O \rightleftharpoons HN_3 + OH^- \quad K_b = \frac{K_w}{K_a} = \frac{1.0\times10^{-14}}{1.9\times10^{-5}} = 5.3 \times 10^{-10}$$

	N_3^-		HN_3	OH^-
Initial	0.010 *M*		0	~0
	x mol/L of N_3^- reacts with H_2O to reach equilibrium			
Change	$-x$	→	$+x$	$+x$
Equil.	$0.010 - x$		x	x

$$K_b = 5.3 \times 10^{-10} = \frac{[HN_3][OH^-]}{[N_3^-]} = \frac{x^2}{0.010 - x} \approx \frac{x^2}{0.010} \quad \text{(assuming } x << 0.010)$$

$$x = [OH^-] = 2.3 \times 10^{-6}\ M;\ [H^+] = \frac{1.0\times10^{-14}}{2.3\times10^{-6}} = 4.3 \times 10^{-9}\ M \quad \text{Assumptions good.}$$

$[HN_3] = [OH^-] = 2.3 \times 10^{-6}\ M$; $[Na^+] = 0.010\ M$; $[N_3^-] = 0.010 - 2.3 \times 10^{-6} = 0.010\ M$

105. a. $CH_3NH_3Cl \rightarrow CH_3NH_3^+ + Cl^-$: $CH_3NH_3^+$ is a weak acid. Cl^- is the conjugate base of a strong acid. Cl^- has no basic (or acidic) properties.

$$CH_3NH_3^+ \rightleftharpoons CH_3NH_2 + H^+ \quad K_a = \frac{[CH_3NH_2][H^+]}{[CH_3NH_3^+]} = \frac{K_w}{K_b} = \frac{1.00\times10^{-14}}{4.38\times10^{-4}} = 2.28 \times 10^{-11}$$

	$CH_3NH_3^+$	$\rightleftharpoons$	CH_3NH_2	+	H^+
Initial	0.10 *M*		0		~0
	x mol/L $CH_3NH_3^+$ dissociates to reach equilibrium				
Change	$-x$	→	$+x$		$+x$
Equil.	$0.10 - x$		x		x

$$K_a = 2.28 \times 10^{-11} = \frac{x^2}{0.10 - x} \approx \frac{x^2}{0.10} \quad \text{(assuming } x << 0.10\text{)}$$

$x = [H^+] = 1.5 \times 10^{-6}\ M$; pH = 5.82   Assumptions good.

b. $NaCN \rightarrow Na^+ + CN^-$: CN^- is a weak base. Na^+ has no acidic (or basic) properties.

$$CN^- + H_2O \rightleftharpoons HCN + OH^- \quad K_b = \frac{K_w}{K_a} = \frac{1.0 \times 10^{-14}}{6.2 \times 10^{-10}} = 1.6 \times 10^{-5}$$

	CN^- + H_2O	$\rightleftharpoons$	HCN	+	OH^-
Initial	0.050 *M*		0		~0
	x mol/L CN^- reacts with H_2O to reach equilibrium				
Change	$-x$	→	$+x$		$+x$
Equil.	$0.050 - x$		x		x

$$K_b = 1.6 \times 10^{-5} = \frac{[HCN][OH^-]}{[CN^-]} = \frac{x^2}{0.050 - x} \approx \frac{x^2}{0.050}$$

$x = [OH^-] = 8.9 \times 10^{-4}\ M$; pOH = 3.05; pH = 10.95   Assumptions good.

107. All these salts contain Na^+, which has no acidic/basic properties, and a conjugate base of a weak acid (except for NaCl where Cl^- is a neutral species.). All conjugate bases of weak acids are weak bases because the K_b values for these species are between 1 and K_w. To identify the species, we will use the data given to determine the K_b value for the weak conjugate base. From the K_b value and data in Table 14.2 of the text, we can identify the conjugate base present by calculating the K_a value for the weak acid. We will use A^- as an abbreviation for the weak conjugate base.

	A^- + H_2O	$\rightleftharpoons$	HA	+	OH^-
Initial	0.100 mol/1.00 L		0		~0
	x mol/L A^- reacts with H_2O to reach equilibrium				
Change	$-x$	→	$+x$		$+x$
Equil.	$0.100 - x$		x		x

$$K_b = \frac{[HA][OH^-]}{[A^-]} = \frac{x^2}{0.100 - x}$$; From the problem, pH = 8.07:

$pOH = 14.00 - 8.07 = 5.93$; $[OH^-] = x = 10^{-5.93} = 1.2 \times 10^{-6}\ M$

$$K_b = \frac{(1.2 \times 10^{-6})^2}{0.100 - 1.2 \times 10^{-6}} = 1.4 \times 10^{-11} = K_b \text{ value for the conjugate base of a weak acid.}$$

The K_a value for the weak acid equals K_w/K_b: $K_a = \dfrac{1.0\times10^{-14}}{1.4\times10^{-11}} = 7.1 \times 10^{-4}$

From Table 14.2 of the text, this K_a value is closest to HF. Therefore, the unknown salt is NaF.

109. Major species present: $Al(H_2O)_6^{3+}$ ($K_a = 1.4 \times 10^{-5}$), NO_3^- (neutral) and H_2O ($K_w = 1.0 \times 10^{-14}$); $Al(H_2O)_6^{3+}$ is a stronger acid than water so it will be the dominant H^+ producer.

	$Al(H_2O)_6^{3+}$	$\rightleftharpoons$	$Al(H_2O)_5(OH)^{2+}$ +	H^+
Initial	0.050 *M*		0	~0
	x mol/L $Al(H_2O)_6^{3+}$ dissociates to reach equilibrium			
Change	$-x$	$\rightarrow$	$+x$	$+x$
Equil.	$0.050 - x$		x	x

$$K_a = 1.4 \times 10^{-5} = \frac{[Al(H_2O)_5(OH)^{2+}][H^+]}{[Al(H_2O)_6^{3+}]} = \frac{x^2}{0.050 - x} \approx \frac{x^2}{0.050}$$

$x = 8.4 \times 10^{-4}\ M = [H^+]$; pH = $-\log(8.4 \times 10^{-4}) = 3.08$; Assumptions good.

111. Reference Table 14.6 of the text and the solution to Exercise 14.99 for some generalizations on acid-base properties of salts.

a. $NaNO_3 \rightarrow Na^+ + NO_3^-$ neutral; Neither species has any acidic/basic properties.

b. $NaNO_2 \rightarrow Na^+ + NO_2^-$ basic; NO_2^- is a weak base and Na^+ has no effect on pH.

$$NO_2^- + H_2O \rightleftharpoons HNO_2 + OH^- \quad K_b = \frac{K_w}{K_{a,\,HNO_2}} = \frac{1.0\times10^{-14}}{4.0\times10^{-4}} = 2.5 \times 10^{-11}$$

c. $C_5H_5NHClO_4 \rightarrow C_5H_5NH^+ + ClO_4^-$ acidic; $C_5H_5NH^+$ is a weak acid and ClO_4^- has no effect on pH.

$$C_5H_5NH^+ \rightleftharpoons H^+ + C_5H_5N \quad K_a = \frac{K_w}{K_{b,\,C_5H_5N}} = \frac{1.0\times10^{-14}}{1.7\times10^{-9}} = 5.9 \times 10^{-6}$$

d. $NH_4NO_2 \rightarrow NH_4^+ + NO_2^-$ acidic; NH_4^+ is a weak acid ($K_a = 5.6 \times 10^{-10}$) and NO_2^- is a weak base ($K_b = 2.5 \times 10^{-11}$). Because $K_{a,\,NH_4^+} > K_{b,\,NO_2^-}$, the solution is acidic.

$NH_4^+ \rightleftharpoons H^+ + NH_3$ $K_a = 5.6 \times 10^{-10}$; $NO_2^- + H_2O \rightleftharpoons HNO_2 + OH^-$ $K_b = 2.5 \times 10^{-11}$

e. $KOCl \rightarrow K^+ + OCl^-$ basic; OCl^- is a weak base and K^+ has no effect on pH.

$$OCl^- + H_2O \rightleftharpoons HOCl + OH^- \quad K_b = \frac{K_w}{K_{a,\,HOCl}} = \frac{1.0\times10^{-14}}{3.5\times10^{-8}} = 2.9 \times 10^{-7}$$

f. $NH_4OCl \rightarrow NH_4^+ + OCl^-$ basic; NH_4^+ is a weak acid and OCl^- is a weak base. Because $K_{b, OCl^-} > K_{a, NH_4^+}$, the solution is basic.

$NH_4^+ \rightleftharpoons NH_3 + H^+$ $K_a = 5.6 \times 10^{-10}$; $OCl^- + H_2O \rightleftharpoons HOCl + OH^-$ $K_b = 2.9 \times 10^{-7}$

Relationships Between Structure and Strengths of Acids and Bases

113. a. $HIO_3 < HBrO_3$; As the electronegativity of the central atom increases, acid strength increases.

b. $HNO_2 < HNO_3$; As the number of oxygen atoms attached to the central nitrogen atom increases, acid strength increases.

c. HOI < HOCl; Same reasoning as in a.

d. $H_3PO_3 < H_3PO_4$; Same reasoning as in b.

115. a. $H_2O < H_2S < H_2Se$; As the strength of the H–X bond decreases, acid strength increases.

b. $CH_3CO_2H < FCH_2CO_2H < F_2CHCO_2H < F_3CCO_2H$; As the electronegativity of neighboring atoms increases, acid strength increases.

c. $NH_4^+ < HONH_3^+$; Same reason as in b.

d. $NH_4^+ < PH_4^+$; Same reason as in a.

117. In general, metal oxides form basic solutions when dissolved in water and nonmetal oxides form acidic solutions in water.

a. basic; $CaO(s) + H_2O(l) \rightarrow Ca(OH)_2(aq)$; $Ca(OH)_2$ is a strong base.

b. acidic; $SO_2(g) + H_2O(l) \rightarrow H_2SO_3(aq)$; H_2SO_3 is a weak diprotic acid.

c. acidic; $Cl_2O(g) + H_2O(l) \rightarrow 2\ HOCl(aq)$; HOCl is a weak acid.

Lewis Acids and Bases

119. A Lewis base is an electron pair donor, and a Lewis acid is an electron pair acceptor.

a. $B(OH)_3$, acid; H_2O, base b. Ag^+, acid; NH_3, base c. BF_3, acid; F^-, base

121. $Al(OH)_3(s) + 3\ H^+(aq) \rightarrow Al^{3+}(aq) + 3\ H_2O(l)$ (Brønsted-Lowry base, H^+ acceptor)

$Al(OH)_3(s) + OH^-(aq) \rightarrow Al(OH)_4^-(aq)$ (Lewis acid, electron pair acceptor)

123. Fe^{3+} should be the stronger Lewis acid. Fe^{3+} is smaller and has a greater positive charge. Because of this, Fe^{3+} will be more strongly attracted to lone pairs of electrons as compared to Fe^{2+}.

Additional Exercises

125. At pH = 2.000, $[H^+] = 10^{-2.000} = 1.00 \times 10^{-2}$ *M*; At pH = 4.000, $[H^+] = 10^{-4.000} = 1.00 \times 10^{-4}$ *M*

$$\text{mol H}^+ \text{ present} = 0.0100 \text{ L} \times \frac{0.0100 \text{ mol H}^+}{\text{L}} = 1.00 \times 10^{-4} \text{ mol H}^+$$

Let V = total volume of solution at pH = 4.000:

$$1.00 \times 10^{-4} \text{ mol/L} = \frac{1.00 \times 10^{-4} \text{ mol H}^+}{\text{V}}, \text{ V} = 1.00 \text{ L}$$

Volume of water added = 1.00 L − 0.0100 L = 0.99 L = 990 mL

127. a. The initial concentrations are halved since equal volumes of the two solutions are mixed.

	$HC_2H_3O_2$	⇌	H^+	+	$C_2H_3O_2^-$
Initial	0.100 *M*		5.00×10^{-4} *M*		0
Equil.	$0.100 - x$		$5.00 \times 10^{-4} + x$		x

$$K_a = 1.8 \times 10^{-5} = \frac{x(5.00 \times 10^{-4} + x)}{(0.100 - x)} \approx \frac{x(5.00 \times 10^{-4})}{(0.100)}$$

$x = 3.6 \times 10^{-3}$; Assumption is horrible. Using the quadratic formula:

$$x^2 + 5.18 \times 10^{-4} x - 1.8 \times 10^{-6} = 0$$

$$x = 1.1 \times 10^{-3} M; \ [H^+] = 5.00 \times 10^{-4} + x = 1.6 \times 10^{-3} M; \ \text{pH} = 2.80$$

b. $x = [C_2H_3O_2^-] = 1.1 \times 10^{-3}$ *M*

129. The light bulb is bright because a strong electrolyte is present, i.e., a solute is present that dissolves to produce a lot of ions in solution. The pH meter value of 4.6 indicates that a weak acid is present. (If a strong acid were present, the pH would be close to zero.) Of the possible substances, only HCl (strong acid), NaOH (strong base) and NH_4Cl are strong electrolytes. Of these three substances, only NH_4Cl contains a weak acid (the HCl solution would have a pH close to zero and the NaOH solution would have a pH close to 14.0). NH_4Cl dissociates into NH_4^+ and Cl^- ions when dissolved in water. Cl^- is the conjugate base of a strong acid, so it has no basic (or acidic properties) in water. NH_4^+, however, is the conjugate acid of the weak base NH_3, so NH_4^+ is a weak acid and would produce a solution with a pH = 4.6 when the concentration is ~1 *M*.

131.

	HBz	⇌	H^+	+	Bz^-
Initial	C		~0		0
	x mol/L HBz dissociates to reach equilibrium				
Change	$-x$	→	$+x$		$+x$
Equil.	$C - x$		x		x

$HBz = C_6H_5CO_2H$

$C = [HBz]_o$ = concentration of HBz that dissolves to give saturated solution.

$$K_a = \frac{[H^+][Bz^-]}{[HBz]} = 6.4 \times 10^{-5} = \frac{x^2}{C - x}, \text{ where } x = [H^+]$$

$$6.4 \times 10^{-5} = \frac{[H^+]^2}{C - [H^+]}; \text{ pH} = 2.80; \ [H^+] = 10^{-2.80} = 1.6 \times 10^{-3}\,M$$

$$C - 1.6 \times 10^{-3} = \frac{(1.6 \times 10^{-3})^2}{6.4 \times 10^{-5}} = 4.0 \times 10^{-2}, \ C = 4.0 \times 10^{-2} + 1.6 \times 10^{-3} = 4.2 \times 10^{-2}\,M$$

The molar solubility of $C_6H_5CO_2H$ is 4.2×10^{-2} mol/L.

133. For $H_2C_6H_6O_6$. $K_{a_1} = 7.9 \times 10^{-5}$ and $K_{a_2} = 1.6 \times 10^{-12}$. Because $K_{a_1} >> K_{a_2}$, the amount of H^+ produced by the K_{a_2} reaction will be negligible.

$$[H_2C_6H_6O_6]_o = \frac{0.500\text{ g} \times \frac{1\text{ mol } H_2C_6H_6O_6}{176.12\text{ g}}}{0.2000\text{ L}} = 0.0142\,M$$

	$H_2C_6H_6O_6(aq)$	$\rightleftharpoons$	$HC_6H_6O_6^-(aq)$	+	$H^+(aq)$	$K_{a_1} = 7.9 \times 10^{-5}$
Initial	0.0142 M		0		~0	
Equil.	0.0142 − x		x		x	

$$K_{a_1} = 7.9 \times 10^{-5} = \frac{x^2}{0.0142 - x} \approx \frac{x^2}{0.0142}, \ x = 1.1 \times 10^{-3}; \text{ Assumption fails the 5\% rule.}$$

Solving by the method of successive approximations:

$$7.9 \times 10^{-5} = \frac{x^2}{0.0142 - 1.1 \times 10^{-3}}, \ x = 1.0 \times 10^{-3}\,M \text{ (consistent answer)}$$

Since H^+ produced by the K_{a_2} reaction will be negligible, $[H^+] = 1.0 \times 10^{-3}$ and pH = 3.00.

135. For this problem we will abbreviate $CH_2{=}CHCO_2H$ as Hacr and $CH_2{=}CHCO_2^-$ as acr^-.

a. Solving the weak acid problem:

	Hacr	$\rightleftharpoons$	H^+	+	acr^-	$K_a = 5.6 \times 10^{-5}$
Initial	0.10 M		~0		0	
Equil.	0.10 − x		x		x	

$$\frac{x^2}{0.10 - x} = 5.6 \times 10^{-5} \approx \frac{x^2}{0.10}, \ x = [H^+] = 2.4 \times 10^{-3}\,M; \text{ pH} = 2.62; \text{ Assumptions good.}$$

b. $\% \text{ dissociation} = \dfrac{[H^+]}{[Hacr]_0} \times 100 = \dfrac{2.4\times10^{-3}}{0.10} \times 100 = 2.4\%$

c. acr^- is a weak base and the major source of OH^- in this solution.

$$acr^- + H_2O \rightleftharpoons Hacr + OH^- \quad K_b = \frac{K_w}{K_a} = \frac{1.0\times10^{-14}}{5.6\times10^{-5}}$$

	acr^-		$Hacr$	OH^-	
Initial	$0.050\ M$		0	~0	$K_b = 1.8 \times 10^{-10}$
Equil.	$0.050 - x$		x	x	

$$K_b = \frac{[Hacr][OH^-]}{[acr^-]} = 1.8 \times 10^{-10} = \frac{x^2}{0.050 - x} \approx \frac{x^2}{0.050}$$

$x = [OH^-] = 3.0 \times 10^{-6}\ M$; pOH = 5.52; pH = 8.48 Assumptions good.

137. a. $Fe(H_2O)_6^{3+} + H_2O \rightleftharpoons Fe(H_2O)_5(OH)^{2+} + H_3O^+$

	$Fe(H_2O)_6^{3+}$	$Fe(H_2O)_5(OH)^{2+}$	H_3O^+
Initial	$0.10\ M$	0	~0
Equil.	$0.10 - x$	x	x

$$K_a = \frac{[Fe(H_2O)_5(OH)^{2+}][H_3O^+]}{[Fe(H_2O)_6^{3+}]} = 6.0 \times 10^{-3} = \frac{x^2}{0.10 - x} \approx \frac{x^2}{0.10}$$

$x = 2.4 \times 10^{-2}$; Assumption is poor (x is 24% of 0.10). Using successive approximations:

$$\frac{x^2}{0.10 - 0.024} = 6.0 \times 10^{-3},\ x = 0.021$$

$$\frac{x^2}{0.10 - 0.021} = 6.0 \times 10^{-3},\ x = 0.022; \quad \frac{x^2}{0.10 - 0.022} = 6.0 \times 10^{-3},\ x = 0.022$$

$x = [H^+] = 0.022\ M$; pH = 1.66

b. Because of the lower charge, $Fe^{2+}(aq)$ will not be as strong an acid as $Fe^{3+}(aq)$. A solution of iron(II) nitrate will be less acidic (have a higher pH) than a solution with the same concentration of iron(III) nitrate.

139. The solution is acidic from $HSO_4^- \rightleftharpoons H^+ + SO_4^{2-}$. Solving the weak acid problem:

	HSO_4^-	H^+	SO_4^{2-}	
				$K_a = 1.2 \times 10^{-2}$
Initial	$0.10\ M$	~0	0	
Equil.	$0.10 - x$	x	x	

$$1.2 \times 10^{-2} = \frac{[H^+][SO_4^{2-}]}{[HSO_4^-]} = \frac{x^2}{0.10 - x} \approx \frac{x^2}{0.10}, \quad x = 0.035$$

Assumption is not good (x is 35% of 0.10). Using successive approximations:

$$\frac{x^2}{0.10 - x} = \frac{x^2}{0.10 - 0.035} = 1.2 \times 10^{-2}, \quad x = 0.028$$

$$\frac{x^2}{0.10 - 0.028} = 1.2 \times 10^{-2}, \quad x = 0.029; \qquad \frac{x^2}{0.10 - 0.029} = 1.2 \times 10^{-2}, \quad x = 0.029$$

$x = [H^+] = 0.029\ M$; pH = 1.54

141. a. In the lungs, there is a lot of O_2 and the equilibrium favors $Hb(O_2)_4$. In the cells, there is a deficiency of O_2, and the equilibrium favors HbH_4^{4+}.

b. CO_2 is a weak acid, $CO_2 + H_2O \rightleftharpoons HCO_3^- + H^+$. Removing CO_2 essentially decreases H^+. $Hb(O_2)_4$ is then favored and O_2 is not released by hemoglobin in the cells. Breathing into a paper bag increases CO_2 in the blood, thus increasing H^+, which shifts the reaction left.

c. CO_2 builds up in the blood and it becomes too acidic, driving the equilibrium to the left. Hemoglobin can't bind O_2 as strongly in the lungs. Bicarbonate ion acts as a base in water and neutralizes the excess acidity.

143. a. H_2SO_3 b. $HClO_3$ c. H_3PO_3

NaOH and KOH are soluble ionic compounds composed of Na^+ and K^+ cations and OH^- anions. All soluble ionic compounds dissolve to form the ions from which they are formed. In oxyacids, the compounds are all covalent compounds in which electrons are shared to form bonds (unlike ionic compounds). When these compounds are dissolved in water, the covalent bond between oxygen and hydrogen breaks to form H^+ ions.

Challenge Problems

145. Since this is a very dilute solution of NaOH, we must worry about the amount of OH^- donated from the autoionization of water.

$NaOH \rightarrow Na^+ + OH^-$

$H_2O \rightleftharpoons H^+ + OH^- \quad K_w = [H^+][OH^-] = 1.0 \times 10^{-14}$

This solution, like all solutions, must be charge balanced, that is [positive charge] = [negative charge]. For this problem, the charge balance equation is:

$[Na^+] + [H^+] = [OH^-]$, where $[Na^+] = 1.0 \times 10^{-7}\ M$ and $[H^+] = \dfrac{K_w}{[OH^-]}$

Substituting into the charge balance equation:

$$1.0 \times 10^{-7} + \frac{1.0\times10^{-14}}{[OH^-]} = [OH^-],\ [OH^-]^2 - 1.0 \times 10^{-7}\ [OH^-] - 1.0 \times 10^{-14} = 0$$

Using the quadratic formula to solve:

$$[OH^-] = \frac{-(-1.0\times10^{-7}) \pm [(-1.0\times10^{-7})^2 - 4(1)(-1.0\times10^{-14})]^{1/2}}{2(1)}$$

$[OH^-] = 1.6 \times 10^{-7}$ *M*; pOH = $-\log(1.6 \times 10^{-7}) = 6.80$; pH = 7.20

147.

	HA	⇌	H^+	+	A^-	$K_a = 1.00 \times 10^{-6}$
Initial	C		~0		0	C = $[HA]_0$; For pH = 4.000,
Equil.	C - 1.00×10^{-4}		1.00×10^{-4}		1.00×10^{-4}	$x = [H^+] = 1.00 \times 10^{-4}$ *M*

$$K_a = \frac{(1.00 \times 10^{-4})^2}{C - 1.00 \times 10^{-4}} = 1.00 \times 10^{-6};\ \text{Solving: } C = 0.0101\ M$$

The solution initially contains 50.0×10^{-3} L × 0.0101 mol/L = 5.05×10^{-4} mol HA. We then dilute to a total volume, V, in liters. The resulting pH = 5.000, so $[H^+] = 1.00 \times 10^{-5}$. In the typical weak acid problem, $x = [H^+]$, so:

	HA	⇌	H^+	+	A^-
Initial	5.05×10^{-4} mol/V		~0		0
Equil.	5.05×10^{-4} /V - 1.00×10^{-5}		1.00×10^{-5}		1.00×10^{-5}

$$K_a = \frac{(1.00 \times 10^{-5})^2}{5.05 \times 10^{-4}/V - 1.00 \times 10^{-5}} = 1.00 \times 10^{-6},\ 1.00 \times 10^{-4} = 5.05 \times 10^{-4}/V - 1.00 \times 10^{-5}$$

V = 4.59 L; 50.0 mL are present initially, so we need to add 4540 mL of water.

149. Major species present are H_2O, $C_5H_5NH^+$ ($K_a = K_w/K_b(C_5H_5N) = 1.0 \times 10^{-14}/1.7 \times 10^{-9} = 5.9 \times 10^{-6}$) and F^- ($K_b = K_w/K_a(HF) = 1.0 \times 10^{-14}/7.2 \times 10^{-4} = 1.4 \times 10^{-11}$). The reaction to consider is the best acid present ($C_5H_5NH^+$) reacting with the best base present (F^-). Solving for the equilibrium concentrations:

	$C_5H_5NH^+(aq)$	+	$F^-(aq)$	⇌	$C_5H_5N(aq)$	+	HF(aq)
Initial	0.200 *M*		0.200 *M*		0		0
Change	$-x$		$-x$	→	$+x$		$+x$
Equil.	$0.200 - x$		$0.200 - x$		x		x

$$K = K_{a,\,C_5H_5NH^+} \times \frac{1}{K_{a,\,HF}} = 5.9 \times 10^{-6}\ (1/7.2 \times 10^{-4}) = 8.2 \times 10^{-3}$$

$$K = \frac{[C_5H_5N][HF]}{[C_5H_5NH^+][F^-]} = 8.2 \times 10^{-3} = \frac{x^2}{(0.200 - x)^2};$$ Taking the square root of both sides:

$$0.091 = \frac{x}{0.200 - x},\quad x = 0.018 - 0.091\,x,\quad x = 0.016\,M$$

From the setup to the problem, $x = [C_5H_5N] = [HF] = 0.016\,M$ and $0.200 - x = 0.200 - 0.016 = 0.184\,M = [C_5H_5NH^+] = [F^-]$. To solve for the $[H^+]$, we can use either the K_a equilibrium for $C_5H_5NH^+$ or the K_a equilibrium for HF. Using $C_5H_5NH^+$ data:

$$K_{a,\,C_5H_5NH^+} = 5.9 \times 10^{-6} = \frac{[C_5H_5N][H^+]}{[C_5H_5NH^+]} = \frac{(0.016)[H^+]}{(0.184)},\quad [H^+] = 6.8 \times 10^{-5}\,M$$

$$pH = -\log(6.8 \times 10^{-5}) = 4.17$$

As one would expect, because the K_a for the weak acid is larger than the K_b for the weak base, a solution of this salt should be acidic.

151. Since NH_3 is so concentrated, we need to calculate the OH^- contribution from the weak base NH_3.

	NH_3 +	⇌	NH_4^+ +	OH^-	$K_b = 1.8 \times 10^{-5}$
Initial	$15.0\,M$		0	$0.0100\,M$	(Assume no volume change.)
Equil.	$15.0 - x$		x	$0.0100 + x$	

$$K_b = 1.8 \times 10^{-5} = \frac{x(0.0100 + x)}{15.0 - x} \approx \frac{x(0.0100)}{15.0},\quad x = 0.027;$$ Assumption is horrible (x is 270% of 0.0100).

Using the quadratic formula:

$$1.8 \times 10^{-5}(15.0 - x) = 0.0100\,x + x^2,\quad x^2 + 0.0100\,x - 2.7 \times 10^{-4} = 0$$

$$x = 1.2 \times 10^{-2},\quad [OH^-] = 1.2 \times 10^{-2} + 0.0100 = 0.022\,M$$

153. $$\text{Molar mass} = \frac{dRT}{P} = \frac{\frac{5.11\,g}{L} \times \frac{0.08206\,L\,atm}{mol\,K} \times 298\,K}{1.00\,atm} = 125\,g/mol$$

$$[HA]_o = \frac{1.50\,g \times \frac{1\,mol}{125\,g}}{0.100\,L} = 0.120\,M;\quad pH = 1.80,\quad [H^+] = 10^{-1.80} = 1.6 \times 10^{-2}\,M$$

	HA	$\rightleftharpoons$	H^+	+	A^-	
Equil.	$0.120 - x$		x		x	$x = [H^+] = 1.6 \times 10^{-2}\ M$

$$K_a = \frac{[H^+][A^-]}{[HA]} = \frac{(1.6 \times 10^{-2})^2}{0.120 - 0.016} = 2.5 \times 10^{-3}$$

155. PO_4^{3-} is the conjugate base of HPO_4^{2-}. The K_a value for HPO_4^{2-} is $K_{a_3} = 4.8 \times 10^{-13}$.

$$PO_4^{3-}(aq) + H_2O(l) \rightleftharpoons HPO_4^{2-}(aq) + OH^-(aq) \quad K_b = \frac{K_w}{K_{a_3}} = \frac{1.0\times10^{-14}}{4.8\times10^{-13}} = 0.021$$

HPO_4^{2-} is the conjugate base of $H_2PO_4^-$ ($K_{a_2} = 6.2 \times 10^{-8}$).

$$HPO_4^{2-} + H_2O \rightleftharpoons H_2PO_4^- + OH^- \quad K_b = \frac{K_w}{K_{a_2}} = \frac{1.0\times10^{-14}}{6.2\times10^{-8}} = 1.6 \times 10^{-7}$$

$H_2PO_4^-$ is the conjugate base of H_3PO_4 ($K_{a_1} = 7.5 \times 10^{-3}$).

$$H_2PO_4^- + H_2O \rightleftharpoons H_3PO_4 + OH^- \quad K_b = \frac{K_w}{K_{a_1}} = \frac{1.0\times10^{-14}}{7.5\times10^{-3}} = 1.3 \times 10^{-12}$$

From the K_b values, PO_4^{3-} is the strongest base. This is expected because PO_4^{3-} is the conjugate base of the weakest acid (HPO_4^{2-}).

157. a. $NH_4(HCO_3) \rightarrow NH_4^+ + HCO_3^-$

$$K_a \text{ for } NH_4^+ = \frac{1.0\times10^{-14}}{1.8\times10^{-5}} = 5.6 \times 10^{-10}; \quad K_b \text{ for } HCO_3^- = \frac{K_w}{K_{a_1}} = \frac{1.0\times10^{-14}}{4.3\times10^{-7}} = 2.3 \times 10^{-8}$$

Solution is basic because HCO_3^- is a stronger base than NH_4^+ is as an acid. The acidic properties of HCO_3^- were ignored because K_{a_2} is very small (5.6×10^{-11}).

b. $NaH_2PO_4 \rightarrow Na^+ + H_2PO_4^-$; Ignore Na^+.

$$K_{a_2} \text{ for } H_2PO_4^- = 6.2 \times 10^{-8}; \quad K_b \text{ for } H_2PO_4^- = \frac{K_w}{K_{a_1}} = \frac{1.0\times10^{-14}}{7.5\times10^{-3}} = 1.3 \times 10^{-12}$$

Solution is acidic because $K_a > K_b$.

c. $Na_2HPO_4 \rightarrow 2\ Na^+ + HPO_4^{2-}$; Ignore Na^+.

$$K_{a_3} \text{ for } HPO_4^{2-} = 4.8 \times 10^{-13}; \quad K_b \text{ for } HPO_4^{2-} = \frac{K_w}{K_{a_2}} = \frac{1.0\times10^{-14}}{6.2\times10^{-8}} = 1.6 \times 10^{-7}$$

Solution is basic because $K_b > K_a$.

d. $NH_4(H_2PO_4) \rightarrow NH_4^+ + H_2PO_4^-$

NH_4^+ is weak acid and $H_2PO_4^-$ is also acidic (see b). Solution with both ions present will be acidic.

e. $NH_4(HCO_2) \rightarrow NH_4^+ + HCO_2^-$; From Appendix 5, K_a for $HCO_2H = 1.8 \times 10^{-4}$.

$$K_a \text{ for } NH_4^+ = 5.6 \times 10^{-10}; \; K_b \text{ for } HCO_2^- = \frac{K_w}{K_a} = \frac{1.0 \times 10^{-14}}{1.8 \times 10^{-4}} = 5.6 \times 10^{-11}$$

Solution is acidic because NH_4^+ is a stronger acid than HCO_2^- is as a base.

159. $$\text{Molality} = m = \frac{0.100 \text{ g} \times \frac{1 \text{ mol}}{100.0 \text{ g}}}{0.5000 \text{ kg}} = 2.00 \times 10^{-3} \text{ mol/kg} \approx 2.00 \times \text{mol/L} \quad \text{(dilute solution)}$$

$\Delta T_f = iK_f m$, $0.0056°C = i(1.86°C/\text{molal})(2.00 \times 10^{-3} \text{ molal})$, $i = 1.5$

If i = 1.0, the percent dissociation of the acid = 0% and if i = 2.0, the percent dissociation of the acid = 100%. Since i = 1.5, the weak acid is 50.% dissociated.

$$HA \rightleftharpoons H^+ + A^- \qquad K_a = \frac{[H^+][A^-]}{[HA]}$$

Because the weak acid is 50.% dissociated:

$$[H^+] = [A^-] = [HA]_o \times 0.50 = 2.00 \times 10^{-3} M \times 0.50 = 1.0 \times 10^{-3} M$$

$$[HA] = [HA]_o - \text{amount HA reacted} = 2.00 \times 10^{-3} M - 1.0 \times 10^{-3} M = 1.0 \times 10^{-3} M$$

$$K_a = \frac{[H^+][A^-]}{[HA]} = \frac{(1.0 \times 10^{-3})(1.0 \times 10^{-3})}{1.0 \times 10^{-3}} = 1.0 \times 10^{-3}$$

Integrative Problems

161. $$[IO^-] = \frac{2.14 \text{ g NaIO} \times \frac{1 \text{ mol NaIO}}{165.89 \text{ g}} \times \frac{1 \text{ mol } IO^-}{\text{mol NaIO}}}{1.25 \text{ L}} = 1.03 \times 10^{-2} M \; IO^-$$

$$IO^- + H_2O \rightleftharpoons HIO + OH^- \qquad K_b = \frac{[HIO][OH^-]}{[IO^-]}$$

	IO^-		HIO	OH^-
Initial	$1.03 \times 10^{-2} M$		0	~0
Equil.	$1.03 \times 10^{-2} - x$		x	x

$$K_b = \frac{x^2}{1.03 \times 10^{-2} - x}$$; From the problem, pOH = 14.00 – 11.32 = 2.68

$$[OH^-] = 10^{-2.68} = 2.1 \times 10^{-3}\,M = x;\ K_b = \frac{(2.1\times10^{-3})^2}{1.03\times10^{-2} - 2.1\times10^{-3}} = 5.4 \times 10^{-4}$$

163. $$\frac{30.0 \text{ mg papH}^+\text{Cl}^-}{\text{mL soln}} \times \frac{1000 \text{ mL}}{\text{L}} \times \frac{1 \text{ g}}{1000 \text{ mg}} \times \frac{1 \text{ mol papH}^+\text{Cl}^-}{378.85 \text{ g}} \times \frac{1 \text{ mol papH}^+}{\text{mol papH}^+\text{Cl}^-} = 0.0792\,M$$

$$\text{papH}^+ \rightleftharpoons \text{pap} + \text{H}^+ \qquad K_a = \frac{K_w}{K_{b,\,pap}} = \frac{2.1\times10^{-14}}{8.33\times10^{-9}} = 2.5 \times 10^{-6}$$

	papH⁺	pap	H⁺
Initial	0.0792 M	0	~0
Equil.	0.0792 – x	x	x

$$K_a = 2.5 \times 10^{-6} = \frac{x^2}{0.0792 - x} \approx \frac{x^2}{0.0792},\quad x = [H^+] = 4.4 \times 10^{-4}\ M$$

pH = -log(4.4 × 10^{-4}) = 3.36; Assumptions good.

CHAPTER FIFTEEN

APPLICATIONS OF AQUEOUS EQUILIBRIA

Questions

13. When an acid dissociates or when a salt dissolves, ions are produced. A common ion effect is observed when one of the product ions in a particular equilibrium is added from an outside source. For a weak acid dissociating to its conjugate base and H^+, the common ion would be the conjugate base; this would be added by dissolving a soluble salt of the conjugate base into the acid solution. The presence of the conjugate base from an outside source shifts the equilibrium to the left so less acid dissociates. For the K_{sp} reaction of a salt dissolving into its respective ions, a common ion would be one of the ions in the salt added from an outside source. When a common ion is present, the K_{sp} equilibrium shifts to the left resulting in less of the salt dissolving into its ions.

15. The more weak acid and conjugate base present, the more H^+ and/or OH^- that can be absorbed by the buffer without significant pH change. When the concentrations of weak acid and conjugate base are equal (so that pH = pK_a), the buffer system is equally efficient at absorbing either H^+ or OH^-. If the buffer is overloaded with weak acid or with conjugate base, then the buffer is not equally efficient at absorbing either H^+ or OH^-.

17. The three key points to emphasize in your sketch are the initial pH, pH at the halfway point to equivalence, and the pH at the equivalence point. For all of the weak bases titrated, pH = pK_a at the halfway point to equivalence (50.0 mL HCl added) because [weak base] = [conjugate acid] at this point. Here, the weak base with $K_b = 10^{-5}$ has a pH = 9.0 at the halfway point and the weak base with $K_b = 10^{-10}$ has a pH = 4.0 at the halfway point to equivalence. For the initial pH, the strong base has the highest pH (most basic), while the weakest base has the lowest pH (least basic). At the equivalence point (100.0 mL HCl added), the strong base titration has pH = 7.0. The weak bases titrated have acidic pHs because the conjugate acids of the weak bases titrated are the major species present. The weakest base has the strongest conjugate acid so its pH will be lowest (most acidic) at the equivalence point.

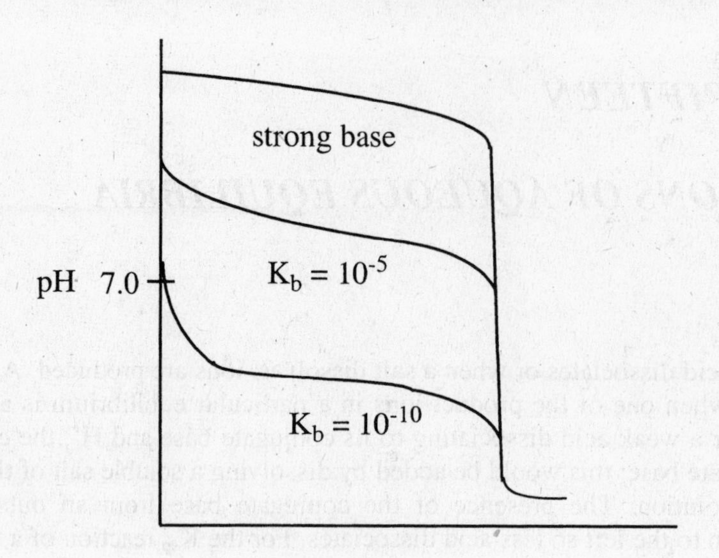

19. i. This is the result when you have a salt that breaks up into two ions. Examples of these salts (but not all) could be AgCl, $SrSO_4$, $BaCrO_4$, and $ZnCO_3$.

ii. This is the result when you have a salt that breaks up into three ions, either two cations and one anion or one cation and two anions. Some examples are SrF_2, Hg_2I_2, and Ag_2SO_4.

iii. This is the result when you have a salt that breaks up into four ions, either three cations and one anion (Ag_3PO_4) or one cation and three anions (ignoring the hydroxides, there are no examples of this type of salt in Table 15.4).

iv. This is the result when you have a salt that breaks up into five ions, either three cations and two anions [$Sr_3(PO_4)_2$] or two cations and three anions (no examples of this type of salt are in Table 15.4).

Exercises

Buffers

21. When strong acid or strong base is added to a bicarbonate/carbonate mixture, the strong acid/base is neutralized. The reaction goes to completion, resulting in the strong acid/base being replaced with a weak acid/base, which results in a new buffer solution. The reactions are:

$$H^+(aq) + CO_3^{2-}(aq) \rightarrow HCO_3^-(aq);\ \ OH^- + HCO_3^-(aq) \rightarrow CO_3^{2-}(aq) + H_2O(l)$$

23. a. This is a weak acid problem. Let $HC_3H_5O_2$ = HOPr and $C_3H_5O_2^-$ = OPr^-.

	HOPr	⇌	H^+	+	OPr^-	$K_a = 1.3 \times 10^{-5}$
Initial	0.100 *M*		~0		0	
	x mol/L HOPr dissociates to reach equilibrium					
Change	$-x$	→	$+x$		$+x$	
Equil.	$0.100 - x$		x		x	

$$K_a = 1.3 \times 10^{-5} = \frac{[H^+][OPr^-]}{[HOPr]} = \frac{x^2}{0.100 - x} \approx \frac{x^2}{0.100}$$

$x = [H^+] = 1.1 \times 10^{-3}\,M$; pH = 2.96   Assumptions good by the 5% rule.

b. This is a weak base problem (Na^+ has no acidic/basic properties).

$$OPr^- + H_2O \rightleftharpoons HOPr + OH^- \qquad K_b = \frac{K_w}{K_a} = 7.7 \times 10^{-10}$$

	OPr^-		$HOPr$	OH^-
Initial	0.100 M		0	~0
	x mol/L OPr^- reacts with H_2O to reach equilibrium			
Change	$-x$	→	$+x$	$+x$
Equil.	$0.100 - x$		x	x

$$K_b = 7.7 \times 10^{-10} = \frac{[HOPr][OH^-]}{[OPr^-]} = \frac{x^2}{0.100 - x} \approx \frac{x^2}{0.100}$$

$x = [OH^-] = 8.8 \times 10^{-6}\,M$; pOH = 5.06; pH = 8.94 Assumptions good.

c. pure H_2O, $[H^+] = [OH^-] = 1.0 \times 10^{-7}\,M$; pH = 7.00

d. This solution contains a weak acid and its conjugate base. This is a buffer solution. We will solve for the pH using the weak acid equilibrium reaction.

$$HOPr \rightleftharpoons H^+ + OPr^- \qquad K_a = 1.3 \times 10^{-5}$$

	$HOPr$		H^+	OPr^-
Initial	0.100 M		~0	0.100 M
	x mol/L HOPr dissociates to reach equilibrium			
Change	$-x$	→	$+x$	$+x$
Equil.	$0.100 - x$		x	$0.100 + x$

$$1.3 \times 10^{-5} = \frac{(0.100 + x)(x)}{0.100 - x} \approx \frac{(0.100)(x)}{0.100} = x = [H^+]$$

$[H^+] = 1.3 \times 10^{-5}\,M$; pH = 4.89 Assumptions good.

Alternatively, we can use the Henderson-Hasselbalch equation to calculate the pH of buffer solutions.

$$pH = pK_a + \log\frac{[Base]}{[Acid]} = pK_a + \log\frac{(0.100)}{(0.100)} = -\log(1.3 \times 10^{-5}) = 4.89$$

The Henderson-Hasselbalch equation will be valid when an assumption of the type $0.1 + x \approx 0.1$ that we just made in this problem is valid. From a practical standpoint, this will almost always be true for useful buffer solutions. If the assumption is not valid, the solution will have such a low buffering capacity that it will be of no use to control the pH. Note: The Henderson-Hasselbalch equation can only be used to solve for the pH of buffer solutions.

25. 0.100 *M* $HC_3H_5O_2$: percent dissociation = $\frac{[H^+]}{[HC_3H_5O_2]_0} \times 100 = \frac{1.1\times10^{-3}\ M}{0.100\ M} = 1.1\%$

0.100 *M* $HC_3H_5O_2$ + 0.100 *M* $NaC_3H_5O_2$: % dissociation = $\frac{1.3\times10^{-5}}{0.100} \times 100 = 1.3 \times 10^{-2}\ \%$

The percent dissociation of the acid decreases from 1.1% to 1.3×10^{-2} % when $C_3H_5O_2^-$ is present. This is known as the common ion effect. The presence of the conjugate base of the weak acid inhibits the acid dissociation reaction.

27. a. We have a weak acid (HOPr = $HC_3H_5O_2$) and a strong acid (HCl) present. The amount of H^+ donated by the weak acid will be negligible as compared to the 0.020 *M* H^+ from the strong acid. To prove it let's consider the weak acid equilibrium reaction:

	HOPr	⇌	H^+	+	OPr^-	$K_a = 1.3 \times 10^{-5}$
Initial	0.100 *M*		0.020 *M*		0	
	x mol/L HOPr dissociates to reach equilibrium					
Change	$-x$	→	$+x$		$+x$	
Equil.	$0.100 - x$		$0.020 + x$		x	

$$K_a = 1.3 \times 10^{-5} = \frac{(0.020 + x)(x)}{0.100 - x} \approx \frac{(0.020)(x)}{0.100}, \quad x = 6.5 \times 10^{-5}\ M$$

$[H^+] = 0.020 + x = 0.020\ M$; pH = 1.70 Assumptions good ($x = 6.5 \times 10^{-5}$ which is $<< 0.020$).

b. Added H^+ reacts completely with the best base present, Opr^-. Since all species present are in the same volume of solution, we can use molarity units to do the stoichiometry part of the problem (instead of moles). The stoichiometry problem is:

	OPr^-	+	H^+	→	HOPr	
Before	0.100 *M*		0.020 *M*		0	
Change	−0.020		−0.020	→	+0.020	Reacts completely
After	0.080		0		0.020 *M*	

After reaction, a weak acid, HOPr , and its conjugate base, OPr^-, are present. This is a buffer solution. Using the Henderson-Hasselbalch equation where $pK_a = -\log(1.3 \times 10^{-5})$ = 4.89:

$$pH = pK_a + \log\frac{[Base]}{[Acid]} = 4.89 + \log\frac{(0.080)}{(0.020)} = 5.49 \qquad \text{Assumptions good.}$$

c. This is a strong acid problem. $[H^+] = 0.020\ M$; pH = 1.70

d. Added H^+ reacts completely with the best base present, OPr^-.

	OPr^-	+ H^+	$\rightarrow$	HOPr	
Before	0.100 *M*	0.020 *M*		0.100 *M*	
Change	−0.020	−0.020	$\rightarrow$	+0.020	Reacts completely
After	0.080	0		0.120	

A buffer solution results (weak acid + conjugate base). Using the Henderson-Hasselbalch equation:

$$pH = pK_a + \frac{[Base]}{[Acid]} = 4.89 + \log\frac{(0.080)}{(0.120)} = 4.71$$

29. a. OH^- will react completely with the best acid present, HOPr.

	HOPr	+ OH^-	$\rightarrow$	OPr^- + H_2O	
Before	0.100 *M*	0.020 *M*		0	
Change	−0.020	−0.020	$\rightarrow$	+0.020	Reacts completely
After	0.080	0		0.020	

A buffer solution results after the reaction. Using the Henderson-Hasselbalch equation:

$$pH = pK_a + \log\frac{[Base]}{[Acid]} = 4.89 + \log\frac{(0.020)}{(0.080)} = 4.29$$

b. We have a weak base and a strong base present at the same time. The amount of OH^- added by the weak base will be negligible compared to the 0.020 *M* OH^- from the strong base. To prove it, let's consider the weak base equilibrium:

$$OPr^- + H_2O \rightleftharpoons HOPr + OH^- \qquad K_b = 7.7 \times 10^{-10}$$

	OPr^-	HOPr	OH^-
Initial	0.100 *M*	0	0.020 *M*

x mol/L OPr^- reacts with H_2O to reach equilibrium

Change	$-x$	$\rightarrow$	$+x$	$+x$
Equil.	$0.100 - x$		x	$0.020 + x$

$$K_b = 7.7 \times 10^{-10} = \frac{(x)(0.020 + x)}{0.100 - x} \approx \frac{(x)(0.020)}{0.100}, \quad x = 3.9 \times 10^{-9}\ M$$

$[OH^-] = 0.020 + x = 0.020\ M$; pOH = 1.70; pH = 12.30 Assumptions good.

c. This is a strong base in water. $[OH^-] = 0.020\ M$; pOH = 1.70; pH = 12.30

d. OH^- will react completely with HOPr, the best acid present.

	HOPr	+ OH^-	$\rightarrow$	OPr^- + H_2O	
Before	0.100 *M*	0.020 *M*		0.100 *M*	
Change	−0.020	−0.020	$\rightarrow$	+0.020	Reacts completely
After	0.080	0		0.120	

Using the Henderson-Hasselbalch equation to solve for the pH of the resulting buffer solution:

$$pH = pK_a + \log\frac{[Base]}{[Acid]} = 4.89 + \frac{(0.120)}{(0.080)} = 5.07$$

31. Consider all of the results to Exercises 15.23, 15.27, and 15.29:

Solution	Initial pH	after added acid	after added base
a	2.96	1.70	4.29
b	8.94	5.49	12.30
c	7.00	1.70	12.30
d	4.89	4.71	5.07

The solution in Exercise 15.23d is a buffer; it contains both a weak acid ($HC_3H_5O_2$) and a weak base ($C_3H_5O_2^-$). Solution d shows the greatest resistance to changes in pH when either strong acid or strong base is added; this is the primary property of buffers.

33. Major species: HNO_2, NO_2^- and Na^+. Na^+ has no acidic or basic properties. One appropriate equilibrium reaction you can use is the K_a reaction of HNO_2 which contains both HNO_2 and NO_2^-. However, you could also use the K_b reaction for NO_2^- and come up with the same answer. Solving the equilibrium problem (called a buffer problem):

	HNO_2	$\rightleftharpoons$	NO_2^-	+	H^+
Initial	1.00 *M*		1.00 *M*		~0
	x mol/L HNO_2 dissociates to reach equilibrium				
Change	$-x$	$\rightarrow$	$+x$		$+x$
Equil.	$1.00 - x$		$1.00 + x$		x

$$K_a = 4.0 \times 10^{-4} = \frac{[NO_2^-][H^+]}{[HNO_2]} = \frac{(1.00 + x)(x)}{1.00 - x} \approx \frac{(1.00)(x)}{1.00} \quad \text{(assuming } x << 1.00\text{)}$$

$x = 4.0 \times 10^{-4}\ M = [H^+]$; Assumptions good ($x$ is 4.0×10^{-2} % of 1.00).

$$pH = -\log(4.0 \times 10^{-4}) = 3.40$$

Note: We would get the same answer using the Henderson-Hasselbalch equation. Use whichever method you prefer.

35. Major species after NaOH added: HNO_2, NO_2^-, Na^+ and OH^-. The OH^- from the strong base will react with the best acid present (HNO_2). Any reaction involving a strong base is assumed to go to completion. Since all species present are in the same volume of solution, we can use molarity units to do the stoichiometry part of the problem (instead of moles). The stoichiometry problem is:

	OH^- +	HNO_2	→	NO_2^- + H_2O	
Before	0.10 mol/1.00 L	1.00 *M*		1.00 *M*	
Change	−0.10 *M*	−0.10 *M*	→	+0.10 *M*	Reacts completely
After	0	0.90		1.10	

After all the OH^- reacts, we are left with a solution containing a weak acid (HNO_2) and its conjugate base (NO_2^-). This is what we call a buffer problem. We will solve this buffer problem using the K_a equilibrium reaction.

	HNO_2	⇌	NO_2^-	+	H^+
Initial	0.90 *M*		1.10 *M*		~0
	x mol/L HNO_2 dissociates to reach equilibrium				
Change	$-x$	→	$+x$		$+x$
Equil.	$0.90 - x$		$1.10 + x$		x

$$K_a = 4.0 \times 10^{-4} = \frac{(1.10 + x)(x)}{0.90 - x} \approx \frac{(1.10)(x)}{0.90}, \; x = [H^+] = 3.3 \times 10^{-4}\,M; \; pH = 3.48;$$

Assumptions good.

Note: The added NaOH to this buffer solution changes the pH only from 3.40 to 3.48. If the NaOH were added to 1.0 L of pure water, the pH would change from 7.00 to 13.00.

Major species after HCl added: HNO_2, NO_2^-, H^+, Na^+, Cl^-; The added H^+ from the strong acid will react completely with the best base present (NO_2^-,).

	H^+ +	NO_2^-	→	HNO_2	
Before	$\frac{0.20\text{ mol}}{1.00\text{ L}}$	1.00 *M*		1.00 *M*	
Change	−0.20 *M*	−0.20 *M*	→	+0.20 *M*	Reacts completely
After	0	0.80		1.20	

After all the H^+ has reacted, we have a buffer solution (a solution containing a weak acid and its conjugate base). Solving the buffer problem:

	HNO_2	⇌	NO_2^-	+	H^+
Initial	1.20 *M*		0.80 *M*		0
Equil.	$1.20 - x$		$0.80 + x$		$+x$

$$K_a = 4.0 \times 10^{-4} = \frac{(0.80 + x)(x)}{1.20 - x} \approx \frac{0.80)(x)}{1.20}, \; x = [H^+] = 6.0 \times 10^{-4}\,M; \; pH = 3.22;$$

Assumptions good.

Note: The added HCl to this buffer solution changes the pH only from 3.40 to 3.22. If the HCl were added to 1.0 L of pure water, the pH would change from 7.00 to 0.70.

37. $$[HC_7H_5O_2] = \frac{21.5 \text{ g } HC_7H_5O_2 \times \frac{1 \text{ mol } HC_7H_5O_2}{122.12 \text{ g}}}{0.2000 \text{ L}} = 0.880 \, M$$

$$[C_7H_5O_2^-] = \frac{37.7 \text{ g } NaC_7H_5O_2 \times \frac{1 \text{ mol } NaC_7H_5O_2}{144.10 \text{ g}} \times \frac{1 \text{ mol } C_7H_5O_2^-}{\text{mol } NaC_7H_5O_2}}{0.2000 \text{ L}} = 1.31 \, M$$

We have a buffer solution since we have both a weak acid and its conjugate base present at the same time. One can use the K_a reaction or the K_b reaction to solve. We will use the K_a reaction for the acid component of the buffer.

	$HC_7H_5O_2$	$\rightleftharpoons$ H^+	+ $C_7H_5O_2^-$
Initial	0.880 M	~0	1.31 M
	x mol/L of $HC_7H_5O_2$ dissociates to reach equilibrium		
Change	$-x$	$\rightarrow$ $+x$	$+x$
Equil.	$0.880 - x$	x	$1.31 + x$

$$K_a = 6.4 \times 10^{-5} = \frac{x(1.31 + x)}{0.880 - x} \approx \frac{x(1.31)}{0.880}, \quad x = [H^+] = 4.3 \times 10^{-5} \, M$$

$pH = -\log(4.3 \times 10^{-5}) = 4.37$; Assumptions good.

Alternatively, we can use the Henderson-Hasselbalch equation to calculate the pH of buffer solutions.

$$pH = pK_a + \log\frac{[\text{Base}]}{[\text{Acid}]} = pK_a + \log\frac{[C_7H_5O_2^-]}{[HC_7H_5O_2]}$$

$$pH = -\log(6.4 \times 10^{-5}) + \log\left(\frac{1.31}{0.880}\right) = 4.19 + 0.173 = 4.36$$

Within round-off error, this is the same answer we calculated solving the equilibrium problem using the K_a reaction.

The Henderson-Hasselbalch equation will be valid when an assumption of the type $1.31 + x \approx 1.31$ that we just made in this problem is valid. From a practical standpoint, this will almost always be true for useful buffer solutions. If the assumption is not valid, the solution will have such a low buffering capacity that it will be of no use to control the pH. Note: The Henderson-Hasselbalch equation can <u>only</u> be used to solve for the pH of buffer solutions.

39. $[H^+] \text{ added} = \frac{0.010 \text{ mol}}{0.25 \text{ L}} = 0.040 \, M$; The added H^+ reacts completely with NH_3 to form NH_4^+.

a.

	NH_3	+ H^+	$\rightarrow$	NH_4^+	
Before	0.050 M	0.040 M		0.15 M	
Change	-0.040	-0.040	$\rightarrow$	$+0.040$	Reacts completely
After	0.010	0		0.19	

A buffer solution still exists after H^+ reacts completely. Using the Henderson-Hasselbalch equation:

$$pH = pK_a + \log\frac{[NH_3]}{[NH_4^+]} = -\log(5.6 \times 10^{-10}) + \log\left(\frac{0.010}{0.19}\right) = 9.25 + (-1.28) = 7.97$$

b.

	NH_3	+ H^+	→	NH_4^+	
Before	0.50 *M*	0.040 *M*		1.50 *M*	
Change	−0.040	−0.040	→	+0.040	Reacts completely
After	0.46	0		1.54	

A buffer solution still exists. $pH = pK_a + \log\frac{[NH_3]}{[NH_4^+]} = 9.25 + \log\left(\frac{0.46}{1.54}\right) = 8.73$

The two buffers differ in their capacity and not in pH (both buffers had an initial pH = 8.77). Solution b has the greater capacity because it has the largest concentration of weak acid and conjugate base. Buffers with greater capacities will be able to absorb more H^+ or OH^- added.

41. $pH = pK_a + \log\frac{[C_2H_3O_2^-]}{[HC_2H_3O_2]}$; $pK_a = -\log(1.8 \times 10^{-5}) = 4.74$

Since the buffer components, $C_2H_3O_2^-$ and $HC_2H_3O_2$, are both in the same volume of water, the concentration ratio of $[C_2H_3O_2^-]/[HC_2H_3O_2]$ will equal the mol ratio of mol $C_2H_3O_2^-$/mol $HC_2H_3O_2$.

$$5.00 = 4.74 + \log\frac{\text{mol } C_2H_3O_2^-}{\text{mol } HC_2H_3O_2};\ \text{mol } HC_2H_3O_2 = 0.5000\ L \times \frac{0.200\ \text{mol}}{L} = 0.100\ \text{mol}$$

$$0.26 = \log\frac{\text{mol } C_2H_3O_2^-}{0.100\ \text{mol}},\ \frac{\text{mol } C_2H_3O_2^-}{0.100} = 10^{0.26} = 1.8,\ \text{mol } C_2H_3O_2^- = 0.18\ \text{mol}$$

$$\text{mass } NaC_2H_3O_2 = 0.18\ \text{mol } NaC_2H_3O_2 \times \frac{82.03\ g}{\text{mol}} = 15\ g\ NaC_2H_3O_2$$

43. $C_5H_5NH^+ \rightleftharpoons H^+ + C_5H_5N \quad K_a = \frac{K_w}{K_b} = \frac{1.0\times10^{-14}}{1.7\times10^{-9}} = 5.9\times10^{-6}$; $pK_a = -\log(5.9\times10^{-6}) = 5.23$

We will use the Henderson-Hasselbalch equation to calculate the concentration ratio necessary for each buffer.

$$pH = pK_a + \log\frac{[\text{Base}]}{[\text{Acid}]},\ pH = 5.23 + \log\frac{[C_5H_5N]}{[C_5H_5NH^+]}$$

a. $4.50 = 5.23 + \log\frac{[C_5H_5N]}{[C_5H_5NH^+]}$

$\log\frac{[C_5H_5N]}{[C_5H_5NH^+]} = -0.73$

$\frac{[C_5H_5N]}{[C_5H_5NH^+]} = 10^{-0.73} = 0.19$

b. $5.00 = 5.23 + \log\frac{[C_5H_5N]}{[C_5H_5NH^+]}$

$\log\frac{[C_5H_5N]}{[C_5H_5NH^+]} = -0.23$

$\frac{[C_5H_5N]}{[C_5H_5NH^+]} = 10^{-0.23} = 0.59$

c. $5.23 = 5.23 + \log\frac{[C_5H_5N]}{[C_5H_5NH^+]}$

$\frac{[C_5H_5N]}{[C_5H_5NH^+]} = 10^{0.0} = 1.0$

d. $5.50 = 5.23 + \log\frac{[C_5H_5N]}{[C_5H_5NH^+]}$

$\frac{[C_5H_5N]}{[C_5H_5NH^+]} = 10^{0.27} = 1.9$

45. A best buffer has large and equal quantities of weak acid and conjugate base. For a best buffer, [acid] = [base], so $pH = pK_a + \log\frac{[Base]}{[Acid]} = pK_a + 0 = pK_a$ ($pH = pK_a$).

The best acid choice for a pH = 7.00 buffer would be the weak acid with a pK_a close to 7.0 or $K_a \approx 1 \times 10^{-7}$. HOCl is the best choice in Table 14.2 ($K_a = 3.5 \times 10^{-8}$; $pK_a = 7.46$). To make this buffer, we need to calculate the [base]/[acid] ratio.

$$7.00 = 7.46 + \log\frac{[Base]}{[Acid]}, \quad \frac{[OCl^-]}{[HOCl]} = 10^{-0.46} = 0.35$$

Any OCl^-/HOCl buffer in a concentration ratio of 0.35:1 will have a pH = 7.00. One possibility is [NaOCl] = 0.35 *M* and [HOCl] = 1.0 *M*.

47. The reaction $OH^- + CH_3NH_3^+ \rightarrow CH_3NH_2 + H_2O$ goes to completion for solutions a, c and d (no reaction occurs between the species in solution because both species are bases). After the OH^- reacts completely, there must be both $CH_3NH_3^+$ and CH_3NH_2 in solution for it to be a buffer. The important components of each solution (after the OH^- reacts completely) is/are:

a. 0.05 *M* CH_3NH_2 (no $CH_3NH_3^+$ remains, no buffer)

b. 0.05 *M* OH^- and 0.1 *M* CH_3NH_2 (two bases present, no buffer)

c. 0.05 *M* OH^- and 0.05 *M* CH_3NH_2 (too much OH^- added, no $CH_3NH_3^+$ remains, no buffer)

d. 0.05 *M* CH_3NH_2 and 0.05 *M* $CH_3NH_3^+$ (a buffer solution results)

Only the combination in mixture d results in a buffer. Note that the concentrations are halved from the initial values. This is because equal volumes of two solutions were added together, which halves the concentrations.

49. Added OH^- converts $HC_2H_3O_2$ into $C_2H_3O_2^-$: $HC_2H_3O_2 + OH^- \rightarrow C_2H_3O_2^- + H_2O$

From this reaction, the moles of $C_2H_3O_2^-$ produced equal the moles of OH^- added. Also, the total concentration of acetic acid plus acetate ion must equal 2.0 *M* (assuming no volume change on addition of NaOH). Summarizing for each solution:

$[C_2H_3O_2^-] + [HC_2H_3O_2] = 2.0\ M$ and $[C_2H_3O_2^-]$ produced = $[OH^-]$ added

a. $pH = pK_a + \log\frac{[C_2H_3O_2^-]}{[HC_2H_3O_2]}$; For $pH = pK_a$, $\log\frac{[C_2H_3O_2^-]}{[HC_2H_3O_2]} = 0$

Therefore, $\frac{[C_2H_3O_2^-]}{[HC_2H_3O_2]} = 1.0$ and $[C_2H_3O_2^-] = [HC_2H_3O_2]$

Because $[C_2H_3O_2^-] + [HC_2H_3O_2] = 2.0\ M$, $[C_2H_3O_2^-] = [HC_2H_3O_2] = 1.0\ M = [OH^-]$ added.

To produce a 1.0 *M* $C_2H_3O_2^-$ solution, we need to add 1.0 mol of NaOH to 1.0 L of the 2.0 *M* $HC_2H_3O_2$ solution. The resultant solution will have $pH = pK_a = 4.74$.

b. $4.00 = 4.74 + \log\frac{[C_2H_3O_2^-]}{[HC_2H_3O_2]}$, $\frac{[C_2H_3O_2^-]}{[HC_2H_3O_2]} = 10^{-0.74} = 0.18$

$[C_2H_3O_2^-] = 0.18\ [HC_2H_3O_2]$ or $[HC_2H_3O_2] = 5.6\ [C_2H_3O_2^-]$; Because $[C_2H_3O_2^-] + [HC_2H_3O_2] = 2.0\ M$, then:

$[C_2H_3O_2^-] + 5.6\ [C_2H_3O_2^-] = 2.0\ M$, $[C_2H_3O_2^-] = \frac{2.0}{6.6} = 0.30\ M = [OH^-]$ added

We need to add 0.30 mol of NaOH to 1.0 L of 2.0 *M* $HC_2H_3O_2$ solution to produce a buffer solution with pH = 4.00.

c. $5.00 = 4.74 + \log\frac{[C_2H_3O_2^-]}{[HC_2H_3O_2]}$, $\frac{[C_2H_3O_2^-]}{[HC_2H_3O_2]} = 10^{0.26} = 1.8$

$1.8\ [HC_2H_3O_2] = [C_2H_3O_2^-]$ or $[HC_2H_3O_2] = 0.56\ [C_2H_3O_2^-]$; Because $[HC_2H_3O_2] + [C_2H_3O_2^-] = 2.0\ M$, then:

$1.56\ [C_2H_3O_2^-] = 2.0\ M$, $[C_2H_3O_2^-] = 1.3\ M = [OH^-]$ added

We need to add 1.3 mol of NaOH to 1.0 L of 2.0 *M* $HC_2H_3O_2$ to produce a buffer solution with pH = 5.00.

Acid-Base Titrations

51.

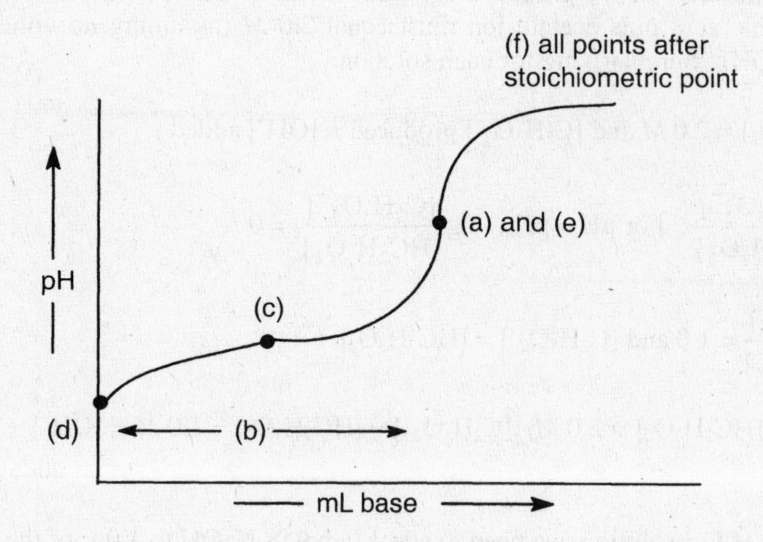

$HA + OH^- \rightarrow A^- + H_2O$; Added OH^- from the strong base converts the weak acid, HA, into its conjugate base, A^-. Initially, before any OH^- is added (point d), HA is the dominant species present. After OH^- is added, both HA and A^- are present and a buffer solution results (region b). At the equivalence point (points a and e), exactly enough OH^- has been added to convert all of the weak acid, HA, into its conjugate base, A^-. Past the equivalence point (region f), excess OH^- is present. For the answer to part b, we included almost the entire buffer region. The maximum buffer region (or the region which is the best buffer solution) is around the halfway point to equivalence (point c). At this point, enough OH^- has been added to convert exactly one-half of the weak acid present initially into its conjugate base so [HA] = $[A^-]$ and pH = pK_a. A best buffer has about equal concentrations of weak acid and conjugate base present.

53. This is a strong acid ($HClO_4$) titrated by a strong base (KOH). Added OH^- from the strong base will react completely with the H^+ present from the strong acid to produce H_2O.

a. Only strong acid present. $[H^+] = 0.200\ M$; pH = 0.699

b. $\text{mmol OH}^- \text{ added} = 10.0 \text{ mL} \times \dfrac{0.100 \text{ mmol OH}^-}{\text{mL}} = 1.00 \text{ mmol OH}^-$

$\text{mmol H}^+ \text{ present} = 40.0 \text{ mL} \times \dfrac{0.100 \text{ mmol H}^+}{\text{mL}} = 8.0 \text{ mmol H}^+$

Note: The units mmoles are usually easier numbers to work with. The units for molarity are moles/L but are also equal to mmoles/mL.

	H^+	+	OH^-	→	H_2O
Before	8.00 mmol		1.00 mmol		
Change	−1.00 mmol		−1.00 mmol		Reacts completely
After	7.00 mmol		0		

The excess H^+ determines pH. $[H^+]_{excess} = \frac{7.00 \text{ mmol } H^+}{40.0 \text{ mL} + 10.0 \text{ mL}} = 0.140\ M$; pH = 0.854

c. mmol OH^- added = 40.0 mL × 0.100 M = 4.00 mmol OH^-

	H^+	+	OH^-	→	H_2O
Before	8.00 mmol		4.00 mmol		
After	4.00 mmol		0		

$$[H^+]_{excess} = \frac{4.00 \text{ mmol}}{(40.0 + 40.0) \text{ mL}} = 0.0500\ M; \text{ pH} = 1.301$$

d. mmol OH^- added = 80.0 mL × 0.100 M = 8.00 mmol OH^-; This is the equivalence point because we have added just enough OH^- to react with all the acid present. For a strong acid-strong base titration, pH = 7.00 at the equivalence point since only neutral species are present (K^+, ClO_4^-, H_2O).

e. mmol OH^- added = 100.0 mL × 0.100 M = 10.0 mmol OH^-

	H^+	+	OH^-	→	H_2O
Before	8.00 mmol		10.0 mmol		
After	0		2.0 mmol		

Past the equivalence point, the pH is determined by the excess OH^- present.

$$[OH^-]_{excess} = \frac{2.0 \text{ mmol}}{(40.0 + 100.0) \text{ mL}} = 0.014\ M; \text{ pOH} = 1.85; \text{ pH} = 12.15$$

55. This is a weak acid ($HC_2H_3O_2$) titrated by a strong base (KOH).

a. Only a weak acid is present. Solving the weak acid problem:

	$HC_2H_3O_2$	⇌	H^+	+	$C_2H_3O_2^-$
Initial	0.200 M		~0		0
	x mol/L $HC_2H_3O_2$ dissociates to reach equilibrium				
Change	$-x$	→	$+x$		$+x$
Equil.	$0.200 - x$		x		x

$$K_a = 1.8 \times 10^{-5} = \frac{x^2}{0.200 - x} = \frac{x^2}{0.200}, \quad x = [H^+] = 1.9 \times 10^{-3}\,M; \quad pH = 2.72;$$

Assumptions good.

b. The added OH^- will react completely with the best acid present, $HC_2H_3O_2$.

$$\text{mmol } HC_2H_3O_2 \text{ present} = 100.0 \text{ mL} \times \frac{0.200 \text{ mmol } HC_2H_3O_2}{\text{mL}} = 20.0 \text{ mmol } HC_2H_3O_2$$

$$\text{mmol } OH^- \text{ added} = 50.0 \text{ mL} \times \frac{0.100 \text{ mmol } OH^-}{\text{mL}} = 5.00 \text{ mmol } OH^-$$

	$HC_2H_3O_2$	+	OH^-	→	$C_2H_3O_2^-$ + H_2O	
Before	20.0 mmol		5.00 mmol		0	
Change	−5.00 mmol		−5.00	→	+5.00 mmol	Reacts completely
After	15.0 mmol		0		5.00 mmol	

After reaction of all the strong base, we will have a buffered solution containing a weak acid ($HC_2H_3O_2$) and its conjugate base ($C_2H_3O_2^-$). We will use the Henderson-Hasselbalch equation to solve for the pH of this buffer.

$$pH = pK_a + \log\frac{[C_2H_3O_2^-]}{[HC_2H_3O_2]} = -\log(1.8 \times 10^{-5}) + \log\left(\frac{5.00\text{mmol}/V_T}{15.0\text{mmol}/V_T}\right)$$

where V_T = total volume

$$pH = 4.74 + \log\left(\frac{5.00}{15.0}\right) = 4.74 + (-0.477) = 4.26$$

Note that the total volume cancels in the Henderson-Hasselbalch equation. For the [base]/ [acid] term, the mole (or mmole) ratio equals the concentration ratio because the components of the buffer are always in the same volume of solution.

c. mmol OH^- added = 100.0 mL × 0.100 mmol OH^-/mL = 10.0 mmol OH^-; The same amount (20.0 mmol) of $HC_2H_3O_2$ is present as before (it never changes). As before, let the OH^- react to completion, then see what remains in solution after the reaction.

	$HC_2H_3O_2$	+	OH^-	→	$C_2H_3O_2^-$ + H_2O
Before	20.0 mmol		10.0 mmol		0
After	10.0 mmol		0		10.0 mmol

A buffered solution results after the reaction. Because $[C_2H_3O_2^-] = [HC_2H_3O_2]$ = 10.0 mmol/ total volume, pH = pK_a. This is always true at the halfway point to equivalence for a weak acid/strong base titration, pH = pK_a.

$$pH = -\log(1.8 \times 10^{-5}) = 4.74$$

d. mmol OH^- added = 150.0 mL × 0.100 *M* = 15.0 mmol OH^-. Added OH^- reacts completely with the weak acid.

	$HC_2H_3O_2$	+	OH^-	→	$C_2H_3O_2^-$	+	H_2O
Before	20.0 mmol		15.0 mmol		0		
After	5.0 mmol		0		15.0 mmol		

We have a buffered solution after all the OH^- reacts to completion. Using the Henderson-Hasselbalch equation:

$$pH = 4.74 + \log\frac{[C_2H_3O_2^-]}{[HC_2H_3O_2]} = 4.74 + \log\left(\frac{15.00\text{ mmol}}{5.0\text{ mmol}}\right)$$ (Total volume cancels, so we can use mmol ratios.)

$pH = 4.74 + 0.48 = 5.22$

e. mmol OH^- added = 200.00 mL × 0.100 *M* = 20.0 mmol OH^-; As before, let the added OH^- react to completion with the weak acid, then see what is in solution after this reaction.

	$HC_2H_3O_2$	+	OH^-	→	$C_2H_3O_2^-$	+	H_2O
Before	20.0 mmol		20.0 mmol		0		
After	0		0		20.0 mmol		

This is the equivalence point. Enough OH^- has been added to exactly neutralize all the weak acid present initially. All that remains that affects the pH at the equivalence point is the conjugate base of the weak acid, $C_2H_3O_2^-$. This is a weak base equilibrium problem because the conjugate bases of all weak acids are weak bases themselves.

$$C_2H_3O_2^- + H_2O \rightleftharpoons HC_2H_3O_2 + OH^- \qquad K_b = \frac{K_w}{K_a} = \frac{1.0 \times 10^{-14}}{1.8 \times 10^{-5}}$$

	$C_2H_3O_2^-$	$HC_2H_3O_2$	OH^-	
Initial	20.0 mmol/300.0 mL	0	0	$K_b = 5.6 \times 10^{-10}$
	x mol/L $C_2H_3O_2^-$ reacts with H_2O to reach equilibrium			
Change	$-x$ →	$+x$	$+x$	
Equil.	$0.0667 - x$	x	x	

$$K_b = 5.6 \times 10^{-10} = \frac{x^2}{0.0667 - x} \approx \frac{x^2}{0.0667}, \quad x = [OH^-] = 6.1 \times 10^{-6}\ M$$

pOH = 5.21; pH = 8.79; Assumptions good.

Note: the pH at the equivalence point for a weak acid-strong base will always be greater than 7.0. This is because a weak base will always be the only major species present having any acidic or basic properties.

f. mmol OH^- added = 250.0 mL × 0.100 *M* = 25.0 mmol OH^-

	$HC_2H_3O_2$	+ OH^-	→	$C_2H_3O_2^-$ + H_2O
Before	20.0 mmol	25.0 mmol		0
After	0	5.0 mmol		20.0 mmol

After the titration reaction, we will have a solution containing excess OH^- and a weak base, $C_2H_3O_2^-$. When a strong base and a weak base are both present, assume the amount of OH^- added from the weak base will be minimal, i.e., the pH past the equivalence point will be determined by the amount of excess strong base.

$$[OH^-]_{excess} = \frac{5.0 \text{ mmol}}{100.0 \text{ mL} + 250.0 \text{ mL}} = 0.014\ M;\ \text{pOH} = 1.85;\ \text{pH} = 12.15$$

57. We will do sample calculations for the various parts of the titration. All results are summarized in Table 15.1 at the end of Exercise 15.59.

At the beginning of the titration, only the weak acid $HC_3H_5O_3$ is present.

	HLac	⇌	H^+	+	Lac^-	$K_a = 10^{-3.86} = 1.4 \times 10^{-4}$
						HLac = $HC_3H_5O_3$
Initial	0.100 *M*		~0		0	$Lac^- = C_3H_5O_3^-$
	x mol/L HLac dissociates to reach equilibrium					
Change	$-x$	→	$+x$		$+x$	
Equil.	$0.100 - x$		x		x	

$$1.4 \times 10^{-4} = \frac{x^2}{0.100 - x} \approx \frac{x^2}{0.100},\ x = [H^+] = 3.7 \times 10^{-3}\ M;\ \text{pH} = 2.43 \quad \text{Assumptions good.}$$

Up to the stoichiometric point, we calculate the pH using the Henderson-Hasselbalch equation because we have buffer solutions that result after the OH^- from the strong base reacts completely with the best acid present, HLac. This is the buffer region. For example, at 4.0 mL of NaOH added:

$$\text{initial mmol HLac present} = 25.0 \text{ mL} \times \frac{0.100 \text{ mmol}}{\text{mL}} = 2.50 \text{ mmol HLac}$$

$$\text{mmol } OH^- \text{ added} = 4.0 \text{ mL} \times \frac{0.100 \text{ mmol}}{\text{mL}} = 0.40 \text{ mmol } OH^-$$

Note: The units mmol are usually easier numbers to work with. The units for molarity are moles/L but are also equal to mmoles/mL.

The 0.40 mmol added OH^- convert 0.40 mmoles HLac to 0.40 mmoles Lac^- according to the equation:

$$\text{HLac} + OH^- \rightarrow Lac^- + H_2O \qquad \text{Reacts completely}$$

mmol HLac remaining = 2.50 - 0.40 = 2.10 mmol; mmol Lac^- produced = 0.40 mmol

We have a buffer solution. Using the Henderson-Hasselbalch equation where $pK_a = 3.86$:

$$pH = pK_a + \log\frac{[Lac^-]}{[HLac]} = 3.86 + \log\frac{(0.40)}{(2.10)}$$ (Total volume cancels, so we can use the mole or mmole ratio.)

$$pH = 3.86 - 0.72 = 3.14$$

Other points in the buffer region are calculated in a similar fashion. Perform a stoichiometry problem first, followed by a buffer problem. The buffer region includes all points up to 24.9 mL OH^- added.

At the stoichiometric (equivalence) point (25.0 mL OH^- added), we have added enough OH^- to convert all of the HLac (2.50 mmol) into its conjugate base, Lac^-. All that is present is a weak base. To determine the pH, we perform a weak base calculation.

$$[Lac^-]_o = \frac{2.50 \text{ mmol}}{25.0 \text{ mL} + 25.0 \text{ mL}} = 0.0500\ M$$

$$Lac^- + H_2O \rightleftharpoons HLac + OH^- \qquad K_b = \frac{1.0 \times 10^{-14}}{1.4 \times 10^{-4}} = 7.1 \times 10^{-11}$$

	Lac^-		HLac	OH^-
Initial	0.0500 *M*		0	0
	x mol/L Lac^- reacts with H_2O to reach equilibrium			
Change	$-x$	→	$+x$	$+x$
Equil.	$0.0500 - x$		x	x

$$K_b = \frac{x^2}{0.0500 - x} \approx \frac{x^2}{0.0500} = 7.1 \times 10^{-11}$$

$x = [OH^-] = 1.9 \times 10^{-6}\ M$; pOH = 5.72; pH = 8.28. Assumptions good.

Past the stoichiometric point, we have added more than 2.50 mmol of NaOH. The pH will be determined by the excess OH^- ion present. An example of this calculation follows.

$$\text{At 25.1 mL: } OH^- \text{ added} = 25.1 \text{ mL} \times \frac{0.100 \text{ mmol}}{\text{mL}} = 2.51 \text{ mmol } OH^-$$

2.50 mmol OH^- neutralizes all the weak acid present. The remainder is excess OH^-.

$$[OH^-]_{excess} = 2.51 - 2.50 = 0.01 \text{ mmol}$$

$$[OH^-]_{excess} = \frac{0.01 \text{ mmol}}{(25.0 + 25.1 \text{ mL})} = 2 \times 10^{-4}\ M; \text{ pOH} = 3.7; \text{ pH} = 10.3$$

All results are listed in Table 15.1 at the end of the solution to Exercise 15.59.

59. At beginning of the titration, only the weak base NH_3 is present. As always, solve for the pH using the K_b reaction for NH_3.

	NH_3 + H_2O	$\rightleftharpoons$	NH_4^+	+	OH^-	$K_b = 1.8 \times 10^{-5}$
Initial	0.100 *M*		0		~0	
Equil.	$0.100 - x$		x		x	

$$K_b = \frac{x^2}{0.100 - x} \approx \frac{x^2}{0.100} = 1.8 \times 10^{-5}$$

$x = [OH^-] = 1.3 \times 10^{-3}\ M$; pOH = 2.89; pH = 11.11 Assumptions good.

In the buffer region (4.0 – 24.9 mL), we can use the Henderson-Hasselbalch equation:

$$K_a = \frac{1.0 \times 10^{-14}}{1.8 \times 10^{-5}} = 5.6 \times 10^{-10};\ pK_a = 9.25;\ pH = 9.25 + \log\frac{[NH_3]}{[NH_4^+]}$$

We must determine the amounts of NH_3 and NH_4^+ present after the added H^+ reacts completely with the NH_3. For example, after 8.0 mL HCl are added:

$$\text{initial mmol } NH_3 \text{ present} = 25.0\ \text{mL} \times \frac{0.100\ \text{mmol}}{\text{mL}} = 2.50\ \text{mmol } NH_3$$

$$\text{mmol } H^+ \text{ added} = 8.0\ \text{mL} \times \frac{0.100\ \text{mmol}}{\text{mL}} = 0.80\ \text{mmol } H^+$$

Added H^+ reacts with NH_3 to completion: $NH_3 + H^+ \rightarrow NH_4^+$

mmol NH_3 remaining = 2.50 – 0.80 = 1.70 mmol; mmol NH_4^+ produced = 0.80 mmol

$$pH = 9.25 + \log\frac{1.70}{0.80} = 9.58$$ (Mole or mmole ratios can be used since the total volume cancels.)

Other points in the buffer region are calculated in similar fashion. Results are summarized in Table 15.1 at the end of Exercise 15.59.

At the stoichiometric point (25.0 mL H^+ added), just enough HCl has been added to convert all of the weak base (NH_3) into its conjugate acid (NH_4^+). Because all conjugate acids of weak bases are weak acids themselves, perform a weak acid calculation to determine the pH. $[NH_4^+]_o$ = 2.50 mmol/50.0 mL = 0.0500 *M*

	NH_4^+	$\rightleftharpoons$	H^+	+	NH_3	$K_a = 5.6 \times 10^{-10}$
Initial	0.0500 *M*		0		0	
Equil.	$0.0500 - x$		x		x	

$$5.6 \times 10^{-10} = \frac{x^2}{0.0500 - x} \approx \frac{x^2}{0.0500},\ x = [H^+] = 5.3 \times 10^{-6}\ M;\ pH = 5.28$$ Assumptions good.

Beyond the stoichiometric point, the pH is determined by the excess H^+. For example, at 28.0 mL of H^+ added:

$$\text{mmol } H^+ \text{ added} = 28.0 \text{ mL} \times \frac{0.100 \text{ mmol}}{\text{mL}} = 2.80 \text{ mmol } H^+$$

$$\text{Excess mmol } H^+ = 2.80 \text{ mmol} - 2.50 \text{ mmol} = 0.30 \text{ mmol excess } H^+$$

$$[H^+]_{excess} = \frac{0.30 \text{ mmol}}{(25.0 + 28.0) \text{ mL}} = 5.7 \times 10^{-3}\ M;\ \text{pH} = 2.24$$

All results are summarized in Table 15.1.

Table 15.1: Summary of pH Results for Exercises 15.57 and 15.60 (Graph follows)

Titrant mL	Exercise 15.57	Exercise 15.59
0.0	2.43	11.11
4.0	3.14	9.97
8.0	3.53	9.58
12.5	3.86	9.25
20.0	4.46	8.65
24.0	5.24	7.87
24.5	5.6	7.6
24.9	6.3	6.9
25.0	8.28	5.28
25.1	10.3	3.7
26.0	11.30	2.71
28.0	11.75	2.24
30.0	11.96	2.04

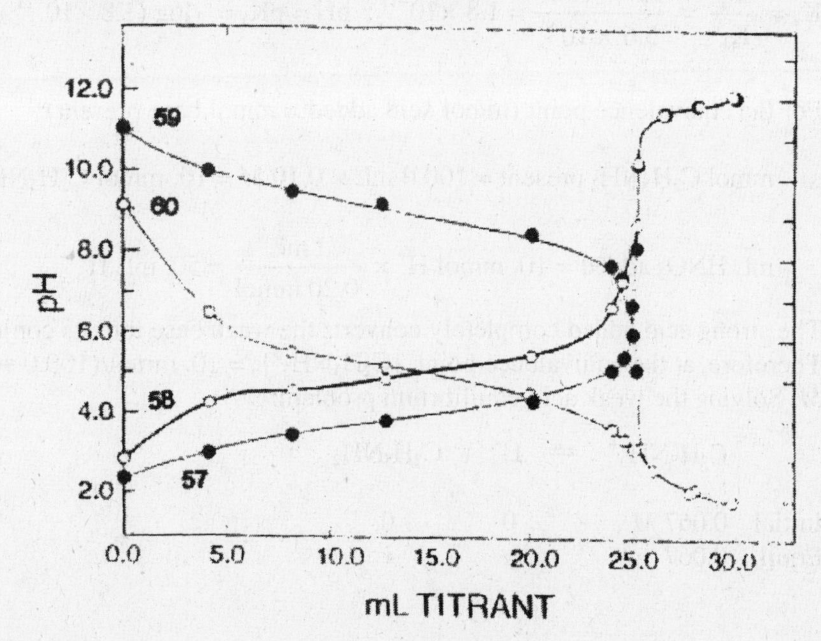

61. a. This is a weak acid/strong base titration. At the halfway point to equivalence, [weak acid] = [conjugate base], so pH = pK_a (always true for a weak acid/strong base titration).

$pH = -\log(6.4 \times 10^{-5}) = 4.19$

mmol $HC_7H_5O_2$ present = 100. mL × 0.10 *M* = 10. mmol $HC_7H_5O_2$. For the equivalence point, 10. mmol of OH^- must be added. The volume of OH^- added to reach the equivalence point is:

$$10.\ \text{mmol OH}^- \times \frac{1\ \text{mL}}{0.10\ \text{mmol OH}^-} = 1.0 \times 10^2\ \text{mL OH}^-$$

At the equivalence point, 10. mmol of $HC_7H_5O_2$ is neutralized by 10. mmol of OH^- to produce 10. mmol of $C_7H_5O_2^-$. This is a weak base. The total volume of the solution is 100.0 mL + 1.0 × 10^2 mL = 2.0 × 10^2 mL. Solving the weak base equilibrium problem:

$$C_7H_5O_2^- + H_2O \rightleftharpoons HC_7H_5O_2 + OH^- \qquad K_b = \frac{K_w}{K_a} = \frac{1.0 \times 10^{-14}}{6.4 \times 10^{-5}}$$

	$C_7H_5O_2^-$	$HC_7H_5O_2$	OH^-	
Initial	10. mmol/2.0 × 10^2 mL	0	0	K_b = 1.6 × 10^{-10}
Equil.	0.050 − x	x	x	

$$K_b = 1.6 \times 10^{-10} = \frac{x^2}{0.050 - x} \approx \frac{x^2}{0.050},\quad x = [OH^-] = 2.8 \times 10^{-6}\ M$$

pOH = 5.55; pH = 8.45 Assumptions good.

b. At the halfway point to equivalence for a weak base/strong acid titration, pH = pK_a since [weak base] = [conjugate acid].

$$K_a = \frac{K_w}{K_b} = \frac{1.0 \times 10^{-14}}{5.6 \times 10^{-4}} = 1.8 \times 10^{-11};\ pH = pK_a = -\log(1.8 \times 10^{-11}) = 10.74$$

For the equivalence point (mmol acid added = mmol base present):

mmol $C_2H_5NH_2$ present = 100.0 mL × 0.10 *M* = 10. mmol $C_2H_5NH_2$

$$\text{mL HNO}_3\ \text{added} = 10.\ \text{mmol H}^+ \times \frac{1\ \text{mL}}{0.20\ \text{mmol}} = 50.\ \text{mL H}^+$$

The strong acid added completely converts the weak base into its conjugate acid. Therefore, at the equivalence point, $[C_2H_5NH_3^+]_o$ = 10. mmol/(100.0 + 50.) mL = 0.067 *M*. Solving the weak acid equilibrium problem:

$$C_2H_5NH_3^+ \rightleftharpoons H^+ + C_2H_5NH_2$$

	$C_2H_5NH_3^+$	H^+	$C_2H_5NH_2$
Initial	0.067 *M*	0	0
Equil.	0.067 − x	x	x

$$K_a = 1.8 \times 10^{-11} = \frac{x^2}{0.067 - x} \approx \frac{x^2}{0.067}, \quad x = [H^+] = 1.1 \times 10^{-6}\ M$$

pH = 5.96; Assumption good.

c. In a strong acid/strong base titration, the halfway point has no special significance other than exactly one-half of the original amount of acid present has been neutralized.

mmol H^+ present = 100.0 mL × 0.50 M = 50. mmol H^+

$$\text{mL } OH^- \text{ added} = 25 \text{ mmol } OH^- \times \frac{1 \text{ mL}}{0.25 \text{ mmol}} = 1.0 \times 10^2 \text{ mL } OH^-$$

	H^+	+	OH^-	→	H_2O
Before	50. mmol		25 mmol		
After	25 mmol		0		

$$[H^+]_{excess} = \frac{25 \text{ mmol}}{(100.0 + 1.0 \times 10^2) \text{ mL}} = 0.13\ M; \ \text{pH} = 0.89$$

At the equivalence point of a strong acid/strong base titration, only neutral species are present (Na^+, Cl^-, H_2O), so the pH = 7.00.

63. $$75.0 \text{ mL} \times \frac{0.10 \text{ mmol}}{\text{mL}} = 7.5 \text{ mmol HA}; \ 30.0 \text{ mL} \times \frac{0.10 \text{ mmol}}{\text{mL}} = 3.0 \text{ mmol } OH^- \text{ added}$$

The added strong base reacts to completion with the weak acid to form the conjugate base of the weak acid and H_2O.

	HA	+	OH^-	→	A^-	+	H_2O
Before	7.5 mmol		3.0 mmol		0		
After	4.5 mmol		0		3.0 mmol		

A buffer results after the OH^- reacts to completion. Using the Henderson-Hasselbalch equation:

$$\text{pH} = \text{p}K_a + \log\frac{[A^-]}{[HA]}, \quad 5.50 = \text{p}K_a + \log\left(\frac{3.0 \text{ mmol}/105.0 \text{ mmol}}{4.5 \text{ mmol}/105.0 \text{ mmol}}\right)$$

$$\text{p}K_a = 5.50 - \log(3.0/4.5) = 5.50 - (-0.18) = 5.68; \ K_a = 10^{-5.68} = 2.1 \times 10^{-6}$$

Indicators

65. $$HIn \rightleftharpoons In^- + H^+ \quad K_a = \frac{[In^-][H^+]}{[HIn]} = 1.0 \times 10^{-9}$$

a. In a very acid solution, the HIn form dominates, so the solution will be yellow.

b. The color change occurs when the concentration of the more dominant form is approximately ten times as great as the less dominant form of the indicator.

$$\frac{[HIn]}{[In^-]} = \frac{10}{1};\ K_a = 1.0 \times 10^{-9} = \left(\frac{1}{10}\right)[H^+],\ [H^+] = 1 \times 10^{-8} M;\ pH = 8.0 \text{ at color change}$$

c. This is way past the equivalence point (100.0 mL OH^- added), so the solution is very basic and the In^- form of the indicator dominates. The solution will be blue.

67. At the equivalence point, P^{2-} is the major species. It is a weak base because it is the conjugate base of a weak acid.

	P^{2-}	+	H_2O	$\rightleftharpoons$	HP^-	+	OH^-	
Initial	$\frac{0.5\text{ g}}{0.1\text{ L}} \times \frac{1\text{ mol}}{204.22\text{ g}} = 0.024\ M$				0		0	(carry extra sig. fig.)
Equil.	$0.024 - x$				x		x	

$$K_b = \frac{[HP^-][OH^-]}{[P^{2-}]} = \frac{K_w}{K_a} = \frac{1.0\times10^{-14}}{10^{-5.51}},\ 3.2\times10^{-9} = \frac{x^2}{0.024 - x} \approx \frac{x^2}{0.024}$$

$x = [OH^-] = 8.8 \times 10^{-6}\ M$; pOH = 5.1; pH = 8.9 Assumptions good.

Phenolphthalein would be a suitable indicator for this titration because it changes color at pH ~9.

69. When choosing an indicator, we want the color change of the indicator to occur approximately at the pH of the equivalence point. Since the pH generally changes very rapidly at the equivalence point, we don't have to be exact. This is especially true for strong acid/strong base titrations. Some choices where color change occurs at about the pH of the equivalence point are:

Exercise	pH at eq. pt.	Indicator
15.53	7.00	bromthymol blue or phenol red
15.55	8.79	o-cresolphthalein or phenolphthalein

71.

Exercise	pH at eq. pt.	Indicator
15.57	8.28	phenolphthalein
15.59	5.28	bromcresol green

73. pH > 5 for bromcresol green to be blue. pH < 8 for thymol blue to be yellow. The pH is between 5 and 8.

Solubility Equilibria

75. a. $AgC_2H_3O_2(s) \rightleftharpoons Ag^+(aq) + C_2H_3O_2^-(aq)$ $K_{sp} = [Ag^+][C_2H_3O_2^-]$

b. $Al(OH)_3(s) \rightleftharpoons Al^{3+}(aq) + 3\ OH^-(aq)$ $K_{sp} = [Al^{3+}][OH^-]^3$

c. $Ca_3(PO_4)_2(s) \rightleftharpoons 3\ Ca^{2+}(aq) + 2\ PO_4^{3-}(aq)$ $K_{sp} = [Ca^{2+}]^3[PO_4^{3-}]^2$

77. In our setup, s = solubility of the ionic solid in mol/L. This is defined as the maximum amount of a salt which can dissolve. Because solids do not appear in the K_{sp} expression, we do not need to worry about their initial and equilibrium amounts.

a.

	$CaC_2O_4(s)$	$\rightleftharpoons$	$Ca^{2+}(aq)$	+	$C_2O_4^{2-}(aq)$
Initial			0		0
	s mol/L of $CaC_2O_4(s)$ dissolves to reach equilibrium				
Change	$-s$	$\rightarrow$	$+s$		$+s$
Equil.			s		s

From the problem, $s = \frac{6.1\times10^{-3}\ g}{L} \times \frac{1\ mol\ CaC_2O_4}{128.10\ g} = 4.8 \times 10^{-5}$ mol/L.

$K_{sp} = [Ca^{2+}][C_2O_4^{2-}] = (s)(s) = s^2$, $K_{sp} = (4.8\times10^{-5})^2 = 2.3 \times 10^{-9}$

b.

	$BiI_3(s)$	$\rightleftharpoons$	$Bi^{3+}(aq)$	+	$3\ I^-(aq)$
Initial			0		0
	s mol/L of $BiI_3(s)$ dissolves to reach equilibrium				
Change	$-s$	$\rightarrow$	$+s$		$+3s$
Equil.			s		$3s$

$K_{sp} = [Bi^{3+}][I^-]^3 = (s)(3s)^3 = 27\ s^4$, $K_{sp} = 27(1.32\times10^{-5})^4 = 8.20 \times 10^{-19}$

79.

	$PbBr_2(s)$	$\rightleftharpoons$	$Pb^{2+}(aq)$	+	$2\ Br^-(aq)$
Initial			0		0
	s mol/L of $PbBr_2(s)$ dissolves to reach equilibrium				
Change	$-s$	$\rightarrow$	$+s$		$+2s$
Equil.			s		$2s$

From the problem, $s = [Pb^{2+}] = 2.14 \times 10^{-2}\ M$. So:

$K_{sp} = [Pb^{2+}][Br^-]^2 = s(2s)^2 = 4s^3$, $K_{sp} = 4(2.14\times10^{-2})^3 = 3.92 \times 10^{-5}$

81. In our setups, s = solubility in mol/L. Because solids do not appear in the K_{sp} expression, we do not need to worry about their initial or equilibrium amounts.

a. $Ag_3PO_4(s) \rightleftharpoons 3\ Ag^+(aq) + PO_4^{3-}(aq)$

Initial			0	0
	s mol/L of $Ag_3PO_4(s)$ dissolves to reach equilibrium			
Change	$-s$	$\rightarrow$	$+3s$	$+s$
Equil.			$3s$	s

$K_{sp} = 1.8 \times 10^{-18} = [Ag^+]^3\ [PO_4^{3-}] = (3s)^3(s) = 27\ s^4$

$27\ s^4 = 1.8 \times 10^{-18}$, $s = (6.7 \times 10^{-20})^{1/4} = 1.6 \times 10^{-5}$ mol/L = molar solubility

b. $CaCO_3(s) \rightleftharpoons Ca^{2+}(aq) + CO_3^{2-}(aq)$

Initial	s = solubility (mol/L)	0	0
Equil.		s	s

$K_{sp} = 8.7 \times 10^{-9} = [Ca^{2+}]\ [CO_3^{2-}] = s^2$, $s = 9.3 \times 10^{-5}$ mol/L

c. $Hg_2Cl_2(s) \rightleftharpoons Hg_2^{2+}(aq) + 2\ Cl^-(aq)$

Initial	s = solubility (mol/L)	0	0
Equil.		s	$2s$

$K_{sp} = 1.1 \times 10^{-18} = [Hg_2^{2+}]\ [Cl^-]^2 = (s)(2s)^2 = 4s^3$, $s = 6.5 \times 10^{-7}$ mol/L

83. $M_2X_3(s) \rightleftharpoons 2\ M^{3+}(aq) + 3\ X^{2-}(aq)$ $K_{sp} = [M^{3+}]^2[X^{2-}]^3$

Initial	s = solubility (mol/L)	0	0
	s mol/L of $M_2X_3(s)$ dissolves to reach equilibrium		
Change	$-s$	$+2s$	$+3s$
Equil.		$2s$	$3s$

$$K_{sp} = (2s)^2(3s)^3 = 108\ s^5;\ \ s = \frac{3.60 \times 10^{-7}\ g}{L} \times \frac{1\ mol\ M_2X_3}{288\ g} = 1.25 \times 10^{-9}\ mol/L$$

$K_{sp} = 108\ (1.25 \times 10^{-9})^5 = 3.30 \times 10^{-43}$

85. Let s = solubility of $Co(OH)_3$ in mol/L. Note: Since solids do not appear in the K_{sp} expression, we do not need to worry about their initial or equilibrium amounts.

$Co(OH)_3(s) \rightleftharpoons Co^{3+}(aq) + 3\ OH^-(aq)$

Initial			0	1.0×10^{-7} M from water
	s mol/L of $Co(OH)_3(s)$ dissolves to reach equilibrium = molar solubility			
Change	$-s$	$\rightarrow$	$+s$	$+3s$
Equil.			s	$1.0 \times 10^{-7} + 3s$

$K_{sp} = 2.5 \times 10^{-43} = [Co^{3+}][OH^-]^3 = (s)(1.0 \times 10^{-7} + 3s)^3 \approx s(1.0 \times 10^{-7})^3$

$s = \dfrac{2.5 \times 10^{-43}}{1.0 \times 10^{-21}} = 2.5 \times 10^{-22}$ mol/L; Assumption good $(1.0 \times 10^{-7} + 3s \approx 1.0 \times 10^{-7})$.

87. a. Both solids dissolve to produce 3 ions in solution, so we can compare the values of K_{sp} to determine relative molar solubility. Because the K_{sp} for CaF_2 is smaller, $CaF_2(s)$ is less soluble (in mol/L).

b. We must calculate molar solubilities since each salt yields a different number of ions when it dissolves.

	$Ca_3(PO_4)_2(s)$	$\rightleftharpoons$	$3\ Ca^{2+}(aq)$	+	$2\ PO_4^{3-}(aq)$	$K_{sp} = 1.3 \times 10^{-32}$
Initial	s = solubility (mol/L)		0		0	
Equil.			$3s$		$2s$	

$K_{sp} = [Ca^{2+}]^3[PO_4^{3-}]^2 = (3s)^3(2s)^2 = 108\ s^5$, $s = (1.3 \times 10^{-32}/108)^{1/5} = 1.6 \times 10^{-7}$ mol/L

	$FePO_4(s)$	$\rightleftharpoons$	$Fe^{3+}(aq)$	+	$PO_4^{3-}(aq)$	$K_{sp} = 1.0 \times 10^{-22}$
Initial	s = solubility (mol/L)		0		0	
Equil.			s		s	

$K_{sp} = [Fe^{3+}][PO_4^{3-}] = s^2$, $s = \sqrt{1.0 \times 10^{-22}} = 1.0 \times 10^{-11}$ mol/L

The molar solubility of $FePO_4$ is smaller, so $FePO_4$ is less soluble (in mol/L).

89. a.

	$Fe(OH)_3(s)$	$\rightleftharpoons$	$Fe^{3+}(aq)$	+	$3\ OH^-(aq)$	
Initial			0		$1 \times 10^{-7}\ M$	from water
	s mol/L of $Fe(OH)_3(s)$ dissolves to reach equilibrium = molar solubility					
Change	$-s$	$\rightarrow$	$+s$		$+3s$	
Equil.			s		$1 \times 10^{-7} + 3s$	

$K_{sp} = 4 \times 10^{-38} = [Fe^{3+}][OH^-]^3 = (s)(1 \times 10^{-7} + 3s)^3 \approx s(1 \times 10^{-7})^3$

$s = 4 \times 10^{-17}$ mol/L Assumption good $(3s << 1 \times 10^{-7})$.

b. $Fe(OH)_3(s) \rightleftharpoons Fe^{3+}(aq) + 3\ OH^-(aq)$ pH = 5.0, $[OH^-] = 1 \times 10^{-9}\ M$

Initial			0	$1 \times 10^{-9}\ M$	(buffered)
	s mol/L dissolves to reach equilibrium				
Change	$-s$	$\rightarrow$	$+s$	–	(assume no pH change in buffer)
Equil.			s	1×10^{-9}	

$K_{sp} = 4 \times 10^{-38} = [Fe^{3+}][OH^-]^3 = (s)(1 \times 10^{-9})^3$, $s = 4 \times 10^{-11}$ mol/L = molar solubility

c. $Fe(OH)_3(s) \rightleftharpoons Fe^{3+}(aq) + 3\ OH^-(aq)$ pH = 11.0 so $[OH^-] = 1 \times 10^{-3}\ M$

Initial			0	0.001 M	(buffered)
	s mol/L dissolves to reach equilibrium				
Change	$-s$	$\rightarrow$	$+s$	$-$	(assume no pH change)
Equil.			s	0.001	

$K_{sp} = 4 \times 10^{-38} = [Fe^{3+}]\ [OH^-]^3 = (s)(0.001)^3,\ s = 4 \times 10^{-29}$ mol/L = molar solubility

Note: As $[OH^-]$ increases, solubility decreases. This is the common ion effect.

91. $Ca_3(PO_4)_2(s) \rightleftharpoons 3\ Ca^{2+}(aq) + 2\ PO_4^{3-}(aq)$

Initial			0	0.20 M
	s mol/L of $Ca_3(PO_4)_2(s)$ dissolves to reach equilibrium			
Change	$-s$	$\rightarrow$	$+3s$	$+2s$
Equil.			$3s$	$0.20 + 2s$

$K_{sp} = 1.3 \times 10^{-32} = [Ca^{2+}]^3\ [PO_4^{3-}]^2 = (3s)^3(0.20 + 2s)^2$

Assuming $0.20 + 2s \approx 0.20$: $1.3 \times 10^{-32} = (3s)^3(0.20)^2 = 27\ s^3(0.040)$

s = molar solubility = 2.3×10^{-11} mol/L; Assumption good.

93. $ZnS(s) \rightleftharpoons Zn^{2+} + S^{2-}$ $K_{sp} = [Zn^{2+}][S^{2-}]$

Initial	s = solubility (mol/L)	0.050 M	0
Equil.		$0.050 + s$	s

$K_{sp} = 2.5 \times 10^{-22} = (0.050 + s)(s) \approx 0.050\ s,\ s = 5.0 \times 10^{-21}$ mol/L; Assumptions good.

$$\text{mass ZnS that dissolves} = 0.3000\ \text{L} \times \frac{5.0 \times 10^{-21}\ \text{mol ZnS}}{\text{L}} \times \frac{97.45\ \text{g ZnS}}{\text{mol}} = 1.5 \times 10^{-19}\ \text{g}$$

95. If the anion in the salt can act as a base in water, the solubility of the salt will increase as the solution becomes more acidic. Added H^+ will react with the base, forming the conjugate acid. As the basic anion is removed, more of the salt will dissolve to replenish the basic anion. The salts with basic anions are Ag_3PO_4, $CaCO_3$, $CdCO_3$ and $Sr_3(PO_4)_2$. Hg_2Cl_2 and PbI_2 do not have any pH dependence since Cl^- and I^- are terrible bases (the conjugate bases of strong acids).

$$Ag_3PO_4(s) + H^+(aq) \rightarrow 3\ Ag^+(aq) + HPO_4^{2-}(aq) \xrightarrow{\text{excess } H^+} 3\ Ag^+(aq) + H_3PO_4(aq)$$

$$CaCO_3(s) + H^+ \rightarrow Ca^{2+} + HCO_3^- \xrightarrow{\text{excess } H^+} Ca^{2+} + H_2CO_3\ [H_2O(l) + CO_2(g)]$$

$$CdCO_3(s) + H^+ \rightarrow Cd^{2+} + HCO_3^- \xrightarrow{\text{excess } H^+} Cd^{2+} + H_2CO_3\ [H_2O(l) + CO_2(g)]$$

$$Sr_3(PO_4)_2(s) + 2\ H^+ \rightarrow 3\ Sr^{2+} + 2\ HPO_4^{2-} \xrightarrow{\text{excess } H^+} 3\ Sr^{2+} + 2\ H_3PO_4$$

97. Potentially, $BaSO_4(s)$ could form if Q is greater than K_{sp}.

$$BaSO_4(s) \rightleftharpoons Ba^{2+}(aq) + SO_4^{2-}(aq) \quad K_{sp} = 1.5 \times 10^{-9}$$

To calculate Q, we need the initial concentrations of Ba^{2+} and SO_4^{2-}.

$$[Ba^{2+}]_o = \frac{\text{mmoles } Ba^{2+}}{\text{total mL solution}} = \frac{75.0\ \text{mL} \times \frac{0.020\ \text{mmoles } Ba^{2+}}{\text{mL}}}{75.0\ \text{mL} + 125\ \text{mL}} = 0.0075\ M$$

$$[SO_4^{2-}]_o = \frac{\text{mmoles } SO_4^{2-}}{\text{total mL solution}} = \frac{125\ \text{mL} \times \frac{0.040\ \text{mmoles } SO_4^{2-}}{\text{mL}}}{200.\ \text{mL}} = 0.025\ M$$

$$Q = [Ba^{2+}]_o[SO_4^{2-}]_o = (0.0075\ M)(0.025\ M) = 1.9 \times 10^{-4}$$

$Q > K_{sp}\ (1.5 \times 10^{-9})$ so $BaSO_4(s)$ will form.

99. The concentrations of ions are large, so Q will be greater than K_{sp} and $BaC_2O_4(s)$ will form. To solve this problem, we will assume that the precipitation reaction goes to completion; then we will solve an equilibrium problem to get the actual ion concentrations. This makes the math reasonable.

$$100.\ \text{mL} \times \frac{0.200\ \text{mmol } K_2C_2O_4}{\text{mL}} = 20.0\ \text{mmol } K_2C_2O_4$$

$$150.\ \text{mL} \times \frac{0.250\ \text{mmol } BaBr_2}{\text{mL}} = 37.5\ \text{mmol } BaBr_2$$

	$Ba^{2+}(aq)$ +	$C_2O_4^{2-}(aq)$	→	$BaC_2O_4(s)$	$K = 1/K_{sp} >> 1$
Before	37.5 mmol	20.0 mmol		0	
Change	-20.0	-20.0	→	+20.0	Reacts completely (K is large)
After	17.5	0		20.0	

New initial concentrations (after complete precipitation) are:

$$[Ba^{2+}] = \frac{17.5 \text{ mmol}}{250. \text{ mL}} = 7.00 \times 10^{-2}\ M;\ [C_2O_4^{2-}] = 0\ M$$

$$[K^+] = \frac{2(20.0 \text{ mmol})}{250. \text{ mL}} = 0.160\ M;\ [Br^-] = \frac{2(37.5 \text{ mmol})}{250. \text{ mL}} = 0.300\ M$$

For K^+ and Br^-, these are also the final concentrations. We can't have 0 *M* $C_2O_4^{2-}$. For Ba^{2+} and $C_2O_4^{2-}$, we need to perform an equilibrium calculation.

	$BaC_2O_4(s)$	$\rightleftharpoons$	$Ba^{2+}(aq)$	+	$C_2O_4^{2-}(aq)$	$K_{sp} = 2.3 \times 10^{-8}$
Initial			0.0700 *M*		0	
	s mol/L of $BaC_2O_4(s)$ dissolves to reach equilibrium					
Equil.			0.0700 + *s*		*s*	

$K_{sp} = 2.3 \times 10^{-8} = [Ba^{2+}][C_2O_4^{2-}] = (0.0700 + s)(s) \approx 0.0700\ s$

$s = [C_2O_4^{2-}] = 3.3 \times 10^{-7}$ mol/L; $[Ba^{2+}] = 0.0700\ M$; Assumption good ($s << 0.0700$).

101. $Ag_3PO_4(s) \rightleftharpoons 3\ Ag^+(aq) + PO_4^{3-}(aq)$; When Q is greater than K_{sp}, precipitation will occur. We will calculate the $[Ag^+]_o$ necessary for $Q = K_{sp}$. Any $[Ag^+]_o$ greater than this calculated number will cause precipitation of $Ag_3PO_4(s)$. In this problem, $[PO_4^{3-}]_o = [Na_3PO_4]_o = 1.0 \times 10^{-5}\ M$.

$K_{sp} = 1.8 \times 10^{-18}$; $Q = 1.8 \times 10^{-18} = [Ag^+]_0^3\ [PO_4^{3-}]_o = [Ag^+]_o^3\ (1.0 \times 10^{-5}\ M)$

$$[Ag^+]_o = \left(\frac{1.8 \times 10^{-18}}{1.0 \times 10^{-5}}\right)^{1/3},\quad [Ag^+]_o = 5.6 \times 10^{-5}\ M$$

When $[Ag^+]_o = [AgNO_3]_o$ is greater than 5.6×10^{-5} *M*, precipitation of $Ag_3PO_4(s)$ will occur.

Complex Ion Equilibria

103. a.

$Ni^{2+} + CN^- \rightleftharpoons NiCN^+$	K_1
$NiCN^+ + CN^- \rightleftharpoons Ni(CN)_2$	K_2
$Ni(CN)_2 + CN^- \rightleftharpoons Ni(CN)_3^-$	K_3
$Ni(CN)_3^- + CN^- \rightleftharpoons Ni(CN)_4^{2-}$	K_4
$Ni^{2+} + 4\ CN^- \rightleftharpoons Ni(CN)_4^{2-}$	$K_f = K_1K_2K_3K_4$

Note: The various K's are included for your information. Each CN^- adds with a corresponding K value associated with that reaction. The overall formation constant, K_f, for the overall reaction is equal to the product of all the stepwise K values.

b.

$$V^{3+} + C_2O_4^{2-} \rightleftharpoons VC_2O_4^{+} \qquad K_1$$
$$VC_2O_4^{+} + C_2O_4^{2-} \rightleftharpoons V(C_2O_4)_2^{-} \qquad K_2$$
$$V(C_2O_4)_2^{-} + C_2O_4^{2-} \rightleftharpoons V(C_2O_3)^{3-} \qquad K_2$$

$$V^{3+} + 3\,C_2O_4^{2-} \rightleftharpoons V(C_2O_4)_3^{3-} \qquad K_f = K_1K_2K_3$$

105.

$$Mn^{2+} + C_2O_4^{2-} \rightleftharpoons MnC_2O_4 \qquad K_1 = 7.9 \times 10^3$$
$$MnC_2O_4 + C_2O_4^{2-} \rightleftharpoons Mn(C_2O_4)_2^{2-} \qquad K_2 = 7.9 \times 10^1$$

$$Mn^{2+}(aq) + 2\,C_2O_4^{2-}(aq) \rightleftharpoons Mn(C_2O_4)_2^{2-}(aq) \qquad K_f = K_1K_2 = 6.2 \times 10^5$$

107. $Hg^{2+}(aq) + 2\,I^-(aq) \rightarrow HgI_2(s)$; $HgI_2(s) + 2\,I^-(aq) \rightarrow HgI_4^{2-}(aq)$

orange ppt — soluble complex ion

109. The formation constant for HgI_4^{2-} is an extremely large number. Because of this, we will let the Hg^{2+} and I^- ions present initially react to completion, and then solve an equilibrium problem to determine the Hg^{2+} concentration.

	Hg^{2+}	+ $4\,I^-$	$\rightleftharpoons$	HgI_4^{2-}	$K = 1.0 \times 10^{30}$
Before	0.010 *M*	0.78 *M*		0	
Change	−0.010	−0.040	→	+0.010	Reacts completely (K large)
After	0	0.74		0.010	New initial
	x mol/L HgI_4^{2-} dissociates to reach equilibrium				
Change	$+x$	$+4x$	←	$-x$	
Equil.	x	$0.74 + 4x$		$0.010 - x$	

$$K = 1.0 \times 10^{30} = \frac{[HgI_4^{2-}]}{[Hg^{2+}][I^-]^4} = \frac{(0.010 - x)}{(x)(0.74 + 4x)^4}; \text{ Making normal assumptions:}$$

$$1.0 \times 10^{30} = \frac{(0.010)}{(x)(0.74)^4}, \quad x = [Hg^{2+}] = 3.3 \times 10^{-32}\,M; \quad \text{Assumptions good.}$$

Note: 3.3×10^{-32} mol/L corresponds to one Hg^{2+} ion per 5×10^7 L. It is very reasonable to approach this problem in two steps. The reaction does essentially go to completion.

111. a.

	AgI(s)	$\rightleftharpoons$	$Ag^+(aq)$ +	$I^-(aq)$	$K_{sp} = [Ag^+][I^-] = 1.5 \times 10^{-16}$
Initial	s = solubility (mol/L)		0	0	
Equil.			s	s	

$K_{sp} = 1.5 \times 10^{-16} = s^2$, $s = 1.2 \times 10^{-8}$ mol/L

b. $AgI(s) \rightleftharpoons Ag^+ + I^-$ $K_{sp} = 1.5 \times 10^{-16}$

$Ag^+ + 2\ NH_3 \rightleftharpoons Ag(NH_3)_2^+$ $K_f = 1.7 \times 10^7$

$AgI(s) + 2\ NH_3(aq) \rightleftharpoons Ag(NH_3)_2^+(aq) + I^-(aq)$ $K = K_{sp} \times K_f = 2.6 \times 10^{-9}$

	$AgI(s)$ +	$2\ NH_3$	$\rightleftharpoons$	$Ag(NH_3)_2^+$	+ I^-
Initial		$3.0\ M$		0	0
	s mol/L of AgBr(s) dissolves to reach equilibrium = molar solubility				
Equil.		$3.0 - 2s$		s	s

$$K = \frac{[Ag(NH_3)_2^+][I^-]}{[NH_3]^2} = \frac{s^2}{(3.0 - 2s)^2} = 2.6\times 10^{-9} \approx \frac{s^2}{(3.0)^2},\ s = 1.5 \times 10^{-4}\ \text{mol/L}$$

Assumption good.

c. The presence of NH_3 increases the solubility of AgI. Added NH_3 removes Ag^+ from solution by forming the complex ion, $Ag(NH_3)_2^+$. As Ag^+ is removed, more AgI(s) will dissolve to replenish the Ag^+ concentration.

113. $AgCl(s) \rightleftharpoons Ag^+ + Cl^-$ $K_{sp} = 1.6 \times 10^{-10}$

$Ag^+ + 2\ NH_3 \rightleftharpoons Ag(NH_3)_2^+$ $K_f = 1.7 \times 10^7$

$AgCl(s) + 2\ NH_3 \rightleftharpoons Ag(NH_3)_2^+(aq) + Cl^-(aq)$ $K = K_{sp} \times K_f = 2.7 \times 10^{-3}$

	$AgCl(s)$ +	$2NH_3$	$\rightleftharpoons$	$Ag(NH_3)_2^+$	+ Cl^-
Initial		$1.0\ M$		0	0
	s mol/L of AgCl(s) dissolves to reach equilibrium = molar solubility				
Equil.		$1.0 - 2s$	$\rightleftharpoons$	s	s

$$K = 2.7 \times 10^{-3} = \frac{[Ag(NH_3)_2^+][Cl^-]}{[NH_3]^2} = \frac{s^2}{(1.0 - 2s)^2},\ \text{Taking the square root:}$$

$$\frac{s}{1.0 - 2s} = (2.7 \times 10^{-3})^{1/2} = 5.2 \times 10^{-2},\ s = 4.7 \times 10^{-2}\ \text{mol/L}$$

In pure water, the solubility of AgCl(s) is $(1.6 \times 10^{-10})^{1/2} = 1.3 \times 10^{-5}$ mol/L. Notice how the presence of NH_3 increases the solubility of AgCl(s) by over a factor of 3500.

115. Test tube 1: added Cl^- reacts with Ag^+ to form a silver chloride precipitate. The net ionic equation is $Ag^+(aq) + Cl^-(aq) \rightarrow AgCl(s)$. Test tube 2: added NH_3 reacts with Ag^+ ions to form a soluble complex ion, $Ag(NH_3)_2^+$. As this complex ion forms, Ag^+ is removed from the solution, which causes the AgCl(s) to dissolve. When enough NH_3 is added, all of the silver

chloride precipitate will dissolve. The equation is $AgCl(s) + 2\ NH_3(aq) \rightarrow Ag(NH_3)_2^+(aq) + Cl^-(aq)$. Test tube 3: added H^+ reacts with the weak base, NH_3, to form NH_4^+. As NH_3 is removed from the $Ag(NH_3)_2^+$ complex ion, Ag^+ ions are released to solution and can then react with Cl^- to reform $AgCl(s)$. The equations are $Ag(NH_3)_2^+(aq) + 2\ H^+(aq) \rightarrow Ag^+(aq) + 2\ NH_4^+(aq)$ and $Ag^+(aq) + Cl^-(aq) \rightarrow AgCl(s)$.

Additional Exercises

117. $NH_3 + H_2O \rightleftharpoons NH_4^+ + OH^-$ $K_b = \dfrac{[NH_4^+][OH^-]}{[NH_3]}$; Taking the −log of the K_b expression:

$$-\log K_b = -\log [OH^-] - \log\frac{[NH_4^+]}{[NH_3]}, \quad -\log [OH^-] = -\log K_b + \log\frac{[NH_4^+]}{[NH_3]}$$

$$pOH = pK_b + \log\frac{[NH_4^+]}{[NH_3]} \text{ or } pOH = pK_b + \log\frac{[Acid]}{[Base]}$$

119. a. $C_2H_5NH_3^+ \rightleftharpoons H^+ + C_2H_5NH_2$ $K_a = \dfrac{K_w}{K_b} = \dfrac{1.0 \times 10^{-14}}{5.6 \times 10^{-4}} = 1.8 \times 10^{-11}$; $pK_a = 10.74$

$$pH = pK_a + \log\frac{[C_2H_5NH_2]}{[C_2H_5NH_3^+]} = 10.74 + \log\frac{0.10}{0.20} = 10.74 - 0.30 = 10.44$$

b. $C_2H_5NH_3^+ + OH^- \rightleftharpoons C_2H_5NH_2$; After 0.050 M OH^- reacts to completion (converting $C_2H_5NH_3^+$ into $C_2H_5NH_2$), a buffer solution still exists where $[C_2H_5NH_3^+] = [C_2H_5NH_2] = 0.15\ M$. Here $pH = pK_a + \log 1.0 = 10.74$ ($pH = pK_a$).

121. A best buffer is when $pH \approx pK_a$; these solutions have about equal concentrations of weak acid and conjugate base. Therefore, choose combinations that yield a buffer where $pH \approx pK_a$, i.e., look for acids whose pK_a is closest to the pH.

a. Potassium fluoride + HCl will yield a buffer consisting of HF ($pK_a = 3.14$) and F^-.

b. Benzoic acid + NaOH will yield a buffer consisting of benzoic acid ($pK_a = 4.19$) and benzoate anion.

c. Sodium acetate + acetic acid ($pK_a = 4.74$) is the best choice for pH = 5.0 buffer since acetic acid has a pK_a value closest to 5.0.

d. HOCl and NaOH: This is the best choice to produce a conjugate acid/base pair with pH = 7.0. This mixture would yield a buffer consisting of HOCl ($pK_a = 7.46$) and OCl^-. Actually, the best choice for a pH = 7.0 buffer is an equimolar mixture of ammonium chloride and sodium acetate. NH_4^+ is a weak acid ($K_a = 5.6 \times 10^{-10}$) and $C_2H_3O_2^-$ is a weak base ($K_b = 5.6 \times 10^{-10}$). A mixture of the two will give a buffer at pH = 7.0 beccause the weak acid and weak base are the same strengths (K_a for NH_4^+ = K_b for

$C_2H_3O_2^-$). $NH_4C_2H_3O_2$ is commercially available, and its solutions are used for pH = 7.0 buffers.

e. Ammonium chloride + NaOH will yield a buffer consisting of NH_4^+ (pK_a = 9.26) and NH_3.

123. a. $HC_2H_3O_2 + OH^- \rightleftharpoons C_2H_3O_2^- + H_2O$

$$K_{eq} = \frac{[C_2H_3O_2^-]}{[HC_2H_3O_2][OH^-]} \times \frac{[H^+]}{[H^+]} = \frac{K_{a,\,HC_2H_3O_2}}{K_w} = \frac{1.8 \times 10^{-5}}{1.0 \times 10^{-14}} = 1.8 \times 10^9$$

b. $C_2H_3O_2^- + H^+ \rightleftharpoons HC_2H_3O_2 \quad K_{eq} = \frac{[HC_2H_3O_2]}{[H^+][C_2H_3O_2^-]} = \frac{1}{K_{a,\,HC_2H_3O_2}} = 5.6 \times 10^4$

c. $HCl + NaOH \rightarrow NaCl + H_2O$

Net ionic equation is: $H^+ + OH^- \rightleftharpoons H_2O; \quad K_{eq} = \frac{1}{K_w} = 1.0 \times 10^{14}$

125. In the final solution: $[H^+] = 10^{-2.15} = 7.1 \times 10^{-3}\,M$

Beginning mmol HCl = 500.0 mL × 0.200 mmol/mL = 100. mmol HCl

Amount of HCl that reacts with NaOH = 1.50×10^{-2} mmol/mL × V

$$\frac{7.1 \times 10^{-3}\text{ mmol}}{\text{mL}} = \frac{\text{final mmol H}^+}{\text{total volume}} = \frac{1.00 - 0.0150\,V}{500.0 + V}$$

$3.6 + 7.1 \times 10^{-3}\,V = 100. - 1.50 \times 10^{-2}\,V$, $2.21 \times 10^{-2}\,V = 100. - 3.6$

$V = 4.36 \times 10^3$ mL = 4.36 L = 4.4 L NaOH

127. $HA + OH^- \rightarrow A^- + H_2O$ where HA = acetylsalicylic acid (assuming a monoprotic acid)

$$\text{mmol HA present} = 27.36\text{ mL OH}^- \times \frac{0.5106\text{ mmol OH}^-}{\text{mL OH}^-} \times \frac{1\text{ mmol HA}}{\text{mmol OH}^-} = 13.97\text{ mmol HA}$$

$$\text{Molar mass of HA} = \frac{\text{grams}}{\text{mol}} \times \frac{2.51\text{ g HA}}{13.97 \times 10^{-3}\text{ mol HA}} = 180.\text{ g/mol}$$

To determine the K_a value, use the pH data. After complete neutralization of acetylsalicylic acid by OH^-, we have 13.97 mmol of OH^- produced from the neutralization reaction. A^- will react completely with the added H^+ and reform acetylsalicylic acid, HA.

$$\text{mmol H}^+\text{ added} = 13.68\text{ mL} \times \frac{0.5106\text{ mmol H}^+}{\text{mL}} = 6.985\text{ mmol H}^+$$

	A^-	+	H^+	→	HA	
Before	13.97 mmol		6.985 mmol		0	
Change	−6.985		−6.985	→	+6.985	Reacts completely
After	6.985 mmol		0		6.985 mmol	

We have back titrated this solution to the halfway point to equivalence where pH = pK_a (assuming HA is a weak acid). We know this because after H^+ reacts completely, equal mmol of HA and A^- are present, which only occurs at the halfway point to equivalence. Assuming acetylsalicylic acid is a weak monoprotic acid, then pH = pK_a = 3.48. $K_a = 10^{-3.48} = 3.3 \times 10^{-4}$

129. $HC_2H_3O_2 \rightleftharpoons H^+ + C_2H_3O_2^-$; Let C_o = initial concentration of $HC_2H_3O_2$

From normal weak acid setup where x = $[H^+]$:

$$K_a = 1.8 \times 10^{-5} = \frac{[H^+][C_2H_3O_2^-]}{[HC_2H_3O_2]} = \frac{[H^+]^2}{C_0 - [H^+]}$$

$$[H^+] = 10^{-2.68} = 2.1 \times 10^{-3}\,M;\ 1.8 \times 10^{-5} = \frac{(2.1 \times 10^{-3})^2}{C_0 - 2.1 \times 10^{-3}},\ C_0 = 0.25\,M$$

25.0 mL × 0.25 mmol/mL = 6.3 mmol $HC_2H_3O_2$; Need 6.3 mol KOH to reach equivalence point.

6.3 mmol KOH = V_{KOH} × 0.0975 mmol/mL, V_{KOH} = 65 mL

131. 50.0 mL × 0.100 M = 5.00 mmol NaOH initially

at pH = 10.50, pOH = 3.50, $[OH^-] = 10^{-3.50} = 3.2 \times 10^{-4}\,M$

mmol OH^- remaining = 3.2×10^{-4} mmol/mL × 73.75 mL = 2.4×10^{-2} mmol

mmol OH^- that reacted = 5.00 − 0.024 = 4.98 mmol

Because the weak acid is monoprotic, 23.75 mL of the weak acid solution contains 4.98 mmol HA.

$$[HA]_o = \frac{4.98\ \text{mmol}}{23.75\ \text{mL}} = 0.210\,M$$

133. a. $Cu(OH)_2 \rightleftharpoons Cu^{2+} + 2\,OH^-$ $\quad K_{sp} = 1.6 \times 10^{-19}$

$Cu^{2+} + 4\,NH_3 \rightleftharpoons Cu(NH_3)_4^{2+}$ $\quad K_f = 1.0 \times 10^{13}$

$Cu(OH)_2(s) + 4\,NH_3(aq) \rightleftharpoons Cu(NH_3)_4^{2+}(aq) + 2\,OH^-(aq)$ $\quad K = K_{sp}K_f = 1.6 \times 10^{-6}$

b. $Cu(OH)_2(s) + 4\,NH_3 \rightleftharpoons Cu(NH_3)_4^{2+} + 2\,OH^-$ $\quad K = 1.6 \times 10^{-6}$

	$4\,NH_3$	$Cu(NH_3)_4^{2+}$	$2\,OH^-$
Initial	5.0 M	0	0.0095 M
	s mol/L $Cu(OH)_2$ dissolves to reach equilibrium		
Equil.	5.0 − 4*s*	*s*	0.0095 + 2*s*

$$K = 1.6 \times 10^{-6} = \frac{[Cu(NH_3)_4^{2+}][OH^-]^2}{[NH_3]^4} = \frac{s(0.0095 + 2s)^2}{(5.0 - 4s)^4}$$

If s is small: $1.6 \times 10^{-6} = \frac{s(0.0095)^2}{(5.0)^4}$, $s = 11.$ mol/L

Assumptions are horrible. We will solve the problem by successive approximations.

$s_{calc} = \frac{1.6 \times 10^{-6}\,(5.0 - 4s_{guess})^4}{(0.0095 + 2s_{guess})^2}$, The results from six trials are:

s_{guess}: 0.10, 0.050, 0.060, 0.055, 0.056

s_{calc}: 1.6×10^{-2}, 0.071, 0.049, 0.058, 0.056

Thus, the solubility of $Cu(OH)_2$ is 0.056 mol/L in 5.0 M NH_3.

135. $Ca_5(PO_4)_3OH(s) \rightleftharpoons 5\,Ca^{2+} + 3\,PO_4^{3-} + OH^-$

		$5\,Ca^{2+}$	$3\,PO_4^{3-}$	OH^-
Initial	s = solubility (mol/L)	0	0	1.0×10^{-7} from water
Equil.		$5s$	$3s$	$s + 1.0 \times 10^{-7} \approx s$

$K_{sp} = 6.8 \times 10^{-37} = [Ca^{2+}]^5\,[PO_4^{3-}]^3\,[OH^-] = (5s)^5(3s)^3(s)$

$6.8 \times 10^{-37} = (3125)(27)s^9$, $s = 2.7 \times 10^{-5}$ mol/L; Assumption is good.

The solubility of hydroxyapatite will increase as the solution gets more acidic because both phosphate and hydroxide can react with H^+.

$Ca_5(PO_4)_3F(s) \rightleftharpoons 5\,Ca^{2+} + 3\,PO_4^{3-} + F^-$

		$5\,Ca^{2+}$	$3\,PO_4^{3-}$	F^-
Initial	s = solubility (mol/L)	0	0	0
Equil.		$5s$	$3s$	s

$K_{sp} = 1 \times 10^{-60} = (5s)^5(3s)^3(s) = (3125)(27)s^9$, $s = 6 \times 10^{-8}$ mol/L

The hydroxyapatite in tooth enamel is converted to the less soluble fluorapatite by fluoride-treated water. The less soluble fluorapatite is more difficult to remove, making teeth less susceptible to decay.

137. a. $Pb(OH)_2(s) \rightleftharpoons Pb^{2+}(aq) + 2\,OH^-(aq)$

		$Pb^{2+}(aq)$	$2\,OH^-(aq)$
Initial	s = solubility (mol/L)	0	1.0×10^{-7} M from water
Equil.		s	$1.0 \times 10^{-7} + 2s$

$K_{sp} = 1.2 \times 10^{-15} = [Pb^{2+}]\,[OH^-]^2 = s(1.0 \times 10^{-7} + 2s)^2 \approx s(2s^2) = 4s^3$

$s = [Pb^{2+}] = 6.7 \times 10^{-6}$ M; Assumption to ignore OH^- from water is good by the 5% rule.

b. $Pb(OH)_2(s) \rightleftharpoons Pb^{2+}(aq) + 2\,OH^-(aq)$

Initial	0	0.10 M	pH = 13.00, $[OH^-] = 0.10\ M$
	s mol/L $Pb(OH)_2(s)$ dissolves to reach equilibrium		
Equil.	s	0.10	(buffered solution)

$1.2 \times 10^{-15} = (s)(0.10)^2$, $s = [Pb^{2+}] = 1.2 \times 10^{-13}\ M$

c. We need to calculate the Pb^{2+} concentration in equilibrium with $EDTA^{4-}$. Because K is large for the formation of $PbEDTA^{2-}$, let the reaction go to completion; then solve an equilibrium problem to get the Pb^{2+} concentration.

	Pb^{2+}	+ $EDTA^{4-}$	$\rightleftharpoons$ $PbEDTA^{2-}$	$K = 1.1 \times 10^{18}$
Before	0.010 M	0.050 M	0	
	0.010 mol/L Pb^{2+} reacts completely (large K)			
Change	−0.010	−0.010 →	+0.010	Reacts completely
After	0	0.040	0.010	New initial
	x mol/L $PbEDTA^{2-}$ dissociates to reach equilibrium			
Equil.	x	$0.040 + x$	$0.010 - x$	

$$1.1 \times 10^{18} = \frac{(0.010 - x)}{(x)(0.040 + x)} \approx \frac{(0.010)}{(x)(0.040)},\quad x = [Pb^{2+}] = 2.3 \times 10^{-19}\ M;$$ Assumptions good.

Now calculate the solubility quotient for $Pb(OH)_2$ to see if precipitation occurs. The concentration of OH^- is 0.10 M because we have a solution buffered at pH = 13.00.

$$Q = [Pb^{2+}]_o[OH^-]_o^2 = (2.3 \times 10^{-19})(0.10)^2 = 2.3 \times 10^{-21} < K_{sp}\ (1.2 \times 10^{-15})$$

$Pb(OH)_2(s)$ will not form since Q is less than K_{sp}.

Challenge Problems

139. $$\text{mmol } HC_3H_5O_2 \text{ present initially} = 45.0 \text{ mL} \times \frac{0.750 \text{ mmol}}{\text{mL}} = 33.8 \text{ mmol } HC_3H_5O_2$$

$$\text{mmol } C_3H_5O_2^- \text{ present initially} = 55.0 \text{ mL} \times \frac{0.700 \text{ mmol}}{\text{mL}} = 38.5 \text{ mmol } C_3H_5O_2^-$$

The initial pH of the buffer is:

$$pH = pK_a + \log\frac{[C_3H_5O_2^-]}{[HC_3H_5O_2]} = -\log(1.3 \times 10^{-5}) + \log\frac{\frac{38.5 \text{ mmol}}{100.0 \text{ mL}}}{\frac{33.8 \text{ mmol}}{100.0 \text{ mL}}}$$

$$pH = 4.89 + \log\frac{38.5}{33.8} = 4.95$$

Note: Because the buffer components are in the same volume of solution, we can use the mol (or mmol) ratio in the Henderson-Hasselbalch equation to solve for pH instead of using the concentration ratio of $[C_3H_5O_2^-]/[HC_3H_5O_2]$. The total volume always cancels for buffer solutions.

When NaOH is added, the pH will increase, and the added OH^- will convert $HC_3H_5O_2$ into $C_3H_5O_2^-$. The pH after addition of OH^- increases by 2.5%, so the resulting pH is:

$$4.95 + 0.025\,(4.95) = 5.07$$

At this pH, a buffer solution still exists and the mmol ratio between $C_3H_5O_2^-$ and $HC_3H_5O_2$ is:

$$pH = pK_a + \log\frac{\text{mmol } C_3H_5O_2^-}{\text{mmol } HC_3H_5O_2}, \quad 5.07 = 4.89 + \log\frac{\text{mmol } C_3H_5O_2^-}{\text{mmol } HC_3H_5O_2}$$

$$\frac{\text{mmol } C_3H_5O_2^-}{\text{mmol } HC_3H_5O_2} = 10^{0.18} = 1.5$$

Let x = mmol OH^- added to increase pH to 5.07. Because OH^- will essentially react to completion with $HC_3H_5O_2$, the setup for the problem using mmol is:

	$HC_3H_5O_2$	+	OH^-	$\rightarrow$	$C_3H_5O_2^- + H_2O$	
Before	33.8 mmol		x mmol		38.5 mmol	
Change	$-x$		$-x$	$\rightarrow$	$+x$	Reacts completely
After	$33.8 - x$		0		$38.5 + x$	

Solving for x:

$$\frac{\text{mmol } C_3H_5O_2^-}{\text{mmol } HC_3H_5O_2} = 1.5 = \frac{38.5 + x}{33.8 - x}, \quad 1.5\,(33.8 - x) = 38.5 + x$$

$$x = 4.9 \text{ mmol } OH^- \text{ added}$$

The volume of NaOH necessary to raise the pH by 2.5% is:

$$4.9 \text{ mmol NaOH} \times \frac{1 \text{ mL}}{0.10 \text{ mmol NaOH}} = 49 \text{ mL}$$

49 mL of 0.10 M NaOH must be added to increase the pH by 2.5%.

141. For HOCl, $K_a = 3.5 \times 10^{-8}$ and $pK_a = -\log(3.5 \times 10^{-8}) = 7.46$; This will be a buffer solution since the pH is close to the pK_a value.

$$pH = pK_a + \log\frac{[OCl^-]}{[HOCl]}, \quad 8.00 = 7.46 + \log\frac{[OCl^-]}{[HOCl]}, \quad \frac{[OCl^-]}{[HOCl]} = 10^{0.54} = 3.5$$

1.00 L × 0.0500 M = 0.0500 mol HOCl initially. Added OH^- converts HOCl into OCl^-. The total moles of OCl^- and HOCl must equal 0.0500 mol. Solving where n = moles:

$$n_{OCl^-} + n_{HOCl} = 0.0500 \text{ and } n_{OCl^-} = 3.5\, n_{HOCl}$$

$$4.5\, n_{HOCl} = 0.0500,\quad n_{HOCl} = 0.011 \text{ mol};\quad n_{OCl^-} = 0.039 \text{ mol}$$

We need to add 0.039 mol NaOH to produce 0.039 mol OCl^-.

0.039 mol OH^- = V × 0.0100 M, V = 3.9 L NaOH

143. The first titration plot (from 0-100.0 mL) corresponds to the titration of H_2A by OH^-. The reaction is $H_2A + OH^- \rightarrow HA^- + H_2O$. After all of the H_2A has been reacted, the second titration (from 100.0 - 200.0 mL) corresponds to the titration of HA^- by OH^-. The reaction is $HA^- + OH^- \rightarrow A^{2-} + H_2O$.

a. At 100.0 mL of NaOH, just enough OH^- has been added to react completely with all of the H_2A present (mol OH^- added = mol H_2A present initially). From the balanced equation, the mol of HA^- produced will equal the mol of H_2A present initially. Because mol HA^- present at 100.0 mL OH^- added equals the mol of H_2A present initially, exactly 100.0 mL more of NaOH must be added to react with all of the HA^-. The volume of NaOH added to reach the second equivalence point equals 100.0 mL + 100.0 mL = 200.0 mL.

b. $H_2A + OH^- \rightarrow HA^- + H_2O$ is the reaction occurring from 0-100.0 mL NaOH added.

i. No reaction has taken place, so H_2A and H_2O are the major species.

ii. Adding OH^- converts H_2A into HA^-. The major species up to 100.0 mL NaOH added are H_2A, HA^-, H_2O, and Na^+.

iii. At 100.0 mL NaOH added, mol of OH^- = mol H_2A, so all of the H_2A present initially has been converted into HA^-. The major species are HA^-, H_2O, and Na^+.

iv. Between 100.0 and 200.0 mL NaOH added, the OH^- converts HA^- into A^{2-}. The major species are HA^-, A^{2-}, H_2O, and Na^+.

v. At the second equivalence point (200.0 mL), just enough OH^- has been added to convert all of the HA^- into A^{2-}. The major species are A^{2-}, H_2O, and Na^+.

vi. Past 200.0 mL NaOH added, excess OH^- is present. The major species are OH^-, A^{2-}, H_2O, and Na^+.

c. 50.0 mL of NaOH added correspond to the first halfway point to equivalence. Exactly one-half of the H_2A present initially has been converted into its conjugate base HA^-, so $[H_2A] = [HA^-]$ in this buffer solution.

$$H_2A \rightleftharpoons HA^- + H^+ \qquad K_{a_1} = \frac{[HA^-][H^+]}{[H_2A]}$$

When $[HA^-] = [H_2A]$, then $K_{a_1} = [H^+]$ or $pK_{a_1} = pH$.

Here, pH = 4.0 so $K_{a_1} = 4.0$ and $K_{a_1} = 10^{-4.0} = 1 \times 10^{-4}$.

150.0 mL of NaOH added correspond to the second halfway point to equivalence where $[HA^-] = [A^{2-}]$ in this buffer solution.

$$HA^- \rightleftharpoons A^{2-} + H^+ \qquad K_{a_2} = \frac{[A^{2-}][H^+]}{[HA^-]}$$

When $[A^{2-}] = [HA^-]$, then $K_{a_2} = [H^+]$ or $pK_{a_2} = pH$.

Here, pH = 8.0 so $pK_{a_2} = 8.0$ and $K_{a_2} = 10^{-8.0} = 1 \times 10^{-8}$.

145. An indicator changes color at pH ≈ pK_a ± 1.The results from each indicator tells us something about the pH. The conclusions are summarized below:

Results from	pH
bromphenol blue	≥ ~ 5.0
bromcresol purple	≤ ~ 5.0
bromcresol green *	pH ~ pK_a ~ 4.8
alizarin	≤ ~ 5.5

*For bromcresol green, the resultant color is green. This is a combination of the extremes (yellow and blue). This occurs when pH ~ pK_a of the indicator.

From the indicator results, the pH of the solution is about 5.0. We solve for K_a by setting up the typical weak acid problem.

	HX	⇌	H^+	+	X^-
Initial	1.0 *M*		~0		0
Equil.	1.0 - *x*		*x*		*x*

$$K_a = \frac{[H^+][X^-]}{[HX]} = \frac{x^2}{1.0 - x}$$; Because pH ~5.0, $[H^+] = x \approx 1 \times 10^{-5}\,M$.

$$K_a \approx \frac{(1 \times 10^{-5})^2}{1.0 - 1 \times 10^{-5}} \approx 1 \times 10^{-10}$$

147. $K_{sp} = [Ni^{2+}][S^{2-}] = 3 \times 10^{-21}$

$$H_2S \rightleftharpoons H^+ + HS^- \qquad K_{a_1} = 1.0 \times 10^{-7}$$
$$HS^- \rightleftharpoons H^+ + S^{2-} \qquad K_{a_2} = 1 \times 10^{-19}$$

$$H_2S \rightleftharpoons 2\,H^+ + S^{2-} \qquad K = K_{a_1} \times K_{a_2} = 1 \times 10^{-26} = \frac{[H^+]^2[S^{2-}]}{[H_2S]}$$

Because K is very small, only a tiny fraction of the H_2S will react. At equilibrium, $[H_2S]$ = 0.10 *M* and $[H^+] = 1 \times 10^{-3}$.

$$[S^{2-}] = \frac{K[H_2S]}{[H^+]^2} = \frac{1\times10^{-26}\,(0.10)}{(1\times10^{-3})^2} = 1 \times 10^{-21}\,M$$

$$NiS(s) \rightleftharpoons Ni^{2+}(aq) + S^{2-}(aq) \qquad K_{sp} = 3.0 \times 10^{-21}$$

Precipitation of NiS will occur when $Q > K_{sp}$. We will calculate $[Ni^{2+}]$ for $Q = K_{sp}$.

$$Q = K_{sp} = [Ni^{2+}][S^{2-}] = 3.0 \times 10^{-21},\ [Ni^{2+}] = \frac{3.0\times10^{-21}}{1\times10^{-21}} = 3\,M$$

149. a. $SrF_2(s) \rightleftharpoons Sr^{2+}(aq) + 2\,F^-(aq)$

		Sr^{2+}	F^-
Initial		0	0
	s mol/L SrF_2 dissolves to reach equilibrium		
Equil.		*s*	2*s*

$[Sr^{2+}]\,[F^-]^2 = K_{sp} = 7.9 \times 10^{-10} = 4s^3$, $s = 5.8 \times 10^{-4}$ mol/L

b. Greater, because some of the F^- would react with water:

$$F^- + H_2O \rightleftharpoons HF + OH^- \qquad K_b = \frac{K_w}{K_a(HF)} = 1.4 \times 10^{-11}$$

This lowers the concentration of F^-, forcing more SrF_2 to dissolve.

c. $SrF_2(s) \rightleftharpoons Sr^{2+} + 2\,F^-$ $K_{sp} = 7.9 \times 10^{-10} = [Sr^{2+}]\,[F^-]^2$

Let *s* = solubility = $[Sr^{2+}]$, then 2*s* = total F^- concentration.

Because F^- is a weak base, some of the F^- is converted into HF. Therefore:

total F^- concentration = 2*s* = $[F^-] + [HF]$.

$$HF \rightleftharpoons H^+ + F^- \quad K_a = 7.2 \times 10^{-4} = \frac{[H^+][F^-]}{[HF]} = \frac{1.0 \times 10^{-2}[F^-]}{[HF]} \quad \text{(pH = 2.00 buffer)}$$

$$7.2 \times 10^{-2} = \frac{[F^-]}{[HF]}, \ [HF] = 14\,[F^-]; \text{ Solving:}$$

$$[Sr^{2+}] = s; \ 2s = [F^-] + [HF] = [F^-] + 14\,[F^-], \ 2s = 15\,[F^-], \ [F^-] = 2s/15$$

$$7.9 \times 10^{-10} = [Sr^{2+}]\,[F^-]^2 = (s)\left(\frac{2s}{15}\right)^2, \quad s = 3.5 \times 10^{-3} \text{ mol/L}$$

Integrative Problems

151. $$\text{pH} = \text{pK}_a + \log\frac{[C_7H_4O_2F^-]}{[C_7H_5O_2F]} = 2.90 + \log\left(\frac{55.0 \text{ mL} \times 0.472\,M / V_T}{75.0 \text{ mL} \times 0.275\,M / V_T}\right)$$

$$\text{pH} = 2.90 + \log\left(\frac{26.0}{20.6}\right) = 2.90 + 0.101 = 3.00$$

153. The added OH^- from the strong base reacts to completion with the best acid present, HF. To determine the pH, see what is in solution after the OH^- reacts to completion.

$$OH^- \text{ added} = 38.7 \text{ g soln} \times \frac{1.50 \text{ g NaOH}}{100.0 \text{ g soln}} \times \frac{1 \text{ mol NaOH}}{40.00 \text{ g}} \times \frac{1 \text{ mol } OH^-}{\text{mol NaOH}} = 0.0145 \text{ mol } OH^-$$

For the 0.174 *m* HF solution, if we had exactly 1 kg of H_2O, then the solution would contain 0.174 mol HF.

$$0.174 \text{ mol HF} \times \frac{20.01 \text{ g}}{\text{mol HF}} = 3.48 \text{ g HF}$$

mass of solution = 1000.00 g H_2O + 3.48 g HF = 1003.48 g

$$\text{volume of solution} = 1003.48 \text{ g} \times \frac{1 \text{ mL}}{1.10 \text{ g}} = 912 \text{ mL}$$

$$\text{mol HF} = 250.\text{ mL} \times \frac{0.174 \text{ mol HF}}{912 \text{ mL}} = 4.77 \times 10^{-2} \text{ mol HF}$$

	OH^-	+ HF	$\rightarrow$ F^-	+ H_2O
Before	0.0145 mol	0.0477 mol	0	
Change	−0.0145	−0.0145	+0.0145	
After	0	0.0332 mol	0.0145 mol	

After reaction, a buffer solution results containing HF, a weak acid, and F^-, its conjugate base. Let V_T = total volume of solution.

$$\text{pH} = \text{pK}_a + \log\frac{[F^-]}{[HF]} = -\log(7.2 \times 10^{-4}) + \log\left(\frac{0.0145/V_T}{0.0332/V_T}\right)$$

$$\text{pH} = 3.14 + \log\left(\frac{0.0145}{0.0332}\right) = 3.14 + (-0.360), \ \text{pH} = 2.78$$

CHAPTER SIXTEEN

SPONTANEITY, ENTROPY, AND FREE ENERGY

Questions

7. Living organisms need an external source of energy to carry out these processes. Green plants use the energy from sunlight to produce glucose from carbon dioxide and water by photosynthesis. In the human body, the energy released from the metabolism of glucose helps drive the synthesis of proteins. For all processes combined, ΔS_{univ} must be greater than zero (2nd law).

9. It appears that the sum of the two processes has no net change. This is not so. By the second law of thermodynamics, ΔS_{univ} must have increased even though it looks as if we have gone through a cyclic process.

11. As a process occurs, ΔS_{univ} will increase; ΔS_{univ} cannot decrease. Time, like ΔS_{univ}, only goes in one direction.

13. Possible arrangements for one molecule:

Both are equally probable.

Possible arrangements for two molecules:

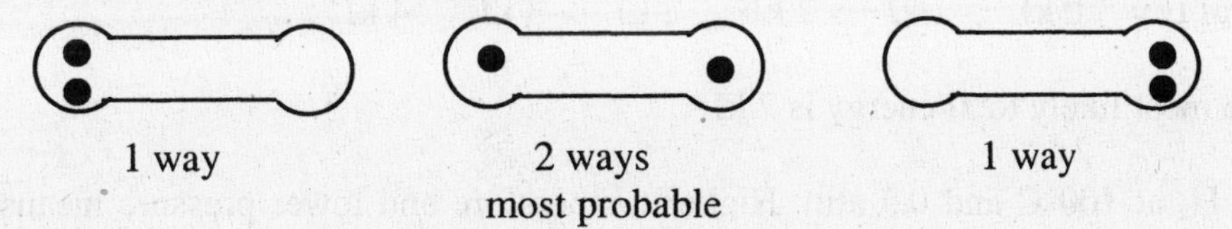

Possible arrangement for three molecules:

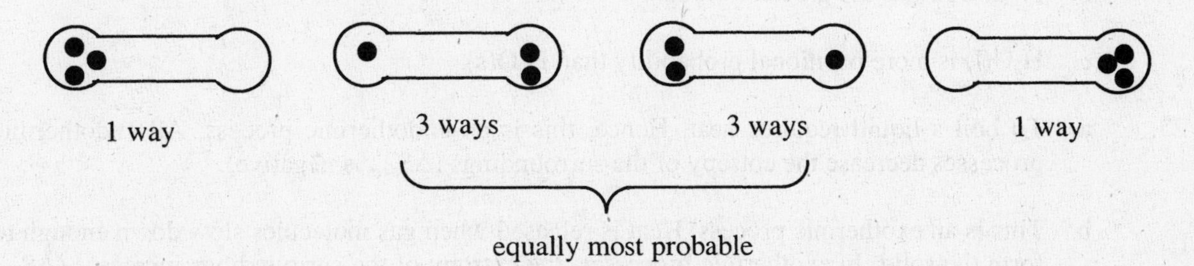

15. Note that these substances are not in the solid state, but are in the aqueous state; water molecules are also present. There is an apparent increase in ordering when these ions are placed in water as compared to the separated state. The hydrating water molecules must be in a highly ordered arrangement when surrounding these anions.

17. One can determine $\Delta S°$ and $\Delta H°$ for the reaction using the standard entropies and standard enthalpies of formation in Appendix 4, then use the equation $\Delta G° = \Delta H° - T\Delta S°$. One can also use the standard free energies of formation in Appendix 4. And finally, one can use Hess's law to calculate $\Delta G°$. Here, reactions having known $\Delta G°$ values are manipulated to determine $\Delta G°$ for a different reaction.

    For temperatures other than 25°C, $\Delta G°$ is estimated using the $\Delta G° = \Delta H° - T\Delta S°$ equation. The assumptions made are that the $\Delta H°$ and $\Delta S°$ values determined from Appendix 4 data are temperature independent. We use the same $\Delta H°$ and $\Delta S°$ values as determined when T = 25°C, then plug in the new temperature in Kelvin into the equation to estimate $\Delta G°$ at the new temperature.

Exercises

Spontaneity, Entropy, and the Second Law of Thermodynamics: Free Energy

19. a, b and c; From our own experiences, salt water, colored water and rust form without any outside intervention. A bedroom, however, spontaneously gets cluttered. It takes an outside energy source to clean a bedroom.

21. We draw all of the possible arrangements of the two particles in the three levels.

2 kJ	___	___	x	___	x	xx
1 kJ	___	x	___	xx	x	___
0 kJ	xx	x	x	___	___	___
Total E =	0 kJ	1 kJ	2 kJ	2 kJ	3 kJ	4 kJ

The most likely total energy is 2 kJ.

23. a. H_2 at 100°C and 0.5 atm; Higher temperature and lower pressure means greater volume and hence, greater positional probability.

    b. N_2 at STP has the greater volume.

    c. $H_2O(l)$ is more positional probability than $H_2O(s)$.

25. a. To boil a liquid requires heat. Hence, this is an endothermic process. All endothermic processes decrease the entropy of the surroundings (ΔS_{surr} is negative).

    b. This is an exothermic process. Heat is released when gas molecules slow down enough to form the solid. In exothermic processes, the entropy of the surroundings increases (ΔS_{surr} is positive).

27. $\Delta G = \Delta H - T\Delta S$; When ΔG is negative, then the process will be spontaneous.

a. $\Delta G = \Delta H - T\Delta S = 25 \times 10^3$ J $-$ (300. K)(5.0 J/K) = 24,000 J, Not spontaneous

b. ΔG = 25,000 J $-$ (300. K)(100. J/K) = $-$5000 J, Spontaneous

c. Without calculating ΔG, we know this reaction will be spontaneous at all temperatures. ΔH is negative and ΔS is positive ($-T\Delta S < 0$). ΔG will always be less than zero with these sign combinations for ΔH and ΔS.

d. $\Delta G = -1.0 \times 10^4$ J $-$ (200. K)($-$40. J/K) = $-$2000 J, Spontaneous

29. At the boiling point, $\Delta G = 0$ so $\Delta H = T\Delta S$.

$$\Delta S = \frac{\Delta H}{T} = \frac{27.5 \text{ kJ/mol}}{(273+35)\text{ K}} = 8.93 \times 10^{-2} \text{ kJ/K•mol} = 89.3 \text{ J/K•mol}$$

31. a. $NH_3(s) \rightarrow NH_3(l)$; $\Delta G = \Delta H - T\Delta S$ = 5650 J/mol $-$ 200. K (28.9 J/K•mol)

ΔG = 5650 J/mol $-$ 5780 J/mol = T = =$-$130 J/mol

Yes, NH_3 will melt because $\Delta G < 0$ at this temperature.

b. At the melting point, $\Delta G = 0$ so $T = \dfrac{\Delta H}{\Delta S} = \dfrac{5650 \text{ J/mol}}{28.9 \text{ J/K•mol}} = 196$ K.

Chemical Reactions: Entropy Changes and Free Energy

33. a. Decrease in disorder; $\Delta S°(-)$

b. Increase in disorder; $\Delta S°(+)$

c. Decrease in disorder ($\Delta n < 0$); $\Delta S°(-)$

d. Increase in disorder ($\Delta n > 0$); $\Delta S°(+)$

For c and d, concentrate on the gaseous products and reactants. When there are more gaseous product molecules than gaseous reactant molecules ($\Delta n > 0$), then $\Delta S°$ will be positive (disorder increases). When Δn is negative, then $\Delta S°$ is negative (disorder decreases or order increases).

35. a. $C_{graphite}(s)$; Diamond has a more ordered structure than graphite.

b. $C_2H_5OH(g)$; The gaseous state is more disordered than the liquid state.

c. $CO_2(g)$; The gaseous state is more disordered than the solid state.

37. a. $2\,H_2S(g) + SO_2(g) \rightarrow 3\,S_{rhombic}(s) + 2\,H_2O(g)$; Because there are more molecules of reactant gases as compared to product molecules of gas ($\Delta n = 2 - 3 < 0$), $\Delta S°$ will be negative.

$$\Delta S° = \sum n_p S°_{products} - \sum n_r S°_{reactants}$$

$\Delta S° = [3 \text{ mol } S_{rhombic}(s) (32 \text{ J/K•mol}) + 2 \text{ mol } H_2O(g) (189 \text{ J/K•mol})]$

$- [2 \text{ mol } H_2S(g) (206 \text{ J/K•mol}) + 1 \text{ mol } SO_2(g) (248 \text{ J/K•mol})]$

$\Delta S° = 474 \text{ J/K} - 660. \text{ J/K} = -186 \text{ J/K}$

b. $2 SO_3(g) \rightarrow 2 SO_2(g) + O_2(g)$; Because Δn of gases is positive ($\Delta n = 3 - 2$), $\Delta S°$ will be positive.

$\Delta S = 2 \text{ mol}(248 \text{ J/K•mol}) + 1 \text{ mol}(205 \text{ J/K•mol}) - [2 \text{ mol}(257 \text{ J/K•mol})] = 187 \text{ J/K}$

c. $Fe_2O_3(s) + 3 H_2(g) \rightarrow 2 Fe(s) + 3 H_2O(g)$; Because Δn of gases $= 0$ ($\Delta n = 3 - 3$), we can't easily predict if $\Delta S°$ will be positive or negative.

$\Delta S = 2 \text{ mol}(27 \text{ J/K•mol}) + 3 \text{ mol}(189 \text{ J/K•mol}) -$

$[1 \text{ mol}(90. \text{ J/K•mol}) + 3 \text{ mol } (141 \text{ J/K•mol})] = 138 \text{ J/K}$

39. $C_2H_2(g) + 4 F_2(g) \rightarrow 2 CF_4(g) + H_2(g)$; $\Delta S° = 2S°_{CF_4} + S°_{H_2} - [S°_{C_2H_2} + 4S°_{F_2}]$

$-358 \text{ J/K} = (2 \text{ mol}) S°_{CF_4} + 131 \text{ J/K} - [201 \text{ J/K} + 4(203 \text{ J/K})]$, $S°_{CF_4} = 262 \text{ J/K•mol}$

41. a. $S_{rhombic} \rightarrow S_{monoclinic}$; This phase transition is spontaneous ($\Delta G < 0$) at temperatures above 95°C. $\Delta G = \Delta H - T\Delta S$; For ΔG to be negative only above a certain temperature, then ΔH is positive and ΔS is positive (see Table 16.5 of text).

b. Because ΔS is positive, $S_{rhombic}$ is the more ordered crystalline structure.

43. a. When a bond is formed, energy is released so ΔH is negative. There are more reactant molecules of gas than product molecules of gas ($\Delta n < 0$), so ΔS will be negative.

b. $\Delta G = \Delta H - T\Delta S$; For this reaction to be spontaneous ($\Delta G < 0$), the favorable enthalpy term must dominate. The reaction will be spontaneous at low temperatures where the ΔH term dominates.

45 a.

	$CH_4(g)$	+ $2 O_2(g)$	$\rightarrow$ $CO_2(g)$	+ $2 H_2O(g)$	
$\Delta H_f°$	−75 kJ/mol	0	−393.5	−242	
$\Delta G_f°$	−51 kJ/mol	0	−394	−229	Data from Appendix 4
S°	186 J/K•mol	205	214	189	

$\Delta H° = \sum n_p \Delta H°_{f,\text{ products}} - \sum n_r \Delta H°_{f,\text{ reactants}}$; $\Delta S° = \sum n_p S°_{\text{products}} - \sum n_r S°_{\text{reactants}}$

$\Delta H° = 2\text{ mol}(-242\text{ kJ/mol}) + 1\text{ mol}(-393.5\text{ kJ/mol}) - [1\text{ mol}(-75\text{ kJ/mol})] = -803\text{ kJ}$

$\Delta S° = 2\text{ mol}(189\text{ J/K·mol}) + 1\text{ mol}(214\text{ J/K•mol})$

$- [1\text{ mol}(186\text{ J/K•mol}) + 2\text{ mol}(205\text{ J/K•mol})] = -4\text{ J/K}$

There are two ways to get $\Delta G°$. We can use $\Delta G° = \Delta H° - T\Delta S°$ (be careful of units):

$$\Delta G° = \Delta H° - T\Delta S° = -803 \times 10^3\text{ J} - (298\text{ K})(-4\text{ J/K}) = -8.018 \times 10^5\text{ J} = -802\text{ kJ}$$

or we can use ΔG_f° values where $\Delta G° = \sum n_p \Delta G^\circ_{f,\ products} - \sum n_r \Delta G^\circ_{f,\ reactants}$:

$\Delta G° = 2\text{ mol}(-229\text{ kJ/mol}) + 1\text{ mol}(-394\text{ kJ/mol}) - [1\text{ mol}(-51\text{ kJ/mol})]$

$\Delta G° = -801\text{ kJ}$ (Answers are the same within round off error.)

b. $6\ CO_2(g) + 6\ H_2O(l) \rightarrow C_6H_{12}O_6(s) + 6\ O_2(g)$

	$CO_2(g)$	$H_2O(l)$	$C_6H_{12}O_6(s)$	$O_2(g)$
ΔH_f°	−393.5 kJ/mol	−286	−1275	0
S°	214 J/K•mol	70.	212	205

$\Delta H° = -1275 - [6(-286) + 6(-393.5)] = 2802\text{ kJ}$

$\Delta S° = 6(205) + 212 - [6(214) + 6(70.)] = -262\text{ J/K}$

$\Delta G° = 2802\text{ kJ} - (298\text{ K})(-0.262\text{ kJ/K}) = 2880.\text{ kJ}$

c. $P_4O_{10}(s) + 6\ H_2O(l) \rightarrow 4\ H_3PO_4(s)$

	$P_4O_{10}(s)$	$H_2O(l)$	$H_3PO_4(s)$
ΔH_f° (kJ/mol)	−2984	−286	−1279
S° (J/K•mol)	229	70.	110.

$\Delta H° = 4\text{ mol}(-1279\text{ kJ/mol}) - [1\text{ mol}(-2984\text{ kJ/mol}) + 6\text{ mol}(-286\text{ kJ/mol})] = -416\text{ kJ}$

$\Delta S° = 4(110.) - [229 + 6(70.)] = -209\text{ J/K}$

$\Delta G° = \Delta H° - T\Delta S° = -416\text{ kJ} - (298\text{ K})(-0.209\text{ kJ/K}) = -354\text{ kJ}$

d.

	$HCl(g)$ +	$NH_3(g)$ →	$NH_4Cl(s)$
ΔH_f° (kJ/mol)	−92	−46	−314
S° (J/K•mol)	187	193	96

$\Delta H^\circ = -314 - [-92 - 46] = -176$ kJ; $\Delta S^\circ = 96 - [187 + 193] = -284$ J/K

$\Delta G^\circ = \Delta H^\circ - T\Delta S^\circ = -176 \text{ kJ} - (298 \text{ K})(-0.284 \text{ kJ/K}) = -91$ kJ

47. $\Delta G^\circ = -58.03 \text{ kJ} - (298 \text{ K})(-0.1766 \text{ kJ/K}) = -5.40$ kJ

$$\Delta G^\circ = 0 = \Delta H^\circ - T\Delta S^\circ,\ T = \frac{\Delta H^\circ}{\Delta S^\circ} = \frac{-58.03 \text{ kJ}}{-0.1766 \text{ kJ/K}} = 328.6 \text{ K}$$

ΔG° is negative below 328.6 K where the favorable ΔH° term dominates.

49. $CH_4(g) + CO_2(g) \rightarrow CH_3CO_2H(l)$

$\Delta H^\circ = -484 - [-75 + (-393.5)] = -16$ kJ; $\Delta S^\circ = 160 - [186 + 214] = -240.$ J/K

$\Delta G^\circ = \Delta H^\circ - T\Delta S^\circ = -16 \text{ kJ} - (298 \text{ K})(-0.240 \text{ kJ/K}) = 56$ kJ

This reaction is spontaneous only at temperatures below $T = \Delta H^\circ/\Delta S^\circ = 67$ K (where the favorable ΔH° term will dominate giving a negative ΔG° value). This is not practical. Substances will be in condensed phases and rates will be very slow at this extremely low temperature.

$CH_3OH(g) + CO(g) \rightarrow CH_3CO_2H(l)$

$\Delta H^\circ = -484 - [-110.5 + (-201)] = -173$ kJ; $\Delta S^\circ = 160 - [198 + 240.] = -278$ J/K

$\Delta G^\circ = -173 \text{ kJ} - (298 \text{ K})(-0.278 \text{ kJ/K}) = -90.$ kJ

This reaction also has a favorable enthalpy and an unfavorable entropy term. This reaction is spontaneous at temperatures below $T = \Delta H^\circ/\Delta S^\circ = 622$ K. The reaction of CH_3OH and CO will be preferred. It is spontaneous at high enough temperatures that the rates of reaction should be reasonable.

51.

Reaction	
$CH_4(g) \rightarrow 2\,H_2(g) + C(s)$	$\Delta G^\circ = -(-51 \text{ kJ})$
$2\,H_2(g) + O_2(g) \rightarrow 2\,H_2O(l)$	$\Delta G^\circ = -2(237 \text{ kJ})$
$C(s) + O_2(g) \rightarrow CO_2(g)$	$\Delta G^\circ = -394$ kJ
$CH_4(g) + 2\,O_2(g) \rightarrow 2\,H_2O(l) + CO_2(g)$	$\Delta G^\circ = -817$ kJ

53. $\Delta G° = \sum n_p \Delta G°_{f,\,products} - \sum n_r \Delta G°_{f,\,reactants}$, $-374\ kJ = -1105\ kJ - \Delta G°_{f,\,SF_4}$

$\Delta G°_{f,\,SF_4} = -731\ kJ/mol$

55. $\Delta G° = \sum n_p \Delta G°_{f,\,products} - \sum n_r \Delta G°_{f,\,reactants}$

$\Delta G° = [-57.37\ kJ + (-68.85\ kJ) + 3(-95.30\ kJ)] - [3(0) + 2(-50.72\ kJ)] = -310.68\ kJ$

For a temperature change from 25°C to ~20°C (room temperature), the magnitude of $\Delta G°$ will not change much. Therefore, $\Delta G°$ will be a negative value at room temperature (~20°C), so the reaction will be spontaneous.

Free Energy: Pressure Dependence and Equilibrium

57. $\Delta G = \Delta G° + RT \ln Q$; For this reaction: $\Delta G = \Delta G° + RT \ln \dfrac{P_{NO_2} \times P_{O_2}}{P_{NO} \times P_{O_3}}$

$\Delta G° = 1\ mol(52\ kJ/mol) + 1\ mol(0) - [1\ mol(87\ kJ/mol) + 1\ mol(163\ kJ/mol)] = -198\ kJ$

$$\Delta G = -198\ kJ + \frac{8.3145\ J/K \bullet mol}{1000\ J/kJ}(298\ K) \ln \frac{(1.00 \times 10^{-7})(1.00 \times 10^{-3})}{(1.00 \times 10^{-6})(1.00 \times 10^{-6})}$$

$\Delta G = -198\ kJ + 9.69\ kJ = -188\ kJ$

59. $\Delta G = \Delta G° + RT \ln Q = \Delta G° + RT \ln \dfrac{P_{N_2O_4}}{P^2_{NO_2}}$

$\Delta G° = 1\ mol(98\ kJ/mol) - 2\ mol(52\ kJ/mol) = -6\ kJ$

a. These are standard conditions so $\Delta G = \Delta G°$ because $Q = 1$ and $\ln Q = 0$. Because $\Delta G°$ is negative, the forward reaction is spontaneous. The reaction shifts right to reach equilibrium.

b. $\Delta G = -6 \times 10^3\ J + 8.3145\ J/K{\bullet}mol\ (298\ K) \ln \dfrac{0.50}{(0.21)^2}$

$\Delta G = -6 \times 10^3\ J + 6.0 \times 10^3\ J = 0$

Because $\Delta G = 0$, this reaction is at equilibrium (no shift).

c. $\Delta G = -6 \times 10^3\ J + 8.3145\ J/K{\bullet}mol\ (298\ K) \ln \dfrac{1.6}{(0.29)^2}$

$\Delta G = -6 \times 10^3\ J + 7.3 \times 10^3\ J = 1.3 \times 10^3\ J = 1 \times 10^3\ J$

Because ΔG is positive, the reverse reaction is spontaneous, and the reaction shifts to the left to reach equilibrium.

61. At 25.0°C: $\Delta G° = \Delta H° - T\Delta S° = -58.03 \times 10^3$ J/mol – (298.2 K)(–176.6 J/K•mol)

$= -5.37 \times 10^3$ J/mol

$$\Delta G° = -RT \ln K, \ \ln K = \frac{-\Delta G°}{RT} = \exp\left(\frac{-(-5.37\ 10^3\ \text{J/mol})}{(8.3145\ \text{J/K} \bullet \text{mol})(298.2\ \text{K})}\right) = 2.166$$

$K = e^{2.166} = 8.72$

At 100.0°C: $\Delta G° = -58.03 \times 10^3$ J/mol – (373.2 K)(–176.6 J/K•mol) = 7.88×10^3 J/mol

$$\ln K = \frac{-(7.88 \times 10^3\ \text{J/mol})}{(8.3145\ \text{J/K} \bullet \text{mol})(373.2\ \text{K})} = -2.540, \ K = e^{-2.540} = 0.0789$$

Note: When determining exponents, we will round off after the calculation is complete. This helps eliminate excessive round off error.

63. When reactions are added together, the equilibrium constants are multiplied together to determine the K value for the final reaction.

$H_2(g) + O_2(g) \rightleftharpoons H_2O_2(g)$	$K = 2.3 \times 10^6$
$H_2O(g) \rightleftharpoons H_2(g) + 1/2O_2(g)$	$K = (1.8 \times 10^{37})^{-1/2}$
$H_2O(g) + 1/2O_2(g) \rightleftharpoons H_2O_2(g)$	$K = 2.3 \times 10^6(1.8 \times 10^{-37})^{-1/2} = 5.4 \times 10^{-13}$

$\Delta G° = -RT \ln K = -8.3145$ J/K•mol (600. K) $\ln(5.4 \times 10^{-13}) = 1.4 \times 10^5$ J/mol = 140 kJ/mol

65. $$K = \frac{P_{NF_3}^2}{P_{N_2} \times P_{F_2}^3} = \frac{(0.48)^2}{0.021(0.063)^3} = 4.4 \times 10^4$$

$\Delta G°_{800} = -RT \ln K = -8.3145$ J/K•mol (800. K) $\ln(4.4 \times 10^4) = -7.1 \times 10^4$ J/mol =

–71 kJ/mol

67. The equation $\ln K = \frac{-\Delta H°}{R}\left(\frac{1}{T}\right) + \frac{\Delta S°}{R}$ is in the form of a straight line equation (y = mx + b). A graph of ln K vs. 1/T will yield a straight line with slope = m = $-\Delta H°/R$ and a y-intercept = b = $\Delta S°/R$.

From the plot:

$$\text{slope} = \frac{\Delta y}{\Delta x} = \frac{0 - 40.}{3.0 \times 10^{-3}\ \text{K}^{-1} - 0} = -1.3 \times 10^4\ \text{K}$$

-1.3×10^4 K = $-\Delta H°/R$, $\Delta H° = 1.3 \times 10^4$ K × 8.3145 J/K•mol = 1.1×10^5 J/mol

y-intercept = 40. = $\Delta S°/R$, $\Delta S°$ = 40. × 8.3145 J/K•mol = 330 J/K•mol

As seen here, when $\Delta H°$ is positive, the slope of the ln K vs. 1/T plot is negative. When $\Delta H°$ is negative as in an exothermic process, then the slope of the ln K vs. 1/T plot will be positive (slope = $-\Delta H°/R$).

Additional Exercises

69. From Appendix 4, S° = 198 J/K•mol for CO(g) and S° = 27 J/K•mol for Fe(s).

Let S_l° = S° for $Fe(CO)_5(l)$ and S_g° = S° for $Fe(CO)_5(g)$.

$$\Delta S° = -677 \text{ J/K} = 1 \text{ mol}(S_l^\circ) - [1 \text{ mol } (27 \text{ J/K•mol}) + 5 \text{ mol}(198 \text{ J/K•mol})]$$

$$S_l^\circ = 340. \text{ J/K•mol}$$

$$\Delta S° = 107 \text{ J/K} = 1 \text{ mol } (S_g^\circ) - 1 \text{ mol } (340. \text{ J/K•mol})$$

$$S_g^\circ = S° \text{ for } Fe(CO)_5(g) = 447 \text{ J/K•mol}$$

71. ΔS will be negative because 2 mol of gaseous reactants form 1 mol of gaseous product. For ΔG to be negative, ΔH must be negative (exothermic). For exothermic reactions, K decreases as T increases. Therefore, the ratio of the partial pressure of PCl_5 to the partial pressure of PCl_3 will decrease when T is raised.

73. solid I → solid II; Equilibrium occurs when $\Delta G = 0$.

$$\Delta G = \Delta H - T\Delta S, \ \Delta H = T\Delta S, \ T = \Delta H/\Delta S = \frac{-743.1 \text{ J/mol}}{-17.0 \text{ J/K•mol}} = 43.7 \text{ K} = -229.5°\text{C}$$

75.

$$HgbO_2 \rightarrow Hgb + O_2 \qquad \Delta G° = -(-70 \text{ kJ})$$
$$Hgb + CO \rightarrow HgbCO \qquad \Delta G° = -80 \text{ kJ}$$

$$HgbO_2 + CO \rightarrow HgbCO + O_2 \qquad \Delta G° = -10 \text{ kJ}$$

$$\Delta G° = -RT \ln K, \ K = \exp\left(\frac{-\Delta G°}{RT}\right) = \exp\left(\frac{-(-10 \times 10^3 \text{ J})}{8.3145 \text{ J/K•mol } (298 \text{ K})}\right) = 60$$

77. $HF(aq) \rightleftharpoons H^+(aq) + F^-(aq); \ \Delta G = \Delta G° + RT \ln\frac{[H^+][F^-]}{[HF]}$

$$\Delta G° = -RT \ln K = -(8.3145 \text{ J/K•mol})(298 \text{ K}) \ln (7.2 \times 10^{-4}) = 1.8 \times 10^4 \text{ J/mol}$$

a. The concentrations are all at standard conditions so $\Delta G = \Delta G = 1.8 \times 10^4$ J/mol (Q = 1.0 and ln Q = 0). Because $\Delta G°$ is positive, the reaction shifts left to reach equilibrium.

b. $\Delta G = 1.8 \times 10^4 \text{ J/mol} + (8.3145 \text{ J/K•mol})(298 \text{ K}) \ln \frac{(2.7 \times 10^{-2})^2}{0.98}$

$\Delta G = 1.8 \times 10^4 \text{ J/mol} - 1.8 \times 10^4 \text{ J/mol} = 0$

$\Delta G = 0$, so the reaction is at equilibrium (no shift).

c. $\Delta G = 1.8 \times 10^4 \text{ J/mol} + 8.3145 (298 \text{ K}) \ln \frac{(1.0 \times 10^{-5})^2}{1.0 \times 10^{-5}} = -1.1 \times 10^4$ J/mol; shifts right

d. $\Delta G = 1.8 \times 10^4 + 8.3145 (298) \ln \frac{7.2 \times 10^{-4} (0.27)}{0.27} = 1.8 \times 10^4 - 1.8 \times 10^4 = 0$; at equilibrium

e. $\Delta G = 1.8 \times 10^4 + 8.3145 (298) \ln \frac{1.0 \times 10^{-3} (0.67)}{0.52} = 2 \times 10^3$ J/mol; shifts left

79. a. $\Delta G° = -RT \ln K$, $K = \exp(-\Delta G°/RT) = \exp\left(\frac{-(-30{,}500 \text{ J})}{8.3145 \text{ J/K} \bullet \text{mol} \times 298 \text{ K}}\right) = 2.22 \times 10^5$

b. $C_6H_{12}O_6(s) + 6\ O_2(g) \rightarrow 6\ CO_2(g) + 6\ H_2O(l)$

$\Delta G° = 6 \text{ mol}(-394 \text{ kJ/mol}) + 6 \text{ mol}(-237 \text{ kJ/mol}) - 1 \text{ mol}(-911 \text{ kJ/mol}) = -2875 \text{ kJ}$

$\frac{2875 \text{ kJ}}{\text{mol glucose}} \times \frac{1 \text{ mol ATP}}{30.5 \text{ kJ}} = \frac{94.3 \text{ mol ATP}}{\text{mol glucose}}$; 94.3 molecules ATP/molecule glucose

This is an overstatement. The assumption that all of the free energy goes into this reaction is false. Actually only 38 moles of ATP are produced by metabolism of one mole of glucose.

c. From Exercise 17.78, $\Delta G = 8.8$ kJ in order to transport 1.0 mol K^+ from the blood to the muscle cells.

$8.8 \text{ kJ} \times \frac{1 \text{ mol ATP}}{30.5 \text{ kJ}} = 0.29 \text{ mol ATP}$

81. ΔS is more favorable for reaction two than for reaction one, resulting in $K_2 > K_1$. In reaction one, seven particles in solution are forming one particle. In reaction two, four particles form one particle which results in a smaller decrease in disorder than for reaction one.

83. $\Delta G° = -RT \ln K$; When $K = 1.00$, $\Delta G° = 0$ since $\ln 1.00 = 0$. $\Delta G° = 0 = \Delta H° - T\Delta S°$

$\Delta H° = 3(-242 \text{ kJ}) - [-826 \text{ kJ}] = 100. \text{ kJ}$; $\Delta S° = [2(27 \text{ J/K}) + 3(189 \text{ J/K})] - [90. \text{ J/K} + 3(131 \text{ J/K})] = 138 \text{ J/K}$

$\Delta H° = T\Delta S°$, $T = \frac{\Delta H°}{\Delta S°} = \frac{100. \text{ kJ}}{0.138 \text{ kJ/K}} = 725 \text{ K}$

Challenge Problems

85. $3\ O_2(g) \rightleftharpoons 2\ O_3(g)$; $\Delta H° = 2(143\ kJ) = 286\ kJ$; $\Delta G° = 2(163\ kJ) = 326\ kJ$

$$\ln K = \frac{-\Delta G°}{RT} = \frac{-326 \times 10^3\ J}{(8.3145\ J/K \bullet mol)(298\ K)} = -131.573,\ K = e^{-131.573} = 7.22 \times 10^{-58}$$

We need the value of K at 230. K. From Section 16.8 of the text:

$$\ln K = \frac{-\Delta G°}{RT} + \frac{\Delta S°}{R}$$

For two sets of K and T:

$$\ln K_1 = \frac{-\Delta H°}{R}\left(\frac{1}{T_1}\right) + \frac{\Delta S}{R};\ \ln K_2 = \frac{-\Delta H°}{R}\left(\frac{1}{T_2}\right) + \frac{\Delta S°}{R}$$

Subtracting the first expression from the second:

$$\ln K_2 - \ln K_1 = \frac{\Delta H°}{R}\left(\frac{1}{T_1} - \frac{1}{T_2}\right) \text{ or } \ln\frac{K_2}{K_1} = \frac{\Delta H°}{R}\left(\frac{1}{T_1} - \frac{1}{T_2}\right)$$

Let $K_2 = 7.22 \times 10^{-58}$, $T_2 = 298$ K; $K_1 = K_{230}$, $T_1 = 230.$ K; $\Delta H° = 286 \times 10^3$ J

$$\ln\frac{7.22 \times 10^{-58}}{K_{230}} = \frac{286 \times 10^3}{8.3145}\left(\frac{1}{230.} - \frac{1}{298}\right) = 34.13$$

$$\frac{7.22 \times 10^{-58}}{K_{230}} = e^{34.13} = 6.6 \times 10^{14},\ K_{230} = 1.1 \times 10^{-72}$$

$$K_{230} = 1.1 \times 10^{-72} = \frac{P_{O_3}^2}{P_{O_2}^3} = \frac{P_{O_3}^2}{(1.0 \times 10^{-3}\ atm)^3},\ P_{O_3} = 3.3 \times 10^{-41}\ atm$$

The volume occupied by one molecule of ozone is:

$$V = \frac{nRT}{P} = \frac{(1/6.022 \times 10^{23}\ mol)(0.08206\ L\ atm/mol \bullet K)(230.\ K)}{(3.3 \times 10^{-41}\ atm)} = 9.5 \times 10^{17}\ L$$

Equilibrium is probably not maintained under these conditions. When only two ozone molecules are in a volume of 9.5×10^{17} L, the reaction is not at equilibrium. Under these conditions, Q > K and the reaction shifts left. But with only 2 ozone molecules in this huge volume, it is extremely unlikely that they will collide with each other. At these conditions, the concentration of ozone is not large enough to maintain equilibrium.

87. a. From the plot, the activation energy of the reverse reaction is $E_a + (-\Delta G°) = E_a - \Delta G°$ ($\Delta G°$ is a negative number as drawn in the diagram).

$$k_f = A\exp\left(\frac{-E_a}{RT}\right) \text{ and } k_r = A\exp\left(\frac{-(E_a - \Delta G°)}{RT}\right), \quad \frac{k_f}{k_r} = \frac{A\exp\left(\frac{-E_a}{RT}\right)}{A\exp\left(\frac{-(E_a - \Delta G°)}{RT}\right)}$$

If the A factors are equal: $\frac{k_f}{k_r} = \exp\left(\frac{-E_a}{RT} + \frac{(E_a - \Delta G°)}{RT}\right) = \exp\left(\frac{-\Delta G°}{RT}\right)$

From $\Delta G° = -RT \ln K$, $K = \exp\left(\frac{-\Delta G°}{RT}\right)$; Because K and $\frac{k_f}{k_r}$ are both equal to the same expression, $K = \frac{k_f}{k_r}$.

b. A catalyst will lower the activation energy for both the forward and reverse reaction (but not change $\Delta G°$). Therefore, a catalyst must increase the rate of both the forward and reverse reactions.

89. a. $\Delta G° = G_B^{\circ} - G_A^{\circ} = 11{,}718 - 8996 = 2722$ J

$$K = \exp\left(\frac{-\Delta G°}{RT}\right) = \exp\left(\frac{-2722\ \text{J}}{(8.3145\ \text{J/K} \bullet \text{mol})(298\ \text{K})}\right) = 0.333$$

b. When Q = 1.00 > K, the reaction shifts left. Let x = atm of B(g) which reacts to reach equilibrium.

	A(g) ⇌	B(g)
Initial	1.00 atm	1.00 atm
Equil.	$1.00 + x$	$1.00 - x$

$$K = \frac{P_B}{P_A} = \frac{1.00 - x}{1.00 + x} = 0.333, \quad 1.00 - x = 0.333 + 0.333\,x, \quad x = 0.50 \text{ atm}$$

$P_B = 1.00 - 0.50 = 0.50$ atm; $P_A = 1.00 + 0.50 = 1.50$ atm

c. $\Delta G = \Delta G° + RT \ln Q = \Delta G° + RT \ln (P_B/P_A)$

$\Delta G = 2722\ \text{J} + (8.3145)(298) \ln (0.50/1.50) = 2722\ \text{J} - 2722\ \text{J} = 0$ (carrying extra sig. figs.)

91. $K = P_{CO_2}$; To insure Ag_2CO_3 from decomposing, P_{CO_2} should be greater than K.

From Exercise 16.67, $\ln K = \frac{\Delta H^o}{RT} + \frac{\Delta S^o}{R}$. For two conditions of K and T, the equation is:

$$\ln\frac{K_2}{K_1} = \frac{\Delta H^o}{R}\left(\frac{1}{T_1} + \frac{1}{T_2}\right)$$

Let $T_1 = 25°C = 298$ K, $K_1 = 6.23 \times 10^{-3}$ torr; $T_2 = 110.°C = 383$ K, $K_2 = ?$

$$\ln\frac{K_2}{6.23 \times 10^{-3}\ \text{torr}} = \frac{79.14 \times 10^3\ \text{J/mol}}{8.3145\ \text{J/K}\bullet\text{mol}}\left(\frac{1}{298\ \text{K}} - \frac{1}{383\ \text{K}}\right)$$

$$\ln\frac{K_2}{6.23 \times 10^{-3}} = 7.1,\quad \frac{K_2}{6.23 \times 10^{-3}} = e^{7.1} = 1.2 \times 10^3,\quad K_2 = 7.5\ \text{torr}$$

To prevent decomposition of Ag_2CO_3, the partial pressure of CO_2 should be greater than 7.5 torr.

93. Use the thermodynamic data to calculate the boiling point of the solvent.

At boiling point: $\Delta G = 0 = \Delta H - T\Delta S$, $T = \frac{\Delta H}{\Delta S} = \frac{33.90 \times 10^3\ \text{J/mol}}{95.95\ \text{J/K}\bullet\text{mol}} = 353.3$ K

$\Delta T = K_b m$, (355.4 K − 353.3 K) = 2.5 K kg/mol (m), $m = \frac{2.1}{2.5} = 0.84$ mol/kg

$$\text{mass solvent} = 150.\ \text{mL} \times \frac{0.879\ \text{g}}{\text{mL}} \times \frac{1\ \text{kg}}{1000\ \text{g}} = 0.132\ \text{kg}$$

$$\text{mass solute} = 0.132\ \text{kg solvent} \times \frac{0.84\ \text{mol solute}}{\text{kg solvent}} \times \frac{142\ \text{g}}{\text{mol}} = 15.7\ \text{g} = 16\ \text{g solute}$$

95. $HX \rightleftharpoons H^+ + X^-$ $\quad K_a = \frac{[H^+][X^-]}{[HX]}$

	HX	H^+	X^-
Initial	0.10 M	~0	0
Equil.	$0.10 - x$	x	x

From problem, $x = [H^+] = 10^{-5.83} = 1.5 \times 10^{-6}$; $K_a = \frac{(1.5 \times 10^{-6})^2}{0.10 - 1.5 \times 10^{-6}} = 2.3 \times 10^{-11}$

$\Delta G° = -RT \ln K = -8.3145$ J/K•mol (298 K) $\ln(2.3 \times 10^{-11}) = 6.1 \times 10^4$ J/mol = 61 kJ/mol

Integrative Problems

97. Because the partial pressure of C(g) decreased, the net change that occurs for this reaction to reach equilibrium is for products to convert to reactants.

	A(g)	+ 2 B(g)	$\rightleftharpoons$	C(g)
Initial	0.100 atm	0.100 atm		0.100 atm
Change	$+x$	$+2x$	$\leftarrow$	$-x$
Equil.	$0.100 + x$	$0.100 + 2x$		$0.100 - x$

From the problem, $P_C = 0.040$ atm $= 0.100 - x$, $x = 0.060$ atm

The equilibrium partial pressures are: $P_A = 0.100 + x = 0.100 + 0.060 = 0.160$ atm, $P_B = 0.100 + 2((0.60) = 0.220$ atm, and $P_C = 0.040$ atm

$$K = \frac{0.040}{0.160(0.220)^2} = 5.2$$

$\Delta G° = -RT \ln K = -8.3145$ J/K•mol (298 K) ln 5.2 $= -4.1 \times 10^3$ J/mol $= -4.1$ kJ/mol

CHAPTER SEVENTEEN

ELECTROCHEMISTRY

Review of Oxidation - Reduction Reactions

13. Oxidation: increase in oxidation number; loss of electrons

    Reduction: decrease in oxidation number; gain of electrons

15. The species oxidized shows an increase in oxidation numbers and is called the reducing agent. The species reduced shows a decrease in oxidation numbers and is called the oxidizing agent. The pertinent oxidation numbers are listed by the substance oxidized and the substance reduced.

	Redox?	Ox. Agent	Red. Agent	Substance Oxidized	Substance Reduced
a.	Yes	H_2O	CH_4	CH_4 (C, $-4 \rightarrow +2$)	H_2O (H, $+1 \rightarrow 0$)
b.	Yes	$AgNO_3$	Cu	Cu ($0 \rightarrow +2$)	$AgNO_3$ (Ag, $+1 \rightarrow 0$)
c.	Yes	HCl	Zn	Zn ($0 \rightarrow +2$)	HCl (H, $+1 \rightarrow 0$)

    d. No; There is no change in any of the oxidation numbers.

Questions

17. Magnesium is an alkaline earth metal; Mg will oxidize to Mg^{2+}. The oxidation state of hydrogen in HCl is +1. To be reduced, the oxidation state of H must decrease. The obvious choice for the hydrogen product is $H_2(g)$ where hydrogen has a zero oxidation state. The balanced reaction is: $Mg(s) + 2HCl(aq) \rightarrow MgCl_2(aq) + H_2(g)$. Mg goes from the 0 to the +2 oxidation state by losing two electrons. Each H atom goes from the +1 to the 0 oxidation state by gaining one electron. Since there are two H atoms in the balanced equation, then a total of two electrons are gained by the H atoms. Hence, two electrons are transferred in the balanced reaction. When the electrons are transferred directly from Mg to H^+, no work is obtained. In order to harness this reaction to do useful work, we must control the flow of electrons through a wire. This is accomplished by making a galvanic cell which separates the reduction reaction from the oxidation reaction in order to control the flow of electrons through a wire to produce a voltage.

19. An extensive property is one that depends directly on the amount of substance. The free energy change for a reaction depends on whether 1 mol of product is produced or 2 mol of product is produced or 1 million mol of product is produced. This is not the case for cell potentials which do not depend on the amount of substance. The equation that relates ΔG to E is $\Delta G = -nFE$. It is the n term that converts the intensive property E into the extensive property ΔG. n is the number of moles of electrons transferred in the balanced reaction that ΔG is associated with.

21. A potential hazard when jump starting a car is the possibility for the electrolysis of $H_2O(l)$ to occur. When $H_2O(l)$ is electrolyzed, the products are the explosive gas mixture of $H_2(g)$ and $O_2(g)$. A spark produced during jump starting a car could ignite any $H_2(g)$ and $O_2(g)$ produced. Grounding the jumper cable far from the battery minimizes the risk of a spark nearby the battery where $H_2(g)$ and $O_2(g)$ could be collecting.

23. You need to know the identity of the metal so you know which molar mass to use. You need to know the oxidation state of the metal ion in the salt so the mol of electrons transferred can be determined. And finally, you need to know the amount of current and the time the current was passed through the electrolytic cell. If you know these four quantities, then the mass of metal plated out can be calculated.

Exercises

Galvanic Cells, Cell Potentials, Standard Reduction Potentials, and Free Energy

25. A typical galvanic cell diagram is:

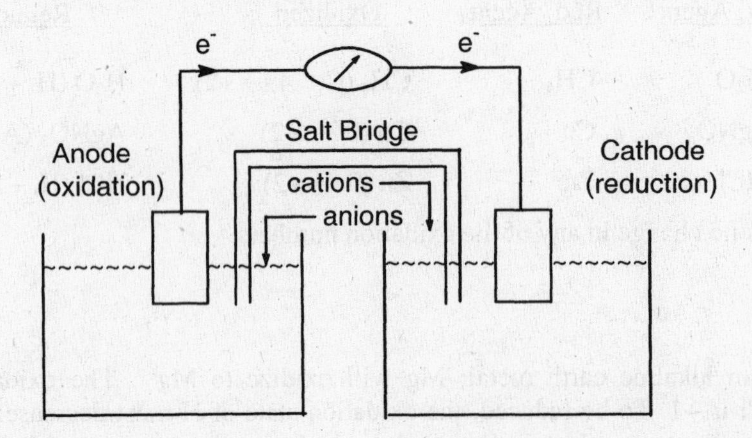

The diagram for all cells will look like this. The contents of each half-cell compartment will be identified for each reaction, with all solute concentrations at 1.0 *M* and all gases at 1.0 atm. For Exercises 17.25 and 17.26, the flow of ions through the salt bridge was not asked for in the questions. If asked, however, cations always flow into the cathode compartment, and anions always flow into the anode compartment. This is required to keep each compartment electrically neutral.

a. Table 17.1 of the text lists balanced reduction half-reactions for many substances. For this overall reaction, we need the Cl_2 to Cl^- reduction half-reaction and the Cr^{3+} to $Cr_2O_7^{2-}$ oxidation half-reaction. Manipulating these two half-reactions gives the overall balanced equation.

$$(Cl_2 + 2\,e^- \rightarrow 2\,Cl^-) \times 3$$
$$7\,H_2O + 2\,Cr^{3+} \rightarrow Cr_2O_7^{2-} + 14\,H^+ + 6\,e^-$$

$$7\,H_2O(l) + 2\,Cr^{3+}(aq) + 3\,Cl_2(g) \rightarrow Cr_2O_7^{2-}(aq) + 6\,Cl^-(aq) + 14\,H^+(aq)$$

The contents of each compartment are:

Cathode: Pt electrode; Cl_2 bubbled into solution, Cl^- in solution

Anode: Pt electrode; Cr^{3+}, H^+, and $Cr_2O_7^{2-}$ in solution

We need a nonreactive metal to use as the electrode in each case, since all the reactants and products are in solution. Pt is a common choice. Another possibility is graphite.

b. $Cu^{2+} + 2\,e^- \rightarrow Cu$

$Mg \rightarrow Mg^{2+} + 2e^-$

$$Cu^{2+}(aq) + Mg(s) \rightarrow Cu(s) + Mg^{2+}(aq)$$

Cathode: Cu electrode; Cu^{2+} in solution; Anode: Mg electrode; Mg^{2+} in solution

27. To determine E° for the overall cell reaction, we must add the standard reduction potential to the standard oxidation potential ($E^o_{cell} = E^o_{red} + E^o_{ox}$). Reference Table 17.1 for values of standard reduction potentials. Remember that $E^o_{ox} = -E^o_{red}$ and that standard potentials are <u>not</u> multiplied by the integer used to obtain the overall balanced equation.

25a. $E^o_{cell} = E^o_{Cl_2 \rightarrow Cl^-} + E^o_{Cr^{3+} \rightarrow Cr_2O_7^{2-}} = 1.36\text{ V} + (-1.33\text{ V}) = 0.03\text{ V}$

25b. $E^o_{cell} = E^o_{Cu^{2+} \rightarrow Cu} + E^o_{Mg \rightarrow Mg^{2+}} = 0.34\text{ V} + 2.37\text{ V} = 2.71\text{ V}$

29. Reference Exercise 17.25 for a typical galvanic cell design. The contents of each half-cell compartment are identified below with all solute concentrations at 1.0 *M* and all gases at 1.0 atm. For each pair of half-reactions, the half-reaction with the largest (most positive) standard reduction potential will be the cathode reaction, and the half-reaction with the smallest (most negative) reduction potential will be reversed to become the anode reaction. Only this combination gives a spontaneous overall reaction, i.e., a reaction with a positive overall standard cell potential. Note that in a galvanic cell as illustrated in Exercise 17.25, the cations in the salt bridge migrate to the cathode, and the anions migrate to the anode.

a. $Cl_2 + 2\,e^- \rightarrow 2\,Cl^-$ $\quad E° = 1.36\text{ V}$

$2\,Br^- \rightarrow Br_2 + 2\,e^-$ $\quad -E° = -1.09\text{ V}$

$$Cl_2(g) + 2\,Br^-(aq) \rightarrow Br_2(aq) + 2\,Cl^-(aq) \qquad E^o_{cell} = 0.27\text{ V}$$

The contents of each compartment are:

Cathode: Pt electrode; $Cl_2(g)$ bubbled in, Cl^- in solution

Anode: Pt electrode; Br_2 and Br^- in solution

b. $(2\ e^- + 2\ H^+ + IO_4^- \rightarrow IO_3^- + H_2O) \times 5$ $E° = 1.60\ V$

$(4\ H_2O + Mn^{2+} \rightarrow MnO_4^- + 8\ H^+ + 5\ e^-) \times 2$ $-E° = -1.51\ V$

$10\ H^+ + 5\ IO_4^- + 8\ H_2O + 2\ Mn^{2+} \rightarrow 5\ IO_3^- + 5\ H_2O + 2\ MnO_4^- + 16\ H^+$ $E°_{cell} = 0.09\ V$

This simplifies to:

$H_2O(l) + 5\ IO_4^-(aq) + 2\ Mn^{2+}(aq) \rightarrow 5\ IO_3^-(aq) + 2\ MnO_4^-(aq) + 6\ H^+(aq)$

$E°_{cell} = 0.09\ V$

Cathode: Pt electrode; IO_4^-, IO_3^-, and H_2SO_4 (as a source of H^+) in solution

Anode: Pt electrode; Mn^{2+}, MnO_4^- and H_2SO_4 in solution

31. In standard line notation, the anode is listed first and the cathode is listed last. A double line separates the two compartments. By convention, the electrodes are on the ends with all solutes and gases towards the middle. A single line is used to indicate a phase change. We also included all concentrations.

25a. Pt | Cr^{3+} (1.0 *M*), $Cr_2O_7^{2-}$ (1.0 *M*), H^+ (1.0 *M*) || Cl_2 (1.0 atm) | Cl^- (1.0 *M*) | Pt

25b. Mg | Mg^{2+} (1.0 *M*) || Cu^{2+} (1.0 *M*) | Cu

29a. Pt | Br^- (1.0 *M*), Br_2 (1.0 *M*) || Cl_2 (1.0 atm) | Cl^- (1.0 *M*) | Pt

29b. Pt | Mn^{2+} (1.0 *M*), MnO_4^- (1.0 *M*), H^+ (1.0 *M*) || IO_4^- (1.0 *M*), H^+ (1.0 *M*), IO_3^- (1.0 *M*) | Pt

33. Locate the pertinent half-reactions in Table 17.1, and then figure which combination will give a positive standard cell potential. In all cases, the anode compartment contains the species with the smallest standard reduction potential. For part a, the copper compartment is the anode, and in part b, the cadmium compartment is the anode.

a. $Au^{3+} + 3\ e^- \rightarrow Au$ $E° = 1.50\ V$

$(Cu^+ \rightarrow Cu^{2+} + e^-) \times 3$ $-E° = -0.16\ V$

$Au^{3+}(aq) + 3\ Cu^+(aq) \rightarrow Au(s) + 3\ Cu^{2+}(aq)$ $E°_{cell} = 1.34\ V$

b. $(VO_2^+ + 2\ H^+ + e^- \rightarrow VO^{2+} + H_2O) \times 2$ $E° = 1.00\ V$

$Cd \rightarrow Cd^{2+} + 2e^-$ $-E° = 0.40\ V$

$2\ VO_2^+(aq) + 4\ H^+(aq) + Cd(s) \rightarrow 2\ VO^{2+}(aq) + 2\ H_2O(l) + Cd^{2+}(aq)$ $E°_{cell} = 1.40\ V$

35. a. $(5\ e^- + 8\ H^+ + MnO_4^- \rightarrow Mn^{2+} + 4\ H_2O) \times 2$ $E° = 1.51\ V$

$(2\ I^- \rightarrow I_2 + 2\ e^-) \times 5$ $-E° = -0.54\ V$

$16\ H^+(aq) + 2\ MnO_4^-(aq) + 10\ I^-(aq) \rightarrow 5\ I_2(aq) + 2\ Mn^{2+}(aq) + 8\ H_2O(l)$ $E°_{cell} = 0.97\ V$

This reaction is spontaneous at standard conditions because $E°_{cell} > 0$.

b. $(5\ e^- + 8\ H^+ + MnO_4^- \rightarrow Mn^{2+} + 4\ H_2O) \times 2$ $E° = 1.51$ V

$(2\ F^- \rightarrow F_2 + 2\ e^-) \times 5$ $-E° = -2.87$ V

$$16\ H^+(aq) + 2\ MnO_4^-(aq) + 10\ F^-(aq) \rightarrow 5\ F_2(aq) + 2\ Mn^{2+}(aq) + 8\ H_2O(l) \quad E^o_{cell} = -1.36\ V$$

This reaction is not spontaneous at standard conditions because $E^o_{cell} < 0$.

37. $Cl_2 + 2\ e^- \rightarrow 2\ Cl^-$ $E° = 1.36$ V

$(ClO_2^- \rightarrow ClO_2 + e^-) \times 2$ $-E° = -0.954$ V

$$2\ ClO_2^-(aq) + Cl_2(g) \rightarrow 2\ ClO_2(aq) + 2\ Cl^-(aq) \quad E^o_{cell} = 0.41\ V = 0.41\ J/C$$

$$\Delta G° = -nFE^o_{cell} = -(2\ mol\ e^-)(96{,}485\ C/mol\ e^-)(0.41\ J/C) = -7.9 \times 10^4\ J = -79\ kJ$$

39. Because the cells are at standard conditions, $w_{max} = \Delta G = \Delta G° = -nFE^o_{cell}$. See Exercise 17.33 for the balanced overall equations and for E^o_{cell}.

33a. $w_{max} = -(3\ mol\ e^-)(96{,}485\ C/mol\ e^-)(1.34\ J/C) = -3.88 \times 10^5\ J = -388\ kJ$

33b. $w_{max} = -(2\ mol\ e^-)(96{,}485\ C/mol\ e^-)(1.40\ J/C) = -2.70 \times 10^5\ J = -270.\ kJ$

41. $CH_3OH(l) + 3/2\ O_2(g) \rightarrow CO_2(g) + 2\ H_2O(l)$ $\Delta G° = 2(-237) + (-394) - [-166] = -702$ kJ

The balanced half-reactions are:

$H_2O + CH_3OH \rightarrow CO_2 + 6\ H^+ + 6\ e^-$ and $O_2 + 4\ H^+ + 4\ e^- \rightarrow 2\ H_2O$

For 3/2 mol O_2, 6 moles of electrons will be transferred (n = 6).

$$\Delta G° = -nFE°,\ E° = \frac{-\Delta G°}{nF} = \frac{-(-702{,}000\ J)}{(6\ mole\ e^-)(96{,}485\ C/mol\ e^-)} = 1.21\ J/C = 1.21\ V$$

43. Good oxidizing agents are easily reduced. Oxidizing agents are on the left side of the reduction half-reactions listed in Table 17.1. We look for the largest, most positive standard reduction potentials to correspond to the best oxidizing agents. The ordering from worst to best oxidizing agents is:

	K^+	<	H_2O	<	Cd^{2+}	<	I_2	<	$AuCl_4^-$	<	IO_3^-
E°(V)	−2.87		−0.83		−0.40		0.54		0.99		1.20

45. a. $2\ H^+ + 2\ e^- \rightarrow H_2$ $E° = 0.00$ V; $Cu \rightarrow Cu^{2+} + 2\ e^-$ $-E° = -0.34$ V

$E^o_{cell} = -0.34$ V; No, H^+ cannot oxidize Cu to Cu^{2+} at standard conditions ($E^o_{cell} < 0$).

b. $Fe^{3+} + e^- \rightarrow Fe^{2+}$ $E° = 0.77$ V; $2\ I^- \rightarrow I_2 + 2\ e^-$ $-E° = -0.54$ V

$E^o_{cell} = 0.77 - 0.54 = 0.23$ V; Yes, Fe^{3+} can oxidize I^- to I_2.

c. $H_2 \rightarrow 2\,H^+ + 2\,e^-$ $-E° = 0.00$ V; $Ag^+ + e^- \rightarrow Ag$ $E° = 0.80$ V

$E°_{cell} = 0.80$ V; Yes, H_2 can reduce Ag^+ to Ag at standard conditions ($E°_{cell} > 0$).

d. $Fe^{2+} \rightarrow Fe^{3+} + e^-$ $-E° = -0.77$ V; $Cr^{3+} + e^- \rightarrow Cr^{2+}$ $E° = -0.50$ V

$E°_{cell} = -0.50 - 0.77 = -1.27$ V; No, Fe^{2+} cannot reduce Cr^{3+} to Cr^{2+} at standard conditions.

47. a. $2\,Br^- \rightarrow Br_2 + 2\,e^-$ $-E° = -1.09$ V; $2\,Cl^- \rightarrow Cl_2 + 2\,e^-$ $-E° = -1.36$ V; $E° > 1.09$ V to oxidize Br^-; $E° < 1.36$ V to not oxidize Cl^-; $Cr_2O_7^{2-}$, O_2, MnO_2, and IO_3^- are all possible since when all of these oxidizing agents are coupled with Br^-, $E°_{cell} > 0$, and when coupled with Cl^-, $E°_{cell} < 0$ (assuming standard conditions).

b. $Mn \rightarrow Mn^{2+} + 2\,e^-$ $-E° = 1.18$; $Ni \rightarrow Ni^{2+} + 2\,e^-$ $-E° = 0.23$ V; Any oxidizing agent with $-0.23\text{ V} > E° > -1.18$ V will work. $PbSO_4$, Cd^{2+}, Fe^{2+}, Cr^{3+}, Zn^{2+} and H_2O will be able to oxidize Mn but not Ni (assuming standard conditions).

49.

$$ClO^- + H_2O + 2\,e^- \rightarrow 2\,OH^- + Cl^- \qquad E° = 0.90\text{ V}$$
$$2\,NH_3 + 2\,OH^- \rightarrow N_2H_4 + 2\,H_2O + 2\,e^- \qquad -E° = 0.10\text{ V}$$

$$ClO^-(aq) + 2\,NH_3(aq) \rightarrow Cl^-(aq) + N_2H_4(aq) + H_2O(l) \qquad E°_{cell} = 1.00\text{ V}$$

Because $E°_{cell}$ is positive for this reaction, at standard conditions ClO^- can spontaneously oxidize NH_3 to the somewhat toxic N_2H_4.

The Nernst Equation

51.

$$H_2O_2 + 2\,H^+ + 2\,e^- \rightarrow 2\,H_2O \qquad E° = 1.78\text{ V}$$
$$(Ag \rightarrow Ag^+ + e^-) \times 2 \qquad -E° = -0.80\text{ V}$$

$$H_2O_2(aq) + 2\,H^+(aq) + 2\,Ag(s) \rightarrow 2\,H_2O(l) + 2\,Ag^+(aq) \qquad E°_{cell} = 0.98\text{ V}$$

a. A galvanic cell is based on spontaneous redox reactions. At standard conditions, this reaction produces a voltage of 0.98 V. Any change in concentration that increases the tendency of the forward reaction to occur will increase the cell potential. Conversely, any change in concentration that decreases the tendency of the forward reaction to occur (increases the tendency of the reverse reaction to occur) will decrease the cell potential. Using Le Chatelier's principle, increasing the reactant concentrations of H_2O_2 and H^+ from 1.0 *M* to 2.0 *M* will drive the forward reaction further to right (will further increase the tendency of the forward reaction to occur). Therefore, E_{cell} will be greater than $E°_{cell}$.

b. Here, we decreased the reactant concentration of H^+ and increased the product concentration of Ag^+ from the standard conditions. This decreases the tendency of the forward reaction to occur which will decrease E_{cell} as compared to $E°_{cell}$ ($E_{cell} < E°_{cell}$).

53. For concentration cells, the driving force for the reaction is the difference in ion concentrations between the anode and cathode. In order to equalize the ion concentrations, the anode always has the smaller ion concentration. The general setup for this concentration cell is:

Cathode: $Ag^+(x\ M) + e^- \to Ag$ $E° = 0.80$ V
Anode: $Ag \to Ag^+\ (y\ M) + e^-$ $-E° = -0.80$ V

$Ag^+(\text{cathode}, x\ M) \to Ag^+\ (\text{anode}, y\ M)$ $E^\circ_{cell} = 0.00$ V

$$E_{cell} = E^\circ_{cell} - \frac{0.0591}{n}\log Q = \frac{-0.0591}{1}\log\frac{[Ag^+]_{anode}}{[Ag^+]_{cathode}}$$

For each concentration cell, we will calculate the cell potential using the above equation. Remember that the anode always has the smaller ion concentration.

a. Both compartments are at standard conditions ($[Ag^+] = 1.0\ M$), so $E_{cell} = E^\circ_{cell} = 0$ V. No voltage is produced since no reaction occurs. Concentration cells only produce a voltage when the ion concentrations are not equal.

b. Cathode = 2.0 *M* Ag^+; Anode = 1.0 *M* Ag^+; Electron flow is always from the anode to the cathode, so electrons flow to the right in the diagram.

$$E_{cell} = \frac{-0.0591}{n}\log\frac{[Ag^+]_{anode}}{[Ag^+]_{cathode}} = \frac{-0.0591}{1}\log\frac{1.0}{2.0} = 0.018\ V$$

c. Cathode = 1.0 *M* Ag^+; Anode = 0.10 *M* Ag^+; Electrons flow to the left in the diagram.

$$E_{cell} = \frac{-0.0591}{n}\log\frac{[Ag^+]_{anode}}{[Ag^+]_{cathode}} = \frac{-0.0591}{1}\log\frac{0.10}{1.0} = 0.059\ V$$

d. Cathode = 1.0 *M* Ag^+; Anode = 4.0×10^{-5} *M* Ag^+; Electrons flow to the left in the diagram.

$$E_{cell} = \frac{-0.0591}{n}\log\frac{4.0\times10^{-5}}{1.0} = 0.26\ V$$

e. The ion concentrations are the same, thus $\log([Ag^+]_{anode}/[Ag^+]_{cathode}) = \log(1.0) = 0$ and $E_{cell} = 0$. No electron flow occurs.

55. n = 2 for this reaction (lead goes from $Pb \to Pb^{2+}$ in $PbSO_4$)

$$E = E° - \frac{-0.0591}{2}\log\frac{1}{[H^+]^2[HSO_4^-]^2} = 2.04\ V - \frac{-0.0591}{2}\log\frac{1}{(4.5)^2(4.5)^2}$$

$E = 2.04\ V + 0.077\ V = 2.12\ V$

57.
$$Cu^{2+} + 2e^- \rightarrow Cu \quad E° = 0.34\ V$$
$$Zn \rightarrow Zn^{2+} + 2e^- \quad -E° = 0.76\ V$$
$$Cu^{2+}(aq) + Zn(s) \rightarrow Zn^{2+}(aq) + Cu(s) \quad E^\circ_{cell} = 1.10\ V$$

Because Zn^{2+} is a product in the reaction, the Zn^{2+} concentration increases from 1.00 *M* to 1.20 *M*. This means that the reactant concentration of Cu^{2+} must decrease from 1.00 *M* to 0.80 *M* (from the 1:1 mol ratio in the balanced reaction).

$$E_{cell} = E^\circ_{cell} - \frac{0.0591}{n}\log Q = 1.10\ V - \frac{0.0591}{2}\log\frac{[Zn^{2+}]}{[Cu^{2+}]}$$

$$E_{cell} = 1.10\ V - \frac{0.0591}{2}\log\frac{1.20}{0.80} = 1.10\ V - 0.0052\ V = 1.09\ V$$

59. $Cu^{2+}(aq) + H_2(g) \rightarrow 2\,H^+(aq) + Cu(s)\quad E^\circ_{cell} = 0.34\ V - 0.00\ V = 0.34\ V$; n = 2 mol electrons

$P_{H_2} = 1.0$ atm and $[H^+] = 1.0\ M$: $E_{cell} = E^\circ_{cell} - \frac{0.0591}{n}\log\frac{1}{[Cu^{2+}]}$

a. $E_{cell} = 0.34\ V - \frac{0.0591}{2}\log\frac{1}{2.5\times10^{-4}} = 0.34\ V - 0.11V = 0.23\ V$

b. $0.195\ V = 0.34\ V - \frac{0.0591}{2}\log\frac{1}{[Cu^{2+}]}$, $\log\frac{1}{[Cu^{2+}]} = 4.91$, $[Cu^{2+}] = 10^{-4.91}$

$= 1.2\times10^{-5}\ M$

Note: When determining exponents, we will carry extra significant figures.

61. $Cu^{2+}(aq) + H_2(g) \rightarrow 2\,H^+(aq) + Cu(s)\quad E^\circ_{cell} = 0.34\ V - 0.00\ V = 0.34\ V$; n = 2

$P_{H_2} = 1.0$ atm and $[H^+] = 1.0\ M$: $E_{cell} = E^\circ_{cell} - \frac{0.0591}{2}\log\frac{1}{[Cu^{2+}]}$

Use the K_{sp} expression to calculate the Cu^{2+} concentration in the cell.

$Cu(OH)_2(s) \rightleftharpoons Cu^{2+}(aq) + 2\,OH^-(aq)\quad K_{sp} = 1.6\times10^{-19} = [Cu^{2+}][OH^-]^2$

From problem, $[OH^-] = 0.10\ M$, so: $[Cu^{2+}] = \frac{1.6\times10^{-19}}{(0.10)^2} = 1.6\times10^{-17}\ M$

$$E_{cell} = E^\circ_{cell} - \frac{0.0591}{2}\log\frac{1}{[Cu^{2+}]} = 0.34\ V - \frac{0.0591}{2}\log\frac{1}{1.6\times10^{-17}} = 0.34 - 0.50 = -0.16\ V$$

Because $E_{cell} < 0$, the forward reaction is not spontaneous, but the reverse reaction is spontaneous. The Cu electrode becomes the anode and $E_{cell} = 0.16$ V for the reverse reaction. The cell reaction is: $2\,H^+(aq) + Cu(s) \rightarrow Cu^{2+}(aq) + H_2(g)$.

63. Cathode: $M^{2+} + 2e^- \rightarrow M(s)$ $E° = -0.31$ V
Anode: $M(s) \rightarrow M^{2+} + 2e^-$ $-E° = 0.31$ V

M^{2+} (cathode) $\rightarrow M^{2+}$ (anode) $E^o_{cell} = 0.00$ V

$$E_{cell} = 0.44\text{ V} = 0.00\text{ V} - \frac{0.0591}{2}\log\frac{[M^{2+}]_{anode}}{[M^{2+}]_{cathode}},\quad 0.44 = -\frac{0.0591}{2}\log\frac{[M^{2+}]_{anode}}{1.0}$$

$$\log[M^{2+}]_{anode} = -\frac{2(0.44)}{0.0591} = -14.89,\ [M^{2+}]_{anode} = 1.3 \times 10^{-15}\ M$$

Because we started with equal numbers of moles of SO_4^{2-} and M^{2+}, $[M^{2+}] = [SO_4^{2-}]$ at equilibrium.

$$K_{sp} = [M^{2+}][SO_4^{2-}] = (1.3 \times 10^{-15})^2 = 1.7 \times 10^{-30}$$

65. See Exercises 17.25, 17.27, and 17.29 for balanced reactions and standard cell potentials. Balanced reactions are necessary to determine n, the moles of electrons transferred.

25a. $7\ H_2O + 2\ Cr^{3+} + 3\ Cl_2 \rightarrow Cr_2O_7^{2-} + 6\ Cl^- + 14\ H^+$ $E^o_{cell} = 0.03\text{ V} = 0.03\text{ J/C}$

$$\Delta G° = -nFE^o_{cell} = -(6\text{ mol e}^-)(96{,}485\text{ C/mol e}^-)(0.03\text{ J/C}) = -1.7 \times 10^4\text{ J} = -20\text{ kJ}$$

$$E_{cell} = E^o_{cell} - \frac{0.0591}{n}\log Q\text{: At equilibrium, } E_{cell} = 0 \text{ and } Q = K\text{, so:}$$

$$E^o_{cell} = \frac{0.0591}{n}\log K,\ \log K = \frac{nE°}{0.0591} = \frac{6(0.03)}{0.0591} = 3.05,\ K = 10^{3.05} = 1 \times 10^3$$

Note: When determining exponents, we will round off to the correct number of significant figures after the calculation is complete in order to help eliminate excessive round-off error.

25b. $\Delta G° = -(2\text{ mol e}^-)(96{,}485\text{ C/mol e}^-)(2.71\text{ J/C}) = -5.23 \times 10^5\text{ J} = -523\text{ kJ}$

$$\log K = \frac{2(2.71)}{0.0591} = 91.709,\ K = 5.12 \times 10^{91}$$

29a. $\Delta G° = -(2\text{ mol e}^-)(96{,}485\text{ C/mol}^-)(0.27\text{ J/C}) = -5.21 \times 10^4\text{ J} = -52\text{ kJ}$

$$\log K = \frac{2(0.27)}{0.0591} = 9.14,\ K = 1.4 \times 10^9$$

29b. $\Delta G° = -(10\text{ mol e}^-)(96{,}485\text{ C/mol e}^-)(0.09\text{ J/C}) = -8.7 \times 10^4\text{ J} = -90\text{ kJ}$

$$\log K = \frac{10(0.09)}{0.0591} = 15.23,\ K = 2 \times 10^{15}$$

67.

$$Cu^{2+} + 2\,e^- \rightarrow Cu \qquad E° = 0.34\ V$$
$$Fe \rightarrow Fe^{2+} + 2\,e^- \qquad -E° = 0.44\ V$$

$$Fe + Cu^{2+} \rightarrow Cu + Fe^{2+} \qquad E°_{cell} = 0.78\ V$$

For this reaction, $K = \dfrac{[Fe^{2+}]}{[Cu^{2+}]}$, so let's solve for K to determine the equilibrium ion ratio.

$$E° = \frac{0.0591}{n}\log K,\ \log K = \frac{2(0.78)}{0.0591} = 26.40,\ K = \frac{[Fe^{2+}]}{[Cu^{2+}]} = 10^{26.40} = 2.5 \times 10^{26}$$

69. a. Possible reaction: $I_2(s) + 2\,Cl^-(aq) \rightarrow 2\,I^-(aq) + Cl_2(g)$ $E°_{cell} = 0.54\ V - 1.36\ V = -0.82\ V$

This reaction is not spontaneous at standard conditions because $E°_{cell} < 0$. No reaction occurs.

b. Possible reaction: $Cl_2(g) + 2\,I^-(aq) \rightarrow I_2(s) + 2\,Cl^-(aq)$ $E°_{cell} = 0.82$ V; This reaction is spontaneous at standard conditions because $E°_{cell} > 0$. The reaction will occur.

$$Cl_2(g) + 2\,I^-(aq) \rightarrow I_2(s) + 2\,Cl^-(aq) \qquad E°_{cell} = 0.82\ V = 0.82\ J/C$$

$$\Delta G° = -nFE°_{cell} = -(2\ mol\ e^-)(96{,}485\ C/mol\ e^-)(0.82\ J/C) = -1.6 \times 10^5\ J = -160\ kJ$$

$$E° = \frac{0.0591}{n}\log K,\ \log K = \frac{nE°}{0.0591} = \frac{2(0.82)}{0.0591} = 27.75,\ K = 10^{27.75} = 5.6 \times 10^{27}$$

c. Possible reaction: $2\,Ag(s) + Cu^{2+}(aq) \rightarrow Cu(s) + 2\,Ag^+(aq)$ $E°_{cell} = -0.46$ V; No reaction occurs.

d. Fe^{2+} can be oxidized or reduced. The other species present are H^+, SO_4^{2-}, H_2O, and O_2 from air. Only O_2 in the presence of H^+ has a large enough standard reduction potential to oxidize Fe^{2+} to Fe^{3+} (resulting in $E°_{cell} > 0$). All other combinations, including the possible reduction of Fe^{2+}, give negative cell potentials. The spontaneous reaction is:

$$Fe^{2+}(aq) + 4\,H^+(aq) + O_2(g) \rightarrow 4\,Fe^{3+}(aq) + 2\,H_2O(l) \qquad E°_{cell} = 1.23 - 0.77 = 0.46\ V$$

$$\Delta G° = -nFE°_{cell} = -(4\ mol\ e^-)(96{,}485\ C/mol\ e^-)(0.46\ J/C)(1\ kJ/1000\ J) = -180\ kJ$$

$$\log K = \frac{4(0.46)}{0.0591} = 31.13,\ K = 1.3 \times 10^{31}$$

71. a.

$$Au^{3+} + 3\,e^- \rightarrow Au \qquad E° = 1.50\ V$$
$$(Tl \rightarrow Tl^+ + e^-) \times 3 \qquad -E° = 0.34\ V$$

$$Au^{3+}(aq) + 3\,Tl(s) \rightarrow Au(s) + 3\,Tl^+(aq) \qquad E°_{cell} = 1.84\ V$$

b. $\Delta G^\circ = -nFE^\circ_{cell} = -(3 \text{ mol } e^-)(96{,}485 \text{ C/mol } e^-)(1.84 \text{ J/C}) = -5.33 \times 10^5 \text{ J} = -533 \text{ kJ}$

$$\log K = \frac{nE^\circ}{0.0591} = \frac{3(1.84)}{0.0591} = 93.401, \; K = 10^{93.401} = 2.52 \times 10^{93}$$

c. $$E^\circ_{cell} = 1.84 \text{ V} - \frac{0.0591}{3}\log\frac{[Tl^+]^3}{[Au^{3+}]} = 1.84 - \frac{0.0591}{3}\log\frac{(1.0\times10^{-4})^3}{1.0\times10^{-2}}$$

$$E^\circ_{cell} = 1.84 - (-0.20) = 2.04 \text{ V}$$

73. The K_{sp} reaction is: $FeS(s) \rightleftharpoons Fe^{2+}(aq) + S^{2-}(aq)$ $K = K_{sp}$. Manipulate the given equations so when added together we get the K_{sp} reaction. Then we can use the value of E°_{cell} for the reaction to determine K_{sp}.

$$FeS + 2\,e^- \rightarrow Fe + S^{2-} \qquad E^\circ = -1.01 \text{ V}$$
$$Fe \rightarrow Fe^{2+} + 2\,e^- \qquad -E^\circ = 0.44 \text{ V}$$

$$Fe(s) \rightarrow Fe^{2+}(aq) + S^{2-}(aq) \qquad E^\circ_{cell} = -0.57 \text{ V}$$

$$\log K_{sp} = \frac{nE^\circ}{0.0591} = \frac{2(-0.57)}{0.0591} = -19.29, \; K_{sp} = 10^{-19.29} = 5.1 \times 10^{-20}$$

75. NO_3^- is a spectator ion. The reaction that occurs is Ag^+ reacting with Zn.

$$(Ag^+ + e^- \rightarrow Ag) \times 2 \qquad E^\circ = 0.80 \text{ V}$$
$$Zn \rightarrow Zn^{2+} + 2\,e^- \qquad -E^\circ = 0.76 \text{ V}$$

$$2\,Ag^+ + Zn \rightarrow 2\,Ag + Zn^{2+} \qquad E^\circ_{cell} = 1.56 \text{ V}$$

$$\log K = \frac{nE}{0.0591} = \frac{2(1.56)}{0.0591}, \; K = 10^{52.792} = 6.19 \times 10^{52}$$

Electrolysis

77. a. $Al^{3+} + 3\,e^- \rightarrow Al$; 3 mol e^- are needed to produce 1 mol Al from Al^{3+}.

$$1.0 \times 10^3 \text{ g Al} \times \frac{1 \text{ mol Al}}{26.98 \text{ g Al}} \times \frac{3 \text{ mol } e^-}{\text{mol Al}} \times \frac{96{,}485 \text{ C}}{\text{mol } e^-} \times \frac{1 \text{ s}}{100.0 \text{ C}} = 1.07 \times 10^5 \text{ s}$$

$$= 30. \text{ hours}$$

b. $$1.0 \text{ g Ni} \times \frac{1 \text{ mol Ni}}{58.69 \text{ g Ni}} \times \frac{2 \text{ mol } e^-}{\text{mol Ni}} \times \frac{96{,}485 \text{ C}}{\text{mol } e^-} \times \frac{1 \text{ s}}{100.0 \text{ C}} = 33 \text{ s}$$

c. $$5.0 \text{ mol Ag} \times \frac{1 \text{ mol } e^-}{\text{mol Ag}} \times \frac{96{,}485 \text{ C}}{\text{mol } e^-} \times \frac{1 \text{ s}}{100.0 \text{ C}} = 4.8 \times 10^3 \text{ s} = 1.3 \text{ hours}$$

79. $15\text{ A} = \frac{15\text{ C}}{\text{s}} \times \frac{60\text{ s}}{\text{min}} \times \frac{60\text{ min}}{\text{h}} = 5.4 \times 10^4$ C of charge passed in 1 hour

a. $5.4 \times 10^4\text{ C} \times \frac{1\text{ mol e}^-}{96{,}485\text{ C}} \times \frac{1\text{ mol Co}}{2\text{ mol e}^-} \times \frac{58.93\text{ g Co}}{\text{mol Co}} = 16\text{ g Co}$

b. $5.4 \times 10^4\text{ C} \times \frac{1\text{ mol e}^-}{96{,}485\text{ C}} \times \frac{1\text{ mol Hf}}{4\text{ mol e}^-} \times \frac{178.5\text{ g Hf}}{\text{mol Hf}} = 25\text{ g Hf}$

c. $2\text{ I}^- \rightarrow I_2 + 2\text{ e}^-$; $5.4 \times 10^4\text{ C} \times \frac{1\text{ mol e}^-}{96{,}485\text{ C}} \times \frac{1\text{ mol I}_2}{2\text{ mol e}^-} \times \frac{253.8\text{ g I}_2}{\text{mol I}_2} = 71\text{ g I}_2$

d. $CrO_3(l) \rightarrow Cr^{6+} + 3\text{ O}^{2-}$; 6 mol e^- are needed to produce 1 mol Cr from molten CrO_3.

$$5.4 \times 10^4\text{ C} \times \frac{1\text{ mol e}^-}{96{,}485\text{ C}} \times \frac{1\text{ mol Cr}}{6\text{ mol e}^-} \times \frac{52.00\text{ g Cr}}{\text{mol Cr}} = 4.9\text{ g Cr}$$

81. $74.1\text{ s} \times \frac{2.00\text{ C}}{\text{s}} \times \frac{1\text{ mol e}^-}{96{,}485\text{ C}} \times \frac{1\text{ mol M}}{3\text{ mol e}^-} = 5.12 \times 10^{-4}$ mol M where M = unknown metal

Molar mass $= \frac{0.107\text{ g M}}{5.12 \times 10^{-4}\text{ mol M}} = \frac{209\text{ g}}{\text{mol}}$; The element is bismuth.

83. F_2 is produced at the anode: $2\text{ F}^- \rightarrow F_2 + 2\text{ e}^-$

$$2.00\text{ h} \times \frac{60\text{ min}}{\text{h}} \times \frac{60\text{ s}}{\text{min}} \times \frac{10.0\text{ C}}{\text{s}} \times \frac{1\text{ mol e}^-}{96{,}485\text{ C}} = 0.746\text{ mol e}^-$$

$0.746\text{ mol e}^- \times \frac{1\text{ mol F}_2}{2\text{ mol e}^-} = 0.373\text{ mol F}_2$; $PV = nRT$, $V = \frac{nRT}{P}$

$$\frac{(0.373\text{ mol})(0.08206\text{ L} \bullet \text{atm/K} \bullet \text{mol})(298\text{ K})}{1.00\text{ atm}} = 9.12\text{ L F}_2$$

K is produced at the cathode: $K^+ + e^- \rightarrow K$

$$0.746\text{ mol e}^- \times \frac{1\text{ mol K}}{\text{mol e}^-} \times \frac{39.10\text{ g K}}{\text{mol K}} = 29.2\text{ g K}$$

85. $\frac{150. \times 10^3\text{ g C}_6H_8N_2}{\text{h}} \times \frac{1\text{ h}}{60\text{ min}} \times \frac{1\text{ min}}{60\text{ s}} \times \frac{1\text{ mol C}_6H_8N_2}{108.14\text{ g C}_6H_8N_2} \times \frac{2\text{ mol e}^-}{\text{mol C}_6H_8N_2} \times \frac{96{,}485\text{ C}}{\text{mol e}^-}$

$= 7.44 \times 10^4$ C/s or a current of 7.44×10^4 A

87. $2.30 \text{ min} \times \frac{60 \text{ s}}{\text{min}} = 138 \text{ s}; \ 138 \text{ s} \times \frac{2.00 \text{ C}}{\text{s}} \times \frac{1 \text{ mol e}^-}{96{,}485 \text{ C}} \times \frac{1 \text{ mol Ag}}{\text{mol e}^-} = 2.86 \times 10^{-3} \text{ mol Ag}$

$[Ag^+] = 2.86 \times 10^{-3} \text{ mol } Ag^+/0.250 \text{ L} = 1.14 \times 10^{-2} \ M$

89. $Au^{3+} + 3\ e^- \rightarrow Au \quad E° = 1.50 \text{ V} \qquad Ni^{2+} + 2\ e^- \rightarrow Ni \quad E° = -0.23 \text{ V}$

$Ag^+ + e^- \rightarrow Ag \quad E° = 0.80 \text{ V} \qquad Cd^{2+} + 2\ e^- \rightarrow Cd \quad E° = -0.40 \text{ V}$

$2\ H_2O + 2e^- \rightarrow H_2 + 2\ OH^- \quad E° = -0.83 \text{ V}$

Au(s) will plate out first since it has the most positive reduction potential, followed by Ag(s), which is followed by Ni(s), and finally Cd(s) will plate out last since it has the most negative reduction potential of the metals listed. Water will not interfere with the plating process.

91. Reduction occurs at the cathode, and oxidation occurs at the anode. First, determine all the species present, then look up pertinent reduction and/or oxidation potentials in Table 17.1 for all these species. The cathode reaction will be the reaction with the most positive reduction potential, and the anode reaction will be the reaction with the most positive oxidation potential.

a. Species present: Ni^{2+} and Br^-; Ni^{2+} can be reduced to Ni, and Br^- can be oxidized to Br_2 (from Table 17.1). The reactions are:

Cathode: $Ni^{2+} + 2e^- \rightarrow Ni \qquad E° = -0.23 \text{ V}$

Anode: $2\ Br^- \rightarrow Br_2 + 2\ e^- \qquad -E° = -1.09 \text{ V}$

b. Species present: Al^{3+} and F^-; Al^{3+} can be reduced, and F^- can be oxidized. The reactions are:

Cathode: $Al^{3+} + 3\ e^- \rightarrow Al \qquad E° = -1.66 \text{ V}$

Anode: $2\ F^- \rightarrow F_2 + 2\ e^- \qquad -E° = -2.87 \text{ V}$

c. Species present: Mn^{2+} and I^-; Mn^{2+} can be reduced, and I^- can be oxidized. The reactions are:

Cathode: $Mn^{2+} + 2\ e^- \rightarrow Mn \qquad E° = -1.18 \text{ V}$

Anode: $2\ I^- \rightarrow I_2 + 2\ e^- \qquad -E° = -0.54 \text{ V}$

Additional Exercises

93. The half-reaction for the SCE is:

$Hg_2Cl_2 + 2\ e^- \rightarrow 2\ Hg + 2\ Cl^- \qquad E_{SCE} = 0.242 \text{ V}$

For a spontaneous reaction to occur, E_{cell} must be positive. Using the standard reduction potentials in Table 17.1 and the given the SCE potential, deduce which combination will produce a positive overall cell potential.

a. $Cu^{2+} + 2\ e^- \rightarrow Cu$ $E° = 0.34\ V$

$E_{cell} = 0.34 - 0.242 = 0.10\ V$; SCE is the anode.

b. $Fe^{3+} + e^- \rightarrow Fe^{2+}$ $E° = 0.77\ V$

$E_{cell} = 0.77 - 0.242 = 0.53\ V$; SCE is the anode.

c. $AgCl + e^- \rightarrow Ag + Cl^-$ $E° = 0.22\ V$

$E_{cell} = 0.242 - 0.22 = 0.02\ V$; SCE is the cathode.

d. $Al^{3+} + 3\ e^- \rightarrow Al$ $E° = -1.66\ V$

$E_{cell} = 0.242 + 1.66 = 1.90\ V$; SCE is the cathode.

e. $Ni^{2+} + 2\ e^- \rightarrow Ni$ $E° = -0.23\ V$

$E_{cell} = 0.242 + 0.23 = 0.47\ V$; SCE is the cathode.

95. $2\ Ag^+(aq) + Cu(s) \rightarrow Cu^{2+}(aq) + 2\ Ag(s)$ $E°_{cell} = 0.80 - 0.34\ V = 0.46\ V$; A galvanic cell produces a voltage as the forward reaction occurs. Any stress that increases the tendency of the forward reaction to occur will increase the cell potential, while a stress that decreases the tendency of the forward reaction to occur will decrease the cell potential.

a. Added Cu^{2+} (a product ion) will decrease the tendency of the forward reaction to occur, which will decrease the cell potential.

b. Added NH_3 removes Cu^{2+} in the form of $Cu(NH_3)_4^{2+}$. Removal of a product ion will increase the tendency of the forward reaction to occur, which will increase the cell potential.

c. Added Cl^- removes Ag^+ in the form of AgCl(s). Removal of a reactant ion will decrease the tendency of the forward reaction to occur, which will decrease the cell potential.

d. $Q_1 = \frac{[Cu^{2+}]_o}{[Ag^+]_o^2}$; As the volume of solution is doubled, each concentration is halved.

$$Q_2 = \frac{1/2[Cu^{2+}]_o}{(1/2[Ag^+]_o)^2} = \frac{[Cu^{2+}]_o}{[Ag^+]_o^2} = 2\ Q_1$$

The reaction quotient is doubled as the concentrations are halved. Because reactions are spontaneous when $Q < K$ and because Q increases when the solution volume doubles, the reaction is closer to equilibrium, which will decrease the cell potential.

e. Because Ag(s) is not a reactant in this spontaneous reaction, and because solids do not appear in the reaction quotient expressions, replacing the silver electrode with a platinum electrode will have no effect on the cell potential.

97. a. $\Delta G^\circ = \sum n_p \Delta G^\circ_{f,\,products} - \sum n_r \Delta G^\circ_{f,\,reactants} = 2(-480.) + 3(86) - [3(-40.)] = -582$ kJ

From oxidation numbers, n = 6. $\Delta G^\circ = -nFE^\circ$, $E^\circ = \frac{-\Delta G^\circ}{nF} = \frac{-(-582{,}000\ J)}{6(96{,}485)\ C} = 1.01$ V

$\log K = \frac{nE^\circ}{0.0591} = \frac{6(1.01)}{0.0591} = 102.538$, $K = 10^{102.538} = 3.45 \times 10^{102}$

b.

$$(2\,e^- + Ag_2S \rightarrow 2\,Ag + S^{2-}) \times 3 \qquad E^\circ_{Ag_2S} = ?$$
$$(Al \rightarrow Al^{3+} + 3\,e^-) \times 2 \qquad -E^\circ = 1.66\ V$$

$$3\,Ag_2S(s) + 2\,Al(s) \rightarrow 6\,Ag(s) + 3\,S^{2-}(aq) + 2\,Al^{3+}(aq) \qquad E^\circ_{cell} = 1.01\ V = E^\circ_{Ag_2S} + 1.66\ V$$

$E^\circ_{Ag_2S} = 1.01\ V - 1.66\ V = -0.65$ V

99. Aluminum has the ability to form a durable oxide coating over its surface. Once the HCl dissolves this oxide coating, Al is exposed to H^+ and is easily oxidized to Al^{3+}, i.e., the Al foil disappears after the oxide coating is dissolved.

101. Consider the strongest oxidizing agent combined with the strongest reducing agent from Table 17.1:

$$F_2 + 2\,e^- \rightarrow 2\,F^- \qquad E^\circ = 2.87\ V$$
$$(Li \rightarrow Li^+ + e^-) \times 2 \qquad -E^\circ = 3.05\ V$$

$$F_2(g) + 2\,Li(s) \rightarrow 2\,Li^+(aq) + 2\,F^-(aq) \qquad E^\circ_{cell} = 5.92\ V$$

The claim is impossible. The strongest oxidizing agent and strongest reducing agent when combined only give an E°_{cell} value of about 6 V.

103.

$$O_2 + 2\,H_2O + 4\,e^- \rightarrow 4\,OH^- \qquad E^\circ = 0.40\ V$$
$$(H_2 + 2\,OH^- \rightarrow 2\,H_2O + 2\,e^-) \times 2 \qquad -E^\circ = 0.83\ V$$

$$2\,H_2(g) + O_2(g) \rightarrow 2\,H_2O(l) \qquad E^\circ_{cell} = 1.23\ V = 1.23\ J/C$$

Because standard conditions are assumed, $w_{max} = \Delta G^\circ$ for 2 mol H_2O produced.

$\Delta G^\circ = -nFE^\circ_{cell} = -(4\text{ mol } e^-)(96{,}485\text{ C/mol } e^-)(1.23\text{ J/C}) = -475{,}000\text{ J} = -475$ kJ

For 1.00×10^3 g H_2O produced, w_{max} is:

$$1.00 \times 10^3\ g\ H_2O \times \frac{1\ mol\ H_2O}{18.02\ g\ H_2O} \times \frac{-475\ kJ}{2\ mol\ H_2O} = -13{,}200\ kJ = w_{max}$$

The work done can be no larger than the free energy change. The best that could happen is that all of the free energy released would go into doing work, but this does not occur in any real process because there is always waste energy in a real process. Fuel cells are more efficient in converting chemical energy into electrical energy; they are also less massive. The major disadvantage is that they are expensive. In addition, $H_2(g)$ and $O_2(g)$ are an explosive mixture if ignited; much more so than fossil fuels.

105. $(CO + O^{2-} \rightarrow CO_2 + 2\,e^-) \times 2$

$O_2 + 4\,e^- \rightarrow 2\,O^{2-}$

$2\,CO + O_2 \rightarrow 2\,CO_2$

$$\Delta G = -nFE,\quad E = \frac{-\Delta G^\circ}{nF} = \frac{-(-380\times10^3\ J)}{(4\ mol\ e^-)(96{,}485\ C/mol\ e^-)} = 0.98\ V$$

107. The oxidation state of gold in $Au(CN)_2^-$ is +1. Each mole of gold produced requires 1 mole of electrons gained (+1 → 0). The only oxygen containing reactant is H_2O. Each mole of oxygen goes from −2 → 0 oxidation states as H_2O is converted into O_2. One mole of O_2 contains 2 moles O, so 4 moles of electrons are lost when 1 mole O_2 is formed. In order to balance the electrons, we need 4.00 moles of gold for every mole of O_2 produced or 0.250 moles O_2 for every 1.00 mole of gold formed.

109. $$mol\ e^- = 50.0\ min \times \frac{60\ s}{min} \times \frac{2.50\ C}{s} \times \frac{1\ mol\ e^-}{96{,}485\ C} = 7.77 \times 10^{-2}\ mol\ e^-$$

$$mol\ Ru = 2.618\ g\ Ru \times \frac{1\ mol\ Ru}{101.1\ g\ Ru} = 2.590 \times 10^{-2}\ mol\ Ru$$

$$\frac{mol\ e^-}{mol\ Ru} = \frac{7.77\times10^{-2}\ mol\ e^-}{2.590\times10^{-2}\ mol\ Ru} = 3.00;$$ The charge on the ruthenium ions is +3. $(Ru^{3+} + 3\,e^- \rightarrow Ru)$

Challenge Problems

111. $$\Delta G^\circ = -nFE^\circ = \Delta H^\circ - T\Delta S^\circ,\ E^\circ = \frac{T\Delta S^\circ}{nF} - \frac{\Delta H^\circ}{nF}$$

If we graph E° vs. T we should get a straight line (y = mx + b). The slope of the line is equal to $\Delta S^\circ/nF$, and the y-intercept is equal to $-\Delta H^\circ/nF$. From the equation above, E° will have a small temperature dependence when ΔS° is close to zero.

113. $(Ag^+ + e^- \rightarrow Ag) \times 2$ $\quad E^\circ = 0.80\ V$

$Pb \rightarrow Pb^{2+} + 2\,e^-$ $\quad -E^\circ = -(-0.13)$

$2\,Ag^+ + Pb \rightarrow 2\,Ag + Pb^{2+}$ $\quad E^\circ_{cell} = 0.93\ V$

$$E = E° - \frac{0.0591}{n}\log\frac{[Pb^{2+}]}{[Ag^+]^2},\ 0.83\ V = 0.93\ V - \frac{0.0591}{n}\log\frac{(1.8)}{[Ag^+]^2}$$

$$\log\frac{(1.8)}{[Ag^+]^2} = \frac{0.10(2)}{0.0591} = 3.4,\ \frac{(1.8)}{[Ag^+]^2} = 10^{3.4},\ [Ag^+] = 0.027\ M$$

$$Ag_2SO_4(s) \rightleftharpoons 2\,Ag^+(aq) + SO_4^{2-}(aq)\quad K_{sp} = [Ag^+]^2[SO_4^{2-}]$$

Initial	s = solubility (mol/L)	0	0
Equil.		$2s$	s

From problem: $2s = 0.027\ M$, $s = 0.027/2$

$K_{sp} = (2s)^2(s) = (0.027)^2(0.027/2) = 9.8 \times 10^{-6}$

115.

$$2\,H^+ + 2\,e^- \rightarrow H_2 \qquad E° = 0.000\ V$$
$$Fe \rightarrow Fe^{2+} + 2\,e^- \qquad -E° = -(-0.440V)$$

$$2\,H^+(aq) + Fe(s) \rightarrow H_2(g) + Fe^{2+}(aq) \qquad E°_{cell} = 0.440\ V$$

$$E_{cell} = E°_{cell} - \frac{0.0591}{n}\log Q,\ \text{where } n = 2 \text{ and } Q = \frac{P_{H_2} \times [Fe^{2+}]}{[H^+]^2}$$

To determine K_a for the weak acid, first use the electrochemical data to determine the H^+ concentration in the half-cell containing the weak acid.

$$0.333\ V = 0.440\ V - \frac{0.0591}{2}\log\frac{1.00(1.00\times10^{-3})}{[H^+]^2}$$

$$\frac{0.107(2)}{0.0591} = \log\frac{1.00\times10^{-3}}{[H^+]^2},\ \frac{1.00\times10^{-3}}{[H^+]^2} = 10^{3.621} = 4.18 \times 10^3,\ [H^+] = 4.89 \times 10^{-4}\ M$$

Now we can solve for the K_a value of the weak acid HA through the normal setup for a weak acid problem.

$$HA \rightleftharpoons H^+ + A^- \qquad K_a = \frac{[H^+][A^-]}{[HA]}$$

	HA	H^+	A^-
Initial	1.00 M	~0	0
Equil.	1.00 - x	x	x

$$K_a = \frac{x^2}{1.00 - x}\ \text{where } x = [H^+] = 4.89 \times 10^{-4}\ M,\ K_a = \frac{(4.89\times10^{-4})^2}{1.00 - 4.89\times10^{-4}} = 2.39 \times 10^{-7}$$

117. a. $E_{cell} = E_{ref} + 0.05916\ pH$, $0.480\ V = 0.250\ V + 0.05916\ pH$

$$pH = \frac{0.480 - 0.250}{0.05916} = 3.888;\ \text{Uncertainty} = \pm 1\ mV = \pm 0.001\ V$$

$$pH_{max} = \frac{0.481 - 0.250}{0.05916} = 3.905; \quad pH_{min} = \frac{0.479 - 0.250}{0.05916} = 3.871$$

If the uncertainty in potential is ± 0.001 V, the uncertainty in pH is ± 0.017 or about ± 0.02 pH units. For this measurement, $[H^+] = 10^{-3.888} = 1.29 \times 10^{-4}\ M$. For an error of +1 mV, $[H^+] = 10^{-3.905} = 1.24 \times 10^{-4}\ M$. For an error of −1 mV, $[H^+] = 10^{-3.871} = 1.35 \times 10^{-4}\ M$. So, the uncertainty in $[H^+]$ is $\pm 0.06 \times 10^{-4}\ M = \pm 6 \times 10^{-6}\ M$.

b. From part a, we will be within ± 0.02 pH units if we measure the potential to the nearest ± 0.001 V (1 mV).

119. a.

$$(Ag^+ + e^- \rightarrow Ag) \times 2 \qquad E° = 0.80\ V$$
$$Cu \rightarrow Cu^{2+} + 2\ e^- \qquad -E° = -0.34\ V$$

$$2\ Ag^+(aq) + Cu(s) \rightarrow 2\ Ag(s) + Cu^{2+}(aq) \qquad E°_{cell} = 0.46\ V$$

$$E_{cell} = E°_{cell} - \frac{0.0591}{n} \log Q \text{ where } n = 2 \text{ and } Q = \frac{[Cu^{2+}]}{[Ag^+]^2}$$

To calculate E_{cell}, we need to use the K_{sp} data to determine $[Ag^+]$.

	$AgCl(s)$	$\rightleftharpoons$	$Ag^+(aq)$	+	$Cl^-(aq)$	$K_{sp} = 1.6 \times 10^{-10}$
Initial	s = solubility (mol/L)		0		0	
Equil.			s		s	

$K_{sp} = 1.6 \times 10^{-10} = [Ag^+][Cl^-] = s^2$, $s = [Ag^+] = 1.3 \times 10^{-5}$ mol/L

$$E_{cell} = 0.46\ V - \frac{0.0591}{2} \log \frac{2.0}{(1.3 \times 10^{-5})^2} = 0.46\ V - 0.30 = 0.16\ V$$

b. $Cu^{2+}(aq) + 4\ NH_3(aq) \rightleftharpoons Cu(NH_4)_4{}^{2+}(aq) \quad K = 1.0 \times 10^{13} = \frac{[Cu(NH_3)_4^{2+}]}{[Cu^{2+}][NH_3]^4}$

Because K is very large for the formation of $Cu(NH_3)_4{}^{2+}$, the forward reaction is dominant. At equilibrium, essentially all of the 2.0 M Cu^{2+} will react to form 2.0 M $Cu(NH_3)_4{}^{2+}$. This reaction requires 8.0 M NH_3 to react with all of the Cu^{2+} in the balanced equation. Therefore, the mol of NH_3 added to 1.0 L solution will be larger than 8.0 mol since some NH_3 must be present at equilibrium. In order to calculate how much NH_3 is present at equilibrium, we need to use the electrochemical data to determine the Cu^{2+} concentration.

$$E_{cell} = E^{\circ}_{cell} - \frac{0.0591}{n}\log Q, \; 0.52\text{ V} = 0.46\text{ V} - \frac{0.0591}{2}\log\frac{[Cu^{2+}]}{(1.3 \times 10^{-5})^2}$$

$$\log\frac{[Cu^{2+}]}{(1.3 \times 10^{-5})^2} = \frac{-0.06(2)}{0.0591} = -2.03, \; \frac{[Cu^{2+}]}{(1.3 \times 10^{-5})^2} = 10^{-2.03} = 9.3\times10^{-3}$$

$[Cu^{2+}] = 1.6 \times 10^{-12} = 2 \times 10^{-12}$ *M* (We carried extra significant figures in the calculation.)

Note: Our assumption that the 2.0 *M* Cu^{2+} essentially reacts to completion is excellent as only 2×10^{-12} *M* Cu^{2+} remains after this reaction. Now we can solve for the equilibrium $[NH_3]$.

$$K = 1.0 \times 10^{13} = \frac{[Cu(NH_3)_4^{2+}]}{[Cu^{2+}][NH_3]^4} = \frac{(2.0)}{(2 \times 10^{-12})\,[NH_3]^4}, \; [NH_3] = 0.6\,M$$

Since 1.0 L of solution is present, then 0.6 mol NH_3 remains at equilibrium. The total mol of NH_3 added is 0.6 mol plus the 8.0 mol NH_3 necessary to form 2.0 *M* $Cu(NH_3)_4^{2+}$. Therefore, 8.0 + 0.6 = 8.6 mol NH_3 were added.

121. $2\,Ag^+ + Ni \rightarrow Ni^{2+} + Ag$; The cell is dead at equilibrium.

$E^{\circ}_{cell} = 0.80\text{ V} + 0.23\text{ V} = 1.03\text{ V}$

$$0 = 1.03\text{ V} - \frac{0.0591}{2}\log K; \text{ Solving: } K = 7.18 \times 10^{34}$$

K is very large. Let the forward reaction go to completion.

$2Ag^+ + Ni \rightarrow Ni^{2+} + 2\,Ag \qquad K = [Ni^{2+}]/[Ag^+]^2 = 7.18 \times 10^{34}$

	Ag^+	Ni^{2+}
Before	1.0 *M*	1.0 *M*
After	0 *M*	1.5 *M*

Now allow reaction to get back to equilibrium.

$2Ag^+ + Ni \rightleftharpoons Ni^{2+} + Ag$

	$2Ag^+$		Ni^{2+}
Initial	0		1.5 *M*
Change	$+2x$	$\leftarrow$	$-x$
Equil.	$2x$		$1.5 - x$

$$K = 7.18 \times 10^{34} = \frac{1.5 - x}{(2x)^2} \approx \frac{1.5}{(2x)^2}; \text{ Solving: } x = 2.3 \times 10^{-18}\,M. \text{ Assumptions good.}$$

$[Ag^+] = 2x = 4.6 \times 10^{-18}\,M$; $[Ni^{2+}] = 1.5 - 2.3 \times 10^{-18}\,M = 1.5\,M$

123. a. $E° = 0$ (concentration cell); $E = 0 - \frac{0.0591}{2}\log\left(\frac{1.0 \times 10^{-4}}{1.00}\right)$, $E = 0.12$ V

b. $Cu^{2+} + 4\ NH_3 \rightleftharpoons Cu(NH_3)_4^{2+}$ $K_{overall} = K_1 \bullet K_2 \bullet K_3 \bullet K_4 = 1.0 \times 10^{13}$

Because K > >1, let the reaction go to completion, then solve the back equilibrium problem.

	Cu^{2+}	+ $4\ NH_3$	→ $Cu(NH_3)_4^{2+}$	$K = 1.0 \times 10^{13}$
Before	1.0×10^{-4} *M*	2.0 *M*	0	
After	0	2.0	1.0×10^{-4}	

Now allow the reaction to reach equilibrium.

	Cu^{2+}	+ $4\ NH_3$	→ $Cu(NH_3)_4^{2+}$
Initial	0	2.0 *M*	1.0×10^{-4} *M*
Equil.	$+x$	$2.0 + 4x$	$1.0 \times 10^{-4} - x$

$$K = 1.0 \times 10^{13} = \frac{(1.0 \times 10^{-4} - x)}{x(2.0 + 4x)^4} \approx \frac{1.0 \times 10^{-4}}{x(2.0)^4} = 1.0 \times 10^{13},\ x = [Cu^{2+}] = 6.3 \times 10^{-19}\ M$$

Assumptions good.

At this Cu^{2+} concentration, the cell potential is:

$$E = 0 - \frac{0.0591}{2}\log\left(\frac{6.3 \times 10^{-19}}{1.00}\right),\quad E = 0.54\ V$$

Integrative Problems

125. a.

$(In^+ + e^- \rightarrow In) \times 2$	$E° = -0.126$ V
$In^+ \rightarrow In^{3+} + 2e^-$	$-E° = 0.444$ V
$3\ In^+ \rightarrow In^{3+} + 2\ In$	$E°_{cell} = 0.318$

$$\log K = \frac{nE°}{0.0591} = \frac{2(0.318)}{0.0591} = 10.761,\ K = 10^{10.761} = 5.77 \times 10^{10}$$

b. $\Delta G° = -nFE° = -(2\ mol\ e^-)(96{,}485\ C/mol\ e^-)(0.318\ J/C) = -6.14 \times 10^5\ J = -61.4\ kJ$

$$\Delta G°_{rxn} = -61.4\ kJ = [2(0) + 1(-97.9\ kJ] - 3\ \Delta G°_{f,In^+},\quad \Delta G°_{f,In^+} = -12.2\ kJ/mol$$

127. Chromium(III) nitrate [$Cr(NO_3)_3$] has chromium in the +3 oxidation state.

$$1.15 \text{ g Cr} \times \frac{1 \text{ mol Cr}}{52.00 \text{ g}} \times \frac{3 \text{ mol e}^-}{\text{mol Cr}} \times \frac{96{,}485 \text{ C}}{\text{mol e}^-} = 6.40 \times 10^3 \text{ C of charge}$$

For the Os cell, 6.40×10^3 C of charge was also passed.

$$3.15 \text{ g Os} \times \frac{1 \text{ mol Os}}{190.2 \text{ g}} = 0.0166 \text{ mol Os}; \quad 6.40 \times 10^3 \text{ C} \times \frac{1 \text{ mol e}^-}{96{,}485 \text{ C}} = 0.0663 \text{ mol e}^-$$

$$\frac{\text{mol e}^-}{\text{mol Os}} = \frac{0.0663}{0.0166} = 3.99 \approx 4$$

This salt is composed of Os^{4+} and NO_3^- ions. The compound is $Os(NO_3)_4$, osmium(IV) nitrate.

For the third cell, identify X by determining its molar mass. Two moles of electrons are transferred when X^{2+} is reduced to X.

$$\text{molar mass} = \frac{2.11 \text{ g X}}{6.40 \times 10^3 \text{ C} \times \frac{1 \text{ mol e}^-}{96{,}485 \text{ C}} \times \frac{1 \text{ mol X}}{2 \text{ mol e}^-}} = 63.6 \text{ g/mol}$$

This is copper, Cu. The electron configuration is: $[Ar]4s^1 3d^{10}$.

CHAPTER EIGHTEEN

THE NUCLEUS: A CHEMIST'S VIEW

Questions

1. Characteristic frequencies of energies emitted in a nuclear reaction suggest that discrete energy levels exist in the nucleus. The extra stability of certain numbers of nucleons and the predominance of nuclei with even numbers of nucleons suggest that the nuclear structure might be described by using quantum numbers.

3. β-particle production has the net effect of turning a neutron into a proton. Radioactive nuclei having too many neutrons typically undergo β-particle decay. Positron production has the net effect of turning a proton into a neutron. Nuclei having too many protons typically undergo positron decay.

5. The transuranium elements are the elements having more protons than uranium. They are synthesized by bombarding heavier nuclei with neutrons and positive ions in a particle accelerator.

7. $\Delta E = \Delta mc^2$; The key difference is the mass change when going from reactants to products. In chemical reactions, the mass change is indiscernible. In nuclear processes, the mass change is discernable. It is the conversion of this discernable mass change into energy that results in the huge energies associated with nuclear processes.

9. Sr-90 is an alkaline earth metal having chemical properties similar to calcium. Sr-90 can collect in bones replacing some of the calcium. Once embedded inside the human body, β particles can do significant damage. Rn-222 is a noble gas so one would expect Rn to be unreactive and pass through the body quickly; it does. The problem with Rn-222 is the rate at which it produces alpha particles. With a short half-life, the few moments that Rn-222 is in the lungs, a significant number of decay events can occur; each decay event produces an alpha particle which is very effective at causing ionization and can produce a dense trail of damage.

Exercises

Radioactive Decay and Nuclear Transformations

11. All nuclear reactions must be charge balanced and mass balanced. To charge balance, balance the sum of the atomic numbers on each side of the reaction, and to mass balance, balance the sum of the mass numbers on each side of the reaction.

a. ${}^{51}_{24}Cr + {}^{0}_{-1}e \rightarrow {}^{51}_{23}V$ b. ${}^{131}_{53}I \rightarrow {}^{0}_{-1}e + {}^{131}_{54}Xe$

13. a. ${}^{68}_{31}Ga + {}^{0}_{-1}e \rightarrow {}^{68}_{30}Zn$ b. ${}^{62}_{29}Cu \rightarrow {}^{0}_{+1}e + {}^{62}_{28}Ni$

c. ${}^{212}_{87}Fr \rightarrow {}^{4}_{2}He + {}^{208}_{85}At$ d. ${}^{129}_{51}Sb \rightarrow {}^{0}_{-1}e + {}^{129}_{52}Te$

15. ${}^{247}_{97}Bk \rightarrow {}^{207}_{82}Pb + ?\ {}^{4}_{2}He + {}^{0}_{-1}e$; The change in mass number (247 − 207 = 40) is due exclusively to the alpha particles. A change in mass number of 40 requires $10\,{}^{4}_{2}He$ particles to be produced. The atomic number only changes by 97 − 82 = 15. The 10 alpha particles change the atomic number by 20, so $5\,{}^{0}_{-1}e$ (5 beta particles) are produced in the decay series of ${}^{247}Bk$ to ${}^{207}Pb$.

17. ${}^{53}_{26}Fe$ has too many protons. It will undergo either positron production, electron capture and/or alpha particle production. ${}^{59}_{26}Fe$ has too many neutrons and will undergo beta particle production. (See Table 18.2 of the text.)

19. a. ${}^{249}_{98}Cf + {}^{18}_{8}O \rightarrow {}^{263}_{106}Sg + 4\,{}^{0}_{1}n$ b. ${}^{259}_{104}Rf$; ${}^{263}_{106}Sg \rightarrow {}^{4}_{2}He + {}^{259}_{104}Rf$

Kinetics of Radioactive Decay

21. All radioactive decay follows first-order kinetics where $t_{1/2} = (\ln 2)/k$.

$$t_{1/2} = \frac{\ln 2}{k} = \frac{0.693}{1.0 \times 10^{-3}\ h^{-1}} = 690\ h$$

23. Kr-81 is most stable because it has the longest half-life while Kr-73 is hottest (least stable) because it has the shortest half-life.

12.5% of each isotope will remain after 3 half-lives:

$$100\% \xrightarrow[t_{1/2}]{} 50\% \xrightarrow[t_{1/2}]{} 25\% \xrightarrow[t_{1/2}]{} 12.5\%$$

For Kr-73: t = 3(27 s) = 81 s; For Kr-74: t = 3(11.5 min) = 34.5 min

For Kr-76: t = 3(14.8 h) = 44.4 h; For Kr-81: t = $3(2.1 \times 10^{5}\ yr) = 6.3 \times 10^{5}$ yr

25. Units for N and N_o are usually the number of nuclei but can also be grams if the units are the same for both N and N_o. In this problem m = the mass of ${}^{32}P$ that remains.

$$175\ mg\ Na_3{}^{32}PO_4 \times \frac{32.0\ mg\ {}^{32}P}{165.0\ mg\ Na_3{}^{32}PO_4} = 33.9\ mg\ {}^{32}P\ \text{initially};\ k = \frac{\ln 2}{t_{1/2}}$$

$\ln\left(\frac{N}{N_0}\right) = -kt = \frac{-0.6931\,t}{t_{1/2}}$, $\ln\left(\frac{m}{33.9\text{ mg}}\right) = \frac{-0.6931(35.0\text{ d})}{14.3\text{ d}}$; Carrying extra sig. figs.:

$\ln(m) = -1.696 + 3.523 = 1.827$, $m = e^{1.827} = 6.22$ mg ^{32}P remains

27. $t = 61.0$ yr; $k = \frac{\ln 2}{t_{1/2}}$; $\ln\left(\frac{N}{N_0}\right) = -kt = \frac{-0.6931 \times 61.0\text{ yr}}{28.8\text{ yr}} = -1.47$, $\left(\frac{N}{N_0}\right) = e^{-1.47} = 0.230$

23.0% of the ^{90}Sr remains as of July 16, 2006.

29. $\ln\left(\frac{N}{N_0}\right) = -kt = \frac{-(\ln 2)\,t}{t_{1/2}}$, $\ln\left(\frac{1.0\text{ g}}{N_0}\right) = \frac{-0.693\left(3.0\text{ d} \times \frac{24\text{ h}}{\text{d}} \times \frac{60\text{ min}}{\text{h}}\right)}{1.0 \times 10^3\text{ min}}$

$\ln\left(\frac{1.0}{N_0}\right) = -3.0$, $\frac{1.0}{N_0} = e^{-3.0}$, $N_0 = 20.$ g ^{82}Br needed

$$20.\text{ g }^{82}\text{Br} \times \frac{1\text{ mol }^{82}\text{Br}}{82.0\text{ g}} \times \frac{1\text{ mol Na}^{82}\text{Br}}{\text{mol }^{82}\text{Br}} \times \frac{105.0\text{ g Na}^{82}\text{Br}}{\text{mol Na}^{82}\text{Br}} = 26\text{ g Na}^{82}\text{Br}$$

31. $k = \frac{\ln 2}{t_{1/2}}$; $\ln\left(\frac{N}{N_0}\right) = -kt = \frac{-0.693\,t}{t_{1/2}}$, $\ln\left(\frac{N}{13.6}\right) = \frac{-0.693\,(15{,}000\text{ yr})}{5730\text{ yr}} = -1.8$

$\frac{N}{13.6} = e^{-1.8} = 0.17$, $N = 13.6 \times 0.17 = 2.3$ counts per minute per g of C

If we had 10. mg C, we would see:

$$10.\text{ mg} \times \frac{1\text{ g}}{1000\text{ mg}} \times \frac{2.3\text{ counts}}{\text{min g}} = \frac{0.023\text{ counts}}{\text{min}}$$

It would take roughly 40 min to see a single disintegration. This is too long to wait, and the background radiation would probably be much greater than the ^{14}C activity. Thus, ^{14}C dating is not practical for very small samples.

33. Assuming 1.000 g ^{238}U present in a sample, then 0.688 g ^{206}Pb is present. Since 1 mol ^{206}Pb is produced per mol ^{238}U decayed, then:

$$^{238}\text{U decayed} = 0.688\text{ g Pb} \times \frac{1\text{ mol Pb}}{206\text{ g Pb}} \times \frac{1\text{ mol U}}{\text{mol Pb}} \times \frac{238\text{ g U}}{\text{mol U}} = 0.795\text{ g }^{238}\text{U}$$

Original mass ^{238}U present = 1.000 g + 0.795 g = 1.795 g ^{238}U

$\ln\left(\frac{N}{N_0}\right) = -kt = \frac{-(\ln 2)\,t}{t_{1/2}}$, $\ln\left(\frac{1.000\text{ g}}{1.795\text{ g}}\right) = \frac{-0.693\,(t)}{4.5 \times 10^9\text{ yr}}$, $t = 3.8 \times 10^9$ yr

Energy Changes in Nuclear Reactions

35. $\Delta E = \Delta mc^2$, $\Delta m = \frac{\Delta E}{c^2} = \frac{3.9\times10^{23}\ \text{kg m}^2/\text{s}^2}{(3.00\times10^8\ \text{m/s})^2} = 4.3\times10^6$ kg

The sun loses 4.3×10^6 kg of mass each second. Note: 1 J = 1 kg m^2/s^2

37. We need to determine the mass defect, Δm, between the mass of the nucleus and the mass of the individual parts that make up the nucleus. Once Δm is known, we can then calculate ΔE (the binding energy) using $E = mc^2$. Note: 1 J = 1 kg m^2/s^2.

For $^{232}_{94}$Pu (94 e, 94 p, 138 n):

mass of ^{232}Pu nucleus = 3.85285×10^{-22} g − mass of 94 electrons

mass of ^{232}Pu nucleus = 3.85285×10^{-22} g − $94(9.10939 \times 10^{-28})$ g = 3.85199×10^{-22} g

$\Delta m = 3.85199 \times 10^{-22}$ g − (mass of 94 protons + mass of 138 neutrons)

$\Delta m = 3.85199 \times 10^{-22}$ g − $[94(1.67262 \times 10^{-24}) + 138(1.67493 \times 10^{-24})]$ g
$= -3.168 \times 10^{-24}$ g

For 1 mol of nuclei: $\Delta m = -3.168 \times 10^{-24}$ g/nuclei $\times 6.0221 \times 10^{23}$ nuclei/mol
$= -1.908$ g/mol

$\Delta E = \Delta mc^2 = (-1.908 \times 10^{-3}\ \text{kg/mol})(2.9979 \times 10^8\ \text{m/s})^2 = -1.715 \times 10^{14} = \text{J/mol}$

For $^{231}_{91}$Pa (91 e, 91 p, 140 n):

mass of ^{231}Pa nucleus = 3.83616×10^{-22} g − $91(9.10939 \times 10^{-28})$ g = 3.83533×10^{-22} g

$\Delta m = 3.83533 \times 10^{-22}$ g − $[91(1.67262 \times 10^{-24}) + 140(1.67493 \times 10^{-24})]$ g
$= -3.166 \times 10^{-24}$ g

$$\Delta E = \Delta mc^2 = \frac{-3.166\times10^{-27}\ \text{kg}}{\text{nuclei}} \times \frac{6.0221\times10^{23}\ \text{nuclei}}{\text{mol}} \times \left(\frac{2.9979\times10^8\ \text{m}}{\text{s}}\right)^2$$
$$= -1.714 \times 10^{14}\ \text{J/mol}$$

39. Let m_e = mass of electron; For ^{12}C (6e, 6p, 6n): mass defect = Δm = mass of ^{12}C nucleus − [mass of 6 protons + mass of 6 neutrons]. Note: the atomic masses of the elements given include the mass of the electrons.

$\Delta m = 12.0000$ amu − 6 m_e − $[6(1.00782 - m_e) + 6(1.00866)]$; Mass of electrons cancel.

$\Delta m = 12.0000 - [6(1.00782) + 6(1.00866)] = -0.0989$ amu

$$\Delta E = \Delta mc^2 = -0.0989 \text{ amu} \times \frac{1.6605 \times 10^{-27} \text{ kg}}{\text{amu}} \times (2.9979 \times 10^8 \text{ m/s})^2 = -1.48 \times 10^{-11} \text{ J}$$

$$\frac{\text{BE}}{\text{nucleon}} = \frac{1.48 \times 10^{-11} \text{ J}}{12 \text{ nucleons}} = 1.23 \times 10^{-12} \text{ J/nucleon}$$

For ^{235}U (92e, 92p, 143n):

$\Delta m = 235.0439 - 92\ m_e - [92(1.00782 - m_e) + 143(1.00866)] = -1.9139$ amu

$$\Delta E = \Delta mc^2 = -1.9139 \text{ amu} \times \frac{1.66054 \times 10^{-27} \text{ kg}}{\text{amu}} \times (2.99792 \times 10^8 \text{ m/s})^2$$

$$= -2.8563 \times 10^{-10} \text{ J}$$

$$\frac{\text{BE}}{\text{nucleon}} = \frac{2.8563 \times 10^{-10} \text{ J}}{235 \text{ nucleons}} = 1.2154 \times 10^{-12} \text{ J/nucleon}$$

Because ^{56}Fe is the most stable known nucleus, the binding energy per nucleon for ^{56}Fe (1.408×10^{-12} J/nucleon) will be larger than that for ^{12}C or ^{235}U (see Figure 18.9 of the text).

41. Let m_{Li} = mass of ^6Li nucleus; A ^6Li nucleus has 3p and 3n.

$-0.03434 \text{ amu} = m_{Li} - (3m_p + 3m_n) = m_{Li} - [3(1.00728 \text{ amu}) + 3(1.00866 \text{ amu})]$

$m_{Li} = 6.01348$ amu

mass of ^6Li atom = 6.01348 amu + 3 m_e = 6.01348 + 3(5.49×10^{-4} amu) = 6.01513 amu (includes mass of 3 e^-)

43. $^1_1H + ^1_1H \rightarrow ^2_1H + ^0_{+1}e$; $\Delta m = (2.01410 \text{ amu} - m_e + m_e) - 2(1.00782 \text{ amu} - m_e)$

$\Delta m = 2.01410 - 2(1.00782) + 2(0.000549) = -4.4 \times 10^{-4}$ amu for two protons reacting

When two mol of protons undergo fusion, $\Delta m = -4.4 \times 10^{-4}$ g.

$\Delta E = \Delta mc^2 = -4.4 \times 10^{-7} \text{ kg} \times (3.00 \times 10^8 \text{ m/s})^2 = -4.0 \times 10^{10}$ J

$$\frac{-4.0 \times 10^{10} \text{ J}}{2 \text{ mol protons}} \times \frac{1 \text{ mol}}{1.01 \text{ g}} = -2.0 \times 10^{10} \text{ J/g of hydrogen nuclei}$$

Detection, Uses, and Health Effects of Radiation

45. The Geiger-Müller tube has a certain response time. After the gas in the tube ionizes to produce a "count," some time must elapse for the gas to return to an electrically neutral state. The response of the tube levels off because at high activities, radioactive particles are entering the tube faster than the tube can respond to them.

47. All evolved oxygen in O_2 comes from water and not from carbon dioxide.

49. $^{235}_{92}U + ^{1}_{0}n \rightarrow ^{144}_{58}Ce + ^{90}_{38}Sr + ?\ ^{1}_{0}n + ?\ ^{0}_{-1}e$; To balance the atomic number, we need 4 β-particles and to balance the mass number, we need 2 neutrons.

51. Release of Sr is probably more harmful. Xe is chemically unreactive. Strontium is in the same family as calcium and could be absorbed and concentrated in the body in a fashion similar to Ca. This puts the radioactive Sr in the bones; red blood cells are produced in bone marrow. Xe would not be readily incorporated into the body.

The chemical properties determine where a radioactive material may be concentrated in the body or how easily it may be excreted. The length of time of exposure and what is exposed to radiation significantly affects the health hazard. (See exercise 18.52 for a specific example.)

Additional Exercises

53. The most abundant isotope is generally the most stable isotope. The periodic table predicts that the most stable isotopes for exercises a - d are ^{39}K, ^{56}Fe, ^{23}Na and ^{204}Tl. (Reference Table 18.2 of the text for potential decay processes.)

a. Unstable; ^{45}K has too many neutrons and will undergo beta particle production.

b. Stable

c. Unstable; ^{20}Na has too few neutrons and will most likely undergo electron capture or positron production. Alpha particle production makes too severe of a change to be a likely decay process for the relatively light ^{20}Na nuclei. Alpha particle production usually occurs for heavy nuclei.

d. Unstable; ^{194}Tl has too few neutrons and will undergo electron capture, positron production and/or alpha particle production.

55. The third-life will be the time required for the number of nuclides to reach one-third of the original value ($N_0/3$).

$$\ln\left(\frac{N}{N_0}\right) = -kt = \frac{-0.6931\,t}{t_{1/2}}, \quad \ln\left(\frac{1}{3}\right) = \frac{-0.6931\,t}{31.4\text{ yr}}, \quad t = 49.8\text{ yr}$$

The third-life of this nuclide is 49.8 yr.

57. $\ln\left(\frac{N}{N_0}\right) = -kt = \frac{-(\ln 2)\,t}{12.3\text{ yr}}$, $\ln\left(\frac{0.17 \times N_0}{N_0}\right) = -5.64 \times 10^{-2}\,t$, $t = 31.4$ yr

It takes 31.4 yr for the tritium to decay to 17% of the original amount. Hence, the watch stopped fluorescing enough to be read in 1975 (1944 + 31.4).

59. $20{,}000\text{ ton TNT} \times \frac{4 \times 10^9\text{ J}}{\text{ton TNT}} \times \frac{1\text{ mol }^{235}\text{U}}{2 \times 10^{13}\text{ J}} \times \frac{235\text{ g }^{235}\text{U}}{\text{mol }^{235}\text{U}} = 940\text{ g }^{235}\text{U} \approx 900\text{ g }^{235}\text{U}$

This assumes that all of the ^{235}U undergoes fission.

61. Mass of nucleus = atomic mass − mass of electron = 2.01410 amu − 0.000549 amu = 2.01355 amu

$$u_{rms} = \left(\frac{3RT}{M}\right)^{1/2} = \left(\frac{3(8.3145\text{ J/K}\bullet\text{mol})(4 \times 10^7\text{ K})}{2.01355\text{ g }(1\text{ kg}/1000\text{ g})}\right)^{1/2} = 7 \times 10^5\text{ m/s}$$

$$KE_{avg} = \frac{1}{2}mu^2 = \frac{1}{2}\left(2.01355\text{ amu} \times \frac{1.66 \times 10^{-27}\text{ kg}}{\text{amu}}\right)(7 \times 10^5\text{ m/s})^2 = 8 \times 10^{-16}\text{ J/nuclei}$$

We could have used $KE_{ave} = (3/2)RT$ to determine the same average kinetic energy.

Challenge Problems

63. Assuming that the radionuclide is long lived enough such that no significant decay occurs during the time of the experiment, the total counts of radioactivity injected are:

$$0.10\text{ mL} \times \frac{5.0 \times 10^3\text{ cpm}}{\text{mL}} = 5.0 \times 10^2\text{ cpm}$$

Assuming that the total activity is uniformly distributed only in the rat's blood, the blood volume is:

$$V \times \frac{48\text{ cpm}}{\text{mL}} = 5.0 \times 10^2\text{ cpm},\ V = 10.4\text{ mL} = 10.\text{ mL}$$

65. a. ^{12}C; It takes part in the first step of the reaction but is regenerated in the last step. ^{12}C is not consumed, so it is not a reactant.

b. ^{13}N, ^{13}C, ^{14}N, ^{15}O, and ^{15}N are the intermediates.

c. $4\,^1_1\text{H} \rightarrow \,^4_2\text{H} + 2\,^0_{+1}\text{e}$; $\Delta m = 4.00260\text{ amu} - 2\,m_e + 2\,m_e - [4(1.00782\text{ amu} - m_e)]$

$\Delta m = 4.00260 - 4(1.00782) + 4(0.000549) = -0.02648$ amu for 4 protons reacting

For 4 mol of protons, $\Delta m = -0.02648$ g and ΔE for the reaction is:

$$\Delta E = \Delta mc^2 = -2.648 \times 10^{-5}\ \text{kg} \times (2.9979 \times 10^8\ \text{m/s})^2 = -2.380 \times 10^{12}\ \text{J}$$

For 1 mol of protons reacting: $\dfrac{-2.380\times10^{12}\ \text{J}}{4\ \text{mol}\ ^1\text{H}} = -5.950 \times 10^{11}\ \text{J/mol}\ ^1\text{H}$

67. $\text{mol I}^- = \dfrac{33\ \text{counts}}{\text{min}} \times \dfrac{1\ \text{mol I} \bullet \text{min}}{5.0\times10^{11}\ \text{counts}} = 6.6 \times 10^{-11}\ \text{mol I}^-$

$$[\text{I}^-] = \frac{6.6\times10^{-11}\ \text{mol I}^-}{0.150\ \text{L}} = 4.4 \times 10^{-10}\ \text{mol/L}$$

	$Hg_2I_2(s)$ →	$Hg_2^{2+}(aq)$	+	$2\ I^-(aq)$	$K_{sp} = [Hg_2^{2+}][I^-]^2$
Initial	s = solubility (mol/L)	0		0	
Equil.		s		$2s$	

From the problem, $2s = 4.4 \times 10^{-10}$ mol/L, $s = 2.2 \times 10^{-10}$ mol/L .

$$K_{sp} = (s)(2s)^2 = (2.2\times10^{-10})(4.4\times10^{-10})^2 = 4.3 \times 10^{-29}$$

Integrative Problems

69. $^{249}_{97}\text{Bk} + ^{22}_{10}\text{Ne} \rightarrow ^{267}_{107}\text{Bh} + ?$; This equation is charge balanced, but it is not mass balanced. The products are off by 4 mass units. The only possibility to account for the 4 mass units is to have 4 neutrons produced. The balanced equation is:

$$^{249}_{97}\text{Bk} + ^{22}_{10}\text{Ne} \rightarrow ^{267}_{107}\text{Bh} + 4\,^1_0\text{n}$$

$$\ln\left(\frac{N}{N_0}\right) = -kt = \frac{-0.6931\ t}{t_{1/2}},\quad \ln\left(\frac{11}{199}\right) = \frac{-0.6931\ t}{15.0\ \text{s}},\quad t = 62.7\ \text{s}\ \ \text{(Assuming 11 is exact.)}$$

Bh: $[\text{Rn}]7s^25f^{14}6d^5$ is the expected electron configuration.

CHAPTER NINETEEN

THE REPRESENTATIVE ELEMENTS: GROUPS 1A THROUGH 4A

Questions

1. The gravity of the earth is not strong enough to keep the light H_2 molecules in the atmosphere.

3. The acidity decreases. Solutions of Be^{2+} are acidic, while solutions of the other M^{2+} ions are neutral.

5. In graphite, planes of carbon atoms slide easily along each other. In addition, graphite is not volatile. The lubricant will not be lost when used in a high-vacuum environment.

7. Group 3A elements have one fewer valence electron than Si or Ge. A p-type semiconductor would form.

9. For groups 1A-3A, the small size of H (as compared to Li), Be (as compared to Mg), and B (as compared to Al) seems to be the reason why these elements have nonmetallic properties, while others in the groups 1A-3A are strictly metallic. The small size of H, Be, and B also causes these species to polarize the electron cloud in nonmetals, thus forcing a sharing of electrons when bonding occurs. For groups 4A-6A, a major difference between the first and second members of a group is the ability to form π bonds. The smaller elements form stable π bonds, while the larger elements are not capable of good overlap between parallel p orbitals and, in turn, do not form strong π bonds. For group 7A, the small size of F as compared to Cl is used to explain the low electron affinity of F and the weakness of the F–F bond.

11. Solids have stronger intermolecular forces than liquids. In order to maximize the hydrogen bonding in the solid phase, ice is forced into an open structure. This open structure is why $H_2O(s)$ is less dense than $H_2O(l)$.

Exercises

Group 1A Elements

13. a. $\Delta H° = -110.5 - [-75 + (-242)] = 207$ kJ; $\Delta S° = 198 + 3(131) - [186 + 189] = 216$ J/K

    b. $\Delta G° = \Delta H° - T\Delta S°$; $\Delta G° = 0$ when $T = \dfrac{\Delta H°}{\Delta S°} = \dfrac{207 \times 10^3 \text{ J}}{216 \text{ J/K}} = 958$ K

    At T > 958 K and standard pressures, the favorable $\Delta S°$ term dominates, and the reaction is spontaneous ($\Delta G° < 0$).

15. a. lithium oxide b. potassium superoxide c. sodium peroxide

17. a. $Li_2O(s) + H_2O(l) \rightarrow 2\ LiOH(aq)$ b. $Na_2O_2(s) + 2\ H_2O(l) \rightarrow 2\ NaOH(aq) + H_2O_2(aq)$

c. $LiH(s) + H_2O(l) \rightarrow H_2(g) + LiOH(aq)$

d. $2\ KO_2(s) + 2H_2O(l) \rightarrow 2\ KOH(aq) + O_2(g) + H_2O_2(aq)$

19. $2\ Li(s) + 2\ C_2H_2(g) \rightarrow 2\ LiC_2H(s) + H_2(g)$; This is an oxidation-reduction reaction.

Group 2A Elements

21. a. magnesium carbonate b. barium sulfate c. strontium hydroxide

23. $CaCO_3(s) + H_2SO_4(aq) \rightarrow CaSO_4(aq) + H_2O(l) + CO_2(g)$

25. In the gas phase, linear molecules would exist.

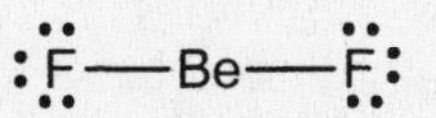

:F̤̈—Be—F̤̈:

In the solid state, BeF_2 has the following extended structure:

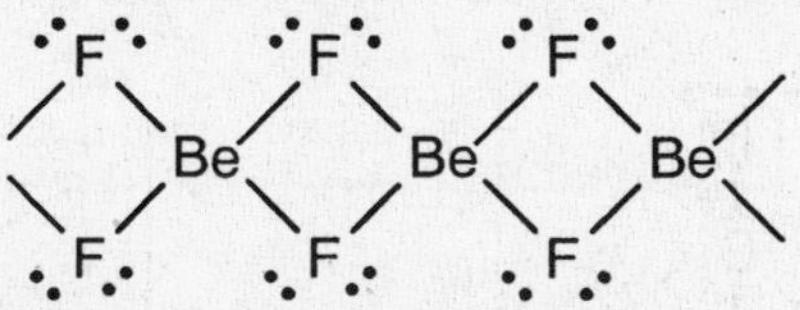

F F F
Be Be Be
F F F

27. $\frac{1\ mg\ F^-}{L} \times \frac{1\ g}{1000\ mg} \times \frac{1\ mol\ F^-}{19.00\ g\ F^-} = 5.3 \times 10^{-5}\ M\ F^- = 5 \times 10^{-5}\ M\ F^-$

$CaF_2(s) \rightleftharpoons Ca^{2+}(aq) + 2\ F^-(aq)$ $K_{sp} = [Ca^{2+}][F^-]^2 = 4.0 \times 10^{-11}$; Precipitation will occur when $Q > K_{sp}$. Let's calculate $[Ca^{2+}]$ so that $Q = K_{sp}$.

$Q = 4.0 \times 10^{-11} = [Ca^{2+}]_o[F^-]_o^2 = [Ca^{2+}]_o(5 \times 10^{-5})^2$, $[Ca^{2+}]_o = 2 \times 10^{-2}\ M$

$CaF_2(s)$ will precipitate when $[Ca^{2+}]_o > 2 \times 10^{-2}\ M$. Therefore, hard water should have a calcium ion concentration of less than $2 \times 10^{-2}\ M$ in order to avoid $CaF_2(s)$ formation.

29. $Ba^{2+} + 2\ e^- \rightarrow Ba$; $6.00\ hr \times \frac{60\ min}{h} \times \frac{60\ s}{min} \times \frac{2.50 \times 10^5\ C}{s} \times \frac{1\ mol\ e^-}{96,485\ C} \times \frac{1\ mol\ Ba}{2\ mol\ e^-}$

$\times \frac{137.3\ g\ Ba}{mol\ Ba} = 3.84 \times 10^6\ g\ Ba$

Group 3A Elements

31. a. AlN b. GaF_3 c. Ga_2S_3

33. $B_2H_6(g) + 3\ O_2(g) \rightarrow 2\ B(OH)_3(s)$

35. $Ga_2O_3(s) + 6\ H^+(aq) \rightarrow 2\ Ga^{3+}(aq) + 3\ H_2O(l)$

$Ga_2O_3(s) + 2\ OH^-(aq) + 3\ H_2O(l) \rightarrow 2\ Ga(OH)_4^-(aq)$

$In_2O_3(s) + 6\ H^+(aq) \rightarrow 2\ In^{3+}(aq) + 3\ H_2O(l)$; $In_2O_3(s) + OH^-(aq) \rightarrow$ no reaction

37. $2\ Ga(s) + 3\ F_2(g) \rightarrow 2\ GaF_3(s)$; $4\ Ga(s) + 3\ O_2(g) \rightarrow 2\ Ga_2O_3(s)$

$16\ Ga(s) + 3\ S_8(s) \rightarrow 8\ Ga_2S_3(s)$; $2\ Ga(s) + N_2(g) \rightarrow 2\ GaN(s)$

Note: GaN would be predicted, but in practice, this reaction does not occur.

$2\ Ga(s) + 6\ HCl(aq) \rightarrow 2\ GaCl_3(aq) + 3\ H_2(g)$

Group 4A Elements

39. CF_4, $4 + 4(7) = 32\ e^-$ GeF_4, $4 + 4(7) = 32\ e^-$ GeF_6^{2-}, $4 + 6(7) + 2 = 48\ e^-$

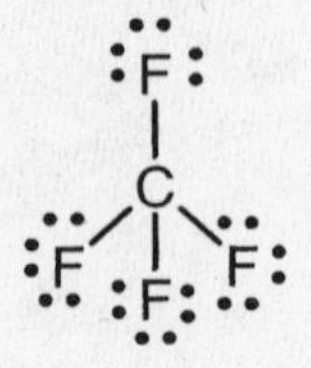

tetrahedral; 109.5°; sp^3 tetrahedral; 109.5°; sp^3 octahedral; 90°; d^2sp^3

In order to form CF_6^{2-}, carbon would have to expand its octet of electrons. Carbon compounds do not expand their octet because of the small atomic size of carbon and because no low energy d-orbitals are available for carbon to accommodate the extra electrons.

41. a. $SiO_2(s) + 2\ C(s) \rightarrow Si(s) + 2\ CO(g)$

b. $SiCl_4(l) + 2\ Mg(s) \rightarrow Si(s) + 2\ MgCl_2(s)$

c. $Na_2SiF_6(s) + 4\ Na(s) \rightarrow Si(s) + 6\ NaF(s)$

43. Lead is very toxic. As the temperature of the water increases, the solubility of lead increases. Drinking hot tap water from pipes containing lead solder could result in higher lead concentrations in the body.

45. $C_6H_{12}O_6(aq) \rightarrow 2\ C_2H_5OH(aq) + 2\ CO_2(g)$

47. The π electrons are free to move in graphite, thus giving it greater conductivity (lower resistance). The electrons in graphite have the greatest mobility within sheets of carbon atoms, resulting in a lower resistance in the plane of the sheets (basal plane). Electrons in diamond are not mobile (high resistance). The structure of diamond is uniform in all directions; thus, resistivity has no directional dependence in diamond.

Additional Exercises

49. Heat released = $0.250 \text{ g Na} \times \frac{1 \text{ mol}}{22.99 \text{ g}} \times \frac{368 \text{ kJ}}{2 \text{ mol}} = 2.00 \text{ kJ}$

To melt 50.0 g of ice requires: $50.0 \text{ g ice} \times \frac{1 \text{ mol } H_2O}{18.02 \text{ g}} \times \frac{6.02 \text{ kJ}}{\text{mol}} = 16.7 \text{ kJ}$

The reaction doesn't release enough heat to melt all of the ice. The temperature will remain at 0°C.

51. $15 \text{ kWh} = \frac{15000 \text{ J h}}{\text{s}} \times \frac{60 \text{ s}}{\text{min}} \times \frac{60 \text{ min}}{\text{h}} = 5.4 \times 10^7 \text{ J or } 5.4 \times 10^4 \text{ kJ}$ (Hall process)

To melt 1.0 kg Al requires: $1.0 \times 10^3 \text{ g Al} \times \frac{1 \text{ mol Al}}{26.98 \text{ g}} \times \frac{10.7 \text{ kJ}}{\text{mol Al}} = 4.0 \times 10^2 \text{ kJ}$

It is feasible to recycle Al by melting the metal because, in theory, it takes less than 1% of the energy required to produce the same amount of Al by the Hall process.

53.
$$HgbO_2 \rightarrow Hgb + O_2 \qquad \Delta G° = -(-70 \text{ kJ})$$
$$Hgb + CO \rightarrow HgbCO \qquad \Delta G° = -80 \text{ kJ}$$
$$HgbO_2 + CO \rightarrow HgbCO + O_2 \qquad \Delta G° = -10 \text{ kJ}$$

$$\Delta G° = -RT \ln K, \; K = \exp\left(\frac{-\Delta G°}{RT}\right) = \exp\left(\frac{-(-10 \times 10^3 \text{ J})}{8.3145 \text{ J/K} \bullet \text{mol } (298 \text{ K})}\right) = 60$$

55. n = 2 for this reaction (lead goes from $Pb \rightarrow Pb^{2+}$ in $PbSO_4$).

$$E = E° - \frac{-0.0591}{2} \log \frac{1}{[H^+]^2[HSO_4^-]^2} = 2.04 \text{ V} - \frac{-0.0591}{2} \log \frac{1}{(4.5)^2(4.5)^2}$$

$$E = 2.04 \text{ V} + 0.077 \text{ V} = 2.12 \text{ V}$$

57. Strontium and calcium are both alkaline earth metals, so they have similar chemical properties. Because milk is a good source of calcium, strontium could replace some calcium in milk without much difficulty.

59. The "inert pair effect" refers to the difficulty of removing the pair of s electrons from some of the elements in the fifth and sixth periods of the periodic table. As a result, multiple oxidation states are exhibited for the heavier elements of Groups 3A and 4A. In^+, In^{3+}, Tl^+ and Tl^{3+} oxidation states are all important to the chemistry of In and Tl.

61. Major species present: $Al(H_2O)_6^{3+}$ ($K_a = 1.4 \times 10^{-5}$), NO_3^- (neutral) and H_2O; $K_w = 1.0 \times 10^{-14}$. $Al(H_2O)_6^{3+}$ is a stronger acid than water so it will be the dominant H^+ producer.

$$Al(H_2O)_6^{3+} \rightleftharpoons Al(H_2O)_5(OH)^{2+} + H^+$$

	$Al(H_2O)_6^{3+}$		$Al(H_2O)_5(OH)^{2+}$	H^+
Initial	0.050 M		0	~0
	x mol/L $Al(H_2O)_6^{3+}$ dissociates to reach equilibrium			
Change	$-x$	→	$+x$	$+x$
Equil.	$0.050 - x$		x	x

$$K_a = 1.4 \times 10^{-5} = \frac{[Al(H_2O)_5(OH)^{2+}][H^+]}{[Al(H_2O)_6^{3+}]} = \frac{x^2}{0.050 - x} \approx \frac{x^2}{0.050}$$

$x = 8.4 \times 10^{-4}$ M = $[H^+]$; pH = $-\log(8.4 \times 10^{-4}) = 3.08$; Assumptions good.

63. Ga(I): $[Ar]4s^23d^{10}$, no unpaired e^-; Ga(III): $[Ar]3d^{10}$, no unpaired e^-

Ga(II): $[Ar]4s^13d^{10}$, 1 unpaired e^-; Note: s electrons are lost before the d electrons.

If the compound contained Ga(II), it would be paramagnetic, and if the compound contained Ga(I) and Ga(III), it would be diamagnetic. This can be determined easily by measuring the mass of a sample in the presence and in the absence of a magnetic field. Paramagnetic compounds will have an apparent increase in mass in a magnetic field.

65. $$750.\text{ mL grape juice} \times \frac{12\text{ mL } C_2H_5OH}{100.\text{ mL juice}} \times \frac{0.79\text{ g } C_2H_5OH}{\text{mL}} \times \frac{1\text{ mol } C_2H_5OH}{46.07\text{ g}}$$

$$\times \frac{2\text{ mol } CO_2}{1\text{ mol } C_2H_5OH} = 1.54\text{ mol } CO_2 \quad \text{(carry extra significant figure)}$$

1.54 mol CO_2 = total mol CO_2 = mol $CO_2(g)$ + mol $CO_2(aq)$ = $n_g + n_{aq}$

$$P_{CO_2} = \frac{n_gRT}{V} = \frac{n_g\left(\frac{0.08206\text{ L atm}}{\text{mol K}}\right)(298\text{ K})}{7.5 \times 10^{-3}\text{ L}} = 326\, n_g$$

$$P_{CO_2} = \frac{C}{k} = \frac{\frac{n_{aq}}{0.750\text{ L}}}{\frac{3.1 \times 10^{-2}\text{ mol}}{\text{L atm}}} = 43.0\, n_{aq}$$

$P_{CO_2} = 326\, n_g = 43.0\, n_{aq}$ and from above, $n_{aq} = 1.54 - n_g$; Solving:

$326\, n_g = 43.0(1.54 - n_g)$, $369\, n_g = 66.2$, $n_g = 0.18$ mol

$P_{CO_2} = 326(0.18) = 59$ atm in gas phase

$$C = kP_{CO_2} = \frac{3.1\times10^{-2}\text{ mol}}{\text{L atm}} \times 59\text{ atm} = 1.8\text{ mol CO}_2\text{/L in wine}$$

67. $Pb(NO_3)_2(aq) + H_3AsO_4(aq) \rightarrow PbHAsO_4(s) + 2\ HNO_3(aq)$

Note: The insecticide used is $PbHAsO_4$ and is commonly called lead arsenate. This is not the correct name, however. Correctly, lead arsenate would be $Pb_3(AsO_4)_2$ and $PbHAsO_4$ should be named lead hydrogen arsenate.

Challenge Problems

69. The reaction is: $X(s) + 2H_2O(l) \rightarrow H_2(g) + X(OH)_2(aq)$

$$\text{mol X} = \text{mol H}_2 = \frac{PV}{RT} = \frac{1.00\text{ atm} \times 6.10\text{ L}}{\frac{0.08206\text{ L atm}}{\text{K mol}} \times 298\text{ K}} = 0.249\text{ mol}$$

$$\text{molar mass X} = \frac{10.00\text{ g X}}{0.249\text{ mol X}} = 40.2\text{ g/mol; X is Ca.}$$

$Ca(s) + 2\ H_2O(l) \rightarrow H_2(g) + Ca(OH)_2(aq)$; $Ca(OH)_2$ is a strong base.

$$[OH^-] = \frac{10.00\text{ g Ca} \times \frac{1\text{ mol Ca}}{40.08\text{ g}} \times \frac{1\text{ mol Ca(OH)}_2}{\text{mol Ca}} \times \frac{2\text{ mol OH}^-}{\text{mol Ca(OH)}_2}}{10.0\text{ L}} = 0.0499\ M$$

pOH = −log(0.0499) = 1.302, pH = 14.000 − 1.302 = 12.698

71. $CO_2(g) + H_2O(l) \rightarrow H_2CO_3(aq)$; $H_2CO_3(aq)$ is a diprotic acid with $K_{a_1} = 4.3 \times 10^{-7}$ and $K_{a_2} = 5.6 \times 10^{-11}$. Because $K_{a_1} >> K_{a_2}$, the H^+ contribution from the K_{a_2} reaction will be insignificant.

	H_2CO_3	⇌	H^+	+	HCO_3^-
Initial	0.50 mol/1.0 L		~0		0
Change	$-x$		$+x$		$+x$
Equil.	0.50 $M - x$		x		x

$$K_{a_1} = 4.3 \times 10^{-7} = \frac{x^2}{0.50 - x} \approx \frac{x^2}{0.50},\ x = [H^+] = 4.6 \times 10^{-4}\ M;\ \text{Assumptions good.}$$

pH = −log(4.6 × 10^{-4}) = 3.34

	HCO_3^-	$\rightleftharpoons$	H^+	+	CO_3^{2-}
Initial	4.6×10^{-4} *M*		4.6×10^{-4} *M*		0
Change	$-x$		$+x$		$+x$
Equil.	$4.6 \times 10^{-4} - x$		$4.6 \times 10^{-4} + x$		x

$$K_{a_2} = 5.6 \times 10^{-11} = \frac{(4.6\times10^{-4} + x)x}{4.6\times10^{-4} - x} \approx \frac{4.6\times10^{-4}x}{4.6\times10^{-4}} = x$$

$x = [CO_3^{2-}] = 5.6 \times 10^{-11}$ *M*; Assumptions good.

73.

	Pb^{2+}	+ H_2EDTA^{2-}	$\rightleftharpoons$	$PbEDTA^{2-}$	+ $2H^+$	
Before	0.0010 *M*	0.050 *M*		0	1.0×10^{-6} *M*	(buffer, $[H^+]$ constant)
Change	-0.0010	-0.0010	→	+0.0010	No change	Reacts completely
After	0	0.049		0.0010	1.0×10^{-6}	New initial conditions
	x mol/L $PbEDTA^{2-}$ dissociates to reach equilibrium					
Change	$+x$	$+x$	←	$-x$		
Equil.	x	$0.049 + x$		$0.0010 - x$	1.0×10^{-6}	(buffer)

$$K = 1.0 \times 10^{23} = \frac{[PbEDTA^{2-}][H^+]^2}{[Pb^{2+}][H_2EDTA^{2-}]} = \frac{(0.0010 - x)(1.0\times10^{-6})^2}{(x)(0.049 + x)}$$

$$1.0 \times 10^{23} \approx \frac{(0.0010)(1.0\times10^{-12})}{(x)(0.049)}, \ x = [Pb^{2+}] = 2.0 \times 10^{-37}\ M;$$ Assumptions good.

75. Carbon cannot form the fifth bond necessary for the transition state because of the small atomic size of carbon and because carbon doesn't have low energy d orbitals available to expand the octet.

77. $PbX_4 \rightarrow PbX_2 + X_2$; From the equation, mol PbX_4 = mol PbX_2. Let x = molar mass of the halogen. Setting up an equation where mol PbX_4 = mol PbX_2:

$$\frac{25.00\text{ g}}{207.2 + 4x} = \frac{16.12\text{ g}}{207.2 + 2x};$$ Solving, x = 127.1; The halogen is iodine, I.

Integrative Problems

79. a. $\text{mol In(CH}_3)_3 = \dfrac{PV}{RT} = \dfrac{2.00\text{ atm} \times 2.56\text{ L}}{0.08206\text{ L atm/K} \bullet \text{mol} \times 900.\text{ K}} = 0.0693\text{ mol}$

$\text{mol PH}_3 = \dfrac{PV}{RT} = \dfrac{3.00\text{ atm} \times 1.38\text{ L}}{0.08206\text{ L atm/K} \bullet \text{mol} \times 900.\text{ K}} = 0.0561\text{ mol}$

Because the reaction requires a 1:1 mole ratio between these reactants, the reactant with the small number of moles (PH_3) is limiting.

$0.0561\text{ mol PH}_3 \times \dfrac{1\text{ mol InP}}{\text{mol PH}_3} \times \dfrac{145.8\text{ g InP}}{\text{mol InP}} = 8.18\text{ g InP}$

The actual yield of InP is: $0.87 \times 8.18\text{ g} = 7.1\text{ g InP}$

b. $\lambda = \dfrac{hc}{E} = \dfrac{6.626 \times 10^{-34}\text{ J s} \times 2.998 \times 10^{8}\text{ m/s}}{2.03 \times 10^{-19}\text{ J}} = 9.79 \times 10^{-7}\text{ m} = 979\text{ nm}$

From the Figure 7.2 of the text, visible light has wavelengths between 4×10^{-7} m and 7×10^{-7} m. Therefore, this wavelength is not visible to humans; it is in the infrared region of the electromagnetic radiation spectrum.

c. $[Kr]5s^2 4d^{10} 5p^4$ is the electron configuration for tellurium, Te. Because Te has more valence electrons than P, this would form an n-type semiconductor (n-type doping).

CHAPTER TWENTY
THE REPRESENTATIVE ELEMENTS: GROUPS 5A THROUGH 8A

Questions

1. This is due to nitrogen's ability to form strong π bonds whereas heavier group 5A elements do not form strong π bonds. Therefore, P_2, As_2, and Sb_2 do not form since two π bonds are required to form these diatomic substances.

3. There are medical studies that have shown an inverse relationship between the incidence of cancer and the selenium levels in soil. The foods grown in these soils and eventually digested are assumed to somehow furnish protection from cancer. Selenium is also involved in the activity of vitamin E and certain enzymes in the human body. In addition, selenium deficiency has been shown to be connected to the occurrence of congestive heart failure.

5. +6 oxidation state: SO_4^{2-}, SO_3, SF_6
   +4 oxidation state: SO_3^{2-}, SO_2, SF_4
   +2 oxidation state: SCl_2
   0 oxidation state: S_8 and all other elemental forms of sulfur
   −2 oxidation state: H_2S, Na_2S

7. a. $H_2(g) + Cl_2(g) \rightarrow 2\ HCl(g)$; this reaction produces a lot of energy which can be used in a cannon apparatus to send a stopper across the room. To initiate this extremely slow reaction, light of specific wavelengths is needed. This is the purpose of lighting the magnesium strip. When magnesium is oxidized to MgO, an intense white light is produced. Some of the wavelengths of this light can break Cl–Cl bonds and get the reaction started.

   b. Br_2 is brown. The disappearance of the brown color indicates that all of the Br_2 has reacted with the alkene (no free Br_2 is remaining).

   c. $2\ Al(s) + 3\ I_2(s) \rightarrow 2\ AlI_3(s)$; This is a highly exothermic reaction, hence the sparks that accompany this reaction. The purple smoke is excess $I_2(s)$ being vaporized [the purple smoke is $I_2(g)$].

Exercises

Group 5A Elements

9. NO_4^{3-}

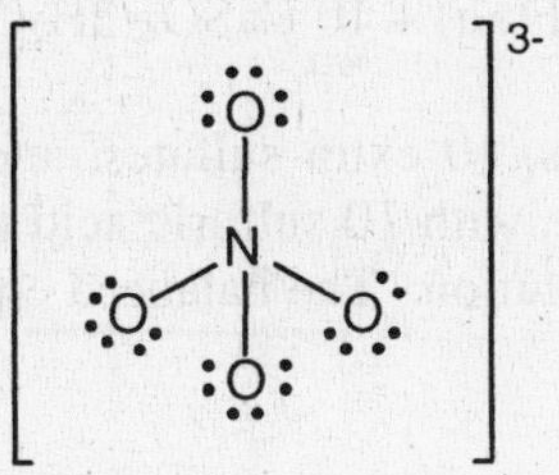

Both NO_4^{3-} and PO_4^{3-} have 32 valence electrons, so both have similar Lewis structures. From the Lewis structure for NO_4^{3-}, the central N atom has a tetrahedral arrangement of electron pairs. N is small. There is probably not enough room for all 4 oxygen atoms around N. P is larger, thus, PO_4^{3-} is stable.

PO_3^-

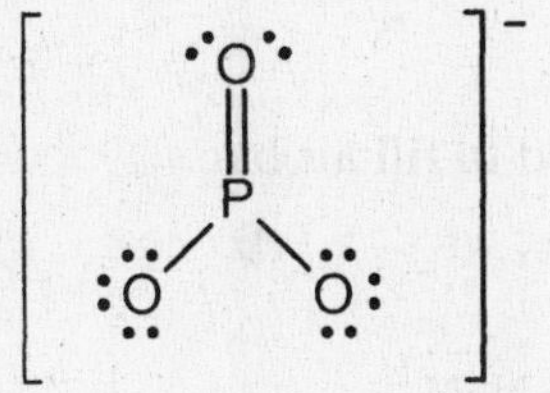

PO_3^- and NO_3^- each have 24 valence electrons so both have similar Lewis structures. From the Lewis structure for PO_3^-, PO_3^- has a trigonal planar arrangement of electron pairs about the central P atom (two single bonds and one double bond). P=O bonds are not particularly stable, while N=O bonds are stable. Thus, NO_3^- is stable.

11. a. NO: $\%N = \frac{14.01\text{ g N}}{30.01\text{ g NO}} \times 100 = 46.68\%\text{ N}$

b. NO_2: $\%N = \frac{14.01\text{ g N}}{46.01\text{ g NO}_2} \times 100 = 30.45\%\text{ N}$

c. N_2O_4: $\%N = \frac{28.02\text{ g N}}{92.02\text{ g N}_2\text{O}_4} \times 100 = 30.45\%\text{ N}$

d. N_2O: $\%N = \frac{28.02\text{ g N}}{44.02\text{ g N}_2\text{O}_4} \times 100 = 63.65\%\text{ N}$

The order from lowest to highest mass percentage of nitrogen is: $NO_2 = N_2O_4 < NO < N_2O$.

13. a. $NH_4NO_3(s) \xrightarrow{\text{heat}} N_2O(g) + 2\,H_2O(g)$

b. $2\,N_2O_5(g) \rightarrow 4\,NO_2(g) + O_2(g)$

c. $2\,K_3P(s) + 6\,H_2O(l) \rightarrow 2\,PH_3(g) + 6\,KOH(aq)$

d. $PBr_3(l) + 3\,H_2O(l) \rightarrow H_3PO_3(aq) + 3\,HBr(aq)$

e. $2\,NH_3(aq) + NaOCl(aq) \rightarrow N_2H_4(aq) + NaCl(aq) + H_2O(l)$

15. Unbalanced equation:

$$CaF_2{\bullet}3Ca_3(PO_4)_2(s) + H_2SO_4(aq) \rightarrow H_3PO_4(aq) + HF(aq) + CaSO_4{\bullet}2H_2O(s)$$

Balancing Ca^{2+}, F^-, and PO_4^{3-}:

$$CaF_2{\bullet}3Ca_3(PO_4)_2(s) + H_2SO_4(aq) \rightarrow 6\ H_3PO_4(aq) + 2\ HF(aq) + 10\ CaSO_4{\bullet}2H_2O(s)$$

On the right hand side, there are 20 extra hydrogen atoms, 10 extra sulfates, and 20 extra water molecules. We can balance the hydrogen and sulfate with 10 sulfuric acid molecules. The extra waters came from the water in the sulfuric acid solution. The balanced equation is:

$$CaF_2{\bullet}3Ca_3(PO_4)_2(s) + 10\ H_2SO_4(aq) + 20\ H_2O(l) \rightarrow$$

$$6\ H_3PO_4(aq) + 2\ HF(aq) + 10\ CaSO_4{\bullet}2H_2O(s)$$

17. $2\ NaN_3(s) \rightarrow 2\ Na(s) + 3\ N_2(g)$

$$n_{N_2} = \frac{PV}{RT} = \frac{1.00\ \text{atm} \times 70.0\ \text{L}}{\frac{0.08206\ \text{L atm}}{\text{mol K}} \times 273\ \text{K}} = 3.12\ \text{mol}\ N_2 \text{ needed to fill air bag.}$$

$$\text{mol}\ NaN_3 \text{ reacted} = 3.12\ \text{mol}\ N_2 \times \frac{2\ \text{mol}\ NaN_3}{3\ \text{mol}\ N_2} = 2.08\ \text{mol}\ NaN_3$$

19. $H_2N{-}NH_2\ (l) + O{=}O\ (g) \longrightarrow N{\equiv}N\ (g) + 2\ H{-}O{-}H\ (g)$

Bonds broken:

1 N–N (160. kJ/mol)

4 N–H (391 kJ/mol)

1 O=O (495 kJ/mol)

Bonds formed:

1 N≡N (941 kJ/mol)

2 × 2 O–H (467 kJ/mol)

$$\Delta H = 160. + 4(391) + 495 - [941 + 4(467)] = 2219\ \text{kJ} - 2809\ \text{kJ} = -590.\ \text{kJ}$$

21. $1/2\ N_2(g) + 1/2\ O_2(g) \rightarrow NO(g)$ $\Delta G° = \Delta G^\circ_{f,\,NO} = 87$ kJ/mol; By definition, ΔG°_f for a compound equals the free energy change that would accompany the formation of 1 mol of that compound from its elements in their standard states. NO (and some other oxides of nitrogen) have weaker bonds as compared to the triple bond of N_2 and the double bond of O_2. Because of this, NO (and some other oxides of nitrogen) have higher (positive) standard free energies of formation as compared to the relatively stable N_2 and O_2 molecules.

23. MO model:

NO^+: $(\sigma_{2s})^2(\sigma_{2s}{*})^2(\pi_{2p})^4(\sigma_{2p})^2$, Bond order = (8 – 2)/2 = 3, 0 unpaired e^- (diamagnetic)

NO: $(\sigma_{2s})^2(\sigma_{2s}{*})^2(\pi_{2p})^4(\sigma_{2p})^2(\pi_{2p}{*})^1$, B.O. = 2.5, 1 unpaired e^- (paramagnetic)

NO^-: $(\sigma_{2s})^2(\sigma_{2s}{*})^2(\pi_{2p})^4(\sigma_{2p})^2(\pi_{2p}{*})^2$, B.O. = 2, 2 unpaired e^- (paramagnetic)

Lewis structures: NO^+: $[:N{\equiv}O:]^+$

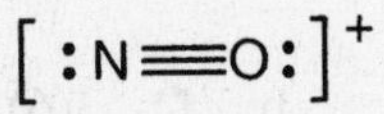

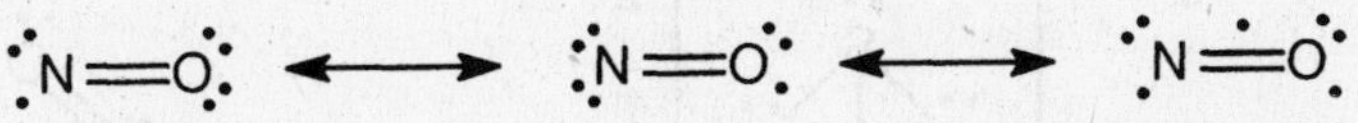

NO: $\dot{N}{=}O \longleftrightarrow N{=}O \longleftrightarrow N{=}\dot{O}$

NO^-: $[\,N{=}O\,]^-$

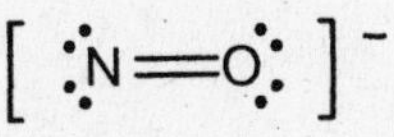

The two models give the same results only for NO^+ (a triple bond with no unpaired electrons). Lewis structures are not adequate for NO and NO^-. The MO model gives a better representation for all three species. For NO, Lewis structures are poor for odd electron species. For NO^-, both models predict a double bond, but only the MO model correctly predicts that NO^- is paramagnetic.

25. a. $H_3PO_4 > H_3PO_3$; The strongest acid has the most oxygen atoms.

b. $H_3PO_4 > H_2PO_4^- > HPO_4^{2-}$; Acid strength decreases as protons are removed.

27. The acidic protons are attached to oxygen.

$H_4P_2O_6$ (50 valence e^-):

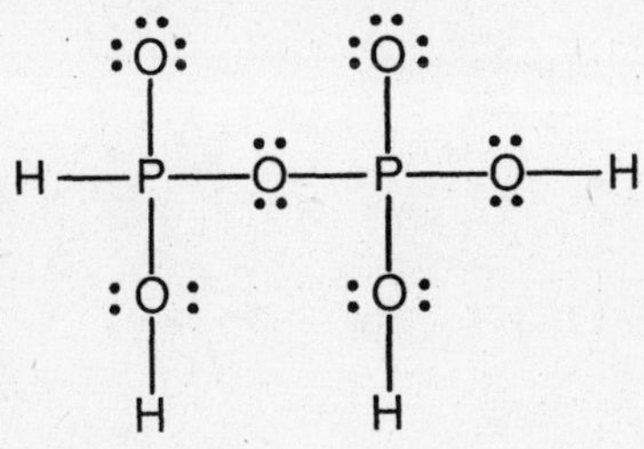

H—P(=O)(—O—H)—O—P(=O)(—O—H)—O—H

$H_4P_2O_5$ (44 valence e^-):

H—O—P(=O)(—H)—O—P(=O)(—H)—O—H

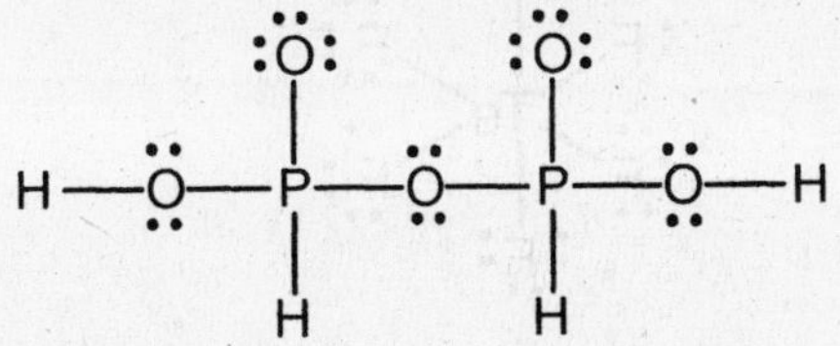

Group 6A Elements

29. O=O–O → O=O + O

Break O–O bond: $\Delta H = \dfrac{146\text{ kJ}}{\text{mol}} \times \dfrac{1\text{ mol}}{6.022 \times 10^{23}} = 2.42 \times 10^{-22}\text{ kJ} = 2.42 \times 10^{-19}\text{ J}$

A photon of light must contain at least 2.42×10^{-19} J to break one O–O bond.

$$E_{photon} = \frac{hc}{\lambda},\ \lambda = \frac{(6.626\times10^{-34}\ J\ s)(2.998\times10^{8}\ m/s)}{2.42\times10^{-19}\ J} = 8.21\times10^{-7}\ m = 821\ nm$$

31. a. $2\ SO_2(g) + O_2(g) \rightarrow 2\ SO_3(g)$

b. $SO_3(g) + H_2O(l) \rightarrow H_2SO_4(aq)$

c. $2\ Na_2S_2O_3(aq) + I_2(aq) \rightarrow Na_2S_4O_6(aq) + 2\ NaI(aq)$

d. $Cu(s) + 2\ H_2SO_4(aq) \rightarrow CuSO_4(aq) + 2\ H_2O(l) + SO_2(aq)$

33. a. SO_3^{2-}, 6 + 3(6) + 2 = 26 e^-

b. O_3, 3(6) = 18 e^-

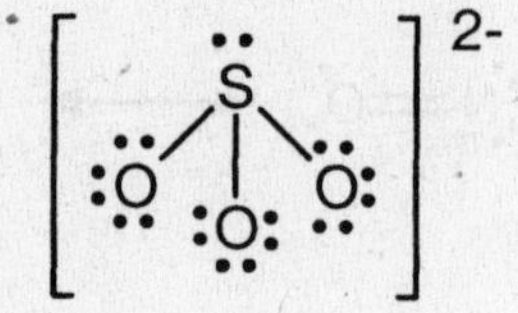

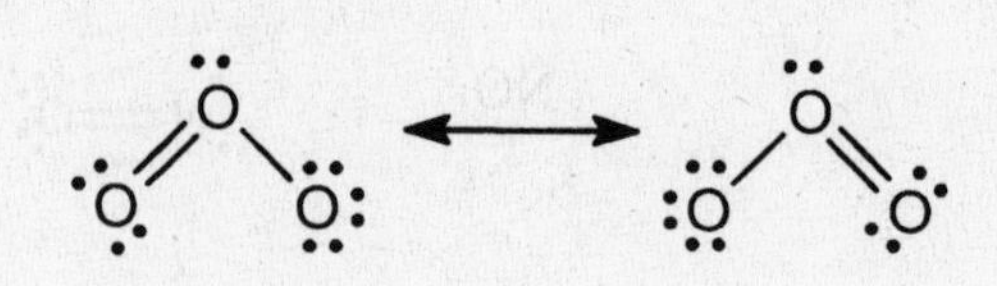

trigonal pyramid; ≈ 109.5°; sp^3

V-shaped; ≈ 120°; sp^2

c. SCl_2, 6 + 2(7) = 20 e^-

d. $SeBr_4$, 6 + 4(7) = 34 e^-

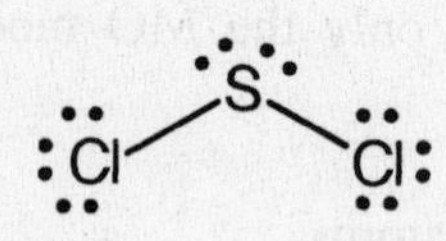

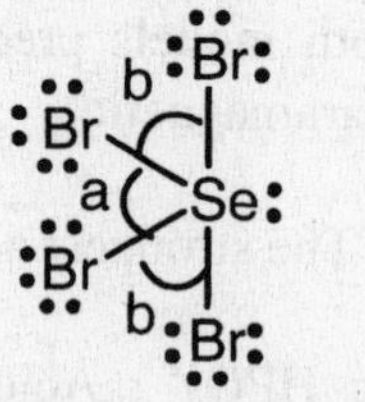

V-shaped; ≈ 109.5°; sp^3

see-saw; a ≈ 120°, b ≈ 90°; dsp^3

e. TeF_6, 6 + 6(7) = 48 e^-

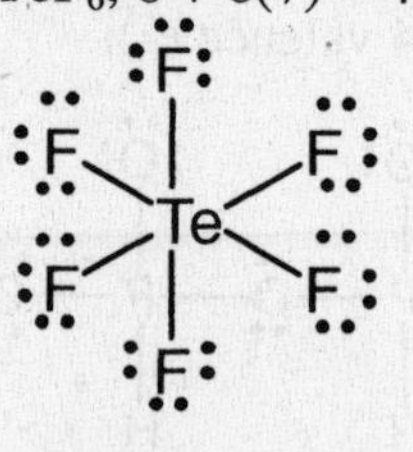

octahedral; 90°; d^2sp^3

35. $$1.50\ g\ BaO_2 \times \frac{1\ mol\ BaO_2}{169.3\ g\ BaO_2} = 8.86\times10^{-3}\ mol\ BaO_2$$

$$25.0\ mL \times \frac{0.0272\ g\ HCl}{mL} \times \frac{1\ mol\ HCl}{36.46\ HCl} = 1.87\times10^{-2}\ mol\ HCl$$

The required mole ratio from the balanced reaction is 2 mol HCl to 1 mol BaO_2. The actual ratio is:

$$\frac{1.87 \times 10^{-2} \text{ mol HCl}}{8.86 \times 10^{-3} \text{ mol BaO}_2} = 2.11$$

Because the actual mole ratio is larger than the required mole ratio, the denominator (BaO_2) is the limiting reagent.

$$8.86 \times 10^{-3} \text{ mol BaO}_2 \times \frac{1 \text{ mol H}_2\text{O}_2}{\text{mol BaO}_2} \times \frac{34.02 \text{ g H}_2\text{O}_2}{\text{mol H}_2\text{O}_2} = 0.301 \text{ g H}_2\text{O}_2$$

The amount of HCl reacted is:

$$8.86 \times 10^{-3} \text{ mol BaO}_2 \times \frac{2 \text{ mol HCl}}{\text{mol BaO}_2} = 1.77 \times 10^{-2} \text{ mol HCl}$$

$$\text{excess mol HCl} = 1.87 \times 10^{-2} \text{ mol} - 1.77 \times 10^{-2} \text{ mol} = 1.0 \times 10^{-3} \text{ mol HCl}$$

$$\text{mass of excess HCl} = 1.0 \times 10^{-3} \text{ mol HCl} \times \frac{36.46 \text{ g HCl}}{\text{mol HCl}} = 3.6 \times 10^{-2} \text{ g HCl}$$

Group 7A Elements

37. O_2F_2 has $2(6) + 2(7) = 26$ valence e^-; From the following Lewis structure, each oxygen atom has a tetrahedral arrangement of electron pairs. Therefore, bond angles ≈ 109.5° and each O is sp^3 hybridized.

:F̤̈—Ö̤—Ö̤—F̤̈:

	F	O	O	F
Formal Charge	0	0	0	0
Oxid. Number	-1	+1	+1	-1

Oxidation numbers are more useful. We are forced to assign +1 as the oxidation number for oxygen. Oxygen is very electronegative, and +1 is not a stable oxidation state for this element.

39. a. $BaCl_2(s) + H_2SO_4(aq) \rightarrow BaSO_4(s) + 2\ HCl(g)$

b. $BrF(s) + H_2O(l) \rightarrow HF(aq) + HOBr(aq)$

c. $SiO_2(s) + 4\ HF(aq) \rightarrow SiF_4(g) + 2\ H_2O(l)$

41.
$$ClO^- + H_2O + 2e^- \rightarrow 2\ OH^- + Cl^- \qquad E° = 0.90 \text{ V}$$
$$2\ NH_3 + 2\ OH^- \rightarrow N_2H_4 + 2\ H_2O + 2e^- \qquad -E° = 0.10 \text{ V}$$

$$ClO^-(aq) + 2\ NH_3(aq) \rightarrow Cl^-(aq) + N_2H_4(aq) + H_2O(l) \qquad E°_{cell} = 1.00 \text{ V}$$

Because E°_{cell} is positive for this reaction, ClO^- , at standard conditions, can spontaneously oxidize NH_3 to the somewhat toxic N_2H_4.

Group 8A Elements

43. Xe has one more valence electron than I. Thus, the isoelectric species will have I plus one extra electron substituted for Xe, giving a species with a net minus one charge.

a. IO_4^- b. IO_3^- c. IF_2^- d. IF_4^- e. IF_6^-

45. XeF_2 can react with oxygen and water to produce explosive xenon oxides and oxyfluorides, and react with water to form HF.

47. Release of Sr is probably more harmful. Xe is chemically unreactive. Strontium is in the same family as calcium and could be absorbed and concentrated in the body in a fashion similar to Ca. This puts the radioactive Sr in the bones, and red blood cells are produced in bone marrow. Xe would not be readily incorporated into the body.

The chemical properties determine where a radioactive material may concentrate in the body or how easily it may be excreted. The length of time of exposure and what is exposed to radiation significantly affects the health hazard.

Additional Exercises

49. As the halogen atoms get larger, it becomes more difficult to fit three halogen atoms around the small nitrogen atom, and the NX_3 molecule becomes less stable.

51. OCN– has 6 + 4 + 5 + 1 = 16 valence electrons.

[:Ö=C=N̈:]⁻ ⟷ [:Ö̤—C≡N:]⁻ ⟷ [:O≡C—N̤̈:]⁻

	O	C	N	O	C	N	O	C	N
Formal charge	0	0	-1	-1	0	0	+1	0	-2

Only the first two resonance structures should be important. The third places a positive formal charge on the most electronegative atom in the ion and a -2 formal charge on N.

CNO^-:

[:C̈=N=Ö:]⁻ ⟷ [:C≡N—Ö̤:]⁻ ⟷ [:C̤̈—N≡O:]⁻

	C	N	O	C	N	O	C	N	O
Formal charge	-2	+1	0	-1	+1	-1	-3	+1	+1

All of the resonance structures for fulminate (CNO^-) involve greater formal charges than in cyanate (OCN^-), making fulminate more reactive (less stable).

53. $1.0 \times 10^4 \text{ kg waste} \times \frac{3.0 \text{ kg } NH_4^+}{100 \text{ kg waste}} \times \frac{1000 \text{ g}}{\text{kg}} \times \frac{1 \text{ mol } NH_4^+}{18.04 \text{ g } NH_4^+} \times \frac{1 \text{ mol } C_5H_7O_2N}{55 \text{ mol } NH_4^+}$

$$\times \frac{113.12 \text{ g } C_5H_7O_2N}{\text{mol } C_5H_7O_2N} = 3.4 \times 10^4 \text{ g tissue if all } NH_4^+ \text{ converted}$$

Since only 95% of the NH_4^+ ions react:

mass of tissue = (0.95) (3.4×10^4 g) = 3.2×10^4 g or 32 kg bacterial tissue

55. TeF_5^- has 6 + 5(7) + 1 = 42 valence electrons.

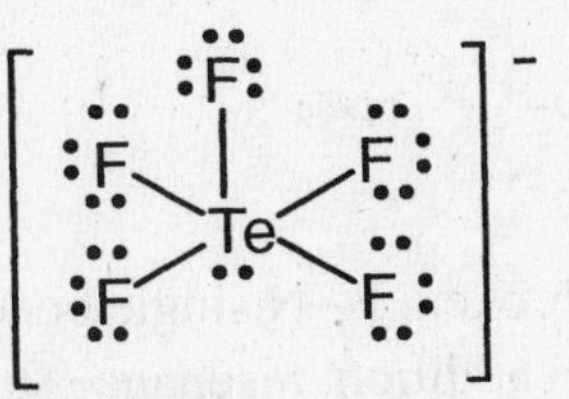

[Lewis structure of TeF_5^-: Te bonded to five F atoms (one axial, four square planar), each F with three lone pairs, one lone pair on Te, bracketed with − charge]

The lone pair of electrons around Te exerts a stronger repulsion than the bonding pairs, pushing the four square planar F's away from the lone pair and thus reducing the bond angles between the axial F atom and the square planar F atoms.

57. As temperature increases, the value of K decreases. This is consistent with an exothermic reaction. In an exothermic reaction, heat is a product and an increase in temperature shifts the equilibrium to the reactant side (as well as lowering the value of K).

59. $MgSO_4(s) \rightarrow Mg^{2+}(aq) + SO_4^{2-}(aq)$; $NH_4NO_3(s) \rightarrow NH_4^+(aq) + NO_3^-(aq)$

Note that the dissolution of $MgSO_4$ used in hot packs is an exothermic process, and the dissolution of NH_4NO_3 in cold packs is an endothermic process.

61. EO_3^- is the formula of the ion. The Lewis structure has 26 valence electrons. Let x = number of valence electrons of element E.

$$26 = x + 3(6) + 1, \quad x = 7 \text{ valence electrons}$$

Element E is a halogen because halogens have 7 valence electrons. Some possible identities are F, Cl, Br and I. The EO_3^- ion has a trigonal pyramid molecular structure with bond angles ≈ 109.5°.

63. $8 \text{ corners} \times \frac{1/8 \text{ Xe}}{\text{corner}} + 1 \text{ Xe inside cell} = 2 \text{ Xe}$; $8 \text{ edges} \times \frac{1/4 \text{ F}}{\text{edge}} + 2 \text{ F inside cell} = 4 \text{ F}$

Empirical formula is XeF_2. This is also the molecular formula.

Challenge Problems

65. For the reaction:

$$O_2N{-}N{=}O \longrightarrow NO_2 + NO$$

the activation energy must in some way involve breaking a nitrogen-nitrogen single bond.

For the reaction:

$$O_2N{-}N{=}O \longrightarrow O_2 + N_2O$$

at some point nitrogen-oxygen bonds must be broken. N–N single bonds (160. kJ/mol) are weaker than N–O single bonds (201 kJ/mol). In addition, resonance structures indicate that there is more double bond character in the N–O bonds than in the N–N bond. Thus, NO_2 and NO are preferred by kinetics because of the lower activation energy.

67. a. NO is the catalyst. NO is present in the first step of the mechanism on the reactant side, but it is not a reactant because it is regenerated in the second step and does not appear in the overall balanced equation.

b. NO_2 is an intermediate. Intermediates also never appear in the overall balanced equation. In a mechanism, intermediates always appear first on the product side while catalysts always appear first on the reactant side.

c. $k = A\exp(-E_a/RT)$; $\dfrac{k_{cat}}{k_{un}} = \dfrac{A\exp[-E_a(cat)/RT]}{A\exp[-E_a(cat)/RT]} = \exp\left(\dfrac{E_a(un) - E_a(cat)}{RT}\right)$

$$\frac{k_{cat}}{k_{un}} = \exp\left(\frac{2100\ J/mol}{8.3145\ J/K \bullet mol \times 298\ K}\right) = e^{0.85} = 2.3$$

The catalyzed reaction is approximately 2.3 times faster than the uncatalyzed reaction at 25°C.

d. The mechanism for the chlorine-catalyzed destruction of ozone is:

$$O_3(g) + Cl(g) \rightarrow O_2(g) + ClO(g) \quad \text{slow}$$
$$ClO(g) + O(g) \rightarrow O_2(g) + Cl(g) \quad \text{fast}$$

$$O_3(g) + O(g) \rightarrow 2\,O_2(g)$$

e. Because the chlorine atom-catalyzed reaction has a lower activation energy, the Cl catalyzed rate is faster. Hence, Cl is a more effective catalyst. Using the activation energy, we can estimate the efficiency that Cl atoms destroy ozone as compared to NO molecules.

$$\text{At 25°C:}\quad \frac{k_{Cl}}{k_{NO}} = \exp\left(\frac{-E_a(Cl)}{RT} + \frac{E_a(NO)}{RT}\right) = \exp\left(\frac{(-2100 + 11{,}900)\ J/mol}{(8.3145 \times 298)\ J/mol}\right)$$

$$= e^{3.96} = 52$$

At 25°C, the Cl–catalyzed reaction is roughly 52 times faster (more efficient) than the NO–catalyzed reaction, assuming the frequency factor A is the same for each reaction and assuming similar rate laws.

69. $NH_3 + NH_3 \rightleftharpoons NH_4^+ + NH_2^- \quad K = [NH_4^+][NH_2^-] = 1.8 \times 10^{-12}$

NH_3 is the solvent, so it is not included in the K expression. In a neutral solution of ammonia:

$$[NH_4^+] = [NH_2^-];\quad 1.8 \times 10^{-12} = [NH_4^+]^2,\quad [NH_4^+] = 1.3 \times 10^{-6}\ M = [NH_2^-]$$

We could abbreviate this autoionization as: $NH_3 \rightleftharpoons H^+ + NH_2^-$, where $[H^+] = [NH_4^+]$.

This abbreviation is synonomous to the abbreviation of the autoionization of water ($H_2O \rightleftharpoons H^+ + OH^-$). So: $pH = pNH_4 = -\log(1.3 \times 10^{-6}) = 5.89$.

71. Let n_{SO_2} = initial mol SO_2 present. The reaction is summarized in the following table (O_2 is in excess).

	$2\ SO_2$	+ $O_2(g)$	$\rightarrow$ $2\ SO_3(g)$
Initial	n_{SO_2}	2.00 mol	0
Change	$-n_{SO_2}$	$-n_{SO_2}/2$	$+n_{SO_2}$
Final	0	$2.00 - n_{SO_2}/2$	n_{SO_2}

d = mass/volume; Let d_i = initial density of gas mixture and d_f = final density of gas mixture after reaction. Because mass is conserved in a chemical reaction, $mass_i = mass_f$.

$$\frac{d_f}{d_i} = \frac{mass_f / V_f}{mass_i / V_i} = \frac{V_i}{V_f}$$

At constant P and T, V ∝ n, so: $\frac{d_f}{d_i} = \frac{V_i}{V_f} = \frac{n_i}{n_f}$; Setting up an equation:

$$\frac{d_f}{d_i} = \frac{0.8471\ g/L}{0.8000\ g/L} = 1.059,\ 1.059 = \frac{n_i}{n_f} = \frac{n_{SO_2} + 2.00}{(2.00 - n_{SO_2}/2) + n_{SO_2}} = \frac{n_{SO_2} + 2.00}{2.00 + n_{SO_2}/2}$$

Solving: $n_{SO_2} = 0.25$ mol; so, 0.25 moles of SO_3 formed

$$0.25\ \text{mol}\ SO_3 \times \frac{80.07\ g}{\text{mol}} = 20.\ g\ SO_3$$

Integrative Problems

73. a. $-307\ kJ = [-1136 + x] - [(-254\ kJ) + 3(-96\ kJ)],\ x = \Delta H^\circ_{f,NI_3} = 287$ kJ/mol

b. IF_2^+, $7 + 2(7) - 1 = 20\ e^-$ BF_4^-, $3 + 4(7) + 1 = 32\ e^-$

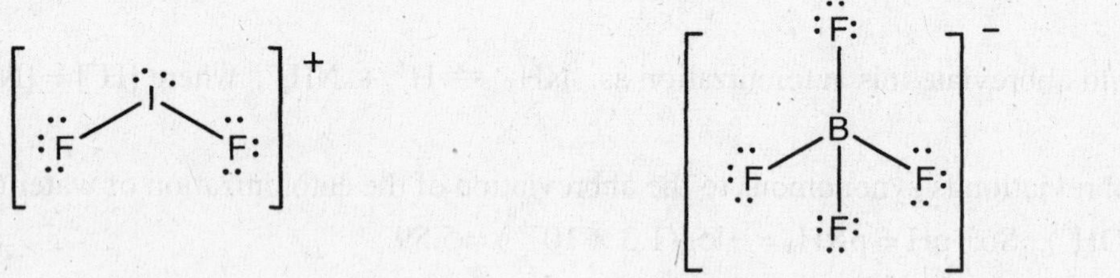

V-shaped; sp^3 tetrahedral; sp^3

CHAPTER TWENTY-ONE

TRANSITION METALS AND COORDINATION CHEMISTRY

Questions

5. $Fe_2O_3(s) + 6\ H_2C_2O_4(aq) \rightarrow 2\ Fe(C_2O_4)_3^{3-}(aq) + 3\ H_2O(l) + 6\ H^+(aq)$; The oxalate anion forms a soluble complex ion with iron in rust (Fe_2O_3), which allows rust stains to be removed.

7. Cl, H_3N, NH_3, Co, H_3N, NH_3, Cl

   Cl, H_3N, Cl, Co, H_3N, NH_3, NH_3 | mirror | Cl, Cl, NH_3, Co, NH_3, NH_3, NH_3

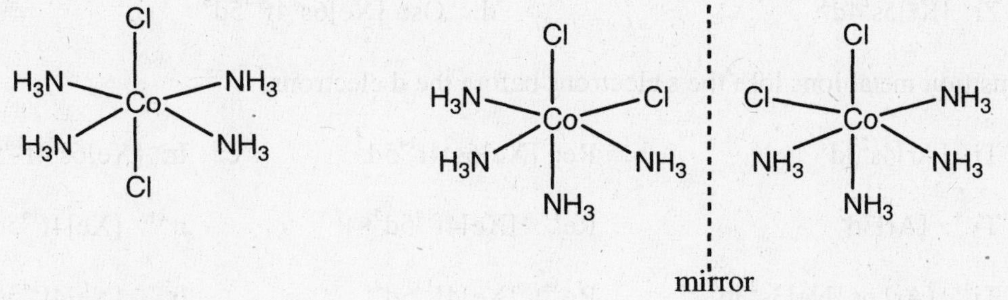

   trans (mirror image is superimposable)

   cis

   The mirror image of the cis isomer is also superimposable.

   No; both the trans or the cis forms of $Co(NH_3)_4Cl_2^+$ have mirror images that are superimposable. For the cis form, the mirror image only needs a 90° rotation to produce the original structure. Hence, neither the trans nor cis forms are optically active.

9. a. $CoCl_4^{2-}$; Co^{2+}: $4s^0 3d^7$; All tetrahedral complexes are a weak field (high-spin).

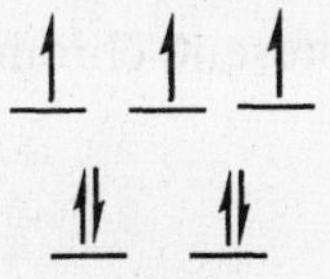

   ↑ ↑ ↑

   ↑↓ ↑↓

   small Δ

   $CoCl_4^{2-}$ is an example of a weak-field case having three unpaired electrons.

   b. $Co(CN)_6^{3-}$: Co^{3+} : $4s^0 3d^6$; Because CN^- is a strong-field ligand, $Co(CN)_6^{3-}$ will be a strong-field case (low-spin case).

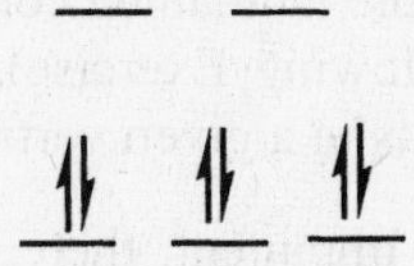

   __ __

   ↑↓ ↑↓ ↑↓

   large Δ

   CN^- is a strong field ligand so $Co(CN)_6^{3-}$ will be a low-spin case having zero unpaired electrons.

11. At high altitudes, the oxygen content of air is lower, so less oxyhemoglobin is formed which diminishes the transport of oxygen in the blood. A serious illness called high-altitude sickness can result from the decrease of O_2 in the blood. High-altitude acclimatization is the phenomenom that occurs in the human body in response to the lower amounts of oxyhemoglobin in the blood. This response is to produce more hemoglobin, and, hence, increase the oxyhemoglobin in the blood. High-altitude acclimatization takes several weeks to take hold for people moving from lower altitudes to higher altitudes.

Exercises

Transition Metals and Coordination Compounds

13. a. Ni: $[Ar]4s^2 3d^8$ b. Cd: $[Kr]5s^2 4d^{10}$

 c. Zr: $[Kr]5s^2 4d^2$ d. Os: $[Xe]6s^2 4f^{14} 5d^6$

15. Transition metal ions lose the s electrons before the d electrons.

 a. Ti: $[Ar]4s^2 3d^2$ b. Re: $[Xe]6s^2 4f^{14} 5d^5$ c. Ir: $[Xe]6s^2 4f^{14} 5d^7$

 Ti^{2+}: $[Ar]3d^2$ Re^{2+}: $[Xe]4f^{14} 5d^5$ Ir^{2+}: $[Xe]4f^{14} 5d^7$

 Ti^{4+}: [Ar] or $[Ne]3s^2 3p^6$ Re^{3+}: $[Xe]4f^{14} 5d^4$ Ir^{3+}: $[Xe]4f^{14} 5d^6$

17. a. With K^+ and CN^- ions present, iron has a +3 charge. Fe^{3+}: $[Ar]3d^5$

 b. With a Cl^- ion and neutral NH_3 molecules present, silver has a +1 charge. Ag^+: [Kr] $4d^{10}$

 c. With Br^- ions and neutral H_2O molecules present, nickel has a +2 charge. Ni^{2+}: $[Ar]3d^8$

 d. With NO_2^- ions, an I^- ion, and neutral H_2O molecules present, chromium has a +3 charge. Cr^{3+}: $[Ar]3d^3$

19. a. molybdenum(IV) sulfide; molybdenum(VI) oxide

 b. MoS_2, +4; MoO_3, +6; $(NH_4)_2Mo_2O_7$, +6; $(NH_4)_6Mo_7O_{24}\bullet 4\ H_2O$, +6

21. The lanthanide elements are located just before the 5d transition metals. The lanthanide contraction is the steady decrease in the atomic radii of the lanthanide elements when going from left to right across the periodic table. As a result of the lanthanide contraction, the sizes of the 4d and 5d elements are very similar (see the following Exercise). This leads to a greater similarity in the chemistry of the 4d and 5d elements in a given vertical group.

23. $CoCl_2(s) + 6\ H_2O(g) \rightleftharpoons CoCl_2\bullet 6\ H_2O(s)$; If rain were imminent, there would be a lot of water vapor in the air causing the reaction to shift to the right. The indicator would take on the color of $CoCl_2\bullet 6\ H_2O$, pink.

25. Test tube 1: added Cl^- reacts with Ag^+ to form a silver chloride precipitate. The net ionic equation is $Ag^+(aq) + Cl^-(aq) \rightarrow AgCl(s)$. Test tube 2: added NH_3 reacts with Ag^+ ions to form the soluble complex ion $Ag(NH_3)_2^+$. As this complex ion forms, Ag^+ is removed from solution, which causes the AgCl(s) to dissolve. When enough NH_3 is added, all of the silver chloride precipitate will dissolve. The equation is $AgCl(s) + 2\ NH_3(aq) \rightarrow Ag(NH_3)_2^+(aq) + Cl^-\ (aq)$. Test tube 3: added H^+ reacts with the weak base NH_3 to form NH_4^+. As NH_3 is removed from the $Ag(NH_3)_2^+$ complex ion equilibrium, Ag^+ ions are released to the solution which can then react with Cl^- to reform AgCl(s). The equations are $Ag(NH_3)_2^+(aq) + 2\ H^+(aq) \rightarrow Ag^+(aq) + 2\ NH_4^+(aq)$ and $Ag^+(aq) + Cl^-(aq) \rightarrow AgCl(s)$.

27. Because each compound contains an octahedral complex ion, the formulas for the compounds are $[Co(NH_3)_6]I_3$, $[Pt(NH_3)_4I_2]I_2$, $Na_2[PtI_6]$ and $[Cr(NH_3)_4I_2]I$. Note that in some cases, the I^- ions are ligands bound to the transition metal ion as required for a coordination number of 6, while in other cases the I^- ions are counterions required to balance the charge of the complex ion. The $AgNO_3$ solution will only precipitate the I^- counterions and will not precipitate the I^- ligands. Therefore, 3 moles of AgI will precipitate per mole of $[Co(NH_3)_6]I_3$, 2 moles of AgI will precipitate per mole of $[Pt(NH_3)_4I_2]I_2$, 0 moles of AgI will precipitate per mole of $Na_2[PtI_6]$, and 1 mole of AgI will precipitate per mole of $[Cr(NH_3)_4I_2]I$.

29. To determine the oxidation state of the metal, you must know the charges of the various common ligands (see Table 21.13 of the text).

a. pentaamminechlororuthenium(III) ion
b. hexacyanoferrate(II) ion
c. tris(ethylenediamine)manganese(II) ion
d. pentaamminenitrocobalt(III) ion

31. a. hexaamminecobalt(II) chloride
b. hexaaquacobalt(III) iodide
c. potassium tetrachloroplatinate(II)
d. potassium hexachloroplatinate(II)
e. pentaamminechlorocobalt(III) chloride
f. triamminetrinitrocobalt(III)

33. a. $K_2[CoCl_4]$
b. $[Pt(H_2O)(CO)_3]Br_2$
c. $Na_3[Fe(CN)_2(C_2O_4)_2]$
d. $[Cr(NH_3)_3Cl(H_2NCH_2CH_2NH_2)]I_2$

35. a.

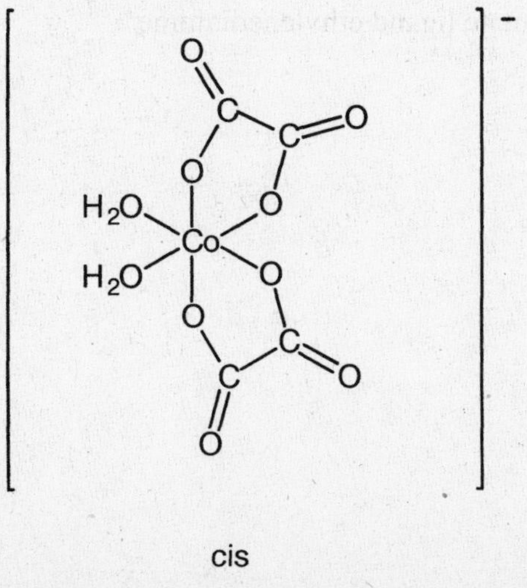

cis

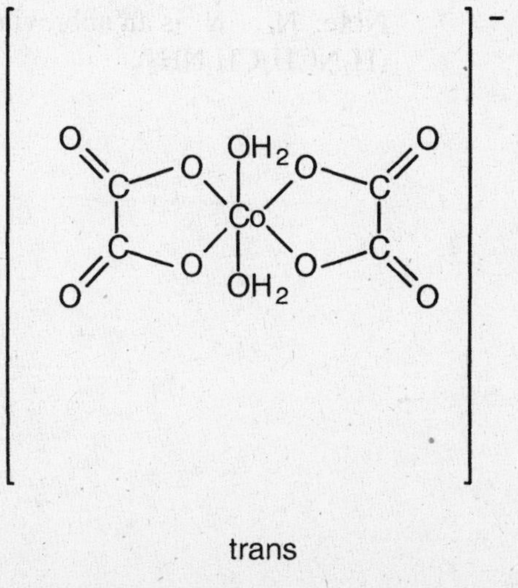

trans

Note: $C_2O_4^{2-}$ is a bidentate ligand. Bidentate ligands bond to the metal at two positions that are 90° apart from each other in octahedral complexes. Bidentate ligands do not bond to the metal at positions 180° apart.

b.

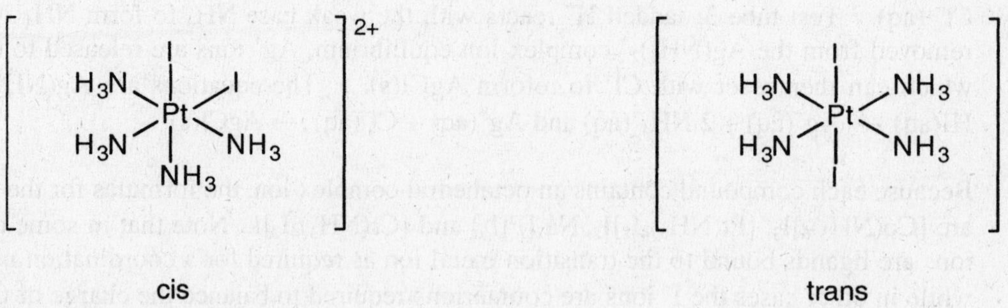

[H_3N, I, Pt, H_3N, NH_3, NH_3]$^{2+}$ [H_3N, NH_3, Pt, H_3N, NH_3]$^{2+}$

cis

trans

c.

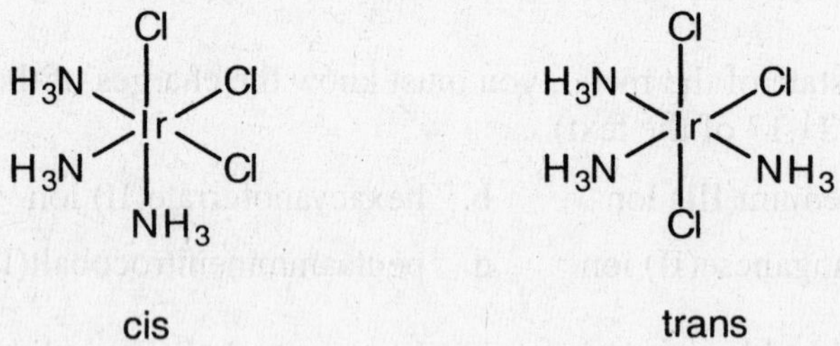

Cl, H_3N, Cl, Ir, H_3N, Cl, NH_3 Cl, H_3N, Cl, Ir, H_3N, NH_3, Cl

cis

trans

d.

[N, H_3N, N, Cr, I, NH_3, I]$^{+}$ [N, I, N, Cr, H_3N, I, NH_3]$^{+}$ [N, I, N, Cr, H_3N, NH_3]$^{+}$

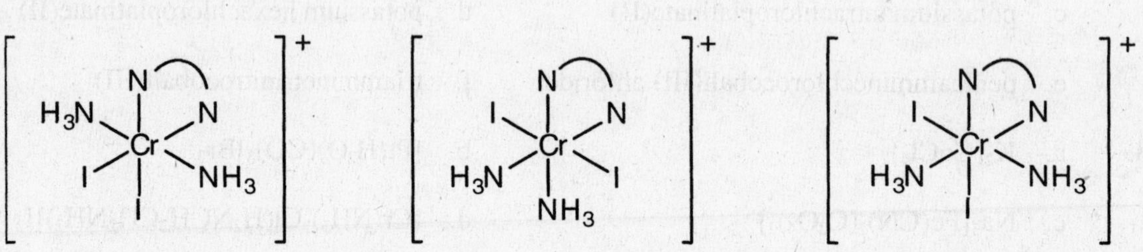

Note: N⌒N is an abbreviation for the bidentate ligand ethylenediamine ($H_2NCH_2CH_2NH_2$).

37.

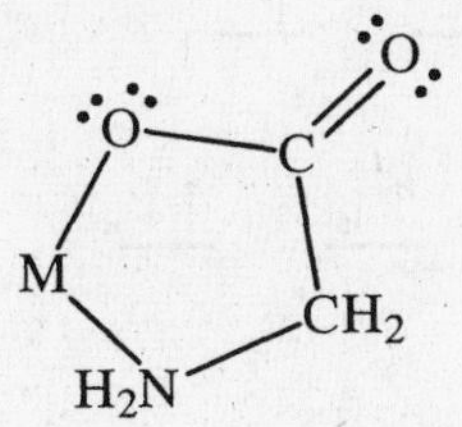

M = transition metal ion

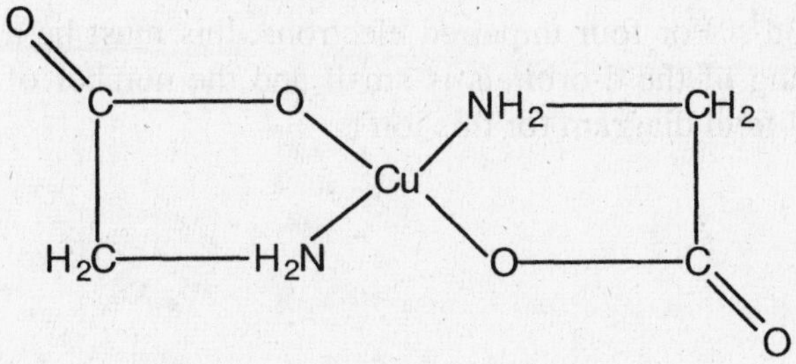

and

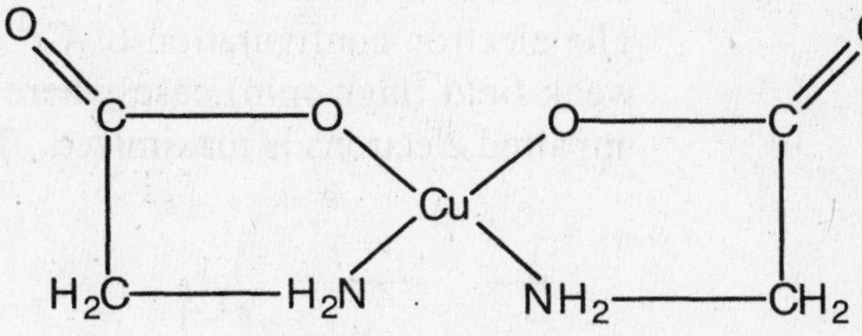

39. Linkage isomers differ in the way the ligand bonds to the metal. SCN^- can bond through the sulfur or through the nitrogen atom. NO_2^- can bond through the nitrogen or through the oxygen atom. OCN^- can bond through the oxygen or through the nitrogen atom. N_3^-, $NH_2CH_2CH_2NH_2$ and I^- are not capable of linkage isomerism.

41. Similar to the molecules discussed in Figures 21.16 and 21.17 of the text, $Cr(acac)_3$ and cis-$Cr(acac)_2(H_2O)_2$ are optically active. The mirror images of these two complexes are nonsuperimposable. There is a plane of symmetry in trans-$Cr(acac)_2(H_2O)_2$, so it is not optically active. A molecule with a plane of symmetry is never optically active because the mirror images are always superimposable. A plane of symmetry is a plane through a molecule where one side reflects the other side of the molecule.

Bonding, Color, and Magnetism in Coordination Compounds

43. a. Fe^{2+}: $[Ar]3d^6$

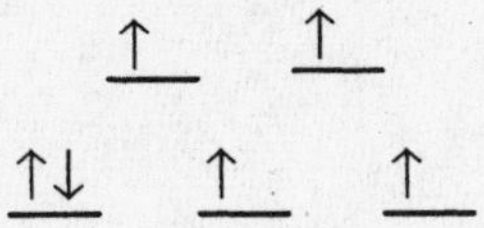

High spin, small Δ

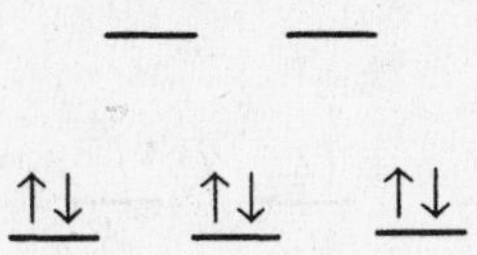

Low spin, large Δ

b. Fe^{3+}: [Ar]$3d^5$

c. Ni^{2+}: [Ar]$3d^8$

45. Because fluorine has a −1 charge as a ligand, chromium has a +2 oxidation state in CrF_6^{4-}. The electron configuration of Cr^{2+} is: [Ar]$3d^4$. For four unpaired electrons, this must be a weak-field (high-spin) case where the splitting of the d-orbitals is small and the number of unpaired electrons is maximized. The crystal field diagram for this ion is:

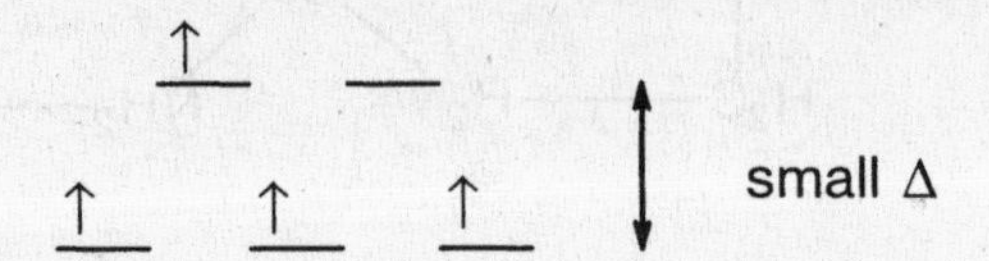

47. To determine the crystal field diagrams, you need to determine the oxidation state of the transition metal, which can only be determined if you know the charges of the ligands (see Table 21.13). The electron configurations and the crystal field diagrams follow.

a. Ru^{2+}: [Kr]$4d^6$, no unpaired e^-

b. Ni^{2+}: [Ar]$3d^8$, 2 unpaired e^-

Low spin, large Δ

c. V^{3+}: [Ar]$3d^2$, 2 unpaired e^-

Note: Ni^{2+} must have 2 unpaired electrons, whether high-spin or low-spin, and V^{3+} must have 2 unpaired electrons, whether high-spin or low-spin.

49. All have octahedral Co^{3+} ions so the difference in d orbital splitting and the wavelength of light absorbed only depends on the ligands. From the spectrochemical series, the order of the ligands from strongest to weakest field is $CN^- >$ en $> H_2O > I^-$. The strongest field ligand produces the greatest d-orbital splitting (Δ) and will absorb light having the smallest

wavelength. The weakest field ligand produces the smallest Δ and absorbs light having the longest wavelength. The order is

$Co(CN)_6^{3-}$ < $Co(en)_3^{3+}$ < $Co(H_2O)_6^{3+}$ < CoI_6^{3-}
shortest λ absorbed — longest λ absorbed

51. From Table 21.16 of the text, the violet complex ion absorbs yellow-green light (λ ~ 570 nm), the yellow complex ion absorbs blue light (λ ~ 450 nm), and the green complex ion absorbs red light (λ ~ 650 nm). The spectrochemical series shows that NH_3 is a stronger-field ligand than H_2O which is a stronger-field ligand than Cl^-. Therefore, $Cr(NH_3)_6^{3+}$ will have the largest d-orbital splitting and will absorb the lowest wavelength electromagnetic radiation (λ ~ 450 nm) since energy and wavelength are inversely related (λ = hc/E). Thus, the yellow solution contains the $Cr(NH_3)_6^{3+}$ complex ion. Similarly, we would expect the $Cr(H_2O)_4Cl_2^+$ complex ion to have the smallest d-orbital splitting since it contains the weakest-field ligands. The green solution with the longest wavelength of absorbed light contains the $Cr(H_2O)_4Cl_2^+$ complex ion. This leaves the violet solution, which contains the $Cr(H_2O)_6^{3+}$ complex ion. This makes sense as we would expect $Cr(H_2O)_6^{3+}$ to absorb light of a wavelength between that of $Cr(NH_3)_6^{3+}$ and $Cr(H_2O)_4Cl_2^+$.

53. $CoBr_6^{4-}$ has an octahedral structure, and $CoBr_4^{2-}$ has a tetrahedral structure (as do most Co^{2+} complexes with four ligands). Coordination complexes absorb electromagnetic radiation (EMR) of energy equal to the energy difference between the split d-orbitals. Because the tetrahedral d-orbital splitting is less than one-half of the octahedral d-orbital splitting, tetrahedral complexes will absorb lower energy EMR, which corresponds to longer wavelength EMR (E = hc/λ). Therefore, $CoBr_6^{2-}$ will absorb EMR having a wavelength shorter than 3.4×10^{-6} m.

55. Because the ligands are Cl^-, iron is in the +3 oxidation state. Fe^{3+}: $[Ar]3d^5$

↑ ↑ ↑

↑ ↑

Since all tetrahedral complexes are high-spin, there are 5 unpaired electrons in $FeCl_4^-$.

Metallurgy

57. a. To avoid fractions, let's first calculate ΔH for the reaction:

$$6\,FeO(s) + 6\,CO(g) \rightarrow 6\,Fe(s) + 6\,CO_2(g)$$

$6\,FeO + 2\,CO_2 \rightarrow 2\,Fe_3O_4 + 2\,CO$	ΔH° = −2(18 kJ)
$2\,Fe_3O_4 + CO_2 \rightarrow 3\,Fe_2O_3 + CO$	ΔH° = −(−39 kJ)
$3\,Fe_2O_3 + 9\,CO \rightarrow 6\,Fe + 9\,CO_2$	ΔH° = 3(−23 kJ)
$6\,FeO(s) + 6\,CO(g) \rightarrow 6\,Fe(s) + 6\,CO_2(g)$	ΔH° = −66 kJ

So for: $FeO(s) + CO(g) \rightarrow Fe(s) + CO_2(g)$ $\Delta H^\circ = \frac{-66 \text{ kJ}}{6} = -11 \text{ kJ}$

b. $\Delta H^\circ = 2(-110.5 \text{ kJ}) - [-393.5 \text{ kJ} + 0] = 172.5 \text{ kJ}$

$\Delta S^\circ = 2(198 \text{ J/K}) - [214 \text{ J/K} + 6 \text{ J/K}] = 176 \text{ J/K}$

$\Delta G^\circ = \Delta H^\circ - T\Delta S^\circ$, $\Delta G^\circ = 0$ when $T = \frac{\Delta H^\circ}{\Delta S^\circ} = \frac{172.5 \text{ kJ}}{0.176 \text{ kJ/K}} = 980.\text{ K}$

Due to the favorable ΔS° term, this reaction is spontaneous at T > 980. K. From Figure 21.36 of the text, this reaction takes place in the blast furnace at temperatures greater than 980. K as required by thermodynamics.

59. Review section 4.10 for balancing reactions in basic solution by the half-reaction method.

$$(2\,CN^- + Ag \rightarrow Ag(CN)_2^- + e^-) \times 4$$
$$4\,e^- + O_2 + 4\,H^+ \rightarrow 2\,H_2O$$

$$8\,CN^- + 4\,Ag + O_2 + 4\,H^+ \rightarrow 4\,Ag(CN)_2^- + 2\,H_2O$$

Adding 4 OH^- to both sides and crossing off 2 H_2O on both sides of the equation gives the balanced equation:

$$8\,CN^-(aq) + 4\,Ag(s) + O_2(g) + 2\,H_2O(l) \rightarrow 4\,Ag(CN)_2^-(aq) + 4\,OH^-(aq)$$

Additional Exercises

61. i. $0.0203 \text{ g } CrO_3 \times \frac{52.00 \text{ g Cr}}{100.0 \text{ g } CrO_3} = 0.0106 \text{ g Cr}$; $\% \text{ Cr} = \frac{0.0106}{0.105} \times 100 = 10.1\% \text{ Cr}$

ii. $32.93 \times 10^{-3} \text{ L HCl} \times \frac{0.100 \text{ mol HCl}}{\text{L}} \times \frac{1 \text{ mol } NH_3}{\text{mol HCl}} \times \frac{17.03 \text{ g } NH_3}{\text{mol}} = 0.0561 \text{ g } NH_3$

$\% NH_3 = \frac{0.0561 \text{ g}}{0.341 \text{ g}} \times 100 = 16.5\% NH_3$

iii. 73.53% I + 16.5% NH_3 + 10.1% Cr = 100.1%; The compound must be composed of only Cr, NH_3, and I.

Out of 100.00 g of compound:

$$10.1 \text{ g Cr} \times \frac{1 \text{ mol}}{52.00 \text{ g}} = 0.194 \qquad \frac{0.194}{0.194} = 1.00$$

$$16.5 \text{ g } NH_3 \times \frac{1 \text{ mol}}{17.03 \text{ g}} = 0.969 \qquad \frac{0.969}{0.194} = 4.99$$

$$73.53 \text{ g I} \times \frac{1 \text{ mol}}{126.9 \text{ g}} = 0.5794 \qquad \frac{0.5794}{0.194} = 2.99$$

$Cr(NH_3)_5I_3$ is the empirical formula. Cr(III) forms octahedral complexes. So, compound A is made of the octahedral $[Cr(NH_3)_5I]^{2+}$ complex ion and two I^- counter ions; the formula is $[Cr(NH_3)_5I]I_2$. Let's check this proposed formula using the freezing point data.

iv. $\Delta T_f = iK_f m$; For $[Cr(NH_3)_5I]I_2$, i = 3.0 (assuming complete dissociation).

$$m = \frac{0.601 \text{ g complex}}{1.000 \times 10^{-2} \text{ kg } H_2O} \times \frac{1 \text{ mol complex}}{517.9 \text{ g complex}} = 0.116 \text{ molal}$$

$$\Delta T_f = 3.0 \times 1.86°C/\text{molal} \times 0.116 \text{ molal} = 0.65°C$$

Because ΔT_f is close to the measured value, this is consistent with the formula $[Cr(NH_3)_5I]I_2$.

63. $Hg^{2+}(aq) + 2\ I^-(aq) \rightarrow HgI_2(s)$, orange ppt.; $HgI_2(s) + 2\ I^-(aq) \rightarrow HgI_4^{2-}(aq)$,
soluble complex ion

Hg^{2+} is a d^{10} ion. Color is the result of electron transfer between split d orbitals. This cannot occur for the filled d orbitals in Hg^{2+}. Therefore, we would not expect Hg^{2+} complex ions to form colored solutions.

65. a. 2; Forms bonds through the lone pairs on the two oxygen atoms.

b. 3; Forms bonds through the lone pairs on the three nitrogen atoms.

c. 4; Forms bonds through the two nitrogen atoms and the two oxygen atoms.

d. 4; Forms bonds through the four nitrogen atoms.

67. a. $Ru(phen)_3^{2+}$ exhibits optical isomerism [similar to $Co(en)_3^{3+}$ in Figure 21.16 of the text].

b. Ru^{2+}: $[Kr]4d^6$; Since there are no unpaired electrons, Ru^{2+} is a strong-field (low-spin) case.

__ __

↑↓ ↑↓ ↑↓ large Δ

69. Octahedral Cr^{2+} complexes should be used. Cr^{2+}: [Ar]$3d^4$; High-spin (weak-field) Cr^{2+} complexes have 4 unpaired electrons and low-spin (strong-field) Cr^{2+} complexes have 2 unpaired electrons. Ni^{2+}: [Ar]$3d^8$; Octahedral Ni^{2+} complexes will always have 2 unpaired electrons, whether high or low-spin. Therefore, Ni^{2+} complexes cannot be used to distinguish weak from strong-field ligands by examining magnetic properties. Alternatively, the ligand field strengths can be measured using visible spectra. Either Cr^{2+} or Ni^{2+} complexes can be used for this method.

71. We need to calculate the Pb^{2+} concentration in equilibrium with $EDTA^{4-}$. Since K is large for the formation of $PbEDTA^{2-}$, let the reaction go to completion; then solve an equilibrium problem to get the Pb^{2+} concentration.

	Pb^{2+}	+ $EDTA^{4-}$	$\rightleftharpoons$ $PbEDTA^{2-}$	$K = 1.1 \times 10^{18}$
Before	0.010 *M*	0.050 *M*	0	
	0.010 mol/L Pb^{2+} reacts completely (large K)			
Change	−0.010	−0.010 →	+0.010	Reacts completely
After	0	0.040	0.010	New initial condition
	x mol/L $PbEDTA^{2-}$ dissociates to reach equilibrium			
Equil.	*x*	0.040 + *x*	0.010 - *x*	

$$1.1 \times 10^{18} = \frac{(0.10 - x)}{(x)(0.040 + x)} \approx \frac{(0.10)}{x(0.040)},\ x = [Pb^{2+}] = 2.3 \times 10^{-19}\ M;\ \text{Assumptions good.}$$

Now calculate the solubility quotient for $Pb(OH)_2$ to see if precipitation occurs. The concentration of OH^- is 0.10 *M* because we have a solution buffered at pH = 13.00.

$$Q = [Pb^{2+}]_o[OH^-]_o^{\ 2} = (2.3 \times 10^{-19})(0.10)^2 = 2.3 \times 10^{-21} < K_{sp}\ (1.2 \times 10^{-15})$$

$Pb(OH)_2(s)$ will not form because Q is less than K_{sp}.

73.

$$HbO_2 \rightarrow Hb + O_2 \qquad \Delta G° = -(-70\text{ kJ})$$
$$Hb + CO \rightarrow HbCO \qquad \Delta G° = -80\text{ kJ}$$

$$HbO_2 + CO \rightarrow HbCO + O_2 \qquad \Delta G° = -10\text{ kJ}$$

$$\Delta G° = -RT\ln K,\ K = \exp\left(\frac{-\Delta G°}{RT}\right) = \exp\left(\frac{-(-10 \times 10^3\text{ J})}{(8.3145\text{ J/K}\bullet\text{mol})(298\text{ kJ})}\right) = 60$$

Challenge Problems

75. $Ni^{2+} = d^8$; If ligands A and B produced very similar crystal fields, the trans-$[NiA_2B_4]^{2+}$ complex ion would give the following octahedral crystal field diagram for a d^8 ion:

↑ ↑

↑↓ ↑↓ ↑↓

This is paramagnetic.

Because it is given that the complex ion is diamagnetic, the A and B ligands must produce different crystal fields giving a unique d-orbital splitting diagram that would result in a diamagnetic species.

77. a. Consider the following electrochemical cell:

$$Co^{3+} + e^- \rightarrow Co^{2+} \qquad E° = 1.82\ V$$

$$Co(en)_3^{2+} \rightarrow Co(en)_3^{3+} + e^- \qquad -E° = ?$$

$$Co^{3+} + Co(en)_3^{2+} \rightarrow Co^{2+} + Co(en)_3^{3+} \qquad E°_{cell} = 1.82 - E°$$

The equilibrium constant for this overall reaction is:

$$Co^{3+} + 3\ en \rightarrow Co(en)_3^{3+} \qquad K_1 = 2.0 \times 10^{47}$$

$$Co(en)_3^{2+} \rightarrow Co^{2+} + 3\ en \qquad K_2 = 1/1.5 \times 10^{12}$$

$$Co^{3+} + Co(en)_3^{2+} \rightarrow Co(en)_3^{3+} + Co^{2+} \qquad K = K_1K_2 = \frac{2.0\times10^{47}}{1.5\times10^{12}} = 1.3 \times 10^{35}$$

From the Nernst equation for the overall reaction:

$$E°_{cell} = \frac{0.0591}{n}\log K = \frac{0.0591}{1}\log(1.3\times10^{35}),\ \log(1.3 \times 10^{35}),\ E°_{cell} = 2.08\ V$$

$$E°_{cell} = 1.82 - E° = 2.08\ V,\ E° = 1.82\ V - 2.08\ V = -0.26\ V$$

b. The stronger oxidizing agent will be the more easily reduced species and will have the more positive standard reduction potential. From the reduction potentials, Co^{3+} ($E° = 1.82$ V) is a much stronger oxidizing agent than $Co(en)_3^{3+}$ ($E° = -0.26$ V).

c. In aqueous solution, Co^{3+} forms the hydrated transition metal complex, $Co(H_2O)_6^{3+}$. In both complexes, $Co(H_2O)_6^{3+}$ and $Co(en)_3^{3+}$, cobalt exists as Co^{3+} which has 6 d electrons. Assuming a strong-field case for each complex ion, the d-orbital splitting diagram for each is:

— — e_g

↑↓ ↑↓ ↑↓ t_{2g}

When each complex gains an electron, the electron enters a higher energy e_g orbital. Since en is a stronger field ligand than H_2O, the d-orbital splitting is larger for $Co(en)_3^{3+}$, and it takes more energy to add an electron to $Co(en)_3^{3+}$ than to $Co(H_2O)_6^{3+}$. Therefore, it is more favorable for $Co(H_2O)_6^{3+}$ to gain an electron than for $Co(en)_3^{3+}$ to gain an electron.

79. No; In all three cases, six bonds are formed between Ni^{2+} and nitrogen, so ΔH values should be similar. ΔS° for formation of the complex ion is most negative for 6 NH_3 molecules reacting with a metal ion (7 independent species become 1). For penten reacting with a metal ion, 2 independent species become 1, so ΔS° is least negative of all three of the reactions. Thus, the chelate effect occurs because the more bonds a chelating agent can form to the metal, the more favorable ΔS° is for the formation of the complex ion and the larger the formation constant.

81.

L, L, M—L, L, L

z-axis

d_{z^2}

$d_{x^2-y^2}$, d_{xy}

d_{xz}, d_{yz}

The d_{z^2} orbital will be destabilized much more than in the trigonal planar case (see Exercise 21.80). The d_{z^2} orbital has electron density on the z-axis directed at the two axial ligands. The $d_{x^2-y^2}$ and d_{xy} orbitals are in the plane of the three trigonal planar ligands and should be destabilized a lesser amount as compared to the d_{z^2} orbital; only a portion of the electron density in the $d_{x^2-y^2}$ and d_{xy} orbitals is directed at the ligands. The d_{xz} and d_{yz} orbitals will be destabilized the least since the electron density is directed between the ligands.

83. The coordinate system for trans-$[Ni(NH_3)_2(CN)_4]^{2-}$ is shown below. Because CN^- produces a much stronger crystal field, it will dominate the d-orbital splitting. From the coordinate system, the CN^- ligands are in a square planar arrangement. Therefore, the diagram will most resemble the square planar diagram. Note that the relative position of d_{z^2} orbital is hard to predict; it could switch positions with the d_{xy} orbital.

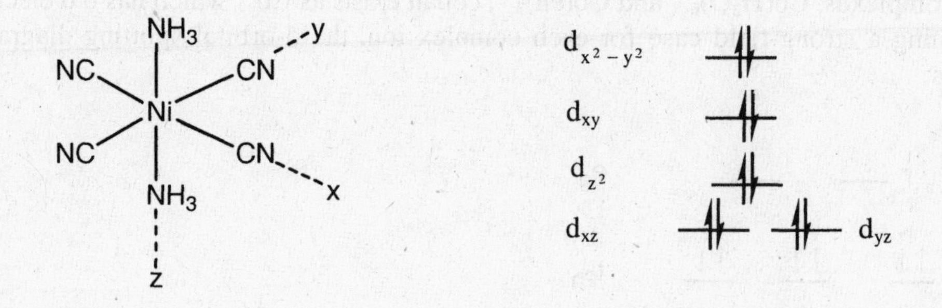

Integrative Problems

85. a. Because O is in the −2 oxidation state, iron must be in the +6 oxidation state. Fe^{6+}: $[Ar]3d^2$.

b. Using the half-reaction method of balancing redox reactions, the balanced equation is:

$$10\ H^+(aq) + 2\ FeO_4^{2-}(aq) + 2\ NH_3(aq) \rightarrow 2\ Fe^{3+}(aq) + N_2(g) + 8\ H_2O(l)$$

$$0.0250\ L \times \frac{0.243\ mol}{L} = 6.08 \times 10^{-3}\ mol\ FeO_4^{2-}$$

$$0.0550\ L \times \frac{1.45\ mol}{L} = 7.98 \times 10^{-2}\ mol\ NH_3$$

$$\frac{mol\ NH_3}{mol\ FeO_4^{2-}} = \frac{7.98 \times 10^{-2}\ mol}{6.08 \times 10^{-3}\ mol} = 13.1$$

The actual mole ratio is larger than the theoretical ratio of 1:1, so FeO_4^{2-} is limiting.

$$V_{N_2} = \frac{nRT}{P} = \frac{(6.08 \times 10^{-3}\ mol\ FeO_4^{2-}) \times \frac{1\ mol\ N_2}{2\ mol\ FeO_4^{2-}} \times \frac{0.08206\ L\ atm}{K\ mol} \times 298\ K}{1.50\ atm}$$

$$V_{N_2} = 0.0496\ L = 49.6\ mL\ N_2$$

CHAPTER TWENTY-TWO

ORGANIC AND BIOLOGICAL MOLECULES

Questions

1. a. 1-sec-butylpropane

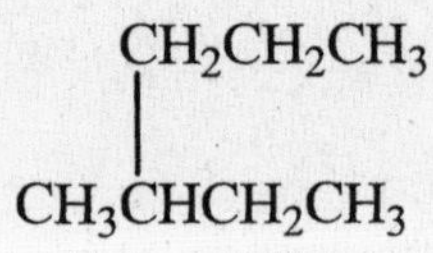

$$\begin{array}{l} CH_2CH_2CH_3 \\ | \\ CH_3CHCH_2CH_3 \end{array}$$

3-methylhexane is correct.

b. 4-methylhexane

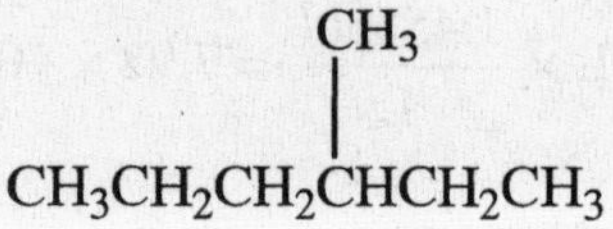

$$\begin{array}{r} CH_3 \\ | \\ CH_3CH_2CH_2CHCH_2CH_3 \end{array}$$

3-methylhexane is correct.

c. 2-ethylpentane

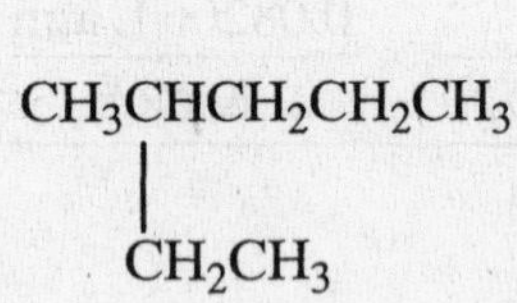

$$\begin{array}{l} CH_3CHCH_2CH_2CH_3 \\ \quad | \\ \quad CH_2CH_3 \end{array}$$

3-methylhexane is correct.

d. 1-ethyl-1-methylbutane

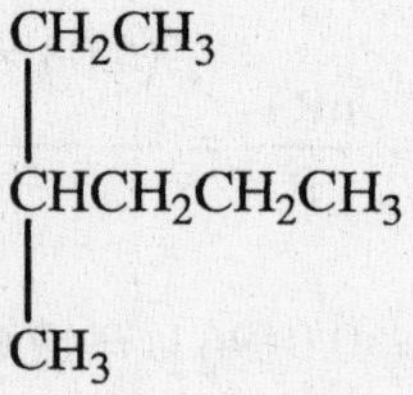

$$\begin{array}{l} CH_2CH_3 \\ | \\ CHCH_2CH_2CH_3 \\ | \\ CH_3 \end{array}$$

3-methylhexane is correct.

e. 3-methylhexane

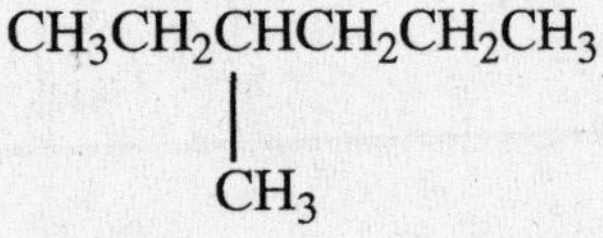

$$\begin{array}{l} CH_3CH_2CHCH_2CH_2CH_3 \\ \qquad\quad | \\ \qquad\quad CH_3 \end{array}$$

f. 4-ethylpentane

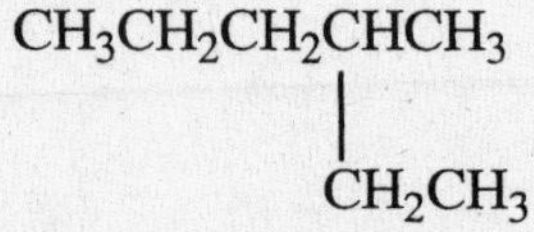

$$\begin{array}{l} CH_3CH_2CH_2CHCH_3 \\ \qquad\qquad\quad | \\ \qquad\qquad\quad CH_2CH_3 \end{array}$$

3-methylhexane is correct.

All six of these compounds are the same. They only differ from each other by rotations about one or more carbon-carbon single bonds. Only one isomer of C_7H_{16} is present in all of these names, 3-methylhexane.

3\. a.

$$\begin{array}{l} CH_3CHCH_3 \\ \quad | \\ \quad CH_2CH_3 \end{array}$$

The longest chain is 4 carbons long. The correct name is 2-methylbutane.

b.

$$\begin{array}{c} \qquad\qquad\qquad\quad I \quad CH_3 \\ \qquad\qquad\qquad\quad | \quad\; | \\ CH_3CH_2CH_2CH_2C — CH_2 \\ \qquad\qquad\qquad\quad | \\ \qquad\qquad\qquad\quad CH_3 \end{array}$$

The longest chain is 7 carbons long and we would start the numbering system at the other end for lowest possible numbers. The correct name is 3-iodo-3-methyl-heptane.

c.

$$\begin{array}{c} \qquad\qquad\quad CH_3 \\ \qquad\qquad\quad | \\ CH_3CH_2CH{=}C — CH_3 \end{array}$$

This compound cannot exhibit cis–trans isomerism since one of the double bonded carbons has the same two groups (CH_3) attached. The numbering system should also start at the other end to give the double bond the lowest possible number. 2-methyl-2-pentene is correct.

d.

$$\begin{array}{c} \;\; Br\; OH \\ \;\; |\quad | \\ CH_3CHCHCH_3 \end{array}$$

The OH functional group gets the lowest number. 3-bromo-2-butanol is correct.

5\. Hydrocarbons are nonpolar substances exhibiting only London dispersion forces. Size and shape are the two most important structural features relating to the strength of London dispersion forces. For size, the bigger the molecule (the larger the molar mass), the stronger the London dispersion forces and the higher the boiling point. For shape, the more branching present in a compound, the weaker the London dispersion forces and the lower the boiling point.

7\. The amide functional group is:

$$\begin{array}{c} O \quad H \\ \| \quad\; | \\ —C — N— \end{array}$$

When the amine end of one amino acid reacts with the carboxylic acid end of another amino acid, the two amino acids link together by forming an amide functional group. A polypeptide has many amino acids linked together, with each linkage made by the formation of an amide functional group. Because all linkages result in the presence of the amide functional group, the resulting polymer is called a polyamide. The correct order of strength is:

$$\left(\begin{matrix} H & H & H & H \\ | & | & | & | \\ -C- & -C- & -C- & -C- \\ | & | & | & | \\ H & H & H & H \end{matrix}\right)_n < \left(\begin{matrix} O & & O & & & \\ \| & & \| & & & \\ -C- & R- & -C- & O- & R'- & O- \end{matrix}\right)_n$$

polyhydrocarbon
weakest fibers

polyester

$$< \left(\begin{matrix} O & & H & O & & H \\ \| & & | & \| & & | \\ -C- & R- & -N- & -C- & R- & -N- \end{matrix}\right)_n$$

polyamide
strongest fibers

The difference in strength is related to the types of intermolecular forces present. All of these types of polymers have London dispersion forces. However, the polar ester group in polyesters and the polar amide group in polya mides give rise to additional dipole forces. The polyamide has the ability to form relatively strong hydrogen bonding interactions, hence why it would form the strongest fibers.

9.

a. $CH_2{=}CH_2 + H_2O \xrightarrow{H^+} \underset{OH}{CH_2}-\underset{H}{CH_2}$ 1° alcohol

b. $CH_3CH{=}CH_2 + H_2O \xrightarrow{H^+} CH_3\overset{OH}{\overset{|}{C}}H-\overset{H}{\overset{|}{C}}H_2$ 2° alcohol

major product

c. $CH_3\underset{CH_3}{\underset{|}{C}}{=}CH_2 + H_2O \xrightarrow{H^+} CH_3\overset{OH}{\overset{|}{\underset{CH_3}{\underset{|}{C}}}}-\overset{H}{\overset{|}{C}}H_2$ 3° alcohol

major product

d. $CH_3CH_2OH \xrightarrow{\text{oxidation}} CH_3\overset{O}{\overset{\|}{C}}H$ aldehyde

e. $CH_3\overset{OH}{\overset{|}{C}}HCH_3 \xrightarrow{\text{oxidation}} CH_3\overset{O}{\overset{\|}{C}}CH_3$ ketone

f. 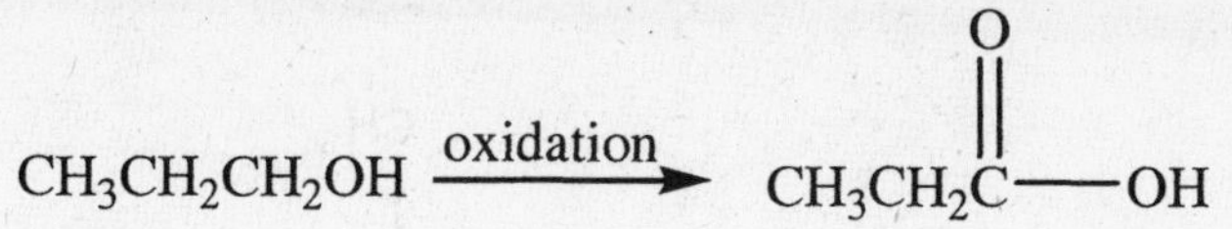 $CH_3CH_2CH_2OH \xrightarrow{\text{oxidation}} CH_3CH_2\overset{\overset{\displaystyle O}{\|}}{C}{-}OH$ carboxylic acid

or

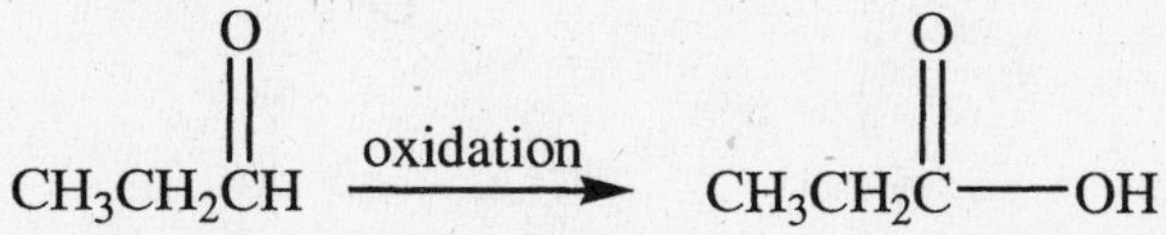

$CH_3CH_2\overset{\overset{\displaystyle O}{\|}}{C}H \xrightarrow{\text{oxidation}} CH_3CH_2\overset{\overset{\displaystyle O}{\|}}{C}{-}OH$

g. 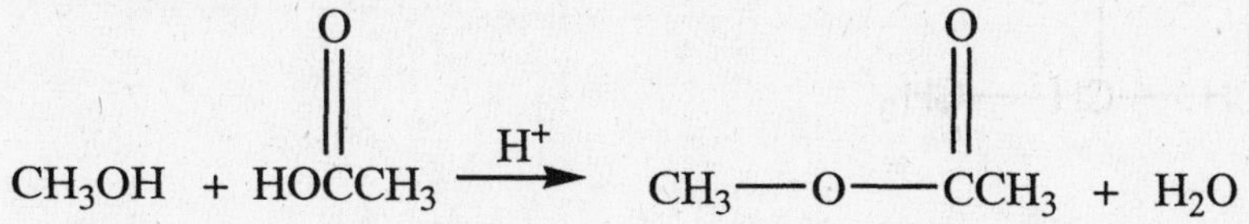 $CH_3OH + HO\overset{\overset{\displaystyle O}{\|}}{C}CH_3 \xrightarrow{H^+} CH_3{-}O{-}\overset{\overset{\displaystyle O}{\|}}{C}CH_3 + H_2O$ ester

11. a. A polyester forms when an alcohol functional group reacts with a carboxylic acid functional group. The monomer for a homopolymer polyester must have an alcohol functional group and a carboxylic acid functional group present in the structure.

b. A polyamide forms when an amine functional group reacts with a carboxylic acid functional group. For a copolymer polyamide, one monomer would have at least two amine functional groups present and the other monomer would have at least two carboxylic acid functional groups present. For polymerization to occur, each monomer must have two reactive functional groups present.

c. To form an addition polymer, a carbon-carbon double bond must be present. To form a polyester, the monomer would need the alcohol and carboxylic acid functional groups present. To form a polyamide, the monomer would need the amine and carboxylic acid functional groups present. The two possibilities are for the monomer to have a carbon-carbon double bond, an alcohol functional group, and a carboxylic acid functional group present, or to have a carbon-carbon double bond, an amine functional group, and a carboxylic acid functional group present.

Exercises

Hydrocarbons

13. i.

$CH_3{-}CH_2{-}CH_2{-}CH_2{-}CH_2{-}CH_3$

ii.

$CH_3{-}\underset{}{\overset{\overset{\displaystyle CH_3}{|}}{C}H}{-}CH_2{-}CH_2{-}CH_3$

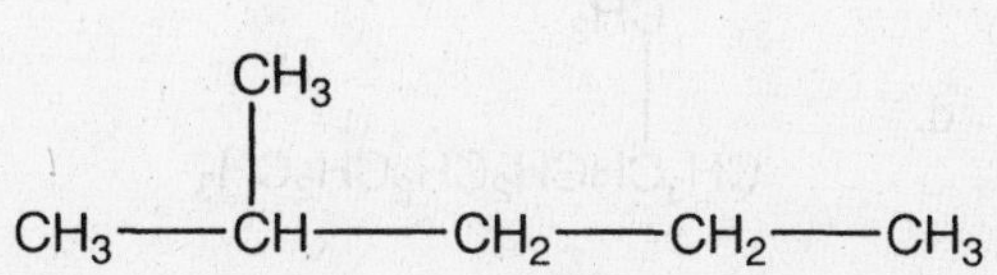

iii.

$$\begin{array}{c} \quad\quad\quad CH_3 \\ \quad\quad\quad | \\ CH_3-CH_2-CH-CH_2-CH_3 \end{array}$$

iv.

$$\begin{array}{c} \quad CH_3 \\ \quad | \\ CH_3-C-CH_2-CH_3 \\ \quad | \\ \quad CH_3 \end{array}$$

v.

$$\begin{array}{c} \quad CH_3 \;\; CH_3 \\ \quad | \quad\;\; | \\ CH_3-CH-CH-CH_3 \end{array}$$

All other possibilities are identical to one of these five compounds.

15. A difficult task in this problem is recognizing different compounds from compounds that differ by rotations about one or more C–C bonds (called conformations). The best way to distinguish different compounds from conformations is to name them. Different name = different compound; same name = same compound, so it is not an isomer, but instead, is a conformation.

a.

$$\begin{array}{c} \quad CH_3 \\ \quad | \\ CH_3CHCH_2CH_2CH_2CH_2CH_3 \end{array}$$

2-methylheptane

$$\begin{array}{c} \quad\quad CH_3 \\ \quad\quad | \\ CH_3CH_2CHCH_2CH_2CH_2CH_3 \end{array}$$

3-methylheptane

$$\begin{array}{c} \quad\quad\quad CH_3 \\ \quad\quad\quad | \\ CH_3CH_2CH_2CHCH_2CH_2CH_3 \end{array}$$

4-methylheptane

b.

$$\begin{array}{c} CH_3 \; CH_3 \\ | \quad | \\ CH_3-C-C-CH_3 \\ | \quad | \\ CH_3 \; CH_3 \end{array}$$

2,2,3,3-tetramethylbutane

17. a.

$$\begin{array}{c} CH_3 \\ | \\ CH_3CHCH_3 \end{array}$$

b.

$$\begin{array}{c} CH_3 \\ | \\ CH_3CHCH_2CH_3 \end{array}$$

c.

$$\begin{array}{c} CH_3 \\ | \\ CH_3CHCH_2CH_2CH_3 \end{array}$$

d.

$$\begin{array}{c} CH_3 \\ | \\ CH_3CHCH_2CH_2CH_2CH_3 \end{array}$$

19. a.

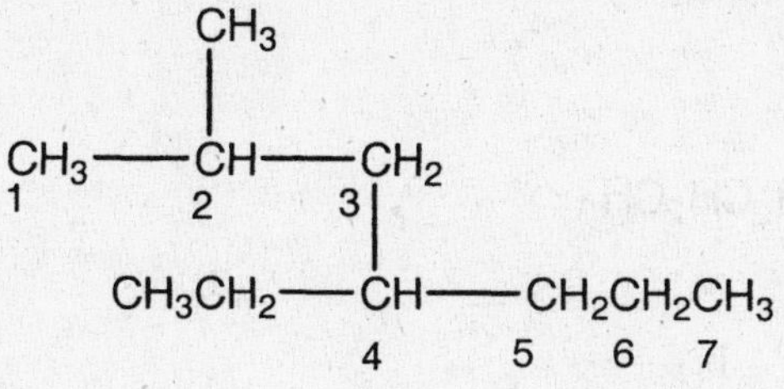

```
                CH3
                 |
CH3 ——— CH ——— CH2
 1       2      3|
                 |
     CH3CH2 ——— CH ——— CH2CH2CH3
                 4      5   6   7
```

b.

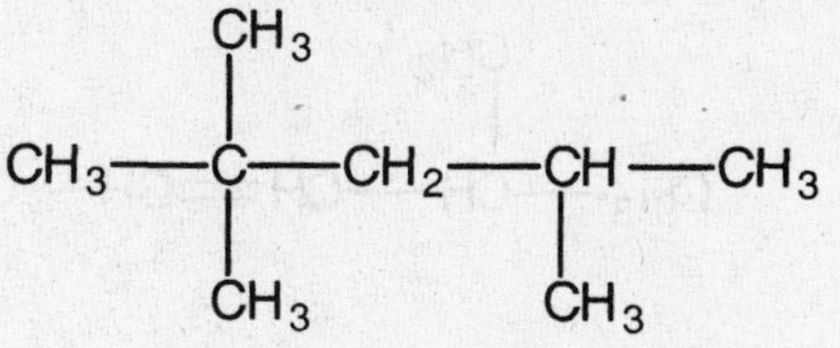

```
        CH3
         |
CH3 ——— C ——— CH2 ——— CH ——— CH3
         |              |
        CH3            CH3
```

c.

```
          CH3
           |
CH3 ——— C ——— CH3
 1       2
           |
   CH3CHCH2CH2CH3
     3  4  5  6
```

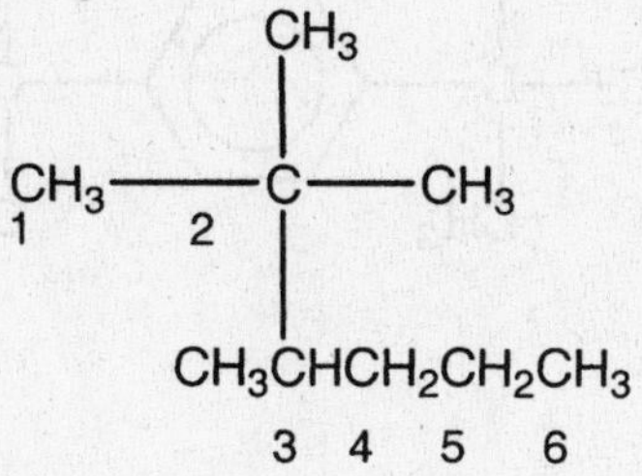

d. For 3-isobutylhexane, the longest chain is 7 carbons long. The correct name is 4-ethyl-2-methylheptane. For 2-tert-butylpentane, the longest chain is 6 carbons long. The correct name is 2,2,3-trimethylhexane.

21. a. 2,2,4-trimethylhexane b. 5-methylnonane c. 2,2,4,4-tetramethylpentane

d. 3-ethyl-3-methyloctane

Note: For alkanes, always identify the longest carbon chain for the base name first, then number the carbons to give the lowest overall numbers for the substituent groups.

23.

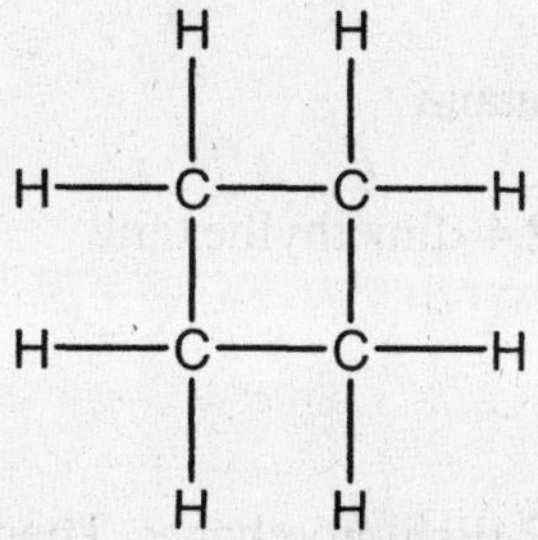

$CH_3—CH_2—CH_2—CH_3$

```
     H    H
     |    |
H —— C —— C —— H
     |    |
H —— C —— C —— H
     |    |
     H    H
```

Each carbon is bonded to four other carbon and/or hydrogen atoms in a saturated hydrocarbon (only single bonds are present).

25. a. 1-butene b. 4-methyl-2-hexene c. 2,5-dimethyl-3-heptene

Note: The multiple bond is assigned the lowest number possible.

27. a. $CH_3–CH_2–CH{=}CH–CH_2–CH_3$ b. $CH_3–CH{=}CH–CH{=}CH–CH_2CH_3$

c.

CH_3

$CH_3-CH-CH=CH-CH_2CH_2CH_2CH_3$

29. a.

CH_3

CH_2CH_3

b.

CH_3 CH_3

H_3C-C- $-C-CH_3$

CH_3 CH_3

c.

CH_2CH_3

CH_2CH_3

d.

$CH_2-CH=CH-CH_3$

31. a. 1,3-dichlorobutane

b. 1,1,1-trichlorobutane

c. 2,3-dichloro-2,4-dimethylhexane

d. 1,2-difluoroethane

Isomerism

33. CH_2Cl-CH_2Cl, 1-2-dichloroethane: There is free rotation about the C–C single bond that doesn't lead to different compounds. CHCl=CHCl, 1,2-dichloroethene: There is no rota-tion about the C=C double bond. This creates the cis and trans isomers, which are different compounds.

35. To exhibit cis-trans isomerism, each carbon in the double bond must have two structurally different groups bonded to it. In Exercise 22.25, this occurs for compounds b and c. The cis isomer has the bulkiest groups on the same side of the double bond while the trans isomer has the bulkiest groups on opposite sides of the double bond. The cis and trans isomers for 25b and 25c are:

25 b.

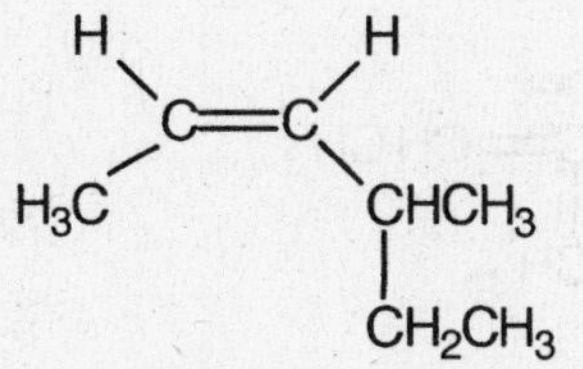

H, H, C=C, H_3C, $CHCH_3$, CH_2CH_3

cis

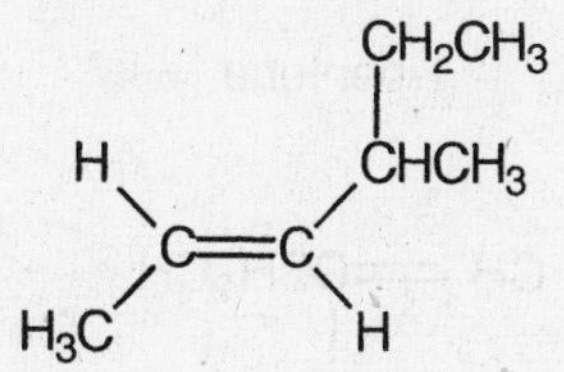

CH_2CH_3, H, $CHCH_3$, C=C, H_3C, H

trans

25 c.

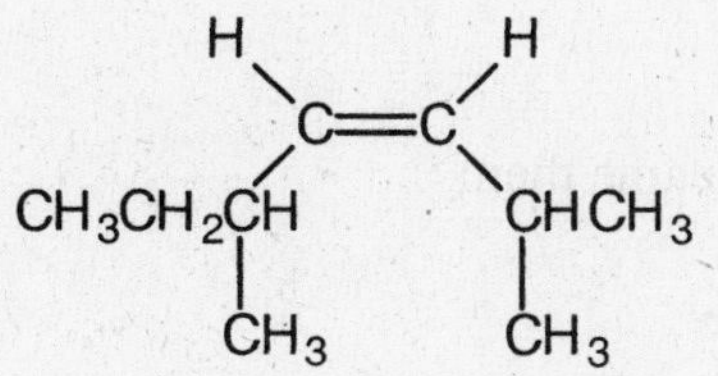

H, H, C=C, CH_3CH_2CH, $CHCH_3$, CH_3, CH_3

cis

CH_3, H, $CHCH_3$, C=C, CH_3CH_2CH, H, CH_3

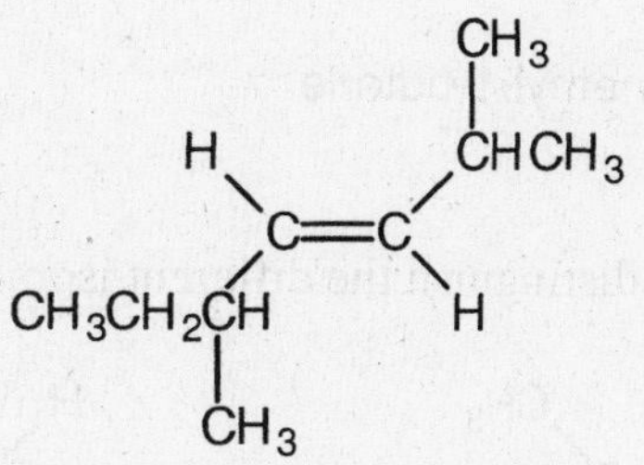

trans

Similarly, all the compounds in Exercise 22.27 exhibit *cis-trans* isomerism.

In compound a of Exercise 22.25, the first carbon in the double bond does not contain two different groups. The first carbon in the double bond contains two H atoms. To illustrate that this compound does not exhibit *cis-trans* isomerism, let's look at the potential *cis-trans* isomers.

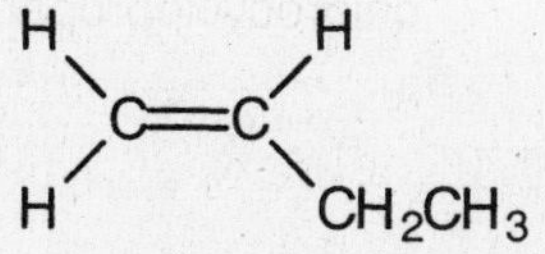

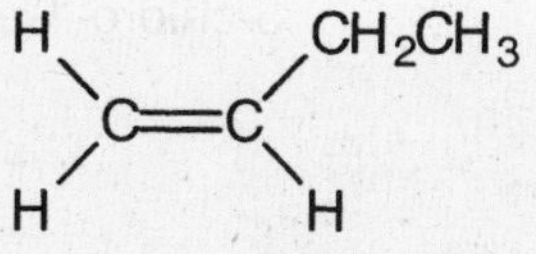

H, H, C=C, H, CH_2CH_3

H, CH_2CH_3, C=C, H, H

These are the same compounds; they only differ by a simple rotation of the molecule. Therefore, they are not isomers of each other, but instead are the same compound.

37. C_5H_{10} has the general formula for alkenes, C_nH_{2n}. To distinguish the different isomers from each other, we will name them. Each isomer must have a different name.

$CH_2{=}CHCH_2CH_2CH_3$

1-pentene

$CH_3CH{=}CHCH_2CH_3$

2-pentene

$CH_2{=}C(CH_3)CH_2CH_3$

2-methyl-1-butene

$CH_3C(CH_3){=}CHCH_3$

2-methyl-2-butene

$CH_3CH(CH_3)CH{=}CH_2$

3-methyl-1-butene

39. To help distinguish the different isomers, we will name them.

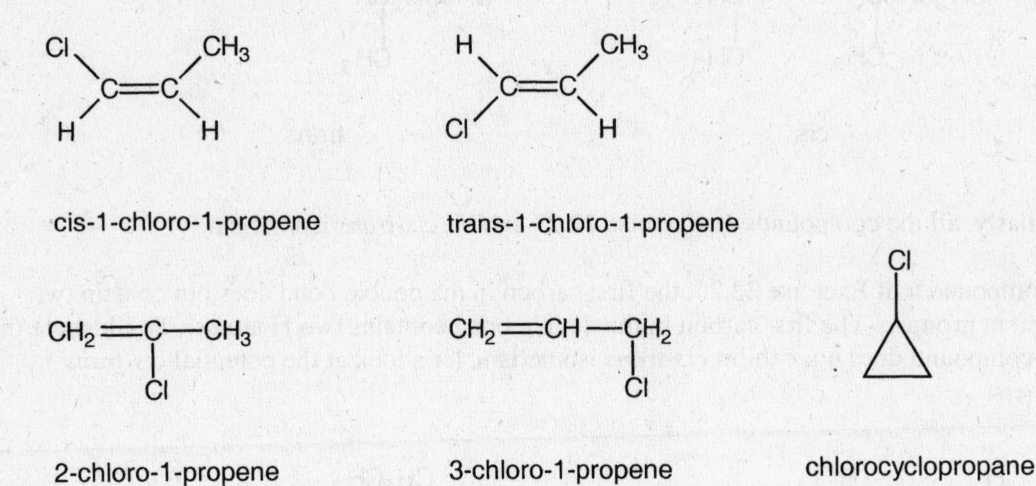

41.

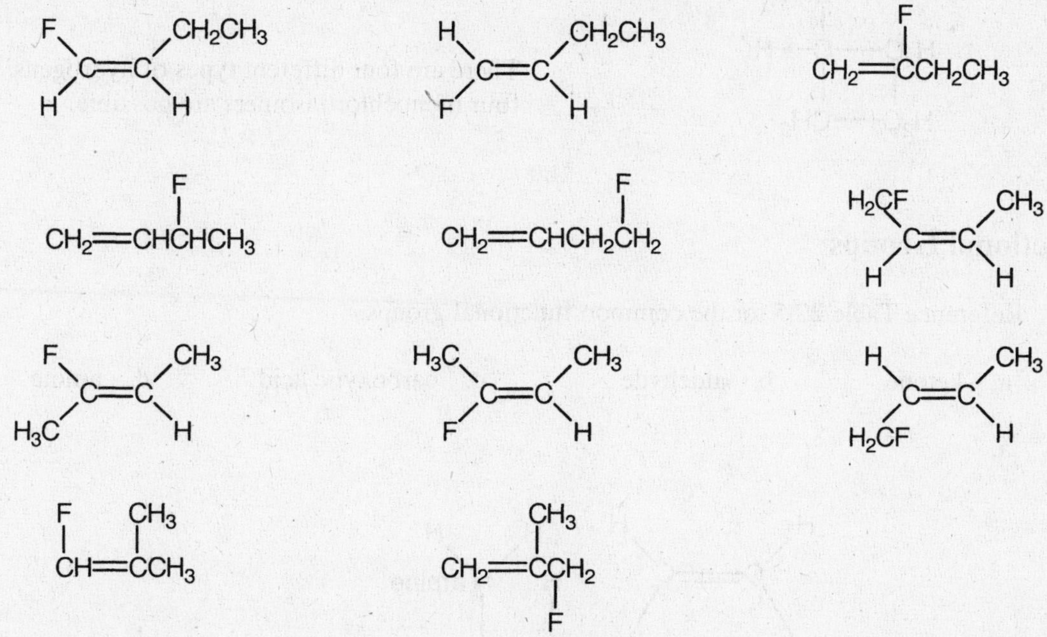

$F(H)C{=}C(H)CH_2CH_3$ $H(F)C{=}C(H)CH_2CH_3$ $CH_2{=}CFCH_2CH_3$

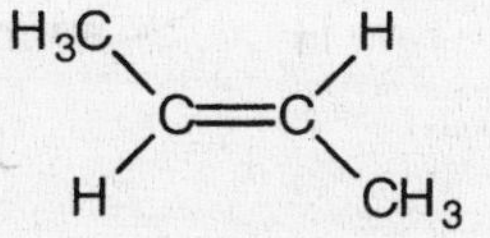

$CH_2{=}CHCHFCH_3$ $CH_2{=}CHCH_2CH_2F$ $H_2CF(H)C{=}C(H)CH_3$

$F(H_3C)C{=}C(H)CH_3$ $H_3C(F)C{=}C(H)CH_3$ $H(H_2CF)C{=}C(H)CH_3$

$CHF{=}C(CH_3)CH_3$ $CH_2{=}C(CH_3)CH_2F$

43.

a. 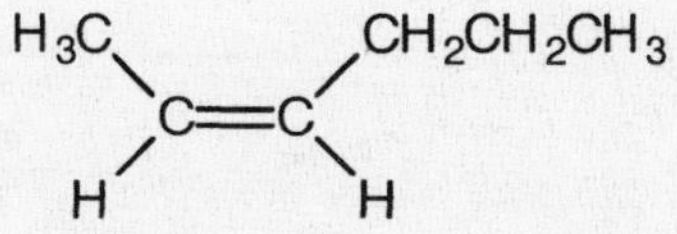 $H_3C(H)C{=}C(H)CH_2CH_2CH_3$ b. $H_3C(H)C{=}C(H)CH_3$ c. $H_3C(Cl)C{=}C(Cl)CH_2CH_3$

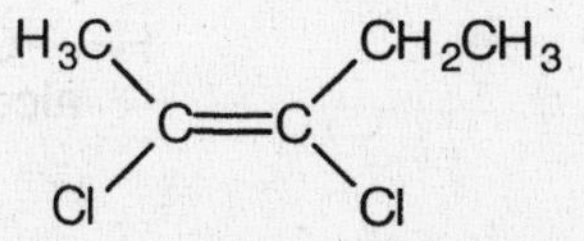

45. a.

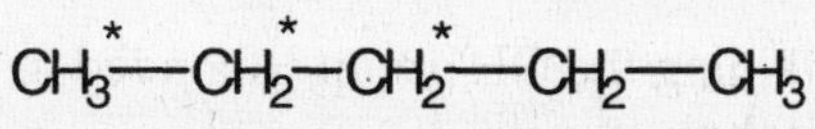

$\overset{*}{C}H_3{-}\overset{*}{C}H_2{-}\overset{*}{C}H_2{-}CH_2{-}CH_3$

There are three different types of hydrogens in n-pentane (see asterisks). Thus there are three monochloro isomers of n-pentane (1-chloropentane, 2-chloropentane and 3-chloropentane).

b.

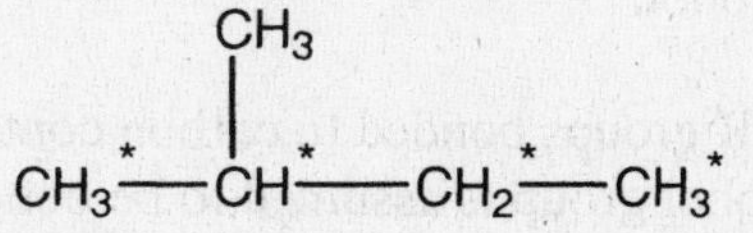

$\overset{*}{C}H_3{-}\overset{*}{C}H(CH_3){-}\overset{*}{C}H_2{-}\overset{*}{C}H_3$

There are four different types of hydrogens in 2-methylbutane, so four monochloro isomers of 2-methylbutane are possible.

c.

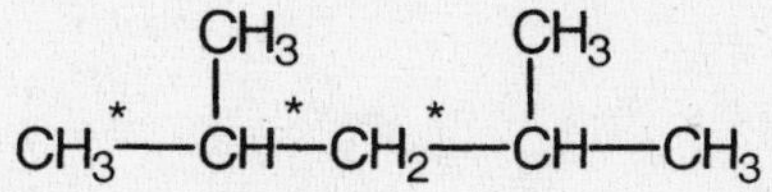

$\overset{*}{C}H_3{-}\overset{*}{C}H(CH_3){-}\overset{*}{C}H_2{-}CH(CH_3){-}CH_3$

There are three different types of hydrogens, so three monochloro isomers are possible.

d.

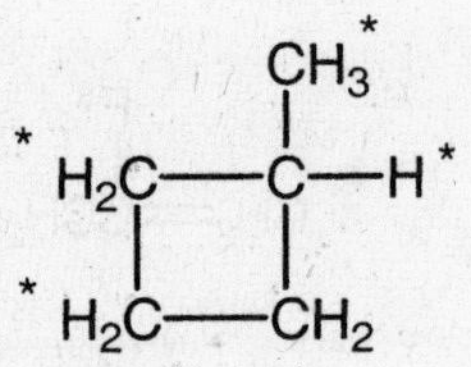

There are four different types of hydrogens, so four monochloro isomers are possible.

Functional Groups

47. Reference Table 22.5 for the common functional groups.

a. ketone b. aldehyde c. carboxylic acid d. amine

49. a.

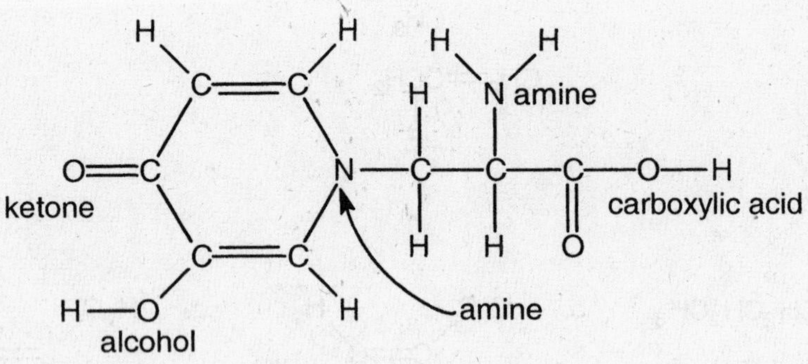

ketone

alcohol

amine

amine

carboxylic acid

b. 5 carbons in the ring and the carbon in $-CO_2H$: sp^2; the other two carbons: sp^3

c. 24 sigma bonds; 4 pi bonds

51. a. 3-chloro-1-butanol: Because the carbon containing the OH group is bonded to just 1 other carbon (1 R group), this is a primary alcohol.

b. 3-methyl-3-hexanol; Because the carbon containing the OH group is bonded to three other carbons (3 R groups), this is a tertiary alcohol.

c. 2-methylcyclopentanol; Secondary alcohol (2 R groups bonded to carbon containing the OH group); Note: In ring compounds, the alcohol group is assumed to be bonded to C_1, so the number designation is commonly omitted for the alcohol group.

53.

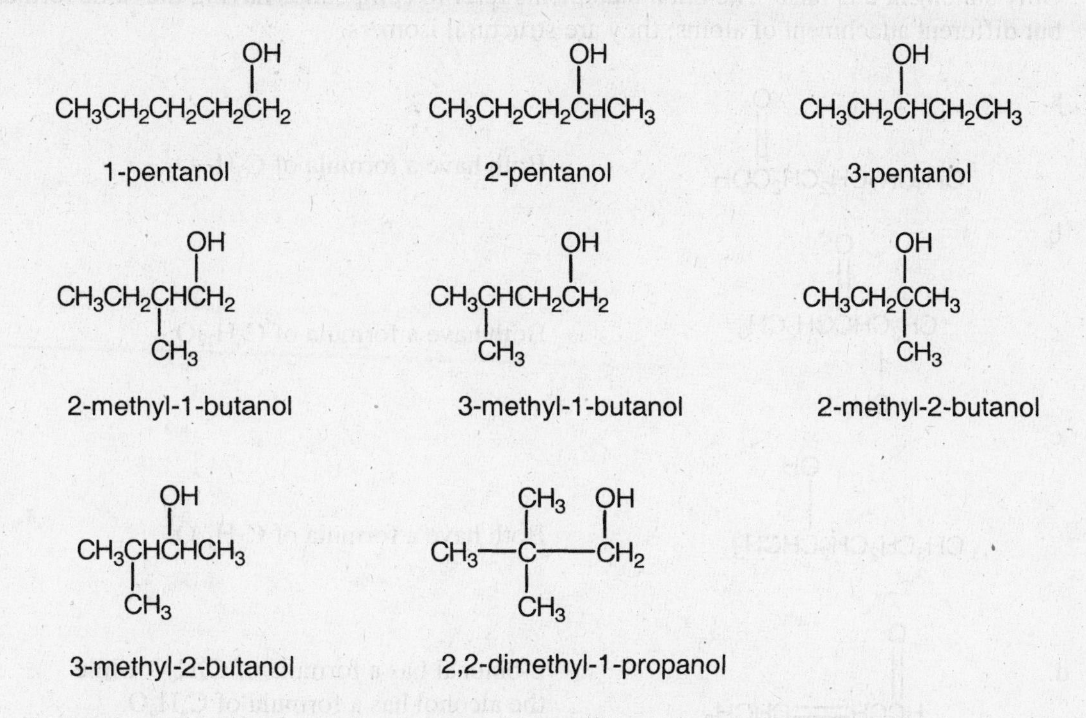

There are six isomeric ethers with formula $C_5H_{12}O$. The structures follow.

$CH_3{-}O{-}CH_2CH_2CH_2CH_3$

$CH_3{-}O{-}CH(CH_3)CH_2CH_3$

$CH_3{-}O{-}CH_2CH(CH_3)CH_3$

$CH_3{-}O{-}C(CH_3)_2{-}CH_3$

$CH_3CH_2{-}O{-}CH_2CH_2CH_3$

$CH_3CH_2{-}O{-}CH(CH_3){-}CH_3$

55. a. 4,5-dichloro-3-hexanone b. 2,3-dimethylpentanal

c. 3-methylbenzaldehyde or m-methylbenzaldehyde

57. a. 4-chlorobenzoic acid or p-chlorobenzoic acid

b. 3-ethyl-2-methylhexanoic acid

c. methanoic acid (common name = formic acid)

59. Only statement d is false. The other statements refer to compounds having the same formula but different attachment of atoms; they are structural isomers.

a. 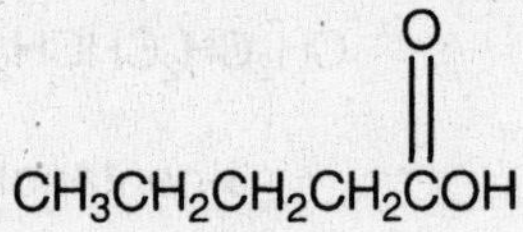 $CH_3CH_2CH_2CH_2\overset{O}{\overset{\|}{C}}OH$ Both have a formula of $C_5H_{10}O_2$.

b. 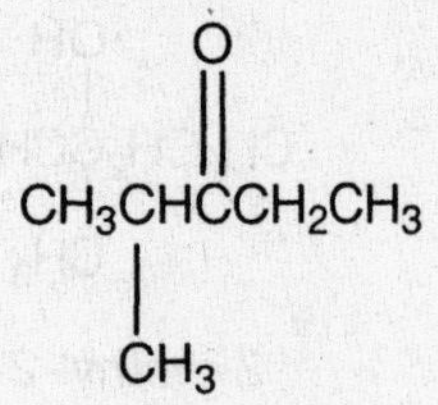 $CH_3CH(CH_3)\overset{O}{\overset{\|}{C}}CH_2CH_3$ Both have a formula of $C_6H_{12}O$.

c. 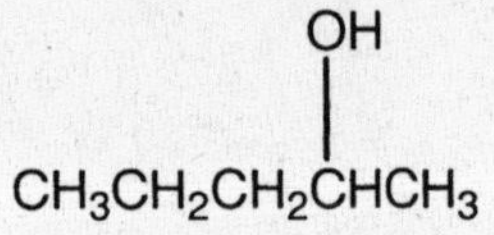 $CH_3CH_2CH_2CH(OH)CH_3$ Both have a formula of $C_5H_{12}O$.

d. 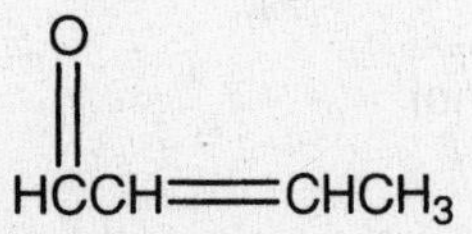 $H\overset{O}{\overset{\|}{C}}CH{=}CHCH_3$ 2-butenal has a formula of C_4H_6O while the alcohol has a formula of C_4H_8O.

e. 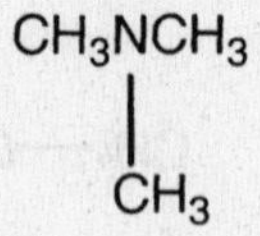 $CH_3N(CH_3)CH_3$ Both have a formula of C_3H_9N.

Reactions of Organic Compounds

61.

a. 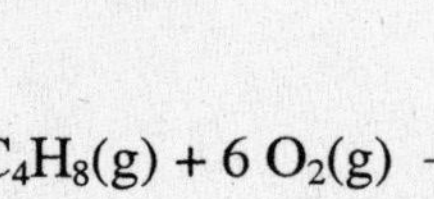 $CH_3CH(H){-}CH(H)CH_3$

b. $CH_2(Cl){-}CH(Cl)(CH_3)CH(Cl){-}CH(Cl)(CH_3)$

c. $C_6H_5{-}Cl + HCl$

d. $C_4H_8(g) + 6\ O_2(g) \rightarrow 4\ CO_2(g) + 4\ H_2O(g)$

63.

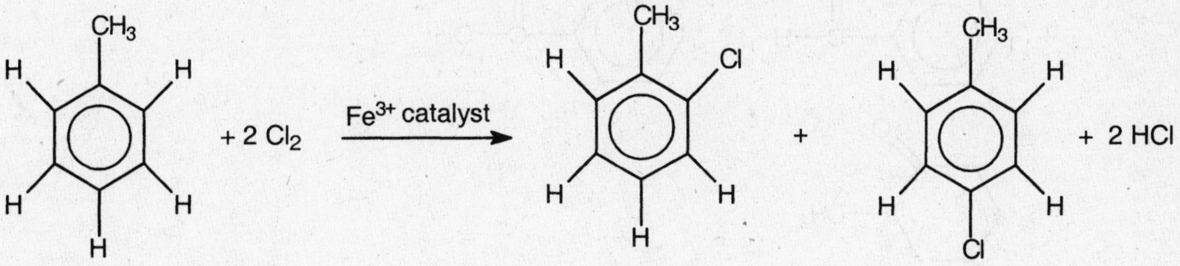

toluene $+ 2\ Cl_2 \xrightarrow{Fe^{3+}\ \text{catalyst}}$ ortho-chlorotoluene + para-chlorotoluene $+ 2\ HCl$

ortho para

toluene $+ Cl_2 \xrightarrow{\text{light}}$ $Cl-CH_2-C_6H_5$ $+ HCl$

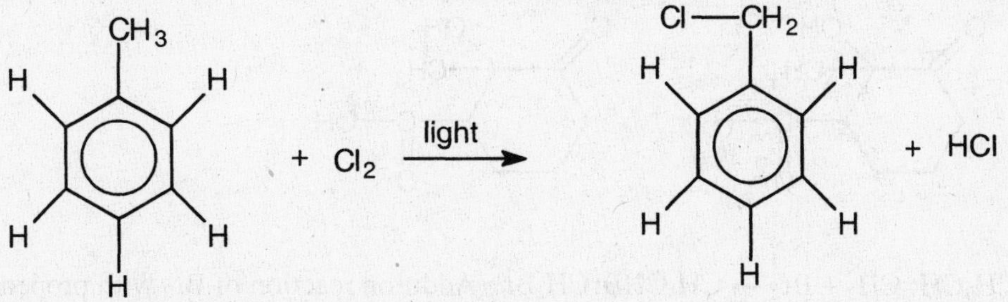

To substitute for the benzene ring hydrogens, an iron(III) catalyst must be present. Without this special iron catalyst, the benzene ring hydrogens are unreactive. To substitute for an alkane hydrogen, light must be present. For toluene, the light-catalyzed reaction substitutes a chlorine for a hydrogen in the methyl group attached to the benzene ring.

65. Primary alcohols (a, d and f) are oxidized to aldehydes, which can be oxidized further to carboxylic acids. Secondary alcohols (b, e and f) are oxidized to ketones, and tertiary alcohols (c and f) do not undergo this type of oxidation reaction. Note that compound f contains a primary, secondary and tertiary alcohol. For the primary alcohols (a, d and f), we listed both the aldehyde and the carboxylic acid as possible products.

a. 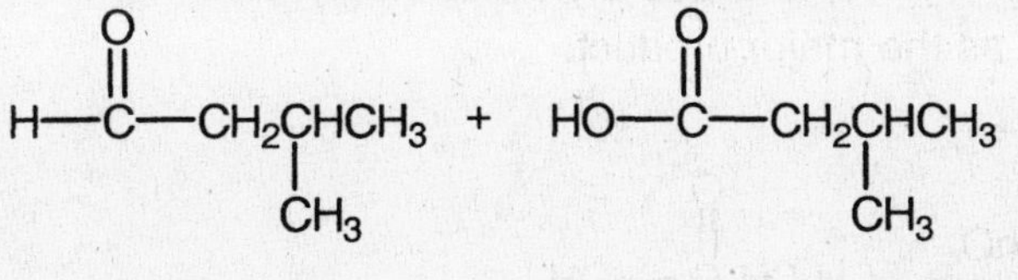 $H-C(=O)-CH_2CH(CH_3)CH_3$ + $HO-C(=O)-CH_2CH(CH_3)CH_3$

b. 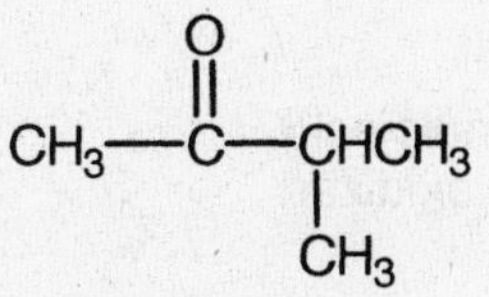 $CH_3-C(=O)-CH(CH_3)CH_3$

c. No reaction

d. $C_6H_5\text{—}\overset{O}{\overset{\|}{C}}\text{—}H \;+\; C_6H_5\text{—}\overset{O}{\overset{\|}{C}}\text{—}OH$

e. 2-methylcyclohexanone (ring C=O with adjacent $\text{—}CH_3$)

f. cyclohexanone ring bearing OH and $\text{—}CH_3$ with $\text{—}\underset{\|}{\underset{O}{C}}\text{—}H$ $\;+\;$ cyclohexanone ring bearing OH and $\text{—}CH_3$ with $\text{—}\underset{\|}{\underset{O}{C}}\text{—}OH$

67. a. $CH_3CH{=}CH_2 + Br_2 \rightarrow CH_3CHBrCH_2Br$ (Addition reaction of Br_2 with propene)

b. $$CH_3\text{—}\overset{OH}{\overset{|}{C}H}\text{—}CH_3 \xrightarrow{\text{oxidation}} CH_3\text{—}\overset{O}{\overset{\|}{C}}\text{—}CH_3$$

Oxidation of 2-propanol yields acetone (2-propanone).

c. $$CH_2{=}\overset{CH_3}{\overset{|}{C}}\text{—}CH_3 + H_2O \xrightarrow{H^+} \underset{H}{\underset{|}{CH_2}}\text{—}\overset{CH_3}{\overset{|}{\underset{OH}{\underset{|}{C}}}}\text{—}CH_3$$

Addition of H_2O to 2-methylpropene would yield tert-butyl alcohol (2-methyl-2-propanol) as the major product.

d. $$CH_3CH_2CH_2OH \xrightarrow{KMnO_4} CH_3CH_2\overset{O}{\overset{\|}{C}}\text{—}OH$$

Oxidation of 1-propanol would eventually yield propanoic acid. Propanal is produced first in this reaction and is then oxidized to propanoic acid.

69. Reaction of a carboxylic acid with an alcohol can produce these esters.

$$CH_3\overset{\overset{O}{\|}}{C}{-}OH + HOCH_2(CH_2)_6CH_3 \longrightarrow CH_3\overset{\overset{O}{\|}}{C}{-}O{-}CH_2(CH_2)_6CH_3 + H_2O$$

ethanoic acid (acetic acid) octanol n-octylacetate

$$CH_3CH_2\overset{\overset{O}{\|}}{C}{-}OH + HOCH_2(CH_2)_4CH_3 \longrightarrow CH_3CH_2\overset{\overset{O}{\|}}{C}{-}O{-}CH_2(CH_2)_4CH_3 + H_2O$$

propanoic acid hexanol

Polymers

71. The backbone of the polymer contains only carbon atoms, which indicates that Kel-F is an addition polymer. The smallest repeating unit of the polymer and the monomer used to produce this polymer are:

$$\left(\begin{array}{cc} F & F \\ | & | \\ C & -\ C \\ | & | \\ Cl & F \end{array} \right)_n \qquad \begin{array}{cc} F & F \\ | & | \\ C & = C \\ | & | \\ Cl & F \end{array}$$

Note: Condensation polymers generally have O or N atoms in the backbone of the polymer.

73.

$$\left(\begin{array}{cccc} CN & & CN & \\ | & & | & \\ C & -CH_2- & C & -CH_2- \\ | & & | & \\ C{-}OCH_3 & & C{-}OCH_3 & \\ \| & & \| & \\ O & & O & \end{array} \right)_n$$

Super glue is an addition polymer formed by reaction of the C=C bond in methyl cyanoacrylate.

75. H_2O is eliminated when Kevlar forms. Two repeating units of Kevlar are:

$$\left(\overset{H}{\overset{|}{N}}{-}C_6H_4{-}\overset{H}{\overset{|}{N}}{-}\overset{O}{\overset{\|}{C}}{-}C_6H_4{-}\overset{O}{\overset{\|}{C}}{-}\overset{H}{\overset{|}{N}}{-}C_6H_4{-}\overset{H}{\overset{|}{N}}{-}\overset{O}{\overset{\|}{C}}{-}C_6H_4{-}\overset{O}{\overset{\|}{C}} \right)_n$$

77. This is a condensation polymer where two molecules of H_2O form when the monomers link together.

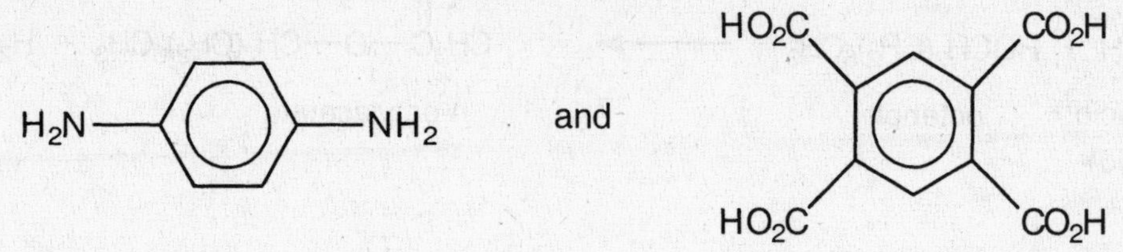

$H_2N-C_6H_4-NH_2$ and $C_6H_2(CO_2H)_4$ (HO₂C, CO₂H, HO₂C, CO₂H substituents on benzene ring)

79. Divinylbenzene has two reactive double bonds that are both used when divinylbenzene inserts itself into two adjacent polymer chains. The chains cannot move past each other because the crosslinks bond adjacent polymer chains together, making the polymer more rigid.

81 a. The polymer formed using 1,2-diaminoethane will exhibit relatively strong hydrogen bonding interactions between adjacent polymer chains. Hydrogen bonding is not present in the ethylene glycol polymer (a polyester polymer forms), so the 1,2-diaminoethane polymer will be stronger.

b. The presence of rigid groups (benzene rings or multiple bonds) makes the polymer stiffer. Hence, the monomer with the benzene ring will produce the more rigid polymer.

c. Polyacetylene will have a double bond in the carbon backbone of the polymer.

$$n\ HC\equiv CH \longrightarrow -(CH=CH)_n-$$

The presence of the double bond in polyacetylene will make polyacetylene a more rigid polymer than polyethylene. Polyethylene doesn't have C=C bonds in the backbone of the polymer (the double bonds in the monomers react to form the polymer).

Natural Polymers

83. a. Serine, tyrosine and threonine contain the -OH functional group in the R group.

b. Aspartic acid and glutamic acid contain the -COOH functional group in the R group.

c. An amine group has a nitrogen bonded to other carbon and/or hydrogen atoms. Histidine, lysine, arginine and tryptophan contain the amine functional group in the R group.

d. The amide functional group is:

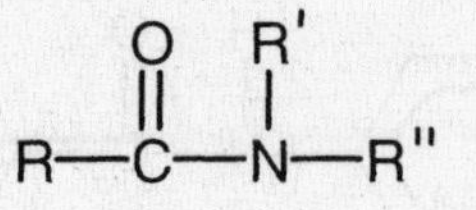

$$R-\overset{\overset{O}{\|}}{C}-\overset{\overset{R'}{|}}{N}-R''$$

This functional group is formed when individual amino acids bond together to form the peptide linkage. Glutamine and asparagine have the amide functional group in the R group.

85. a. Aspartic acid and phenylalanine make up aspartame.

H O N H O
$H_2N—C—C—OH$ H—N—C—COH
CH_2 amide bond forms here CH_2
CO_2H

b. Aspartame contains the methyl ester of phenylalanine. This ester can hydrolyze to form methanol:

$$R–CO_2CH_3 + H_2O \rightleftharpoons RCO_2H + HOCH_3$$

87.

$H_2NCHC(=O)—NHCHCO_2H$ with CH_2OH and CH_3 side chains: ser - ala

$H_2NCHC(=O)—NHCHCO_2H$ with CH_3 and CH_2OH side chains: ala - ser

89. a. Six tetrapeptides are possible. From NH_2 to CO_2H end:

phe-phe-gly-gly, gly-gly-phe-phe, gly-phe-phe-gly,

phe-gly-gly-phe, phe-gly-phe-gly, gly-phe-gly-phe

b. Twelve tetrapeptides are possible. From NH_2 to CO_2H end:

phe-phe-gly-ala, phe-phe-ala-gly, phe-gly-phe-ala,

phe-gly-ala-phe, phe-ala-phe-gly, phe-ala-gly-phe,

gly-phe-phe-ala, gly-phe-ala-phe, gly-ala-phe-phe

ala-phe-phe-gly, ala-phe-gly-phe, ala-gly-phe-phe

91. a. Ionic: Need NH_2 on side chain of one amino acid with CO_2H on side chain of the other amino acid. The possibilities are:

NH_2 on side chain = His, Lys or Arg; CO_2H on side chain = Asp or Glu

b. Hydrogen bonding: Need N–H or O–H bond present in side chain. The hydrogen bonding interaction occurs between the X– H bond and a carbonyl group from any amino acid.

X–H · · · · · · · O = C (carbonyl group)

Ser	Asn	Any amino acid
Glu	Thr	
Tyr	Asp	
His	Gln	
Arg	Lys	

c. Covalent: Cys–Cys (forms a disulfide linkage)

d. London dispersion: All amino acids with nonpolar R groups. They are:

Gly, Ala, Pro, Phe, Ile, Trp, Met, Leu and Val

e. Dipole-dipole: Need side chain with OH group. Tyr, Thr and Ser all could form this specific dipole-dipole force with each other since all contain an OH group in the side chain.

93. Glutamic acid: R = $-CH_2CH_2CO_2H$; Valine: R = $-CH(CH_3)_2$; A polar side chain is replaced by a nonpolar side chain. This could affect the tertiary structure of hemoglobin and the ability of hemoglobin to bind oxygen.

95. See Figures 22.29 and 22.30 of the text for examples of the cyclization process.

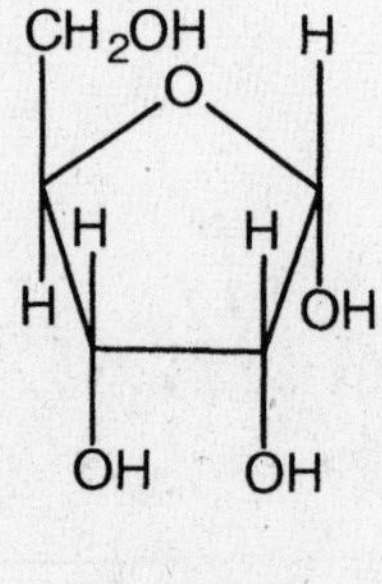

CH_2OH H O H H H OH OH OH

D-Ribose

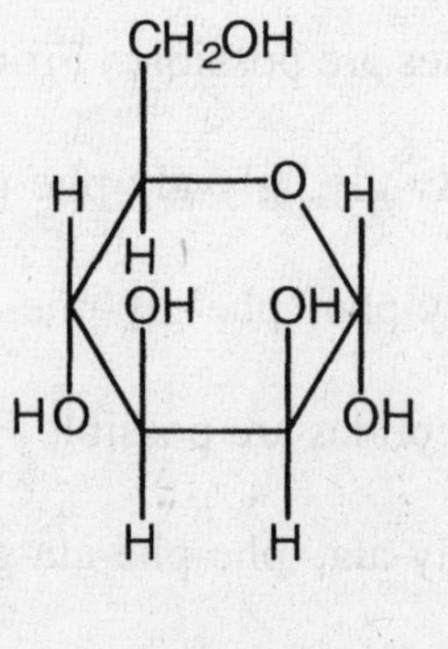

CH_2OH H O H H OH OH HO OH H H

D-Mannose

97. The aldohexoses contain 6 carbons and the aldehyde functional group. Glucose, mannose and galactose are aldohexoses. Ribose and arabinose are aldopentoses since they contain 5 carbons with the aldehyde functional group. The ketohexose (6 carbons + ketone functional group) is fructose and the ketopentose (5 carbons + ketone functional group) is ribulose.

99. The α and β forms of glucose differ in the orientation of a hydroxy group on one specific carbon in the cyclic forms (see Figure 22.30 of the text). Starch is a polymer composed of only α-D-glucose, and cellulose is a polymer composed of only β-D-glucose.

101. A chiral carbon has four different groups attached to it. A compound with a chiral carbon is optically active. Isoleucine and threonine contain more than the one chiral carbon atom (see asterisks).

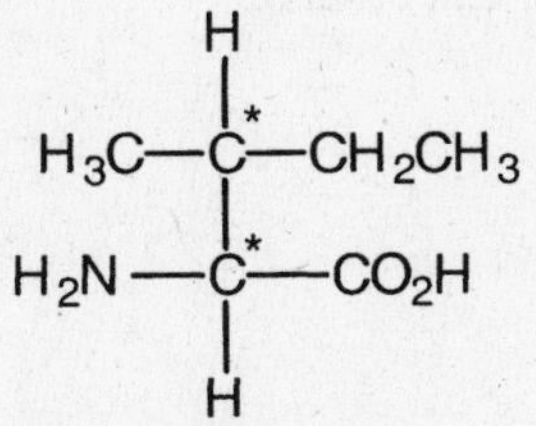

H
H_3C—C*—CH_2CH_3
H_2N—C*—CO_2H
H

isoleucine

H
H_3C—C*—OH
H_2N—C*—CO_2H
H

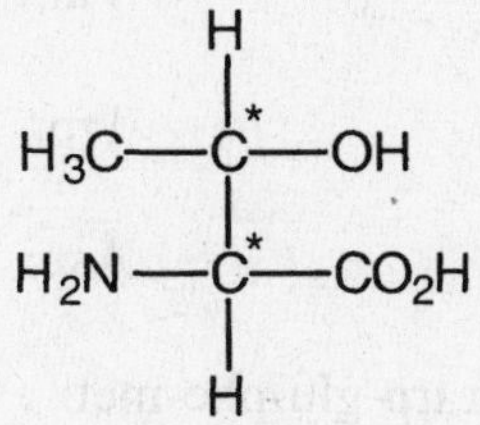

threonine

103. Only one of the isomers is optically active. The chiral carbon in this optically active isomer is marked with an asterisk.

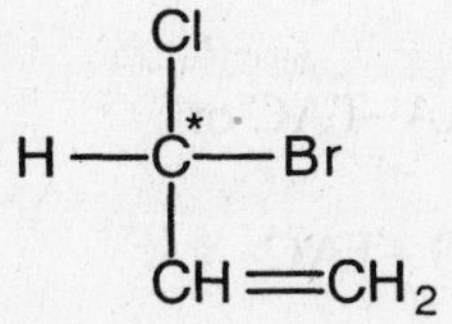

Cl
H—C*—Br
CH=CH_2

105. The complimentary base pairs in DNA are cytosine (C) and guanine (G), and thymine (T) and adenine (A). The complimentary sequence is: C-C-A-G-A-T-A-T-G

107. Uracil will hydrogen bond to adenine.

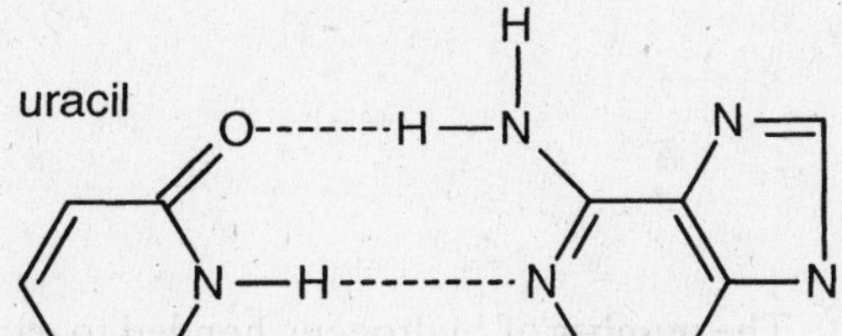

uracil
O------H—N
N—H--------N
sugar
sugar
adenine

109. Base pair:

RNA	DNA
A	T
G	C
C	G
U	A

a. Glu: CTT, CTC   Val: CAA, CAG, CAT, CAC

Met: TAC   Trp: ACC

Phe: AAA, AAG   Asp: CTA, CTG

b. DNA sequence for trp-glu-phe-met:

ACC –CTT –AAA –TAC
        or   or
      CTC AAG

c. Due to glu and phe, there is a possibility of four different DNA sequences. They are:

ACC–CTT–AAA–TAC or ACC–CTC–AAA–TAC or

ACC–CTT–AAG–TAC or ACC–CTC–AAG –TAC

d.

T—A—C—C—T—G—A—A—G

met (T—A—C), asp (C—T—G), phe (A—A—G)

e. TAC–CTA–AAG; TAC–CTA–AAA; TAC–CTG–AAA

Additional Exercises

111. We omitted the hydrogens for clarity. The number of hydrogens bonded to each carbon is the number necessary to form four bonds.

a.

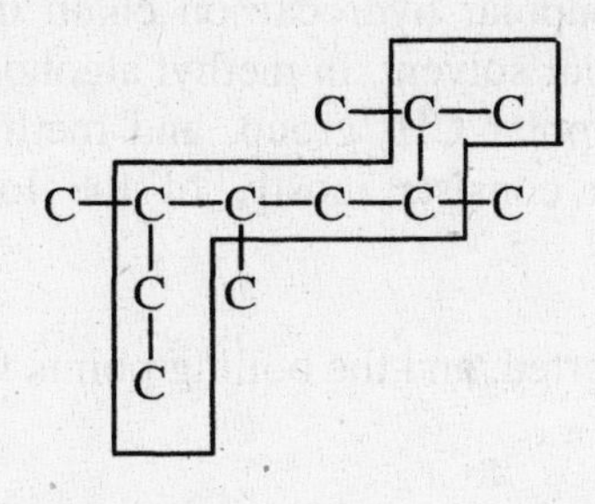

2,3,5,6-tetramethyloctane

b.

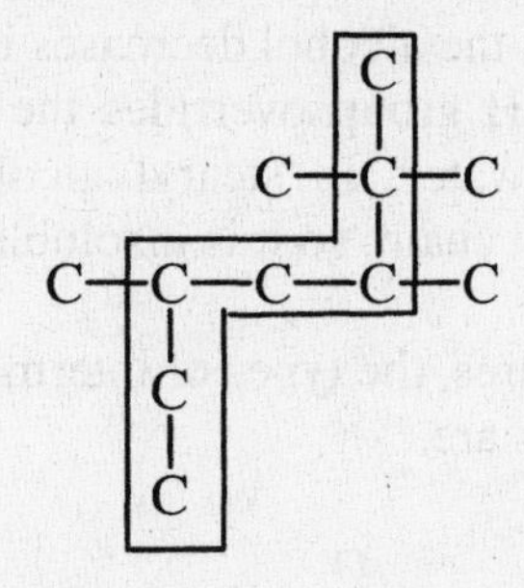

2,2,3,5-tetramethylheptane

c.

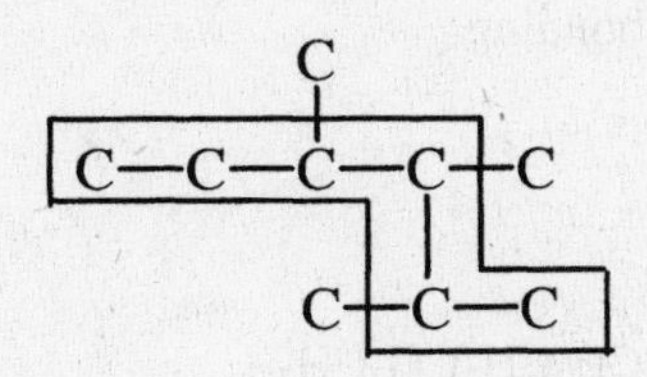

2,3,4-trimethylhexane

d.

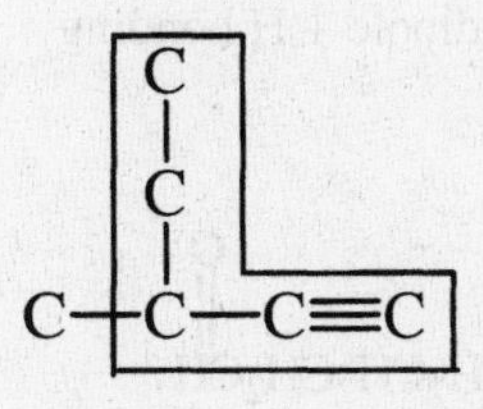

3-methyl-1-pentyne

113.

There are many possibilities for isomers. Any structure with four chlorines replacing four hydrogens in any four of the numbered positions would be an isomer, i.e., 1,2,3,4-tetra-chloro-dibenzo-p-dioxin is a possible isomer.

115. The isomers are:

$CH_3—O—CH_3$ CH_3CH_2OH

dimethyl ether, −23°C ethanol, 78.5°C

Ethanol, with its ability to form the relatively strong hydrogen bonding interactions, boils at the higher temperature.

117. Alcohols consist of two parts, the polar OH group and the nonpolar hydrocarbon chain attached to the OH group. As the length of the nonpolar hydrocarbon chain increases, the solubility of the alcohol decreases in water, a very polar solvent. In methyl alcohol (methanol), the polar OH group overrides the effect of the nonpolar CH_3 group, and methyl alcohol is soluble in water. In stearyl alcohol, the molecule consists mostly of the long nonpolar hydrocarbon chain, so it is insoluble in water.

119. The structures, the types of intermolecular forces exerted, and the boiling points for the compounds are:

$CH_3CH_2CH_2\overset{\overset{O}{\|}}{C}OH$	$CH_3CH_2CH_2CH_2CH_2OH$
butanoic acid, 164°C LD + dipole + H bonding	1-pentanol, 137°C LD + H bonding
$CH_3CH_2CH_2CH_2\overset{\overset{O}{\|}}{C}H$	$CH_3CH_2CH_2CH_2CH_2CH_3$
pentanal, 103°C LD + dipole	n-hexane, 69°C LD only

All these compounds have about the same molar mass. Therefore, the London dispersion (LD) forces in each are about the same. The other types of forces determine the boiling point order. Since butanoic acid and 1-pentanol both exhibit hydrogen bonding inter-actions, these two compounds will have the two highest boiling points. Butanoic acid has the highest boiling point since it exhibits H bonding along with dipole-dipole forces due to the polar C=O bond.

121. $85.63\text{ g C} \times \frac{1\text{ mol C}}{12.01\text{ g C}} = 7.130\text{ mol C}$; $14.37\text{ g H} \times \frac{1\text{ mol H}}{1.008\text{ g H}} = 14.26\text{ mol H}$

Because the mol H to mol C ratio is 2:1 (14.26/7.130 = 2.000), the empirical formula is CH_2. The empirical formula mass ≈ 12 + 2(1) = 14. Since 4 × 14 = 56 puts the molar mass between 50 and 60, the molecular formula is C_4H_8. The isomers of C_4H_8 are:

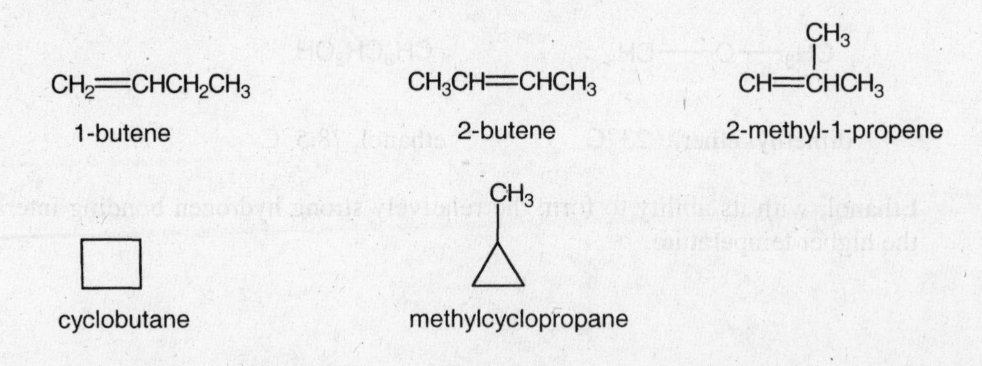

Only the alkenes will react with H_2O to produce alcohols, and only 1-butene will produce a secondary alcohol for the major product and a primary alcohol for the minor product.

$$CH_2{=}CHCH_2CH_3 + H_2O \longrightarrow \underset{\text{2° alcohol, major product}}{\overset{\text{H}\quad\ \text{OH}}{CH_2{-}CHCH_2CH_3}}$$

$$CH_2{=}CHCH_2CH_3 + H_2O \longrightarrow \underset{\text{1° alcohol, minor product}}{\overset{\text{OH}\quad\ \text{H}}{CH_2{-}CHCH_2CH_3}}$$

2-butene will produce only a secondary alcohol when reacted with H_2O, and 2-methyl-1-propene will produce a tertiary alcohol as the major product and a primary alcohol as the minor product.

123. $KMnO_4$ will oxidize primary alcohols to aldehydes and then to carboxylic acids. Secondary alcohols are oxidized to ketones by $KMnO_4$. Tertiary alcohols and ethers are not oxidized by $KMnO_4$.

The three isomers and their reactions with $KMnO_4$ are:

$$\underset{\text{ether}}{CH_3{-}O{-}CH_2CH_3} \xrightarrow{KMnO_4} \text{no reaction}$$

$$\underset{\text{2° alcohol}}{CH_3{-}\overset{\text{OH}}{CH}{-}CH_3} \xrightarrow{KMnO_4} \underset{\text{2-propanone (acetone)}}{CH_3{-}\overset{\text{O}}{\overset{\|}{C}}{-}CH_3}$$

$$\underset{\text{1° alcohol}}{CH_3CH_2\overset{\text{OH}}{CH_2}} \xrightarrow{KMnO_4} \underset{\text{propanal}}{CH_3CH_2\overset{\text{O}}{\overset{\|}{C}}H} \xrightarrow{KMnO_4} \underset{\text{propanoic acid}}{CH_3CH_2\overset{\text{O}}{\overset{\|}{C}}{-}OH}$$

The products of the reactions with excess $KMnO_4$ are 2-propanone and propanoic acid.

125. In nylon, hydrogen-bonding interactions occur due to the presence of N–H bonds in the polymer. For a given polymer chain length, there are more N–H groups in Nylon-46 as compared to Nylon-6. Hence, Nylon-46 forms a stronger polymer compared to Nylon-6 due to the increased hydrogen-bonding interactions.

127. a.

$H_2N-C_6H_4-NH_2$ and $HO_2C-C_6H_4-CO_2H$

b. Repeating unit:

$$\left(-N(H)-C_6H_4-N(H)-C(=O)-C_6H_4-C(=O)- \right)_n$$

The two polymers differ in the substitution pattern on the benzene rings. The Kevlar chain is straighter, and there is more efficient hydrogen bonding between Kevlar chains than between Nomex chains.

129. a. The bond angles in the ring are about 60°. VSEPR predicts bond angles close to 109°. The bonding electrons are closer together than they prefer, resulting in strong electron-electron repulsions. Thus, ethylene oxide is unstable (reactive).

b. The ring opens up during polymerization; the monomers link together through the formation of O–C bonds.

$$\left(O-CH_2CH_2-O-CH_2CH_2-O-CH_2CH_2 \right)_n$$

131. Glutamic acid:

$H_2N-CH(CH_2CH_2CO_2H)-CO_2H$

Monosodium glutamate:

$H_2N-CH(CH_2CH_2CO_2^-Na^+)-CO_2H$

One of the two acidic protons in the carboxylic acid groups is lost to form MSG. Which proton is lost is impossible for you to predict.

In MSG, the acidic proton from the carboxylic acid in the R group is lost, allowing formation of the ionic compound.

133. $\Delta G = \Delta H - T\Delta S$; For the reaction, we break a P–O and O–H bond and form a P–O and O–H bond, so $\Delta H \approx 0$. ΔS for this process is negative because order increases. Thus, $\Delta G > 0$ and the reaction is not spontaneous.

135. Alanine can be thought of as a diprotic acid. The first proton to leave comes from the carboxylic acid end with $K_a = 4.5 \times 10^{-3}$. The second proton to leave comes from the protonated amine end (K_a for $R{-}NH_3^+ = K_w/K_b = 1.0 \times 10^{-14}/7.4 \times 10^{-5} = 1.4 \times 10^{-10}$).

In 1.0 *M* H^+, both the carboxylic acid and the amine end will be protonated since H^+ is in excess. The protonated form of alanine is below. In 1.0 *M* OH^-, the dibasic form of alanine will be present because the excess OH^- will remove all acidic protons from alanine. The dibasic form of alanine follows.

1.0 *M* H+ : $H_3\overset{+}{N}{-}CH(CH_3){-}C(=O){-}OH$

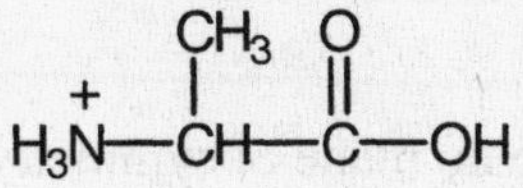

protonated form

1.0 *M* OH^- : $H_2N{-}CH(CH_3){-}C(=O){-}O^-$

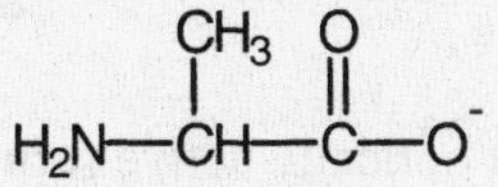

dibasic form

137. For denaturation, heat is added so it is an endothermic process. Because the highly ordered secondary structure is disrupted, entropy (disorder) will increase. Thus, ΔH and ΔS are both positive for protein denaturation.

Challenge Problems

139. For the reaction:

$$^+H_3NCH_2CO_2H \rightleftharpoons 2\,H^+ + H_2NCH_2CO_2^- \quad K_{eq} = 7.3 \times 10^{-13} = K_a\,(\text{-}CO_2H) \times K_a\,(\text{-}NH_3^+)$$

$$7.3 \times 10^{-13} = \frac{[H^+]^2[H_2NCH_2CO_2^-]}{[^+H_3NCH_2CO_2H]} = [H^+]^2, \; [H^+] = (7.3 \times 10^{-13})^{1/2}$$

$[H^+] = 8.5 \times 10^{-7}$ *M*; pH = −log $[H^+]$ = 6.07 = isoelectric point

141. a. Even though this form of tartaric acid contains 2 chiral carbon atoms (see asterisks in the following structure), the mirror image of this form of tartaric acid is superim-posable. Therefore, it is not optically active. An easier way to identify optical activity in molecules with two or more chiral carbon atoms is to look for a plane of symmetry in the molecule. If a molecule has a plane of symmetry, then it is never optically active. A plane of symmetry is a plane that bisects the molecule where one side exactly reflects on the other side.

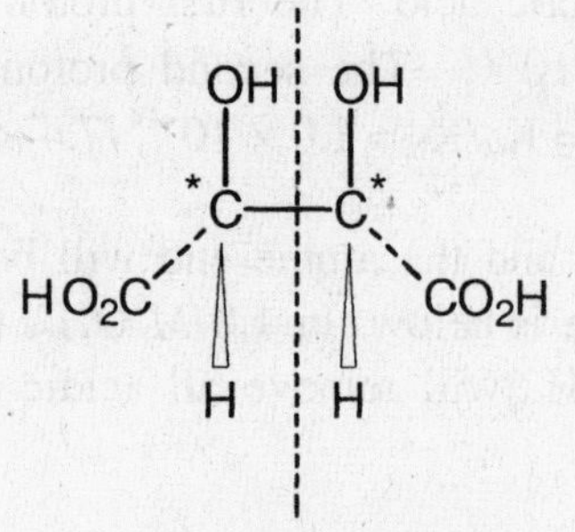

OH OH
C—C
HO_2C CO_2H
H H

symmetry plane

b. The optically active forms of tartaric acid have no plane of symmetry. The structures of the optically active forms of tartaric acid are:

OH OH | OH OH
C—C | C—C
H CO_2H | HO_2C H
CO_2H H | H CO_2H

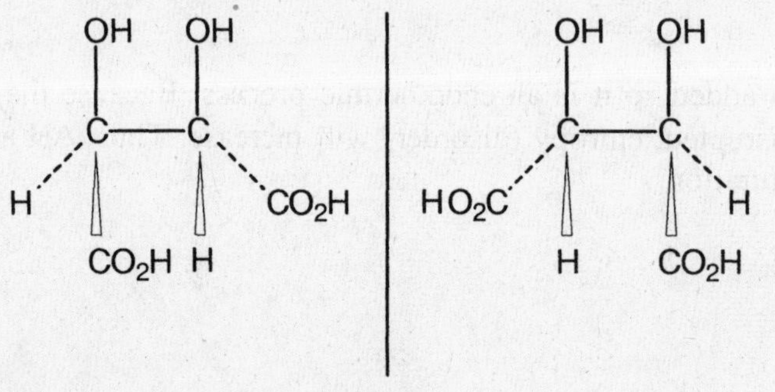

mirror

These two forms of tartaric acid are nonsuperimposable.

143.

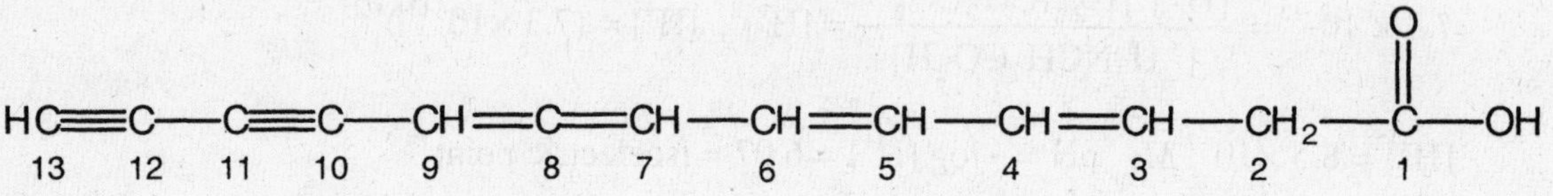

$$\underset{13}{HC}\equiv\underset{12}{C}-\underset{11}{C}\equiv\underset{10}{C}-\underset{9}{CH}=\underset{8}{C}=\underset{7}{CH}-\underset{6}{CH}=\underset{5}{CH}-\underset{4}{CH}=\underset{3}{CH}-\underset{2}{CH_2}-\underset{1}{\overset{O}{\overset{\|}{C}}}-OH$$

145. a. The three structural isomers of C_5H_{12} are:

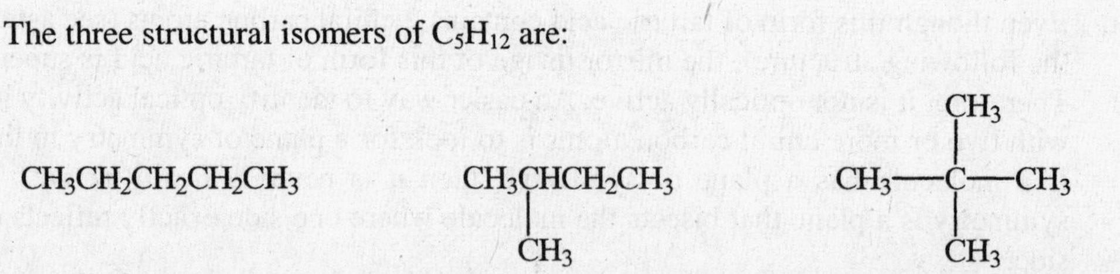

$CH_3CH_2CH_2CH_2CH_3$	$CH_3CH(CH_3)CH_2CH_3$	$CH_3-C(CH_3)_2-CH_3$
n-pentane	2-methylbutane	2,2-dimethylpropane

n-pentane will form three different monochlorination products: 1-chloropentane, 2-chloropentane and 3-chloropentane (the other possible monochlorination products differ by a simple rotation of the molecule; they are not different products from the ones listed). 2-2,dimethylpropane will only form one monochlorination product: 1-chloro-2,2-dimethylpropane. 2-methylbutane is the isomer of C_5H_{12} that forms four different monochlorination products: 1-chloro-2-methylbutane, 2-chloro-2-methyl-butane, 3-chloro-2-methylbutane (or we could name this compound 2-chloro-3-methylbutane), and 1-chloro-3-methylbutane.

b. The isomers of C_4H_8 are:

$CH_2{=}CHCH_2CH_3$ — 1-butene

$CH_3CH{=}CHCH_3$ — 2-butene

$CH_2{=}C(CH_3)CH_3$ — 2-methyl-1-propene or 2-methylpropene

cyclobutane

methylcyclopropane

The cyclic structures will not react with H_2O; only the alkenes will add H_2O to the double bond. From Exercise 22.62, the major product of the reaction of 1-butene and H_2O is 2-butanol (a 2° alcohol). 2-butanol is also the major (and only) product when 2-butene and H_2O react. 2-methylpropene forms 2-methyl-2-propanol as the major product when reacted with H_2O; this product is a tertiary alcohol. Therefore, the C_4H_8 isomer is 2-methylpropene.

$$CH_2{=}C(CH_3){-}CH_3 + HOH \longrightarrow CH_3{-}C(CH_3)(OH){-}CH_3$$

2-methyl-2-propanol (a 3° alcohol, 3 R groups)

c. The structure of 1-chloro-1-methylcyclohexane is:

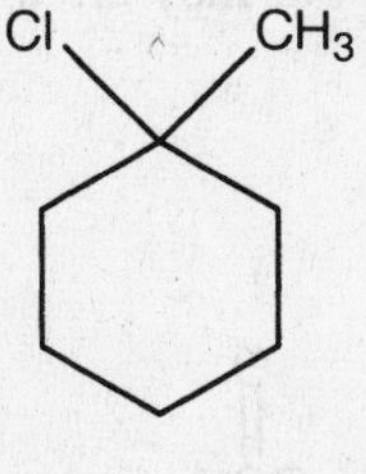

(cyclohexane ring with Cl and CH_3 on the same carbon)

The addition reaction of HCl with an alkene is a likely choice for this reaction (see Exercise 22.62). The two isomers of C_7H_{12} that produce 1-chloro-1-methylcyclohexane as the major product are:

d. Working backwards, 2° alcohols produce ketones when they are oxidized (1° alco-hols produce aldehydes, then carboxylic acids). The easiest way to produce the 2° alcohol from a hydrocarbon is to add H_2O to an alkene. The alkene reacted is 1-propene (or propene).

$$CH_2{=}CHCH_3 + H_2O \longrightarrow CH_3CH(OH)CH_3 \xrightarrow{\text{oxidation}} CH_3C(O)CH_3$$

propene — acetone

e. The $C_5H_{12}O$ formula has too many hydrogens to be anything other than an alcohol (or an unreactive ether). 1° alcohols are first oxidized to aldehydes, then to carboxylic acids. Therefore, we want a 1° alcohol. The 1° alcohols with formula $C_5H_{12}O$ are:

1-pentanol: $HOCH_2CH_2CH_2CH_2CH_3$

2-methyl-1-butanol: $HOCH_2CH(CH_3)CH_2CH_3$

3-methyl-1-butanol: $CH_3CH(CH_3)CH_2CH_2OH$

2,2-dimethyl-1-propanol: $HOCH_2C(CH_3)_2CH_3$

There are other alcohols with formula $C_5H_{12}O$, but they are all 2° or 3° alcohols, which do not produce carboxylic acids when oxidized.

147.

$$\left(\!-OCH_2CH_2OC(O)NH-C_6H_4-NHC(O)OCH_2CH_2OC(O)NH-C_6H_4-NHC(O)-\right)_n$$

149. a. The temperature of the rubber band increases when it is stretched.

b. Exothermic because heat is released.

c. As the chains are stretched, they line up more closely together, resulting in stronger London dispersion forces between the chains. Heat is released as the strength of the intermolecular forces increases.

d. Stretching is not spontaneous so, ΔG is positive. $\Delta G = \Delta H - T\Delta S$; Since ΔH is negative then ΔS must be negative in order to give a positive ΔG.

e.

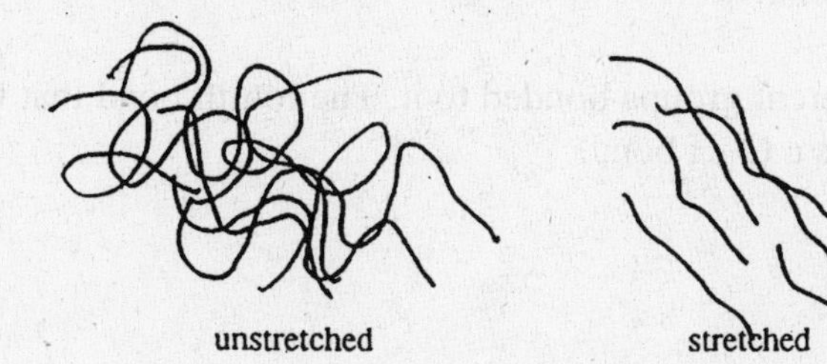

The structure of the stretched polymer is more ordered (lower S).

151. 4.2×10^{-3} g $K_2CrO_7 \times \frac{1 \text{ mol } K_2Cr_2O_7}{294.20 \text{ g}} \times \frac{1 \text{ mol } Cr_2O_7^{2-}}{\text{mol } K_2Cr_2O_7} \times \frac{3 \text{ mol } C_2H_5OH}{2 \text{ mol } Cr_2O_7^{2-}}$

$= 2.1 \times 10^{-5}$ mol C_2H_5OH

$$n_{\text{breath}} = \frac{PV}{RT} = \frac{\left(750.\text{ mm Hg} \times \frac{1 \text{ atm}}{760 \text{ mm Hg}}\right) \times 0.500 \text{ L}}{\frac{0.08206 \text{ L atm}}{\text{K mol}} \times 303 \text{ K}} = 0.0198 \text{ mol breath}$$

$$\text{mol \% } C_2H_5OH = \frac{2.1 \times 10^{-5} \text{ mol } C_2H_5OH}{0.0198 \text{ mol total}} \times 100 = 0.11\% \text{ alcohol}$$

153. The five chiral carbons are marked with an asterisk.

Each of these five carbons have four different groups bonded to it. The fourth bond that is not shown for any of the five chiral carbons is a C–H bond.

Integrative Problems

155. a. Zn^{2+} has the $[Ar]3d^{10}$ electron configuration and zinc does form +2 charged ions.

$$\text{mass \%Zn} = \frac{\text{mass of 1 mol Zn}}{\text{mass of 1 mol } CH_3CH_2ZnBr} \times 100 = \frac{65.38 \text{ g}}{174.34 \text{ g}} \times 100 = 37.50\% \text{ Zn}$$

b. The reaction is:

$$CH_3CH_2CH(CH_3)-\overset{*}{C}(=O)-CH_3 \longrightarrow CH_3CH_2CH(CH_3)-\overset{*}{C}(OH)(CH_2CH_3)-CH_3$$

The hybridization changes from sp^2 to sp^3.

c. 3,4-dimethyl-3-hexanol